Contemporary Bayesian
Econometrics and Statistics

WILEY SERIES IN PROBABILITY AND STATISTICS

Established by WALTER A. SHEWHART and SAMUEL S. WILKS

Editors: *David J. Balding, Noel A. C. Cressie, Nicholas I. Fisher, Iain M. Johnstone, J. B. Kadane, Geert Molenberghs, Louise M. Ryan, David W. Scott, Adrian F. M. Smith, Jozef L. Teugels*
Editors Emeriti: *Vic Barnett, J. Stuart Hunter, David G. Kendall*

A complete list of the titles in this series appears at the end of this volume.

Contemporary Bayesian Econometrics and Statistics

JOHN GEWEKE

University of Iowa
Departments of Economics and Statistics
Iowa City, Iowa

WILEY-INTERSCIENCE

A JOHN WILEY & SONS, INC., PUBLICATION

Published by John Wiley & Sons, Inc., Hoboken, New Jersey.
Published simultaneously in Canada.

Library of Congress Cataloging-in-Publication Data:

Geweke, John.
Contemporary Bayesian econometrics and statistics / John Geweke.
p. cm.
Includes bibliographical references and index.
ISBN-13 978-0-471-67932-5 (cloth)
ISBN-10 0-471-67932-1 (cloth)
1. Econometrics. 2. Bayesian statistical decision theory. 3. Decision making—Mathematical models. I. Title.

HB139.G478 2005
330′.01′519542—dc22 2005043948

Printed in the United States of America.

10 9 8 7 6 5 4 3 2

Contents

Preface

Bayesian analysis provides a unified and coherent way of thinking about decision problems and how to solve them using data and other information. The goal of this book is to acquaint the reader in a serious way with this approach and its problem-solving potential, and to this end it has two objectives. The first is to provide a clear understanding of Bayesian analysis, grounded in the theory of inference and optimal decisionmaking, which will enable the reader to confidently analyze real problems. The second is to equip the reader with state-of-the-art simulation methods that can be used to solve these problems.

This book is written for research professionals who use econometrics and similar statistical methods in their work, and for Ph.D. students in disciplines that do the same. These disciplines include economics and statistics, as well as the many social sciences and fields in business and public policy schools that study decisionmaking on the basis of data and other information. The book assumes the same knowledge of mathematical statistics as most Ph.D. courses in econometrics, and familiarity with linear models at the level of a graduate applied econometrics course or a master's statistics course. The entire book was developed through a decade of teaching at this level, all of the material having been presented at least twice and some more than a half-dozen times. This vetting process has afforded the opportunity to minimize the barriers to entry to a sound and practical grasp of Bayesian analysis for the intended audience.

Loosely speaking, the first three chapters address the objective of a clear understanding of Bayesian analysis—how to think—and the next five, the objective of presenting and applying simulation methods—how to act. There is no sharp distinction between these two objectives. In particular, as one gains greater confidence with "hands on" methods, it is natural to rethink the formulation of problems at hand with the knowledge that what was not long ago impossible is now practical. The text has many examples and exercises that follow this path, ranging from questions that have been used in examinations to substantial projects that extend or apply the methods presented. Some of these examples and exercises use the Bayesian analysis, computation, and communication (BACC) extension of the mathematical applications Matlab, Splus, R, and Gauss. The reader will find the software and documentation, along with data and code for examples, in the online appendix for this text at http://www.biz.uiowa.edu/cbes.

The book takes up specific models as vehicles for understanding Bayesian analysis and applying simulation methods. This entails solving problems in a practical way and at the level of detail required by research professionals whose work must withstand subsequent scrutiny. In some cases these solutions did not exist only a few years ago (prior to 2005), and are not yet widely known among econometricians and statisticians. Therefore the book concentrates on a handful of models in some depth, rather than attempting to survey all models with a scope similar to that of leading (and much longer) graduate econometrics texts. The coverage here should not be taken as a judgment that other models are somehow less important or significant, or cannot be approached using Bayesian analysis. Just the opposite is true. The approaches and methods in this book are being used to improve models and decisionmaking at an accelerating rate, as perusal of the tables of contents of leading journals such as the *Journal of the American Statistical Association*, the *Journal of the Royal Statistical Society*, and the *Journal of Econometrics* will verify. The reader of this book will be well equipped to understand this research, to appreciate its relevance to problems at hand, and to tailor existing methods to these problems.

The organization is designed to meet a variety of uses in graduate education. All begin with Chapter 1, which provides an overview of the rest of the text at a lower technical level than is used subsequently. This material, which can be covered in 1–2 weeks in a traditional setting or in the first day of an intensive course, provides the reader with motivation for the more technical work that follows. A full-year graduate course can cover the first four chapters in the first semester, perhaps using the material on discrete-state Markov processes in Chapter 7 as an entrée to the theory of Markov chain Monte Carlo (MCMC) methods in Chapter 4. The second semester then begins with hands-on computing and applications and proceeds through the rest of the book. One can base a one-semester course on Chapters 1 and 2, the first three sections of Chapter 4, Section 5.1, plus other parts of Chapters 5, 6, and 7 as time and interests dictate. For example, completion of Chapter 5 will concentrate on linear models. Chapter 6 concentrates on latent variable models, and for this concentration the material on hierarchical priors at the start of Chapter 3 may also be of interest. An intensive applications-oriented course of 1–2 weeks can be based on Chapter 1, Section 2.1, Section 4.3, and Section 5.1, plus other parts of Chapters 5, 6, and 7 consistent with time and interests. The online appendix provides ample material for computing laboratory sessions in such a course.

I am very grateful to a number of people who contributed, in one way or another, to the book. Scores of graduate students were involved since the mid-1990s as material was developed, discarded, modified, and redeveloped in graduate courses at the Universities of Minnesota and Iowa. Of these former graduate or postdoctoral students, Gianni Amisano, Pat Bajari, Hulya Eraslan, Merrell Hora, John Landon-Lane, Lea Petrella, Arnie Quinn, Hisashi Tanizaki, and Nobuhiko Terui all played roles in improving the text, computing code, or examples. I owe a special debt to my former student Bill McCausland, who also conceived the BACC software and brought it into being. I am grateful to the National Science Foundation for support of software development and research incorporated here. For nurturing many aspects of the Bayesian approach to thinking reflected in these pages, I am especially grateful to Jim Berger, Jay Kadane, Dennis Lindley, Dale Poirier, Christopher Sims, Luke Tierney, and Arnold Zellner. Finally, for advice and comments on many specific aspects of the book I thank Siddhartha Chib, Bill Griffiths, Gary Koop, Peter Rossi, Christopher Sims, Mark Steel, and Herman van Dijk.

JOHN GEWEKE

Iowa City, Iowa

CHAPTER 1

Introduction

The evolution of modern society is driven by decisions that affect the welfare and choices of large groups of individuals. Of the scores of examples, a few will illustrate the characteristics of decisionmaking that motivate our approach:

1. A new drug has been developed in the laboratories of a private firm over a period of several years and at a cost of tens of millions of dollars. It has been tested in animals, and in increasingly larger groups of human beings in a succession of highly structured clinical trials. If the drug is approved by the Food and Drug Administration (FDA), it will be available for all licensed physicians to use at their discretion. The FDA must decide whether to approve the drug.
2. Since the mid-1980s evidence from many different sources, taken together, clearly indicates that the earth's climate is warming. The evidence that this warming is due to human activities, in particular the emission of carbon dioxide, is not as compelling but becomes stronger every year. The economic activities responsible for increases in the emission of carbon dioxide are critical to the aspirations of billions of people, and to the political order that would be needed to sustain a policy that would limit emissions. How should the evidence be presented to political leaders who are able to make and enforce decisions about emissions policy? What should their decision be?
3. A multi-billion-dollar firm is seeking to buy a firm of similar size. The two firms have documented cost reductions that will be possible because of the merger. On the other hand, joint ownership of the two firms will likely increase market power, making it in the interests of the merged firm to set higher price cost margins than did the two firms separately. How should lawyers and economists—whether disinterested or not—document and synthesize the evidence on both points for the regulatory authorities who decide whether to permit the merger? How should the regulatory authorities

---

Contemporary Bayesian Econometrics and Statistics, by John Geweke

make their decision? If they deny the merger, the firms must decide whether to appeal the decision to the courts.

4. A standard petroleum refining procedure produces two-thirds unleaded gasoline and one-third heating oil (or jet aviation fuel, its near equivalent). Refinery management buys crude oil, and produces and sells gasoline and heating oil. The wholesale prices of these products are volatile. Management can guarantee the difference between selling and buying prices, by means of futures contracts in which speculators (risk takers) commit to purchasing specified amounts of gasoline or heating oil, and selling agreed-on amounts of crude oil, at fixed prices. Should management lock in some or all of its net return in this way? If some, then how much?

These decisions differ in many ways. The second and third will appear prominently in the media; the first might, the last rarely will. The second is a matter of urgent global public policy, and the last is entirely private. The other two are mixtures; in each case the final decision is a matter of public policy, but in both the matter is raised to the level of public policy through a sequence of private decisions, in which anticipation of the ultimate public policy decision is quite important.

Yet these decisions have many features in common:

1. The decision must be made on the basis of less-than-perfect information. By "perfect information" is meant all the information the decisionmaker(s) would requisition if information were free, that is, immediately available at no cost in resources diverted from other uses.
2. The decision must be made at a specified time. Either waiting is prohibited by law or regulation (examples 1 and 3), is denied by the definition of the decision (example 4), or "wait" amounts to making a critical choice that may circumscribe future options (example 2).
3. The information bearing on the decision, and the consequences of the decision, are primarily quantitative. The relationship between information and outcome, mediated by working hypotheses about the connection between the two, is nondeterministic.
4. There are multiple sources of information bearing on each decision. Whether the information is highly structured and derived from controlled experiments (example 1), consists of numerous studies using different approaches and likely reaching different conclusions (examples 2 and 3), or originates in different time periods and settings whose relation to the decision at hand must be assessed repeatedly (example 4), this information must be aggregated, explicitly or implicitly, in the decision.

We will often refer to "investigators" and "clients," terms due to Hildreth (1963). The investigator is the applied statistician or econometrician whose function is to convey quantitative information in a manner that facilitates and thereby improves decisions. The client may be the actual decisionmaker, or—more often—another scientist working to support the decision with information. The client's identity and preferences may be well known to the investigator (example: an expert witness hired by any interested party), or many clients may be unknown to the investigator (example: the readers of a subsequently well-cited academic paper reporting the investigator's work).

The objective of this book is to provide investigators with understanding and technical tools that will enable them to communicate effectively with clients, including decisionmakers and other investigators. Several themes emerge:

1. Make all assumptions explicit.
2. Explicitly quantify all of the essentials, including the assumptions.
3. Synthesize, or provide the means to synthesize, different approaches and models.
4. Represent the inevitable uncertainty in ways that will be useful to the client.

The understanding of effective communication is grounded in Bayesian inference and decision theory. The grounding emerges not from any single high-minded principle, but rather from the fact that this foundation is by far the most coherent and comprehensive one that presently exists. It may eventually be superseded by a superior model, but for the foreseeable future it is the foundation of economics and rational quantitative decisionmaking.

The reader grounded in non-Bayesian methods need not take any of this for granted. To these readers, the utility of the approach taken here will emerge as successive real problems succumb to effective treatment using Bayesian methods, while remaining considerably more difficult, if not entirely intractable, using non-Bayesian approaches.

Simulation methods provide an indispensable link between principles and practice. These methods, essentially unavailable before the late 1980s, represent uncertainty in terms of a large but finite number of synthetic random drawings from the distribution of unobservables (examples: parameters and latent variables), conditional on what is known (examples: data and the constraints imposed by economic theory) and the model(s) used to relate unobservables to what is known. Algorithms for the generation of the synthetic random drawings are governed by this representation of uncertainty. The investigator who masters these tools not only becomes a more fluent communicator of results but also greatly expands the choices of contexts, or models, in which to represent uncertainty and provide useful information to decisionmakers.

1.1 TWO EXAMPLES

This chapter is an overview of the chapters that follow. It provides much of what is needed for the reader to be a knowledgeable client, that is, a receiver of information communicated in the way just discussed. Being an effective investigator requires the considerably more detailed and technical understanding that the other chapters convey.

1.1.1 Public School Class Sizes

The determination of class size in public schools is a political and fiscal decision whose details vary from state to state and district to district. Regardless of the details, the decision ultimately made balances the fact that, given the number of students in the district, a lower student : teacher ratio is more costly, against the perception that a lower student : teacher ratio also increases the quality of education. Moreover, quality is difficult to measure. The most readily available measures are test scores. Changes made in federal funding of locally controlled public education since 2001 emphasize test scores as indicators of quality, and create fiscal incentives for local school boards to maintain and improve the test scores of students in their districts.

In this environment, there are several issues that decisionmaking clients must address and in which Bayesian investigation is important:

1. What is the relationship between the student : teacher ratio and test scores? Quite a few other factors, all of them measurable, may also affect test scores. We are uncertain about how to model the relationship, and for any one model there is uncertainty about the parameters in this model. Even if we were certain of both the model and the parameters, there would still be uncertainty about the resulting test scores. Full reporting and effective decisionmaking require that all these aspects of uncertainty be expressed.
2. The tradeoff between costs, on one hand, and quality of education, on the other hand, needs to be expressed. "Funding formulas" that use test scores to determine revenues available to school administrators (the clients) express at least part of this relationship quantitatively. In addition, a client may wish to see the implications of alternative valuations of educational quality, as expressed in test scores, for decisions about class size. Funding formulas may be expressed in terms of targets that make this an analytically challenging problem. The simulation methods that are an integral part of contemporary Bayesian econometrics and statistics make it practical to solve such problems routinely.
3. Another set of prospective clients consists of elected and appointed policymakers who determine funding formulas. Since these policymakers are distinct from school administrators, any funding formula anticipates (at least implicitly) the way that these administrators will handle tradeoffs between the costs of classroom staffing and the incentives created in the funding formulas. Depending on administrators' behavior, different policies may incur higher, or lower, costs to attain the same outcome as measured by test scores.

Bayesian analysis provides a coherent and practical framework for combining information and data in a useful way in this and other decisionmaking situations. Chapters 2 and 3 take up the critical technical steps in integrating data and other sources of information and representing the values of the decisionmaking client. Chapter 4 provides the simulation methods that make it practical and routine to undertake the required analysis. The remaining chapters return to this particular decision problem at several points.

1.1.2 Value at Risk

Financial institutions (banks, brokerage firms, insurance companies) own a variety of financial assets, often with total value in the many billions of dollars. They may include debt issued by businesses, loans to individuals, and government bonds. These firms also have financial liabilities: for example, deposit accounts in the case of private banks and life insurance policies in the case of insurance companies. Taken together, the holdings of financial assets or liabilities by a firm are known as its "portfolio."

The value of an institution's portfolio, or of a particular part of it, is constantly changing. This is the case even if the institution initiates no change in its holdings, because the market price of the institution's assets or liabilities change from day to day and even minute to minute. Thus every such institution is involved in a risky business. In general, the larger the institution, the more difficult it is to assess this risk because of both the large variety of assets and liabilities and the number of individuals within the institution who have authority to change specified holdings in the institution's portfolio.

Beginning about 1990 financial institutions, and government agencies with oversight and regulatory responsibility for these institutions, developed measures of the risk inherent in institutions' portfolios. One of the simplest and most widely used is value at risk. To convey the essentials of the idea, let p_t be the market value of an institution's entire portfolio, or of a defined portion of it. In the former case, p_t is the net worth of the institution—what would remain in the hypothetical situation that the institution were to sell all its assets and meet all of its liabilities. In the latter case it might be (for example) the institution's holding of conventional mortgages, or of U.S. government bonds.

The value p_t is constantly changing. This is in part a consequence of holdings by the institution, but it is also a result of changes in market prices. Value at risk is more concerned with the latter, so p_t is taken to be the portfolio value assuming that its composition remains fixed. "Value at risk" is defined with respect to a future time period, say, t^*, relative to the current period t, where $t^* > t$ and $t^* - t$ may range from less than a day to as long as a month. A typical definition of value at risk is that it is the loss in portfolio value v_{t,t^*} that satisfies

$$P(p_t - p_{t^*} \geq v_{t,t^*}) = .05. \tag{1.1}$$

Thus value at risk is a hypothetical decline in value, such that the probability of an even greater decline is 5%. The choice of .05 appears arbitrary, since other values could be used, but .05 is by far the most common, and in fact some regulatory authorities establish limits of v_{t,t^*} in relation to p_t based on (1.1).

The precise notion of probability in (1.1) is important. Models for establishing value at risk provide a distribution for p_{t^*}, conditional on p_t and, perhaps, other information available at time t. From this distribution we can then determine v_{t,t^*}. Most models used for this purpose are formulated in terms of the period-to-period return on the portfolio

$$r_t = (p_t - p_{t-1})/p_{t-1},$$

and statistical modeling usually directly addresses the behavior of the time series

$$y_t = \log(1 + r_t) = \log(p_t/p_{t-1}). \tag{1.2}$$

One of the simplest models is

$$y_t \sim N(\mu, \sigma^2). \tag{1.3}$$

Even this simple model leaves open a number of questions. For example, is it really intended that the same model (including the same mean and variance) pertains today for "high tech" stocks as it did in 1999, before the rapid decline in their value? In any event, the parameters μ and σ^2 are unknown, so how is this fact to be handled in the context of (1.1)? This problem is especially vexing if μ and σ^2 are subject to periodic changes, as the high-tech example suggests at least sometimes must be the case if we insist on proceeding with (1.3).

One of the biggest difficulties with (1.3) is that it is demonstrably bad as a description of returns that are relatively large in absolute value, at least with fixed μ and σ^2. If we take as the fixed values of μ and σ^2 their conventional estimates based on daily stock price indices for the entire twentieth century, then the model implies that "crashes" like the one that occurred in October 1987, are events that are so rare as to be impossible for all practical purposes. [For the daily *Standard and Poors 500* stock returns for January 3, 1928–April 30, 1991, from Ryden et al. (1998) used in Sections 7.3 and 8.3, the mean is .000182, the standard deviation is .0135, and the largest return in absolute value is $-.228$, which is 16.9 standard deviations from the mean. If $z \sim N(0,1)$ then $P(z \leq -16.9) = 2.25 \times 10^{-64}$. The inverse of this probability is 4.44×10^{63}. Dividing by 260 trading days in the year yields 1.71×10^{61} years. The estimated age of the universe is 1.2×10^{10} years. Chapter 8 takes up Bayesian specification analysis, which is the systematic and constructive assessment of this sort of incongruence of a model with reality.] This, of course, makes explicit the fact that we are uncertain about more than just the unknown parameters μ and σ^2 in (1.3). In fact we are also uncertain about the functional form of the distribution, and our notion of "probability" in (1.1) should account for this, too.

Section 1.4 introduces an alternative to (1.3), which is developed in detail in Section 7.3. An important variant on the value at risk problem arises when a decisionmaker (say, a vice president of an investment bank) selects the value .05, as opposed to some other probability, in (1.1). This integration of behavior with probability is the foundation of Bayesian decision theory, as well as of important parts of modern economics and finance. We shall return to this theme repeatedly, for example, in Sections 2.4 and 4.1.

1.2 OBSERVABLES, UNOBSERVABLES, AND OBJECTS OF INTEREST

A model is a simplified description of reality that is at least potentially useful in decisionmaking. Since models are simplified, they are never literally true; whatever the "data-generating process" may be, it is not the model. Since models are constructed for the purpose of decisionmaking, different decision problems can appropriately lead to different models despite the fact that the reality they simplify is the same. A well-known example is Newtonian physics, which is inadequate when applied to cosmology or subatomic interactions but works quite well in launching satellites and sending people to the moon. In the development of positron emission tomography and other kinds of imaging based on the excitation of subatomic particles, on other hand, quantum mechanics (a different model) functions quite well whereas Newtonian mechanics is inapplicable.

All scientific models have certain features in common. One is that they often reduce an aspect of reality to a few quantitative concepts that are unobservable but organize observables in a way that is useful in decisionmaking. The gravitational constant or the charge of an electron in physics, and the variance of asset returns or the equation of a demand function in the examples in the previous section are all examples of unobservables. Observables can be measured directly; the acceleration of an object when dropped, the accumulation of charge on an electrode, average test scores in different school districts, and sample means of asset returns are all examples.

A model posits certain relationships between observables and unobservables; without these relationships the concepts embodied in the unobservables would be vacuous. A scientific model takes the form "Given the values of the unobservables, the observables will behave in the following way." The relationship may or may not be deterministic. Thus a model may be cast in the form

$$p(\mathbf{y} \mid \boldsymbol{\theta}),$$

in which $\boldsymbol{\theta}$ is a vector of unobservables and $\mathbf{y}$ is a vector of observables. The unobservables $\boldsymbol{\theta}$ are typically parameters or latent variables. It is important to distinguish between the observables $\mathbf{y}$, a random vector, and their values after they are observed, which we shall denote $\mathbf{y}^o$ and are commonly called "data." The functional form of the probability density p gives the model some of its content. In the simple example of Section 1.1.1 the observables might be pairs of student : teacher ratios and test score averages in a sample of school districts, and the unobservables the slope and intercept parameters of a normal linear regression model linking the two. In the simple example of Section 1.1.2, the observables might be asset returns $y_1, \ldots, y_T$, and the unobservable is $\sigma^2 = \text{var}(y_t)$.

The relationship $p(\mathbf{y} \mid \boldsymbol{\theta})$ between observables and unobservables is central, but it is not enough for decisionmaking. The relationship between the gravitational constant g and the acceleration that results when a force is applied to a mass is not enough to deliver a communications satellite into orbit—we had better know quite a lot about the value of g. Likewise, in assessing value at risk using the simplified model of Section 1.1.2, we must know something about σ^2. In general, the density $p(\mathbf{y} \mid \boldsymbol{\theta})$ may restrict the behavior of $\mathbf{y}$ regardless of $\boldsymbol{\theta}$ (e.g., when dropped, everyday objects accelerate at rates that differ negligibly with their mass) but for decisionmaking we must know something about $\boldsymbol{\theta}$. (An object will fall about how many meters per second squared at sea level?) A very general way to represent knowledge about $\boldsymbol{\theta}$ is by means of a density $p(\boldsymbol{\theta})$. Formally, we may combine $p(\boldsymbol{\theta})$ and $p(\mathbf{y} \mid \boldsymbol{\theta})$ to produce information about the observables:

$$p(\mathbf{y}) = \int p(\boldsymbol{\theta}) p(\mathbf{y} \mid \boldsymbol{\theta}) \, d\boldsymbol{\theta}.$$

How we obtain information about $\boldsymbol{\theta}$, and how $p(\boldsymbol{\theta})$ changes in response to new information are two of the central topics of this book. In particular, we shall turn shortly to the question of how information about $\boldsymbol{\theta}$ changes when $\mathbf{y}$ is observed.

In any decision there is typically more than one model at hand that is at least potentially useful. In fact, much of the work of actual decisionmakers lies in sorting through and weighing the implications of different models. To recognize this fact, we shall further index the relation between observables and unobservables by A to denote the model: $p(\mathbf{y} \mid \boldsymbol{\theta})$ becomes $p(\mathbf{y} \mid \boldsymbol{\theta}_A, A)$, and $p(\boldsymbol{\theta})$ becomes $p(\boldsymbol{\theta}_A \mid A)$. The vector of unobservables (in many cases, the parameters of the model A) $\boldsymbol{\theta}_A$ belongs to the set $\Theta_A \subseteq \mathbb{R}^{k_A}$. Alternative models will be denoted $A_1, A_2, \ldots$. Note that the unobservables need not be the same in the models, but the observables $\mathbf{y} \in Y$ are. When several models have the same set of observables, and then we obtain observations (which we call "data"), it becomes possible to discriminate among models. We shall return to this topic in Section 1.5, where we will see that with a bit more effort we can actually use the data to assign probabilities to competing models.

More generally, however, the models relevant to the decision at hand need not all have the same set of observables. A classic example is the work of Friedman (1957) on the marginal propensity to consume. One model (A_1) used aggregate time series data on income and consumption, while another model (A_2) used income and consumption measures for different households at the same point in time. The sets of models addressed the same unobservable—marginal propensity to consume—but reached different conclusions. Friedman's contribution was to show that the models A_1 and A_2 did, indeed, have different unobservables ($\boldsymbol{\theta}_{A_1}$ and $\boldsymbol{\theta}_{A_2}$), and that the differences in $\boldsymbol{\theta}_{A_1}$ and $\boldsymbol{\theta}_{A_2}$ were consistent with a third, more appropriate, concept of marginal propensity to consume. We shall denote the object of interest on which decisionmaking depends, and which all models relevant to the decision have something to say, by the vector $\boldsymbol{\omega}$. We shall denote the implications of model A for $\boldsymbol{\omega}$ by $p(\boldsymbol{\omega} \mid \mathbf{y}, \boldsymbol{\theta}_A, A)$. The models at hand *must* specify this density; if they do not, then they are not pertinent to the decision at hand.

We can apply this idea to the two examples in the previous section. In the case of the class size decision, $\boldsymbol{\omega}$ might be a $q \times 1$ vector of average test scores conditional on q alternative decisions that might be made about class size. In the case of value at risk, $\boldsymbol{\omega}$ might be a 5×1 vector, the value of the portfolio at the end of each of the next 5 business days.

In summary, we have identified three components of a *complete model*, A, involving unobservables (often parameters) $\boldsymbol{\theta}_A$, observables $\mathbf{y}$, and a vector of interest $\boldsymbol{\omega}$:

$$p(\boldsymbol{\theta}_A \mid A), \tag{1.4}$$

$$p(\mathbf{y} \mid \boldsymbol{\theta}_A, A), \tag{1.5}$$

$$p(\boldsymbol{\omega} \mid \mathbf{y}, \boldsymbol{\theta}_A, A). \tag{1.6}$$

The ordering of (1.4)–(1.6) emphasizes the fact that the model A specifies the joint distribution

$$p(\boldsymbol{\theta}_A, \mathbf{y}, \boldsymbol{\omega} \mid A) = p(\boldsymbol{\theta}_A \mid A)p(\mathbf{y} \mid \boldsymbol{\theta}_A, A)p(\boldsymbol{\omega} \mid \mathbf{y}, \boldsymbol{\theta}_A, A). \tag{1.7}$$

It is precisely this joint distribution that makes it possible to use data to inform decisions in an internally consistent manner, and—with more structure to be introduced in Section 1.6—addresses the question of which decision would be optimal.

Exercise 1.2.1 Conditional Probability. A test for the presence of a disease can be administered by a nurse. A result "positive" (+) indicates disease present; a result "negative" (−) indicates disease absent. However, the test is not perfect. The *sensitivity* of the test is the probability of a "positive" result conditional on the disease being present; it is .98. The *specificity* of the test is the probability of a "negative" result conditional on the disease being absent; it is .90. The *incidence* of the disease is the probability that the disease is present in a randomly selected individual; it is .005.

Denoting specificity by p, sensitivity by q, incidence by π, and test outcome by $+$ or $-$, develop an expression for the probability of disease conditional on a "positive" outcome and one for the probability of disease conditional on a "negative" outcome, if the test is administered to a randomly selected individual. Evaluate these expressions using the values given above.

Exercise 1.2.2 Non-Bayesian Statistics. Suppose the model A is $y \sim N(\mu, 1)$, $\mu \geq 0$, and the sample consists of a single observation $y = y^o$.

(a) Show that $S = (\max(y - 1.96, 0), \max(y + 1.96, 0))$ is a 95% classical confidence interval for μ, that is, $P(\mu \in S \mid \mu, A) = .95$.

(b) Show that if $y^o = -2.0$ is observed, then the 95% classical confidence interval is the empty set.

Exercise 1.2.3 Ex Ante and Ex Post Tests. Let y have a uniform distribution on the interval $(\theta, \theta + 1)$, and suppose that it is desired to test the null hypothesis $H_0 : \theta = 0$ versus the alternative hypothesis $H_1 : \theta = 0.9$ (which are the only two values of θ that are possible). A single observation x is available. Consider the test that rejects H_0 if $y \geq 0.95$, and accepts H_0 otherwise.

(a) Calculate the probabilities of type I and type II errors for this test.

(b) Explain why it does not make common sense, for decisionmaking purposes, to accept mechanically the outcome of this test when the observed y^o lies in the interval (0.9, 1.0).

1.3 CONDITIONING AND UPDATING

Because a complete model provides a joint density $p(\boldsymbol{\theta}_A, \mathbf{y}, \boldsymbol{\omega} \mid A)$, it is in principle possible to address the entire range of possible marginal and conditional distributions involving the unobservables, observables, and vector of interest. Let $\mathbf{y}^o$ denote the actual value of the observable—the data, "$\mathbf{y}$ observed." Then with the data in hand, the relevant probability density for a decision based on the model A is $p(\boldsymbol{\omega} \mid \mathbf{y}^o, A)$. This is the single most important principle in Bayesian inference in support of decisionmaking. The principle, however, subsumes a great many details taken up in subsequent chapters.

It is useful to break up the process of obtaining $p(\boldsymbol{\omega} \mid \mathbf{y}^o, A)$ into a number of steps, and to introduce some more terminology. The distribution corresponding to the density $p(\boldsymbol{\theta}_A \mid A)$ is usually known as the *prior distribution* and that corresponding to $p(\mathbf{y} \mid \boldsymbol{\theta}_A, A)$, as the *observables distribution*. The distribution of the unobservable $\boldsymbol{\theta}_A$, conditional on the observed $\mathbf{y}^o$, has density

$$p(\boldsymbol{\theta}_A \mid \mathbf{y}^o, A) = \frac{p(\boldsymbol{\theta}_A, \mathbf{y}^o \mid A)}{p(\mathbf{y}^o \mid A)} = \frac{p(\boldsymbol{\theta}_A \mid A)p(\mathbf{y}^o \mid \boldsymbol{\theta}_A, A)}{p(\mathbf{y}^o \mid A)} \propto p(\boldsymbol{\theta}_A \mid A)p(\mathbf{y}^o \mid \boldsymbol{\theta}_A, A). \tag{1.8}$$

Expression (1.8) is usually called the *posterior density* of the unobservable $\boldsymbol{\theta}_A$. The corresponding distribution is the *posterior distribution*.

The distinction between the prior and posterior distributions of $\boldsymbol{\theta}_A$ is not quite as tidy as this widely used notation and terminology suggests, however. To see this, define $\mathbf{Y}'_t = (\mathbf{y}'_1, \ldots, \mathbf{y}'_t)$, for $t = 0, \ldots, T$ with the understanding that $\mathbf{Y}_0 = \{\varnothing\}$, and consider the decomposition of the probability density of the observables $\mathbf{y} = \mathbf{Y}_T$:

$$p(\mathbf{y} \mid \boldsymbol{\theta}_A, A) = \prod_{t=1}^{T} p(\mathbf{y}_t \mid \mathbf{Y}_{t-1}, \boldsymbol{\theta}_A, A). \tag{1.9}$$

In fact, densities of observables are usually constructed in exactly this way, because when there is dependence between observations, a recursive model is typically the natural representation.

Suppose that $\mathbf{Y}_t^{o\prime} = (\mathbf{y}_1^{o\prime}, \ldots, \mathbf{y}_t^{o\prime})$ is available but $(\mathbf{y}_{t+1}^{o\prime}, \ldots, \mathbf{y}_T^{o\prime})$ is not. (If "t" denotes time, then we are between periods t and $t+1$). Then

$$p(\boldsymbol{\theta}_A \mid \mathbf{Y}_t^o, A) \propto p(\boldsymbol{\theta}_A \mid A)p(\mathbf{Y}_t^o \mid \boldsymbol{\theta}_A, A) = p(\boldsymbol{\theta}_A \mid A)\prod_{s=1}^{t} p(\mathbf{y}_s^o \mid \mathbf{Y}_{s-1}^o, \boldsymbol{\theta}_A, A).$$

When $\mathbf{y}_{t+1}^{o}$ becomes available, then

$$p(\boldsymbol{\theta}_A \mid \mathbf{Y}_{t+1}^{o}, A) \propto p(\boldsymbol{\theta}_A \mid A) \prod_{s=1}^{t+1} p(\mathbf{y}_s^{o} \mid \mathbf{Y}_{s-1}^{o}, \boldsymbol{\theta}_A, A)$$
$$\propto p(\boldsymbol{\theta}_A \mid \mathbf{Y}_t^{o}, A) p(\mathbf{y}_{t+1}^{o} \mid \mathbf{Y}_t^{o}, \boldsymbol{\theta}_A, A). \qquad (1.10)$$

The change in the distribution of $\boldsymbol{\theta}_A$ brought about by the introduction of $\mathbf{y}_{t+1}^{o}$, made clear in (1.10), is usually known as *Bayesian updating*. Comparing (1.10) with (1.8), note that $p(\boldsymbol{\theta}_A \mid \mathbf{Y}_t^{o}, A)$ plays the same role in (1.10) as does the prior density $p(\boldsymbol{\theta}_A \mid A)$ in (1.8), and that $p(\mathbf{y}_{t+1}^{o} \mid \mathbf{Y}_t^{o}, \boldsymbol{\theta}_A, A)$ plays the same role in (1.10) as does $p(\mathbf{y}^{o} \mid \boldsymbol{\theta}_A, A)$ in (1.8). Indeed, from the perspective of what happens at "time" $t + 1$, $p(\boldsymbol{\theta}_A \mid \mathbf{Y}_t^{o}, A)$ is the prior density of $\boldsymbol{\theta}_A$, and $p(\boldsymbol{\theta}_A \mid \mathbf{Y}_{t+1}^{o}, A)$ is the posterior density of $\boldsymbol{\theta}_A$. This emphasizes the fact that "prior" and "posterior" distributions (or densities, or moments, or other properties of unobservables) are always with respect to an incremental information set. In (1.8) this information is the entire data set $\mathbf{y}^{o} = \mathbf{Y}_T^{o}$, whereas in (1.10) it is $\mathbf{y}_{t+1}^{o}$.

From the posterior density (1.8), the density relevant for decisionmaking is

$$p(\boldsymbol{\omega} \mid \mathbf{y}^{o}, A) = \int_{\Theta_A} p(\boldsymbol{\theta}_A \mid \mathbf{y}^{o}, A) p(\boldsymbol{\omega} \mid \boldsymbol{\theta}_A, \mathbf{y}^{o}, A) \, d\boldsymbol{\theta}_A. \qquad (1.11)$$

It is important to acknowledge that we are proceeding in a way that is different from most non-Bayesian statistics, generally termed "classical" statistics. The key difference between Bayesian and non-Bayesian statistics is, in fact, in conditioning. Likelihood-based non-Bayesian statistics conditions on A and $\boldsymbol{\theta}_A$, and compares the implication $p(\mathbf{y} \mid \boldsymbol{\theta}_A, A)$ with $\mathbf{y}^{o}$. This avoids the need for any statement about the prior density $p(\boldsymbol{\theta}_A \mid A)$, at the cost of conditioning on what is unknown. Bayesian statistics conditions on $\mathbf{y}^{o}$, and utilizes the full density $p(\boldsymbol{\theta}_A, \mathbf{y}, \boldsymbol{\omega} \mid A)$ to build up coherent tools for decisionmaking, but demands specification of $p(\boldsymbol{\theta}_A \mid A)$.

The strategic advantage of Bayesian statistics stems from the fact that its conditioning is driven by the actual availability of information and by its complete integration with the theory of economic behavior under uncertainty, achieved by Friedman and Savage (1948, 1952). We shall return to this point in Section 1.6 and subsequently in this book.

Two additional matters need to be addressed, as well. The first is that (1.8) and (1.11) are mere formalities as stated; actually representing the densities $p(\boldsymbol{\theta}_A \mid \mathbf{y}^{o}, A)$ and $p(\boldsymbol{\omega} \mid \mathbf{y}^{o}, A)$ in practical ways for decisionmaking is a technical challenge of high order. Indeed, the principles stated here have been recognized since at least the mid-1950s, but it was not until the application of simulation methods in the 1980s that they began to take on the practical significance that they have today. We return to these developments in Section 1.4 and Chapter 4.

The other matter ignored is explicit attention to multiple models $A_1, \ldots, A_J$. In fact, it is not necessary to confine attention to a single model, and the developments here may be extended to several models simultaneously. We do this in Section 1.5.

Exercise 1.3.1 A Simple Posterior Distribution. Suppose that $y \sim N(\mu, 1)$ and the sample consists of a single observation y^o. Suppose that an investigator has a prior distribution for μ that is uniform on (0, 4).

(a) Derive the investigator's posterior distribution for μ.

(b) Suppose that $y^o = -2$. Find an interval (μ_1, μ_2) such that

$$P[\mu \in (\mu_1, \mu_2) \mid y^o] = 0.95.$$

(The answer consists of a pair of real numbers.)

(c) Do the same for the case $y^o = 1$.

(d) Are your intervals in (b) and (c) the shortest possible in each case? (You need not use a formal argument. A sketch is enough.)

Exercise 1.3.2 Applied Conditioning and Updating. On a popular, nationally televised game show the guest is shown three doors. Behind one door there is a valuable prize (e.g., a new luxury automobile), and behind the other two doors there are trivial prizes (perhaps a new toaster). The host of the game show knows which prizes are behind which doors. The guest, who cannot see the prizes, chooses one door for the host to open. But before he opens the door selected by the guest, the host always opens one of the two doors not chosen by the guest, and this always reveals a trivial prize. (The guest and the television audience, having watched the show many times, know that this always happens.) The guest is then given the opportunity to change her selected door. After the guest makes her final choice, that door is opened and the guest receives the prize behind her chosen door.

If you were the guest, would you change your door selection when given the opportunity to do so? Would you be indifferent about changing your selection? Defend your answer with a formal probability argument.

Exercise 1.3.3 Prior Distributions. Two graduate students play the following game. An amount of money W is placed in a sealed envelope. An amount $2W$ is placed in another sealed envelope. Student A is given one envelope, and student B is given the other envelope. (The assignment of envelopes is random, and the students do not know which envelope they have received.) Before student A opens his envelope and keeps the money inside, he may exchange envelopes with student B, if B is willing to do this. (At this point, B has not opened her envelope, either; the game is symmetric.) In either case, each student keeps the money in the envelope finally accepted. Both students are rational and risk-neutral; that is, they behave so as to maximize the expected value of the money they keep at the end of the game.

Student A reasons as follows. "There is an unknown amount of money, x, in my envelope. It is just as likely that B's envelope has $2x$ as it is that it has $x/2$. Conditional on x, my expected gain from switching envelopes is $.5(2x + .5x) - x = .25x$. Since this is positive for all x, I should offer to switch envelopes."

Student B says that the expected gain from switching envelopes is zero.

Explain the fallacy in A's argument, and provide the details of B's argument. In each case use the laws of probability carefully.

1.4 SIMULATORS

Decisionmaking requires specific tasks involving posterior distributions. The financial manager in Section 1.1.2 is concerned about the distribution of values of an asset 5 days from now $\omega = p_{T+5} = p_T \exp(\sum_{s=1}^{5} y_{T+s})$. She has at hand observations on returns through the present time period, T, of the form $\mathbf{y}^o = (y_1^o, \ldots, y_T^o)'$, and is using a model with a parameter vector $\boldsymbol{\theta}_A$. The value at risk she seeks to determine is the number c with the property

$$\int_{-\infty}^{p_T - c} p(\omega \mid \mathbf{y}^o, A)\, d\omega = 0.05.$$

The manager might recognize that she can decompose this problem into two parts. First, if she knows the value of $\boldsymbol{\theta}_A$—or, more precisely, if the model A specifies the value of $\boldsymbol{\theta}_A$ with no uncertainty—then finding c amounts to deriving the inverse cumulative distribution function (cdf) of ω from $p(y_{T+1}, \ldots, y_{T+5} \mid \mathbf{y}^o, \boldsymbol{\theta}_A, A)$. This task can be completed analytically for the model (1.3) with known μ and σ^2, but for realistic models with uncertainty about parameters this is at best tedious and in general impossible.

At this point the financial manager, or one of her staff, might point out that it is relatively easy to simulate most models of financial time series. One such model is the Markov mixture of normals model, discussed in more detail in Section 7.3, in which each y_t is drawn from one of L alternative normal distributions $N(\mu_j, \sigma_j^2)$. Each day t is characterized by an unobserved state variable s_t that assumes one of the values 1, 2, ... or L, and then

$$s_t = j \Rightarrow y_t \sim N(\mu_j, \sigma_j^2). \tag{1.12}$$

The state variables themselves obey a first-order Markov process in which

$$P(s_t = j \mid s_{t-1} = i) = p_{ij}. \tag{1.13}$$

In applications to financial modeling it is reasonable that the values of σ_j^2 vary substantially depending on the state, for example, $\sigma_1^2/\sigma_2^2 \approx 3$, and the state variable is persistent as indicated by $p_{ii} \gg \sum_{j \neq i} p_{ij}$. Such a structure gives rise to episodes of high and low volatility, a feature seen in most financial returns data.

Widely available mathematical applications software makes it easy to simulate this and many other models. Given the current state $s_t = i$, the next period's state is drawn from the distribution (1.13), and then y_{t+1} is drawn from the selected normal distribution in (1.12). Our firm manager can exploit this fact if she knows the parameters of the model and the current state $s_T = j$. She repeatedly simulates the model forward from the current day T, obtaining in simulation m the returns $y_{T+s}^{(m)}$ $(s = 1, \ldots, 5)$ and the corresponding simulated asset price 5 days hence, $\omega^{(m)} = p_T^o \exp(\sum_{s=1}^{5} y_{T+s}^{(m)})$. At the end she can sort the M simulations of ω, and find a number $c^{(M)}$ such that 5% of the draws are below and 95% are above $p_T^o - c^{(M)}$. It turns out that $c^{(M)} \stackrel{\text{a.s.}}{\to} c$ as M increases.

This solves only part of the manager's problem. The model, in fact, has many unobservables, not only the unknown parameters μ_j, σ_j^2 and p_{ij} but also the states s_t. Together they constitute the unobservables vector $\boldsymbol{\theta}_A$ in this model. The simulation just described requires all of the parameters and the current state s_T. Noting that

$$p(\omega \mid \mathbf{y}^o, A) = \int_{\Theta_A} p(\omega \mid \mathbf{y}^o, \boldsymbol{\theta}_A, A) p(\boldsymbol{\theta}_A \mid \mathbf{y}^o, A) \, d\boldsymbol{\theta}_A, \tag{1.14}$$

the manager might well recognize that if she could simulate

$$\boldsymbol{\theta}_A^{(m)} \backsim p(\boldsymbol{\theta}_A \mid \mathbf{y}^o, A) \tag{1.15}$$

and next apply the algorithm just described to draw

$$\omega^{(m)} \backsim p(\omega \mid \mathbf{y}^o, \boldsymbol{\theta}_A^{(m)}, A), \tag{1.16}$$

then the distribution of $\omega^{(m)}$ would be that corresponding to the density (1.14).

This strategy is valid, but producing the draws in (1.15) is much more challenging than was developing the algorithm behind (1.16). The latter simulation was relatively easy because it corresponds to the recursion in the natural expression of the model; recall (1.4)–(1.6). Given $\boldsymbol{\theta}_A$, the model tells us how $\mathbf{y}_1$, then $\mathbf{y}_2$, and so on, are produced, and as a consequence simulating into the future is typically straightforward. The distribution (1.15), on the other hand, asks us to reverse this process: given that a set of observables was produced by the model A, with prior distribution $p(\boldsymbol{\theta}_A \mid A)$ and observables distribution $p(\mathbf{y} \mid \boldsymbol{\theta}_A, A)$, make drawings from the distribution with posterior density $p(\boldsymbol{\theta}_A \mid \mathbf{y}^o, A)$. The formal definition (1.8) is not much help in this task.

This impasse is typical if we attempt to use simulation to unravel the actual distribution corresponding to $p(\omega \mid \mathbf{y}^o, A)$ in a useful way. Until the late 1980s this problem had succumbed to solution in only a few simple cases, and these did not go very far beyond the even smaller set of cases that could be solved analytically from start to finish. Geweke (1989a) pointed out that importance sampling methods described in Hammersly and Handscomb (1964) could be used together with standard optimization methods to simulate $\boldsymbol{\theta}_A^{(m)} \backsim p(\boldsymbol{\theta}_A \mid \mathbf{y}^o, A)$. The following year Gelfand and Smith (1990) published their discovery that methods then being used in image reconstruction could be adapted to construct a Markov chain G such that if

$$\boldsymbol{\theta}_A^{(m)} \backsim p(\boldsymbol{\theta}_A \mid \boldsymbol{\theta}_A^{(m-1)}, \mathbf{y}^o, G)$$

then $\boldsymbol{\theta}_A^{(m)} \xrightarrow{d} p(\boldsymbol{\theta}_A \mid \mathbf{y}^o, A)$. This work in short order burgeoned into an even more general set of procedures, known as *Markov chain Monte Carlo* (MCMC), which achieves the same result for almost any complete model. Section 7.3 shows how to apply these methods to the Markov mixture of normals model used in this example.

All of these methods, including importance sampling, produce what are known as *posterior simulators*. These algorithms make it practical to address quantitative decisionmaking problems, using a rich variety of models. Posterior simulators are the focus of Chapter 4.

1.5 MODELING

To this point we have taken the complete model (1.4)–(1.6) as given. In fact, the investigator begins with much less. Typically the vector of interest $\boldsymbol{\omega}$ is specified (at least implicitly) by the client making the decision. The composition of the observables vector is sometimes obvious, but in general the question of which observables are best used to inform quantitative decisionmaking is itself an important, interesting, and sometimes difficult question.

This leaves almost all of (1.4)–(1.6) to be specified by the investigator. There is, of course, no algorithm mapping reality into models. The ability to isolate the important features of an actual decision problem, and organize them into a model that is workable and brings to bear all the important features of the decision is an acquired and well-rewarded skill. However this process does involve some specific technical steps that themselves can be cast as intermediate decision problems addressed by the investigator.

One such step is to incorporate competing models $A_1, A_2, \ldots, A_J$ in the process of inference and decisionmaking. In Section 1.2 we constructed a joint probability distribution for the unobservables $\boldsymbol{\theta}_A$, the observables $\mathbf{y}$, and the vector of interest $\boldsymbol{\omega}$, in the context of model A. Suppose that we have done that for each of models $A_1, \ldots, A_J$ and that the vector of observables is the same for each of these models. Then we have

$$p(\boldsymbol{\theta}_{A_j} \mid A_j), \quad p(\mathbf{y} \mid \boldsymbol{\theta}_{A_j}, A_j), \quad p(\boldsymbol{\omega} \mid \boldsymbol{\theta}_{A_j}, \mathbf{y}, A_j) \quad (j = 1, \ldots, J).$$

If we now provide a prior probability $p(A_j)$ for each model, with $\sum_{j=1}^{J} p(A_j) = 1$, there is a complete probability distribution over models, unobservables, observables, and the vector of interest. Let $A = \bigcup_{j=1}^{J} A_j$. In each model the density (1.14), built up from (1.8) and (1.6), provides $p(\boldsymbol{\omega} \mid \mathbf{y}^o, A_j)$. Then

$$p(\boldsymbol{\omega} \mid \mathbf{y}, A) = \sum_{j=1}^{J} p(\boldsymbol{\omega} \mid \mathbf{y}, A_j) p(A_j \mid \mathbf{y}, A). \tag{1.17}$$

The posterior density of $\boldsymbol{\omega}$ is given by (1.17) with the data $\mathbf{y}^o$ replacing the observable $\mathbf{y}$. It is a weighted average of the posterior densities of $\boldsymbol{\omega}$ in the various models;

indeed, (1.17) is sometimes called *model averaging*. The weights are

$$p(A_j \mid \mathbf{y}^o, A) = \frac{p(A_j)p(\mathbf{y}^o \mid A_j)}{p(\mathbf{y}^o \mid A)} = \frac{p(A_j)p(\mathbf{y}^o \mid A_j)}{\sum_{j=1}^{J} p(A_j)p(\mathbf{y}^o \mid A_j)}. \tag{1.18}$$

The data therefore affect the weights by means of

$$\begin{aligned} p(\mathbf{y}^o \mid A_j) &= \int_{\Theta_{A_j}} p(\boldsymbol{\theta}_{A_j}, \mathbf{y}^o \mid A_j)\, d\boldsymbol{\theta}_{A_j} \\ &= \int_{\Theta_{A_j}} p(\boldsymbol{\theta}_{A_j} \mid A_j)p(\mathbf{y}^o \mid \boldsymbol{\theta}_{A_j}, A_j)\, d\boldsymbol{\theta}_{A_j}. \end{aligned} \tag{1.19}$$

The number $p(\mathbf{y}^o \mid A_j)$ is known as the *marginal likelihood* of model A_j. The technical obstacles to the computation, or approximation, of $p(\mathbf{y}^o \mid A_j)$ are at least as severe as those for simulating $\boldsymbol{\theta}_A$, but rapid progress on this problem was made during the 1990s, and this is becoming an increasingly routine procedure.

For any pair of models (A_i, A_j), we obtain

$$\frac{p(A_i \mid \mathbf{y}^o)}{p(A_j \mid \mathbf{y}^o)} = \frac{p(A_i)}{p(A_j)} \cdot \frac{p(\mathbf{y}^o \mid A_i)}{p(\mathbf{y}^o \mid A_j)}. \tag{1.20}$$

Note that the ratio is independent of the composition of the full complement of models in A. It is therefore a useful summary of the evidence in the data $\mathbf{y}^o$ about the relative posterior probabilities of the two models. The left side of (1.20) is known as the *posterior odds ratio*, and it is decomposed on the right side into the product of the *prior odds ratio* and the *Bayes factor*. Expressions (1.17) and (1.18) imply that providing the marginal likelihood of a model is quite useful for the subsequent work, including decisionmaking, with several models.

Expression (1.19) for the marginal likelihood makes plain that the bearing of a model on decisionmaking—its weight in the model averaging process (1.17)—depends on the prior density $p(\boldsymbol{\theta}_{A_i} \mid A_i)$ as well as the observables density $p(\mathbf{y} \mid \boldsymbol{\theta}_{A_i}, A_i)$. In particular, a model A_i may be an excellent representation of the data in the sense that for some value(s) of $\boldsymbol{\theta}_{A_i}$, $p(\mathbf{y}^o \mid \boldsymbol{\theta}_{A_i}, A_i)$ is large relative to the best fit $p(\mathbf{y}^o \mid \boldsymbol{\theta}_{A_j}, A_j)$ in other models, but if $p(\boldsymbol{\theta}_{A_i} \mid A_i)$ places low (even zero) probability on those values, then the posterior odds ratio (1.20) may run heavily against model A_i.

The investigator's problem in specifying $p(\boldsymbol{\theta}_{A_i} \mid A_i)$ is no more (or less) difficult than that of designing the observables density $p(\mathbf{y} \mid \boldsymbol{\theta}_{A_i}, A_i)$. The two are inseparable: $p(\boldsymbol{\theta}_{A_i} \mid A_i)$ has no implications for observables without $p(\mathbf{y} \mid \boldsymbol{\theta}_{A_i}, A_i)$, and $p(\mathbf{y} \mid \boldsymbol{\theta}_{A_i}, A_i)$ says little about $p(\mathbf{y} \mid A_i)$ until we have $p(\boldsymbol{\theta}_{A_i} \mid A_i)$ in hand. The first two components of any complete model, (1.4) and (1.5), combined with some relatively simple simulation, can help in these steps of the investigator's problem. Suppose that one or more aspects of the observables $\mathbf{y}$, which we can represent quite generally as $g(\mathbf{y})$, are thought to be important aspects of reality bearing on a decision, that therefore should be well represented by the model. In the case of our financial decisionmaker from Section 1.1.2, one concern might focus on the model's stance on "crashes" in the value of financial assets like the one day return of worse than -20% experienced during October 1987, for many assets; then $g(\mathbf{y}) = 1$ if $\mathbf{y}$ exhibits such a day and $g(\mathbf{y}) = 0$ if not. For any specified prior and observables densities, it is generally straightforward to simulate

$$\boldsymbol{\theta}_A^{(m)} \backsim p(\boldsymbol{\theta}_A \mid A), \quad \mathbf{y}^{(m)} \backsim p(\mathbf{y} \mid \boldsymbol{\theta}_A^{(m)}, A)$$

and then construct $g(\mathbf{y}^{(m)})$. The resulting $g(\mathbf{y}^{(m)})$ $(m = 1, \ldots,)$ is an independent identically distributed (i.i.d.) sample from $p[g(\mathbf{y}) \mid A]$.

This process enables the investigator to understand key properties of a model A before undertaking the more demanding task of developing a posterior simulator $\boldsymbol{\theta}_A^{(m)} \backsim p(\boldsymbol{\theta}_A \mid \mathbf{y}^o, A)$. It provides guidance in choosing the prior density $p(\boldsymbol{\theta}_A \mid A)$ corresponding to $p(\mathbf{y} \mid \boldsymbol{\theta}_A, A)$, and can reveal that an observables density $p(\mathbf{y} \mid \boldsymbol{\theta}_A, A)$ fails to capture important aspects of reality no matter what the value of $\boldsymbol{\theta}_A$. These tasks are all part of what is generally referred to as "model specification" in econometrics. We shall return to them in detail in Chapter 8.

1.6 DECISIONMAKING

The key property of the vector of interest $\boldsymbol{\omega}$ is that it mediates aspects of reality that are relevant for the decision that motivates the econometric or statistical modeling in the first place. To illustrate this point, return again to the decision of school administrators about the class sizes described in Section 1.1.1. School administrators prefer certain outcomes to others; for example, it is quite likely that they prefer high test scores and small teaching budgets to low test scores and large expenditures for teachers' salaries. Suppose, for sake of simplicity, that the teaching budget can be controlled with certainty by hiring more or fewer teachers. In Bayesian decision theory such a decision is known as an *action*, and represented generically by a vector $\mathbf{a}$. The vector of interest $\boldsymbol{\omega}$ includes all the uncertain factors that matter to administrators in evaluating the outcome; it could be a single summary of test scores, or it might disaggregate to measure test outcomes for different groups of students. The expected utility paradigm, associated with von Neumann and Morgenstern (1944), states that decisions are made so as to maximize the expected value of a *utility function* $U(\mathbf{a}, \boldsymbol{\omega})$ defined over all possible outcomes and decisions. The term "utility" is universal in economics, whereas in Bayesian decision theory the concept of "loss" prevails; the loss function $L(\mathbf{a}, \boldsymbol{\omega})$ is used in place of the utility function $U(\mathbf{a}, \boldsymbol{\omega})$. The only distinction is that the decisionmaker seeks to minimize, not maximize, $E[L(\mathbf{a}, \boldsymbol{\omega})]$. We can always take $L(\mathbf{a}, \boldsymbol{\omega}) = -U(\mathbf{a}, \boldsymbol{\omega})$.

This paradigm fits naturally into the relationship between the model A with parameter vector $\boldsymbol{\theta}_A$, the observable vector $\mathbf{y}$, and the vector of interest $\boldsymbol{\omega}$. Expression (1.11) provides the distribution relevant to the decision in the use of a single model A—that is, the distribution relevant for the expectation $E[L(\mathbf{a}, \boldsymbol{\omega})]$, which therefore may be written

$$E[L(\mathbf{a}, \boldsymbol{\omega}) \mid \mathbf{y}^o, A] = \int_{\Theta_A} L(\mathbf{a}, \boldsymbol{\omega}) p(\boldsymbol{\omega} \mid \mathbf{y}^o, A)\, d\boldsymbol{\omega}$$

$$= \int_{\Omega} \int_{\Theta_A} L(\mathbf{a}, \boldsymbol{\omega}) p(\boldsymbol{\theta}_A \mid \mathbf{y}^o, A) p(\boldsymbol{\omega} \mid \boldsymbol{\theta}_A, \mathbf{y}^o, A)\, d\boldsymbol{\theta}_A\, d\boldsymbol{\omega}.$$

Section 1.4 outlines how, in principle, we might obtain drawings $\boldsymbol{\omega}^{(m)}$ from (1.11). Typically those drawings can be used to solve the formal decision problem. In the simplest case, there are only two possible actions ($a = 0$, $a = 1$), and the drawings $\boldsymbol{\omega}^{(m)}$ are i.i.d. Then, so long as $E[L(0, \boldsymbol{\omega})]$ and $E[L(1, \boldsymbol{\omega})]$ both exist—a requirement for the expected utility paradigm to be applicable—the strong law of large numbers implies

$$M^{-1} \sum_{m=1}^{M} L(a, \boldsymbol{\omega}^{(m)}) \stackrel{\text{a.s.}}{\to} E[L(a, \boldsymbol{\omega}) \mid \mathbf{y}^o, A]$$

for $a = 0$ and $a = 1$. More generally, if $\mathbf{a}$ is continuous and $E[L(\mathbf{a}, \boldsymbol{\omega})]$ is twice differentiable, then typically

$$M^{-1} \sum_{m=1}^{M} \partial L(\mathbf{a}, \boldsymbol{\omega}^{(m)})/\partial \mathbf{a} \stackrel{\text{a.s.}}{\to} \partial E[L(\mathbf{a}, \boldsymbol{\omega}) \mid \mathbf{y}^o, A]/\partial \mathbf{a}$$

and this feature may be exploited to solve for the value $\mathbf{a} = \widehat{\mathbf{a}}$ that minimizes expected loss, using a steepest-descent algorithm. More often, the draws $\boldsymbol{\omega}^{(m)}$ from (1.11) are serially dependent, but this complication turns out not to be essential. We revisit these issues at the level of technical detail required for their application subsequently in Chapter 4. This formalization of the decisionmaking process can be extended to the case of several competing models, using the setup developed in Sections 2.6 and 8.2.

Decisionmaking plays, or should play, an important role in modeling and inference. It focuses attention, first, on the vector of interest $\boldsymbol{\omega}$ that is relevant to the decision problem—namely, the unobservables that will ultimately drive the subjective evaluation of the decision ex post. Given $\boldsymbol{\omega}$, we may then consider the observables $\mathbf{y}$ that are most likely to be useful in providing information about $\boldsymbol{\omega}$ before the decision is made. The observables then govern consideration of the relevant models A, their vectors of unobservables $\boldsymbol{\theta}_A$, and the associated prior densities $p(\boldsymbol{\theta}_A \mid A)$. Note that this amounts to stepping backward through the marginal–conditional decomposition (1.7), a process that is often informal.

In practice, formal decisionmaking is most useful for the structure that it places on the research endeavor from start to finish. Rarely, if ever, do decisionmakers think and talk about decisions entirely and explicitly within the formal framework we have laid out here. However, the discipline of formal decision theory combined with Bayesian inference can, when well executed, earn the respect of real decisionmakers, and therefore a "seat at the table." In many ways, this is the ultimate goal of Bayesian inference, and achieving it is a high reward to applied econometrics and statistics.

CHAPTER 2

Elements of Bayesian Inference

This chapter systematically develops the principles of Bayesian inference that are used repeatedly in the rest of the book. The purpose is threefold: to set up notation, to provide an introduction for statisticians and econometricians unfamiliar with Bayesian methods, and to set forth some technical challenges addressed in subsequent chapters. The development emphasizes the eventual application of Bayesian inference in decisionmaking contexts.

The introduction here is concise, concentrating on analytic essentials and touching lightly on some concepts of greater depth. Those versed in Bayesian methods at the level of Berger (1985) or Bernardo and Smith (1994) can easily skip to the fourth chapter and beyond, consulting Section 2.1 as required for notation. Those seeking a complete introduction can consult these references as well as the next chapter, perhaps supplemented by DeGroot (1970), Berger and Wolpert (1988), and Poirier (1988) on the distinction between Bayesian and non-Bayesian methods. On Bayesian econometrics in particular, see Zellner (1971), Poirier (1995), Koop (2003), and Lancaster (2004). All the concepts introduced in this chapter are illustrated using the normal linear regression model.

The results presented in this chapter are not operational. In particular, they all involve integrals that rarely can be evaluated analytically, and the dimensions of integration are typically greater than the four or five for which deterministic numerical methods are practical. Chapter 4 provides the analytical development of posterior simulators, which are then used in the practical procedures developed subsequently.

2.1 BASICS

Bayesian inference takes place in the context of one or more statistical or econometric models. A model describes the behavior of a $p \times 1$ vector of observable random vectors $\mathbf{y}_t$. The index t has one of a number of interpretations determined by the application including time, individuals in a random sample, location, as

Contemporary Bayesian Econometrics and Statistics, by John Geweke

well as combinations of these and other relevant attributes of the observables. Let $\mathbf{Y}_t = \{\mathbf{y}_s\}_{s=1}^t$ denote the subsample consisting of the first t observables. The sample space for $\mathbf{y}_t$ is ψ_t, that for $\mathbf{Y}_t$ is Ψ_t, and $\psi_0 = \Psi_0 = \{\emptyset\}$. A model, A, specifies a corresponding sequence of probability density functions

$$p(\mathbf{y}_t \mid \mathbf{Y}_{t-1}, \boldsymbol{\theta}_A, A), \tag{2.1}$$

in which $\boldsymbol{\theta}_A$ is a $k_A \times 1$ vector of unobservables and $\boldsymbol{\theta}_A \in \Theta_A \subseteq \mathbb{R}^{k_A}$. The interpretation of $\boldsymbol{\theta}_A$ depends on the model A. In many contexts $\boldsymbol{\theta}_A$ is a vector of a fixed number of unknown parameters. However, $\boldsymbol{\theta}_A$ may also include unobservables that economists and other social scientists call "latent variables," and in that case k_A typically depends on the size of the sample. We return to this interpretation in more detail in Section 3.1.

The notation $p(\cdot)$ will be used to denote a generic probability density function (pdf) with respect to a generic measure $\nu(\cdot)$. The measure ν permits the random vector to be continuous (Lebesgue measure), discrete (point mass) or a mixture of the two. Thus, for example, if $\nu(\cdot)$ assigns ordinary (Lebesgue) measure to the unit interval and the measure one to the point $x = 0.7$, and $p(x) = \left(\frac{1}{2}\right) I_{(0,1)}(x)$, then $P(A) = \int_A p(x)\, d\nu(x)$ is the probability function corresponding to a random variable that is uniformly distributed on the unit interval with probability $\frac{1}{2}$, and takes on the value 0.7 with probability $\frac{1}{2}$.

The pdf of $\mathbf{Y}_T$, conditional on the model and the vector of unobservables $\boldsymbol{\theta}_A$, is

$$p(\mathbf{Y}_T \mid \boldsymbol{\theta}_A, A) = \prod_{t=1}^{T} p(\mathbf{y}_t \mid \mathbf{Y}_{t-1}, \boldsymbol{\theta}_A, A). \tag{2.2}$$

If the model specifies that the $\mathbf{y}_t$ are independent and identically distributed, then

$$p(\mathbf{y}_t \mid \mathbf{Y}_{t-1}, \boldsymbol{\theta}_A, A) = p(\mathbf{y}_t \mid \boldsymbol{\theta}_A, A)$$

and in this case

$$p(\mathbf{Y}_T \mid \boldsymbol{\theta}_A, A) = \prod_{t=1}^{T} p(\mathbf{y}_t \mid \boldsymbol{\theta}_A, A).$$

More generally, the index t may pertain to cross sections, to time series, or both. Time series models and language preserve this generality.

When used alone, expressions like $\mathbf{y}_t$ and $\mathbf{Y}_T$ denote random vectors. In equations (2.1) and (2.2) $\mathbf{y}_t$ and $\mathbf{Y}_T$ are arguments of functions. These uses are distinct from the observed values themselves. To preserve this distinction explicitly, denote observed $\mathbf{y}_t$ by $\mathbf{y}_t^o$ and observed $\mathbf{Y}_T$ by $\mathbf{Y}_T^o$. In general, the superscript o will denote the observed value of a random vector. For example, if the observed value of the random vector $\mathbf{Y}_T$ is $\mathbf{Y}_T^o$, then the *likelihood function* is any function $L(\boldsymbol{\theta}_A; \mathbf{Y}_T^o, A) \propto p(\mathbf{Y}_T^o \mid \boldsymbol{\theta}_A, A)$. Unless the number of observations, T, is important to the topic at hand, we shall simply denote the observables by $\mathbf{y}$, their observed values (the data) by $\mathbf{y}^o$, and the sample space by Ψ.

The pdf (2.2) is the first component of the model A introduced in Section 1.2. The second component is the prior density $p(\boldsymbol{\theta}_A \mid A)$. The prior density is a formal representation of the values of the vector of unobservables $\boldsymbol{\theta}_A$ that are reasonable in the model A. It reflects everything that is known, or believed, about $\boldsymbol{\theta}_A$ prior to learning the observed values $\mathbf{y}^o$. For example, if Θ_A has finite Lebesgue measure, then the prior density could assign probability uniformly in Θ_A: $p(\boldsymbol{\theta}_A) = [\nu(\Theta_A)]^{-1}$. The notation extends to the extreme case in which the model assigns exact values $\boldsymbol{\theta}_A^*$ to all unobservables, $\boldsymbol{\theta}_A = \boldsymbol{\theta}_A^*$. In that case ν places point mass at $\boldsymbol{\theta}_A^*$ and $p(\boldsymbol{\theta}_A^*) = 1$. In general, the functional forms of the observables densities $p(\mathbf{y}_t \mid \mathbf{Y}_{t-1}, \boldsymbol{\theta}_A, A)$ $(t = 1, \ldots, T)$ and the prior density $p(\boldsymbol{\theta}_A \mid A)$ are chosen simultaneously as part of the model A as discussed in Section 1.2.

Given (2.2) and the prior density, we obtain

$$p(\mathbf{y}, \boldsymbol{\theta}_A \mid A) = p(\boldsymbol{\theta}_A \mid A) p(\mathbf{y} \mid \boldsymbol{\theta}_A, A). \tag{2.3}$$

Thus model A provides a joint density of the observables, $\mathbf{y}$, and unobservables, $\boldsymbol{\theta}_A$. Expression (2.3) decomposes this density as a marginal density in $\boldsymbol{\theta}_A$ (the prior) and a density in $\mathbf{y}$ conditional on $\boldsymbol{\theta}_A$ (the data density). The joint density can also be expressed as the product of the marginal density in $\mathbf{y}$ and the conditional density in $\boldsymbol{\theta}_A$:

$$p(\mathbf{y}, \boldsymbol{\theta}_A \mid A) = p(\mathbf{y} \mid A) p(\boldsymbol{\theta}_A \mid \mathbf{y}, A). \tag{2.4}$$

Both terms on the right side of (2.4) may be written in terms of the prior density and observables density. The marginal density in $\mathbf{y}$ is

$$p(\mathbf{y} \mid A) = \int_{\Theta_A} p(\mathbf{y}, \boldsymbol{\theta}_A \mid A)\, d\nu(\boldsymbol{\theta}_A) = \int_{\Theta_A} p(\boldsymbol{\theta}_A \mid A) p(\mathbf{y} \mid \boldsymbol{\theta}_A, A)\, d\nu(\boldsymbol{\theta}_A). \tag{2.5}$$

Note that the integral appearing in this expression is absolutely convergent for almost all $\mathbf{y}$, because $p(\mathbf{y}, \boldsymbol{\theta}_A \mid A)$ is the joint distribution of $\mathbf{y}$ and $\boldsymbol{\theta}_A$. Expression (2.5) is the density of the observable $\mathbf{y}$ implied by the model a priori. It is a prediction of what the data will be, before they are observed. It makes explicit the predictive content of the model A, and indicates the instrumental role of the vector of unobservables $\boldsymbol{\theta}_A$ in expressing this prediction. This expression also emphasizes that since the observables density and prior density are complementary in forming the predictions of the model, they should be chosen together. If $\mathbf{y}$ is replaced with $\mathbf{y}^o$ in (2.5), then $p(\mathbf{y}^o \mid A)$ is a real number.

Definition 2.1.1 The *marginal likelihood* of the model A is $p(\mathbf{y}^o \mid A)$.

This terminology [which dates at least to Raiffa and Schlaifer (1961), Section 2.1] reflects the fact that with $\mathbf{y}^o$ in place of $\mathbf{y}$, (2.5) can be interpreted as "marginalizing" the vector of unobservables in the likelihood function, that is, $p(\mathbf{y}^o \mid A) = \int_{\Theta_A} L(\boldsymbol{\theta}_A; \mathbf{y}^o, A)\, d\nu(\boldsymbol{\theta}_A)$. But note that this is true only if $L(\boldsymbol{\theta}_A; \mathbf{y}^o, A)$ carries forward all constants of integration in $p(\mathbf{y}^o, \boldsymbol{\theta}_A \mid A)$, that is, $L(\boldsymbol{\theta}_A; \mathbf{y}^o, A) = p(\mathbf{y}^o, \boldsymbol{\theta}_A \mid A)$ and not simply $L(\boldsymbol{\theta}_A; \mathbf{y}^o, A) \propto p(\mathbf{y}^o, \boldsymbol{\theta}_A \mid A)$.

The second conditional density on the right side of (2.4) is

$$p(\boldsymbol{\theta}_A \mid \mathbf{y}, A) = \frac{p(\boldsymbol{\theta}_A \mid A)p(\mathbf{y} \mid \boldsymbol{\theta}_A, A)}{p(\mathbf{y} \mid A)}. \tag{2.6}$$

If $\mathbf{y}$ is replaced with $\mathbf{y}^o$ in (2.6), then this expression provides the distribution of the vector of unobservables $\boldsymbol{\theta}_A$ conditional on the data $\mathbf{y}^o$ in the context of the model A. Since $p(\boldsymbol{\theta}_A \mid A)$ conveys what is known about $\boldsymbol{\theta}_A$ before ("a priori") learning $\mathbf{y}^o$, $p(\boldsymbol{\theta}_A \mid \mathbf{y}^o, A)$ communicates what is known after ("a posteriori") learning $\mathbf{y}^o$.

Definition 2.1.2 The *posterior density* of the vector of unobservables $\boldsymbol{\theta}_A$ in the model A is

$$p(\boldsymbol{\theta}_A \mid \mathbf{y}^o, A) = \frac{p(\boldsymbol{\theta}_A \mid A)p(\mathbf{y}^o \mid \boldsymbol{\theta}_A, A)}{p(\mathbf{y}^o \mid A)}. \tag{2.7}$$

The expression in the denominator of (2.7) is the marginal likelihood. In many circumstances it suffices to know just the shape of the posterior density $p(\boldsymbol{\theta}_A \mid \mathbf{y}^o, A)$ and it is costly to evaluate $p(\mathbf{y}^o \mid A)$. In this case it is useful to exploit the fact that

$$p(\boldsymbol{\theta}_A \mid \mathbf{y}^o, A) \propto p(\boldsymbol{\theta}_A \mid A)p(\mathbf{y}^o \mid \boldsymbol{\theta}_A, A). \tag{2.8}$$

Definition 2.1.3 Any nonnegative function $k(\mathbf{x})$ proportional to a probability density function $p(\mathbf{x})$ is a *kernel* of $p(\mathbf{x})$.

The expression on the right side of (2.8) is a kernel of the posterior density. In general, any finitely integrable nonnegative function is the kernel of some probability density function. To emphasize the distinction between the posterior density function proper and a kernel of that function, we shall sometimes refer to (2.7) as the *normalized posterior density*, whereas the right side of (2.8) is the *posterior density kernel in standard form*. Any function $k(\boldsymbol{\theta}_A \mid \mathbf{y}^o, A) \propto p(\boldsymbol{\theta}_A \mid \mathbf{y}^o, A)$ is a kernel of the posterior density.

The third and final component of the model A is a vector of interest $\boldsymbol{\omega} \in \Omega \subseteq \mathbb{R}^q$ representing entities the model is intended to describe, together with a conditional density $p(\boldsymbol{\omega} \mid \mathbf{y}, \boldsymbol{\theta}_A, A)$. Whereas $\boldsymbol{\theta}_A$ is specific to the model A, $\boldsymbol{\omega}$ remains the same across models. This includes a wide range of possibilities.

Example 2.1.1 Vector of Interest in the Value at Risk Example If we are interested only in the portfolio value 5 days hence, we can take $\omega = p_T \exp(\sum_{s=1}^{5} y_{T+s})$. A model of asset returns, A, provides $p(y_t \mid \mathbf{Y}_{t-1}, \boldsymbol{\theta}_A, A)$, and thereby $p(\omega \mid \mathbf{Y}_T^o, \boldsymbol{\theta}_A, A)$. The simple, but unrealistic, model (1.3) implies $\log(\omega|p_T) \mid (\mu, \sigma^2) \sim N(5\mu, 5\sigma^2)$ conditional on $\boldsymbol{\theta}_A = (\mu, \sigma^2)'$. As we shall see (Examples 2.1.2 and 2.3.3), for certain prior distributions we may remove the conditioning on μ and σ^2 and obtain similar compact expressions for the distribution of ω. In general, however, this will not be possible, and in particular it cannot be done in the more realistic model (1.12)–(1.13).

Definition 2.1.4 A *complete model* A consists of three components: the observables density $p(\mathbf{y} \mid \boldsymbol{\theta}_A, A)$, the prior density $p(\boldsymbol{\theta}_A \mid A)$, and the vector of interest density, $p(\boldsymbol{\omega} \mid \mathbf{y}, \boldsymbol{\theta}_A, A)$.

The objective of inference can generally be expressed as the posterior density of the vector of interest $\boldsymbol{\omega}$. When there is just one model, A, this is

$$p(\boldsymbol{\omega} \mid \mathbf{y}^o, A) = \int_{\Theta_A} p(\boldsymbol{\omega} \mid \mathbf{y}^o, \boldsymbol{\theta}_A, A) p(\boldsymbol{\theta}_A \mid \mathbf{y}^o, A)\, d\nu(\boldsymbol{\theta}_A). \tag{2.9}$$

By means of (2.8) and (2.9), $p(\boldsymbol{\omega} \mid \mathbf{y}^o, A)$ is expressed in terms of the three components of the model A. Quite often—but by no means always—the objective of inference can be expressed $E[h(\boldsymbol{\omega}) \mid \mathbf{y}^o, A]$ for suitably chosen $h(\cdot)$. This formulation includes several special cases of interest.

If a hypothesis restricts $\boldsymbol{\theta}_A$ to a set $\Theta_{A0} \subset \Theta_A$, then, by taking $h(\boldsymbol{\omega}) = I_{\Theta_{A0}}(\boldsymbol{\theta}_A)$, we have $E[h(\boldsymbol{\omega}) \mid \mathbf{y}^o, A] = P(\boldsymbol{\theta}_A \in \Theta_{A0} \mid \mathbf{y}^o, A)$, the posterior probability that the hypothesis is true in the context of model A. For example, suppose that in the class size example (Section 1.1.1) the investigator uses a normal linear model in which class size is the second component of $\mathbf{x}_t$, and wishes to ascertain whether increasing class size lowers test scores. Then Θ_{A0} might consist of all those sets of parameters $(\boldsymbol{\beta}, h)$ for which $\beta_2 < 0$.

Another important class of cases arises from prediction problems, $\boldsymbol{\omega}' = (y_{T+1}, \ldots, y_{T+q})$. The appropriate choice of $h(\boldsymbol{\omega})$ may include expected values, turning point probabilities, and predictive intervals. In the value at risk example (Section 1.1.2) let $\omega_s = p_T \exp(\sum_{r=1}^{s} y_{T+r})$ $(s = 1, \ldots, 5)$ denote the portfolio values over the next 5 days. The maximum value over the period corresponds to $h(\boldsymbol{\omega}) = \sup_{s=1,\ldots,5} \omega_s$. To assess the probability of a turning point in portfolio value on day $T+3$, we would set $h(\boldsymbol{\omega}) = 1$ if $\omega_2 < \omega_3 > \omega_4$ or $\omega_2 > \omega_3 < \omega_4$ and $h(\boldsymbol{\omega}) = 0$ otherwise. To assess the probability that $p_T - p_{T+5} \geq c$ define $h(\boldsymbol{\omega}; c) = 1$ if $\omega_5 \leq p_T - c$ and $h(\boldsymbol{\omega}; c) = 0$ otherwise. The value at risk problem is to find that c for which $E[h(\boldsymbol{\omega}; c) \mid \mathbf{Y}_T^o, A] = .05$, a relatively easy task if we can compute $E[h(\boldsymbol{\omega}; c) \mid \mathbf{Y}_T^o, A]$ for any value of c.

Yet another useful class of functions arises whenever a decisionmaker must take one of two actions, $\mathbf{a}_1$ or $\mathbf{a}_2$. Then, $h(\boldsymbol{\omega}) = L(\mathbf{a}_1, \boldsymbol{\omega}) - L(\mathbf{a}_2, \boldsymbol{\omega})$, in which $L(\mathbf{a}, \boldsymbol{\omega})$ denotes the loss incurred if action $\mathbf{a}$ is taken and then the realization of the vector of interest is $\boldsymbol{\omega}$. In the drug approval example at the beginning of Chapter 1, the FDA must decide whether to approve a drug; $\boldsymbol{\omega}$ might be a vector of health outcomes. In the merger example, the regulatory authority must either approve or disapprove the proposed merger; $\boldsymbol{\omega}$ might be a vector of prices of products produced by the firms involved.

Example 2.1.2 Normal Linear Regression Model For an observable $T \times 1$ vector of dependent variables $\mathbf{y}$ and $T \times k$ matrix of fixed covariates $\mathbf{X}$, assume

$$\mathbf{y} \mid (\boldsymbol{\beta}, h, \mathbf{X}, A) \sim N(\mathbf{X}\boldsymbol{\beta}, h^{-1}\mathbf{I}_T);\ \operatorname{rank}(\mathbf{X}) = k. \tag{2.10}$$

The fixed and observed covariates $\mathbf{X}$ are part of the model specification A, but it will prove convenient to include them explicitly in the notation for the conditional distribution of observables (2.10). We shall sometimes write $\mathbf{X}' = [\mathbf{x}_1, \ldots, \mathbf{x}_T]$ and let $\boldsymbol{\varepsilon}$ denote the $T \times 1$ vector of disturbances, with $\varepsilon_t = y_t - \boldsymbol{\beta}'\mathbf{x}_t$ $(t = 1, \ldots, T)$. The parameter h is the *precision* of each of the i.i.d. disturbances ε_t; it is the inverse of $\mathrm{var}(\varepsilon_t) = \sigma^2$. (More generally, the precision of any random variable is the inverse of its variance.) The vector of unobservables is the parameter vector $\boldsymbol{\theta}'_A = (\boldsymbol{\beta}', h)$. The coefficient vector $\boldsymbol{\beta}$ and the precision h are independent in the prior. The prior distribution of $\boldsymbol{\beta}$ is

$$\boldsymbol{\beta} \mid A \sim N(\underline{\boldsymbol{\beta}}, \underline{\mathbf{H}}^{-1}). \tag{2.11}$$

In (2.11) the mean $\underline{\boldsymbol{\beta}}$ is a $k \times 1$ vector of constants. This vector is specified as part of the prior distribution. The precision $\underline{\mathbf{H}}$ is a $k \times k$ positive definite matrix of constants, also specified as part of the prior distribution. (In general, an underscore will denote constants in prior distributions.) The prior distribution of h is

$$\underline{s}^2 h \mid A \sim \chi^2(\underline{\nu}). \tag{2.12}$$

The formulation (2.12) is a concise way of expressing a gamma distribution for h. In general, the two-parameter gamma distribution can be represented in this form [see Johnson et al. (1994), Section 17.3]. In a given application, (2.11)–(2.12) is not necessarily an adequate representation of prior information and beliefs, and other prior distributions could be used. However, (2.11)–(2.12) has attractive analytical properties that will become clear in due course, and Section 8.4 discusses methods for modifying this and other prior distributions.

In (2.11) $\underline{\boldsymbol{\beta}}$, $\underline{\mathbf{H}}$, and $\underline{\mathbf{H}}^{-1}$ are respectively the prior mean, prior precision, and prior variance of $\boldsymbol{\beta}$. Thus

$$p(\boldsymbol{\beta} \mid A) = (2\pi)^{-k/2} \left|\underline{\mathbf{H}}\right|^{1/2} \exp[-(\boldsymbol{\beta} - \underline{\boldsymbol{\beta}})'\underline{\mathbf{H}}(\boldsymbol{\beta} - \underline{\boldsymbol{\beta}})/2]. \tag{2.13}$$

To derive the probability density corresponding to (2.12) recall that if $w \sim \chi^2(\nu)$ then

$$p(w) = [2^{\nu/2}\Gamma(\nu/2)]^{-1} w^{(\nu-2)/2} \exp(-w/2), \tag{2.14}$$

$E(w) = \nu$, and $\mathrm{var}(w) = 2\nu$. Hence the prior mean and variance of h are $E(h \mid A) = \underline{\nu}/\underline{s}^2$ and $\mathrm{var}(h \mid A) = 2\underline{\nu}/\underline{s}^4$, respectively. Through the usual change of variable, the pdf of h is

$$p(h \mid A) = [2^{\underline{\nu}/2}\Gamma(\underline{\nu}/2)]^{-1} (\underline{s}^2)^{\underline{\nu}/2} h^{(\underline{\nu}-2)/2} \exp(-\underline{s}^2 h/2). \tag{2.15}$$

Another change of variable yields the prior pdf of $\sigma^2 = h^{-1}$:

$$p(\sigma^2 \mid A) = [2^{\underline{\nu}/2}\Gamma(\underline{\nu}/2)]^{-1} (\underline{s}^2)^{\underline{\nu}/2} (\sigma^2)^{-(\underline{\nu}+2)/2} \exp(-\underline{s}^2/2\sigma^2). \tag{2.16}$$

From the specification (2.10), we obtain

$$p(\mathbf{y} \mid \boldsymbol{\beta}, h, \mathbf{X}, A) = (2\pi)^{-T/2} h^{T/2} \exp[-h(\mathbf{y} - \mathbf{X}\boldsymbol{\beta})'(\mathbf{y} - \mathbf{X}\boldsymbol{\beta})/2]. \tag{2.17}$$

The posterior density kernel in standard form is the product of (2.13), (2.15), and (2.17) evaluated at $\mathbf{y}^o$:

$$(2\pi)^{-(T+k)/2}[2^{\underline{\nu}/2}\Gamma(\underline{\nu}/2)]^{-1} \tag{2.18a}$$

$$\cdot \left|\underline{\mathbf{H}}\right|^{1/2} (\underline{s}^2)^{\underline{\nu}/2} \tag{2.18b}$$

$$\cdot\, h^{(T+\underline{\nu}-2)/2} \exp(-\underline{s}^2 h/2) \tag{2.18c}$$

$$\cdot \exp\{-[(\boldsymbol{\beta} - \underline{\boldsymbol{\beta}})'\underline{\mathbf{H}}(\boldsymbol{\beta} - \underline{\boldsymbol{\beta}}) + h(\mathbf{y}^o - \mathbf{X}\boldsymbol{\beta})'(\mathbf{y}^o - \mathbf{X}\boldsymbol{\beta})]/2\}. \tag{2.18d}$$

To interpret this expression, it is useful to begin with some algebra. Complete the square in $\boldsymbol{\beta}$ of the term in brackets in (2.18d) to obtain

$$(\boldsymbol{\beta} - \underline{\boldsymbol{\beta}})'\underline{\mathbf{H}}(\boldsymbol{\beta} - \underline{\boldsymbol{\beta}}) + h(\mathbf{y}^o - \mathbf{X}\beta)'(\mathbf{y}^o - \mathbf{X}\beta) = (\boldsymbol{\beta} - \overline{\boldsymbol{\beta}})'\overline{\mathbf{H}}(\boldsymbol{\beta} - \overline{\boldsymbol{\beta}}) + Q,$$

where

$$\overline{\mathbf{H}} = \underline{\mathbf{H}} + h\mathbf{X}'\mathbf{X}, \tag{2.19}$$

$$\overline{\beta} = \overline{\mathbf{H}}^{-1}(\underline{\mathbf{H}}\underline{\beta} + h\mathbf{X}'\mathbf{y}) = \overline{\mathbf{H}}^{-1}(\underline{\mathbf{H}}\underline{\boldsymbol{\beta}} + h\mathbf{X}'\mathbf{X}\mathbf{b}), \tag{2.20}$$

$$Q = h\mathbf{y}^{o\prime}\mathbf{y}^o + \underline{\boldsymbol{\beta}}'\underline{\mathbf{H}}\underline{\boldsymbol{\beta}} - \overline{\boldsymbol{\beta}}'\overline{\mathbf{H}}\overline{\boldsymbol{\beta}}, \tag{2.21}$$

where $\mathbf{b}$ denotes the coefficients in the ordinary least-squares fit of $\mathbf{y}^o$ to $\mathbf{X}$, $\mathbf{b} = (\mathbf{X}'\mathbf{X})^{-1}\mathbf{X}'\mathbf{y}^o$.

If (2.18a)–(2.18d) is interpreted as a function of $\boldsymbol{\beta}$ only, then (2.18d) is a posterior density kernel for $\boldsymbol{\beta}$ conditional on h, and our square completion shows that

$$p(\boldsymbol{\beta} \mid h, \mathbf{y}^o, \mathbf{X}, A) \propto \exp[-(\boldsymbol{\beta} - \overline{\boldsymbol{\beta}})'\overline{\mathbf{H}}(\boldsymbol{\beta} - \overline{\boldsymbol{\beta}})/2]. \tag{2.22}$$

Consequently

$$\boldsymbol{\beta} \mid (h, \mathbf{y}^o, \mathbf{X}, A) \sim N(\overline{\boldsymbol{\beta}}, \overline{\mathbf{H}}^{-1}). \tag{2.23}$$

The conditional posterior distribution of $\boldsymbol{\beta}$ is normal because the prior distribution of $\boldsymbol{\beta}$ is normal and the likelihood function in $\boldsymbol{\beta}$, (2.18d), is a kernel of a multivariate normal distribution. Note the symmetry of the prior and observables distributions as they are combined in (2.19), (2.20), and (2.23). The precision of the posterior distribution is the sum of the prior precision and the term $h\mathbf{X}'\mathbf{X}$. The latter is the posterior precision in the limit as the prior precision $\underline{\mathbf{H}} \rightarrow 0$, and might therefore be called the *observables precision matrix*. The mean of the posterior distribution is the matrix weighted average of the prior mean $\underline{\boldsymbol{\beta}}$ and the vector $\mathbf{b}$. The latter is the limiting posterior mean as $\underline{\mathbf{H}} \rightarrow \mathbf{0}$. The respective matrix weights are the prior precision matrix $\underline{\mathbf{H}}$ and the observables precision matrix $h\mathbf{X}'\mathbf{X}$.

Interpreting the product of (2.18a)–(2.18d) as a function of h alone, we see that a kernel of $h \mid (\boldsymbol{\beta}, \mathbf{y}^o, \mathbf{X}, A)$ is the product of (2.18c) and (2.18d). Hence the posterior density of h conditional on $\boldsymbol{\beta}$ and the data is

$$p(h \mid \boldsymbol{\beta}, \mathbf{y}^o, \mathbf{X}, A) \propto h^{(T+\underline{\nu}-2)/2} \exp\{-[\underline{s}^2 + (\mathbf{y}^o - \mathbf{X}\boldsymbol{\beta})'(\mathbf{y}^o - \mathbf{X}\boldsymbol{\beta})]h/2\}. \quad (2.24)$$

Comparing this expression with (2.12) and (2.15), it is evident that

$$\overline{s}^2 h \mid (\boldsymbol{\beta}, \mathbf{y}^o, \mathbf{X}, A) \sim \chi^2(\overline{\nu}) \quad (2.25)$$

where

$$\overline{s}^2 = \underline{s}^2 + (\mathbf{y}^o - \mathbf{X}\boldsymbol{\beta})'(\mathbf{y}^o - \mathbf{X}\boldsymbol{\beta}) \text{ and } \overline{\nu} = T + \underline{\nu}. \quad (2.26)$$

There is again an evident symmetry between the prior and the observables, and it again arises because the kernel of the prior density for h, (2.15), and the kernel of the observables density function in h, (2.17) have the same functional form.

The prior distributions (2.11) and (2.12) are attractive because they lead to the simple and interpretable results (2.23) and (2.25). We shall study this property more generally and systematically in the consideration of conjugate and conditionally conjugate priors, in Section 2.3. The results (2.23) and (2.25) are not immediately useful, for they do not provide distributions conditional only on the data and prior information. We cannot obtain these distributions analytically, although this is possible using different priors. Alternatively, a numerical approach may be taken. We will pursue the former strategy in Section 2.3, and the latter method will be developed in Section 4.3.

Example 2.1.3 Geometric Interpretation of the Normal Linear Regression Model with Two Covariates There is an informative geometric interpretation of the posterior mean $\overline{\boldsymbol{\beta}}$, conditional on h, due to Leamer (1973). A geometric representation of the level contours of the prior pdf of $\boldsymbol{\beta}$ (2.13) consists of the ellipses

$$\boldsymbol{\beta} : (\boldsymbol{\beta} - \underline{\boldsymbol{\beta}})'\underline{\mathbf{H}}(\boldsymbol{\beta} - \underline{\boldsymbol{\beta}}) = c_1$$

for various positive constants c_1. Because (2.13) implies $(\boldsymbol{\beta} - \underline{\boldsymbol{\beta}})'\underline{\mathbf{H}}(\boldsymbol{\beta} - \underline{\boldsymbol{\beta}}) \mid A \sim \chi^2(k)$, the prior probability that $\boldsymbol{\beta}$ is in the interior of the ellipse is $1 - \alpha$ if $c_1 = \chi^2_\alpha(k)$. Interpreting the pdf of $\mathbf{y}$ given in (2.17) as a density kernel for $\boldsymbol{\beta}$ and substituting $\mathbf{y}^o$ for $\mathbf{y}$, the level contours of that density are the ellipses $\boldsymbol{\beta}$: $(\boldsymbol{\beta} - \mathbf{b})'h\mathbf{X}'\mathbf{X}(\boldsymbol{\beta} - \mathbf{b}) = c_2$.

Now consider the set of points $\boldsymbol{\beta}$ such that there is no point $\boldsymbol{\beta}^* \in \mathbb{R}^k$ for which both $p(\boldsymbol{\beta}^* \mid A) > p(\boldsymbol{\beta} \mid A)$ and $p(\mathbf{y}^o \mid \boldsymbol{\beta}^*, h, \mathbf{X}, A) > p(\mathbf{y}^o \mid \boldsymbol{\beta}, h, \mathbf{X}, A)$. Through the usual constrained optimization calculus, this is the set of points that solves the first-order condition for the objective function

$$(\boldsymbol{\beta} - \mathbf{b})'h\mathbf{X}'\mathbf{X}(\boldsymbol{\beta} - \mathbf{b}) + \lambda(\boldsymbol{\beta} - \underline{\boldsymbol{\beta}})'\underline{\mathbf{H}}(\boldsymbol{\beta} - \underline{\boldsymbol{\beta}}). \quad (2.27)$$

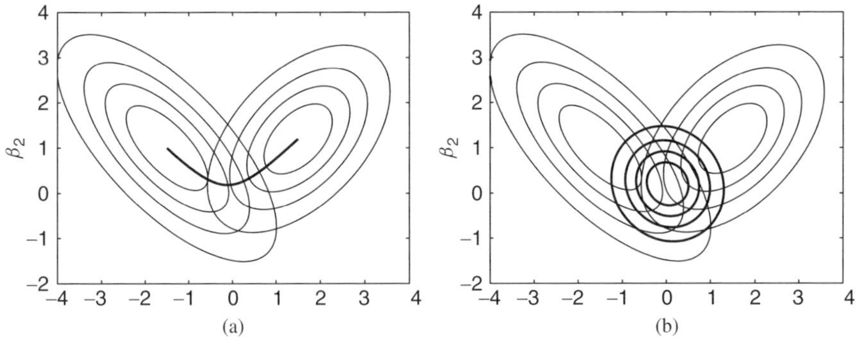

Figure 2.1. Normal linear model with two regressors—geometric interpretation: (a) prior and likelihood contours with locus of posterior means; (b) prior, likelihood, and posterior contours.

This entire set is indexed by the Lagrange multiplier λ. Setting the first derivative of (2.27) with respect to $\boldsymbol{\beta}$ to zero, this curve may be expressed

$$\boldsymbol{\beta}(\lambda; h) = [h\mathbf{X}'\mathbf{X} + \lambda\underline{\mathbf{H}}]^{-1}(h\mathbf{X}'\mathbf{X}\mathbf{b} + \lambda\underline{\mathbf{H}}\underline{\boldsymbol{\beta}})$$

Observe that $\boldsymbol{\beta}(1; h) = \overline{\boldsymbol{\beta}}$.

An example for the case $k = 2$ is presented in Figure 2.1. The level contours of the prior distribution are represented by the lighter ellipses; those for the likelihood, by the darker ellipses. The curve $\boldsymbol{\beta}(1; h)$ is the solid curve shown in Figure 2.1a. This locus of tangencies between the ellipses is indexed by h and traces the conditional posterior mean $\overline{\boldsymbol{\beta}}$ as a function of h. In particular, $\lim_{h\to 0} \boldsymbol{\beta}(1; h) = \underline{\boldsymbol{\beta}}$, and $\lim_{h\to\infty} \boldsymbol{\beta}(1; h) = b$. In the example portrayed, $\beta_2(1; h) < \underline{\beta}_2$ and $\beta_2(1; h) < b_2$, for most values of h. This illustrates some subtleties of the matrix weighted average of $\underline{\boldsymbol{\beta}}$ and $\mathbf{b}$ in (2.20). The darkest ellipses in Figure 2.1b indicate the level contours of the posterior distribution of $\boldsymbol{\beta}$ for the case $h = 1$. For more on the question of the sensitivity of the posterior mean to the prior mean, in this setting, see Leamer (1982).

Exercise 2.1.1 Combining Information "Suppose that a random sample of size T is drawn from a normal population with unknown mean μ and known precision h. There is always a normal prior distribution for μ that will lead to a normal posterior distribution for μ with mean $\overline{\mu}$ and precision $\overline{h}$, for any real $\overline{\mu}$ and positive $\overline{h}$."

Is this statement true or false? If true, prove it. If false, provide a counterexample, a correct version of the statement, and a proof of the correct version.

Exercise 2.1.2 Probability Density Kernels Let $p(\mathbf{y})$ denote the probability density of an $n \times 1$ random vector $\mathbf{y}$ and suppose that

$$\log p(\mathbf{y}) = \mathbf{y}'\mathbf{A}\mathbf{y} + \mathbf{b}'\mathbf{y} + c.$$

The $n \times n$ matrix $\mathbf{A}$ is negative definite.

(a) Show that $\mathbf{y}$ has a multivariate normal distribution. Express its precision $\mathbf{H}$ in terms of $\mathbf{A}$ and its mean $\boldsymbol{\mu}$ in terms of $\mathbf{A}$ and $\mathbf{b}$.

(b) Express c in terms of $\mathbf{H}$ and $\boldsymbol{\mu}$.

Exercise 2.1.3 The Generalized Normal Linear Regression Model Suppose that in model A of Example 2.1.2, $\mathbf{y} \mid (\boldsymbol{\beta}, \mathbf{X}, A) \sim N(\mathbf{X}\boldsymbol{\beta}, \mathbf{V})$, where $\mathbf{V}$ is a known $T \times T$ positive definite matrix. Show that given the prior distribution (2.11) the posterior distribution of $\boldsymbol{\beta} \mid (\mathbf{y}^o, \mathbf{X}, A)$ is then $\boldsymbol{\beta} \sim N(\overline{\boldsymbol{\beta}}, \overline{\mathbf{H}}^{-1})$, and provide expressions for $\overline{\mathbf{H}}$ and $\overline{\boldsymbol{\beta}}$.

Exercise 2.1.4 The Short Rank Normal Linear Model Suppose that in the normal linear model of Example 2.1.2, rank($\mathbf{X}$) $< k$. All other aspects of the model remain the same as in Example 2.1.2.

(a) Show that $\boldsymbol{\beta} \mid (h = 1, \mathbf{y}^o, \mathbf{X}, A) \sim N(\overline{\boldsymbol{\beta}}, \overline{\mathbf{H}}^{-1})$, and provide expressions for $\overline{\mathbf{H}}$ and $\overline{\boldsymbol{\beta}}$. [Of course, the least-squares estimate $\mathbf{b} = (\mathbf{X}'\mathbf{X})^{-1}\mathbf{X}'\mathbf{y}^o$ does not exist in this case.]

(b) Show that if $k = 6$, then

$$P[(\boldsymbol{\beta}-\overline{\boldsymbol{\beta}})'\overline{\mathbf{H}}(\boldsymbol{\beta}-\overline{\boldsymbol{\beta}}) < 12.5922 \mid (h = 1, \mathbf{y}^o, \mathbf{X}, A)] = .95.$$

(c) Suppose that for the $k \times 1$ vector $\mathbf{a}$, it is the case that $\mathbf{Xa} = \mathbf{0}$. Suppose also that $\underline{\mathbf{H}} = \mathbf{I}_k$. Show that the prior distribution and the posterior distribution of $\mathbf{a}'\boldsymbol{\beta}$ are the same. For $k = 2$, draw a sketch in β_1 and β_2 space similar to Figure 2.1a to illustrate what is going on. [*Hint*: $(\underline{\mathbf{H}} + \mathbf{X}'\mathbf{X})^{-1} = \underline{\mathbf{H}}^{-1} - \underline{\mathbf{H}}^{-1}\mathbf{X}'(\mathbf{X}\underline{\mathbf{H}}^{-1}\mathbf{X}' + \mathbf{I}_T)^{-1}\mathbf{X}\underline{\mathbf{H}}^{-1}$.]

(d) Rework (c) for the case of any nonsingular prior variance matrix $\underline{\mathbf{H}}^{-1}$. For some linear combinations $\mathbf{a}'\boldsymbol{\beta}$, the prior and posterior distribution are the same. What can you say about the vectors $\mathbf{a}$ for which this is true? How does the sketch you drew in (c) change for this more general case?

Exercise 2.1.5 Distribution of σ^2 and Related Parameters This exercise is about the prior distribution of σ^2 in (2.16), its properties, and related distributions.

(a) Derive (2.16) from (2.15).

(b) Suppose that σ^2 has the pdf (2.16). Derive the mean and variance of σ^2, indicating any additional assumptions that are necessary for these moments to exist. [*Hints*:

- **(i)** This is not an elaborate integration problem. What does (2.16) tell you about the integral of the kernel of the pdf of σ^2?
- **(ii)** $\Gamma(x + 1) = x\Gamma(x)$ for all $x > 0$.]

(c) If σ^2 has the pdf (2.16), what is the pdf of σ? Express the median of the distribution of σ in terms of the median $\chi^2_{.50}(\underline{\nu})$ of the chi-squared distribution with $\underline{\nu}$ degrees of freedom.

2.2 SUFFICIENCY, ANCILLARITY, AND NUISANCE PARAMETERS

The steps that are undertaken to derive the posterior distribution $p(\boldsymbol{\theta}_A \mid \mathbf{y}^o, A)$ or the marginal likelihood $p(\mathbf{y}^o \mid A)$ depend on the relations between $\mathbf{y}^o$ and $\boldsymbol{\theta}_A$ in these expressions. In particular circumstances these expressions can be simplified. Two of the most useful arise when the data can be reduced to a smaller set of statistics (called *sufficient statistics*) for the purpose of inference, and again when a subset of this set (called *ancillary statistics*) can be regarded as fixed for the same purpose.

2.2.1 Sufficiency

Definition 2.2.1 The vector $\mathbf{s} = \mathbf{s}(\mathbf{y}; A)$ is a *sufficient statistic* in a model with the observables density $p(\mathbf{y} \mid \boldsymbol{\theta}_A, A)$ if

$$p[\mathbf{y} \mid \mathbf{s}(\mathbf{y}; A), \boldsymbol{\theta}_A, A] = p[\mathbf{y} \mid \mathbf{s}(\mathbf{y}; A), A] \; \forall \; \boldsymbol{\theta}_A \in \Theta_A. \tag{2.28}$$

Heuristically, (2.28) implies that there is no information about $\mathbf{y}$ originating in $\boldsymbol{\theta}_A$ in the density $p(\mathbf{y} \mid \boldsymbol{\theta}_A, A)$, beyond that conveyed by $\mathbf{s}$ ex ante. This suggests that in learning about $\boldsymbol{\theta}_A$, nothing would be lost by confining attention to $\mathbf{s}(\mathbf{y}; A)$ rather than $\mathbf{y}$. This is indeed the case.

Theorem 2.2.1 Ex Post and Ex Ante Equivalence of Sufficiency The vector $\mathbf{s}(\mathbf{y}; A)$ is a sufficient statistic in a model with observables density $p(\mathbf{y} \mid \boldsymbol{\theta}_A, A)$ if and only if for all $\boldsymbol{\theta}_A \in \Theta_A$ and for all $\mathbf{y} \in \Psi$, we have

$$p(\boldsymbol{\theta}_A \mid \mathbf{y}, A) = p[\boldsymbol{\theta}_A \mid \mathbf{s}(\mathbf{y}; A), A]. \tag{2.29}$$

Proof: Suppose that (2.28) is true. Then

$$\begin{aligned} p(\boldsymbol{\theta}_A \mid \mathbf{y}, A) = p(\boldsymbol{\theta}_A \mid \mathbf{y}, \mathbf{s}, A) &= \frac{p(\mathbf{y} \mid \boldsymbol{\theta}_A, \mathbf{s}, A)\, p(\boldsymbol{\theta}_A \mid \mathbf{s}, A)}{p(\mathbf{y} \mid \mathbf{s}, A)} \\ &= \frac{p(\mathbf{y} \mid \mathbf{s}, A)\, p(\boldsymbol{\theta}_A \mid \mathbf{s}, A)}{p(\mathbf{y} \mid \mathbf{s}, A)} = p(\boldsymbol{\theta}_A \mid \mathbf{s}, A). \end{aligned}$$

Conversely, if (2.29) is true, then

$$\begin{aligned} p(\mathbf{y} \mid \mathbf{s}, \boldsymbol{\theta}_A, A) &= \frac{p(\boldsymbol{\theta}_A \mid \mathbf{y}, \mathbf{s}, A)\, p(\mathbf{y} \mid \mathbf{s}, A)}{p(\boldsymbol{\theta}_A \mid \mathbf{s}, A)} \\ &= \frac{p(\boldsymbol{\theta}_A \mid \mathbf{y}, A)\, p(\mathbf{y} \mid \mathbf{s}, A)}{p(\boldsymbol{\theta}_A \mid \mathbf{y}, A)} = p(\mathbf{y} \mid \mathbf{s}, A). \end{aligned}$$

■

Note that the conditions in Theorem 2.2.1 hold for any choice of the prior density $p(\boldsymbol{\theta}_A \mid A)$ and vector of interest $\boldsymbol{\omega}$. This is because sufficiency is a property of the observables density $p(\mathbf{y} \mid \boldsymbol{\theta}_A, A)$ alone. In demonstrating the sufficiency of $\mathbf{s}(\mathbf{y}; A)$ in an observables density, it is usually easiest to use a third, equivalent condition.

Theorem 2.2.2 Factorization Criterion The vector $\mathbf{s}(\mathbf{y}; A)$ is a sufficient statistic in a model with data density $p(\mathbf{y} \mid \boldsymbol{\theta}_A, A)$ if and only if

$$p(\mathbf{y} \mid \boldsymbol{\theta}_A, A) = p[\mathbf{s}(\mathbf{y}; A) \mid \boldsymbol{\theta}_A, A]r(\mathbf{y}; A) \tag{2.30}$$

for some function $r(\mathbf{y}; A)$.

Proof: It suffices to show that (2.29) and (2.30) are equivalent. First suppose that (2.30) is true. Then

$$\begin{aligned} p(\boldsymbol{\theta}_A \mid \mathbf{y}, A) &= \frac{p(\mathbf{y} \mid \boldsymbol{\theta}_A, A)p(\boldsymbol{\theta}_A \mid A)}{p(\mathbf{y} \mid A)} \\ &= \frac{p(\mathbf{s} \mid \boldsymbol{\theta}_A, A)r(\mathbf{y}; A)p(\boldsymbol{\theta}_A \mid A)}{p(\mathbf{y} \mid A)} = p(\boldsymbol{\theta}_A \mid \mathbf{s}, A). \end{aligned}$$

On the other hand, given (2.29), it follows that

$$\begin{aligned} p(\mathbf{y} \mid \boldsymbol{\theta}_A, A) &= \frac{p(\boldsymbol{\theta}_A \mid \mathbf{y}, A)p(\mathbf{y} \mid A)}{p(\boldsymbol{\theta}_A \mid A)} \\ &= p(\boldsymbol{\theta}_A \mid \mathbf{s}, A)\frac{p(\mathbf{y} \mid A)}{p(\boldsymbol{\theta}_A \mid A)} \\ &= \frac{p(\mathbf{s} \mid \boldsymbol{\theta}_A, A)p(\boldsymbol{\theta}_A \mid A)}{p(\mathbf{s} \mid A)} \cdot \frac{p(\mathbf{y} \mid A)}{p(\boldsymbol{\theta}_A \mid A)} \\ &= \frac{p(\mathbf{s} \mid \boldsymbol{\theta}_A, A)p(\mathbf{y} \mid A)}{p(\mathbf{s} \mid A)} = p(\mathbf{s} \mid \boldsymbol{\theta}_A, A)r(\mathbf{y}; A), \end{aligned}$$

where in the last equation $r(\mathbf{y}; A) = p(\mathbf{y} \mid A)/p(\mathbf{s} \mid A)$. ■

The factorization criterion is particularly useful in demonstrating that $\mathbf{s}(\mathbf{y}; A)$ is a sufficient statistic. This is because it is often relatively easy to demonstrate that for the likelihood function

$$L(\boldsymbol{\theta}_A; \mathbf{y}, A) = L[\boldsymbol{\theta}_A; \mathbf{s}(\mathbf{y}; A), A]. \tag{2.31}$$

It is always the case that

$$L(\boldsymbol{\theta}_A; \mathbf{y}, A) = p(\mathbf{y} \mid \boldsymbol{\theta}_A, A)r_1(\mathbf{y}),$$

where $r_1(\mathbf{y})$ absorbs any constants excluded from the likelihood function, and

$$L[\boldsymbol{\theta}_A; \mathbf{s}(\mathbf{y}; A), A] = p[\mathbf{s}(\mathbf{y}; A) \mid \boldsymbol{\theta}_A, A]r_2(\mathbf{y}),$$

where $r_2(\mathbf{y})$ absorbs any such constants, including the Jacobian of transformation between $\mathbf{y}$ and $\mathbf{s}(\mathbf{y}; A)$. Hence, if (2.31) is true for all $\boldsymbol{\theta}_A \in \Theta_A$ and $\mathbf{y} \in Y$, then $\mathbf{s}(\mathbf{y}; A)$ is a sufficient statistic.

Example 2.2.1 Sufficient Statistics in the Normal Linear Regression Model From (2.17), we have

$$p(\mathbf{y} \mid \boldsymbol{\beta}, h, \mathbf{X}, A) \propto h^{T/2} \exp[-h(\mathbf{y} - \mathbf{X}\boldsymbol{\beta})'(\mathbf{y} - \mathbf{X}\boldsymbol{\beta})/2].$$

Completing the square, we obtain

$$\begin{aligned}(\mathbf{y} - \mathbf{X}\boldsymbol{\beta})'(\mathbf{y} - \mathbf{X}\boldsymbol{\beta}) &= (\mathbf{y} - \mathbf{X}\mathbf{b})'(\mathbf{y} - \mathbf{X}\mathbf{b}) + (\boldsymbol{\beta} - \mathbf{b})'\mathbf{X}'\mathbf{X}(\boldsymbol{\beta} - \mathbf{b}) \\ &= s^2 + (\boldsymbol{\beta} - \mathbf{b})'\mathbf{X}'\mathbf{X}(\boldsymbol{\beta} - \mathbf{b}),\end{aligned}$$

where $\mathbf{b} = (\mathbf{X}'\mathbf{X})^{-1}\mathbf{X}'\mathbf{y}$, and $s^2 = (\mathbf{y} - \mathbf{X}\mathbf{b})'(\mathbf{y} - \mathbf{X}\mathbf{b})$. By the factorization criterion, $[\mathbf{b}, s^2, \mathbf{X}'\mathbf{X}, T]$ is a sufficient statistic in the normal linear regression model A. This is equivalent to $[\mathbf{Z}'\mathbf{Z}, T]$, where $\mathbf{Z} = [\mathbf{X}, \mathbf{y}]$.

2.2.2 Ancillarity

Definition 2.2.2 Suppose that $\mathbf{s}(\mathbf{y}; A)$ is a sufficient statistic in the observables density $p(\mathbf{y} \mid \boldsymbol{\theta}_A, A)$. If there exist partitions $\mathbf{s}' = (\mathbf{s}_1', \mathbf{s}_2')$ and $\boldsymbol{\theta}_A' = (\boldsymbol{\theta}_{A1}', \boldsymbol{\theta}_{A2}')$ such that

$$p(\boldsymbol{\theta}_A \mid A) = p(\boldsymbol{\theta}_{A1} \mid A)p(\boldsymbol{\theta}_{A2} \mid A), \tag{2.32}$$

$$p(\mathbf{s}_1 \mid \boldsymbol{\theta}_A, A) = p(\mathbf{s}_1 \mid \boldsymbol{\theta}_{A1}, A), \tag{2.33}$$

$$p(\mathbf{s}_2 \mid \mathbf{s}_1, \boldsymbol{\theta}_A, A) = p(\mathbf{s}_2 \mid \mathbf{s}_1, \boldsymbol{\theta}_{A2}, A), \tag{2.34}$$

then $\mathbf{s}_1$ is an *ancillary statistic* with respect to $\boldsymbol{\theta}_{A2}$.

Ancillarity implies

$$\begin{aligned}p(\boldsymbol{\theta}_A \mid \mathbf{y}, A) &\propto p(\boldsymbol{\theta}_A \mid \mathbf{s}, A) \\ &\propto p(\boldsymbol{\theta}_{A1} \mid A)p(\boldsymbol{\theta}_{A2} \mid A)p(\mathbf{s}_1 \mid \boldsymbol{\theta}_A, A)p(\mathbf{s}_2 \mid \mathbf{s}_1, \boldsymbol{\theta}_A, A) \\ &= p(\boldsymbol{\theta}_{A1} \mid A)p(\mathbf{s}_1 \mid \boldsymbol{\theta}_{A1}, A)p(\boldsymbol{\theta}_{A2} \mid A)p(\mathbf{s}_2 \mid \mathbf{s}_1, \boldsymbol{\theta}_{A2}, A). \end{aligned} \tag{2.35}$$

It simplifies inference when the vector of interest $\boldsymbol{\omega}$ depends on $\boldsymbol{\theta}_{A2}$ and $\mathbf{y}$, but not $\boldsymbol{\theta}_{A1}$, or, alternatively, when $\boldsymbol{\omega}$ depends on $\boldsymbol{\theta}_{A1}$ and $\mathbf{y}$, but not $\boldsymbol{\theta}_{A2}$. In the first case $p(\boldsymbol{\omega} \mid \mathbf{y}, \boldsymbol{\theta}_A, A) = p(\boldsymbol{\omega} \mid \mathbf{y}, \boldsymbol{\theta}_{A2}, A)$ and (2.35) implies

$$p(\boldsymbol{\theta}_{A2} \mid \mathbf{y}, A) \propto p(\boldsymbol{\theta}_{A2} \mid A)p(\mathbf{s}_2 \mid \mathbf{s}_1, \boldsymbol{\theta}_{A2}, A).$$

Then

$$p(\boldsymbol{\omega} \mid \mathbf{y}, A) = \int_{\Theta_A} p(\boldsymbol{\omega} \mid \mathbf{y}, \boldsymbol{\theta}_A, A) p(\boldsymbol{\theta}_A \mid \mathbf{y}, A)\, d\nu(\boldsymbol{\theta}_A)$$

$$\propto \int_{\Theta_{A2}} p(\boldsymbol{\omega} \mid \mathbf{y}, \boldsymbol{\theta}_{A2}, A) p(\boldsymbol{\theta}_{A2} \mid A) p(\mathbf{s}_2 \mid \mathbf{s}_1, \boldsymbol{\theta}_{A2}, A)\, d\nu(\boldsymbol{\theta}_{A2}).$$

This means that for purposes of learning about $\boldsymbol{\omega}$ from the data $\mathbf{y}^o$ it is necessary only to use the prior density of $\boldsymbol{\theta}_{A2}$, $p(\boldsymbol{\theta}_{A2} \mid A)$, and the conditional density of $\mathbf{s}_2$, $p(\mathbf{s}_2 \mid \mathbf{s}_1, \boldsymbol{\theta}_{A2}, A)$. Since $p(\boldsymbol{\omega} \mid \mathbf{y}, A)$ is not affected by the prior distribution of $\boldsymbol{\theta}_{A1}$ or the marginal density $p(\mathbf{s}_1 \mid \boldsymbol{\theta}_A, A)$, it is not necessary to develop these distributions beyond establishing the properties (2.32)–(2.34). The random vector $\mathbf{s}_1(\mathbf{y}_T; A)$ can be treated as fixed, and the parameter vector $\boldsymbol{\theta}_{A1}$ can be ignored.

Example 2.2.2 Ancillarity in the Normal Linear Regression Model Modify the assumptions made in Example 2.1.2 by making $\mathbf{X}$ random, with pdf $p(\mathbf{X} \mid \boldsymbol{\eta}, A), \boldsymbol{\eta} \in H$. If $p(\boldsymbol{\beta}, h, \boldsymbol{\eta} \mid A) = p(\boldsymbol{\beta}, h \mid A) p(\boldsymbol{\eta} \mid A)$, then $\mathbf{X}$ is ancillary with respect to $(\boldsymbol{\beta}, h)$. If the distribution of the vector of interest depends only on $\boldsymbol{\beta}$ and h, the matrix $\mathbf{X}$ can be treated as fixed and the parameter vector $\boldsymbol{\eta}$ ignored. That is exactly what was done in Example 2.1.2. Therefore that treatment of the normal linear regression model with $\mathbf{X}$ fixed is also appropriate when $\mathbf{X}$ is random but ancillary with respect to $(\boldsymbol{\beta}, h)$, and the distribution of the vector of interest depends only on $\boldsymbol{\beta}$ and h. This happens often in applied work.

If $p(\boldsymbol{\omega} \mid \mathbf{y}, \boldsymbol{\theta}_A, A) = p(\boldsymbol{\omega} \mid \mathbf{y}, \boldsymbol{\theta}_{A1}, A)$, the factorization (2.35) is also useful, because then

$$p(\boldsymbol{\omega} \mid \mathbf{y}, A) \propto \int_{\Theta_{A_1}} p(\boldsymbol{\omega} \mid \mathbf{y}, \boldsymbol{\theta}_{A1}, A) p(\boldsymbol{\theta}_{A1} \mid A) p(\mathbf{s}_1 \mid \boldsymbol{\theta}_{A1}, A)\, d\nu(\boldsymbol{\theta}_{A1}).$$

Since $p(\boldsymbol{\omega} \mid \mathbf{y}, A)$ is not affected by the prior distribution of $\boldsymbol{\theta}_{A2}$ or the conditional density $p(\mathbf{s}_2 \mid \mathbf{s}_1, \boldsymbol{\theta}_{A2}, A)$, it is not necessary to develop these distributions beyond establishing the properties (2.32)–(2.34). The random vector $\mathbf{s}_2$ can simply be ignored.

Example 2.2.3 Missing Data It is sometimes the case that not all of the observables $\mathbf{y}$ are, in fact, observed. For example, in a survey some respondents may not reply to some questions; time series data may be quarterly before a certain date and monthly thereafter. Let $\mathbf{y}' = (\mathbf{y}_o', \mathbf{y}_m')$, where $\mathbf{y}_o$ denotes the observables subsequently observed and $\mathbf{y}_m$ denotes those that are subsequently missing. Let the inclusion indicator $\mathbf{I}$ be isomorphic to $\mathbf{y}$ with $I_t = 1$ if $y_t \in \mathbf{y}_o$ and $I_t = 0$ if $y_t \in \mathbf{y}_m$. Without further assumptions a complete model must specify $p(\mathbf{y}, \mathbf{I} \mid \boldsymbol{\theta}_A, A)$. But suppose further that $\boldsymbol{\theta}_A' = (\boldsymbol{\theta}_{A1}', \boldsymbol{\theta}_{A2}')$ and

$$p(\mathbf{y}, \mathbf{I} \mid \boldsymbol{\theta}_{A1}, \boldsymbol{\theta}_{A2}, A) = p(\mathbf{y} \mid \boldsymbol{\theta}_{A1}, A) p(\mathbf{I} \mid \mathbf{y}, \boldsymbol{\theta}_{A2}, A) \tag{2.36}$$

and $p(\boldsymbol{\theta}_A \mid A) = p(\boldsymbol{\theta}_{A1} \mid A)\ p(\boldsymbol{\theta}_{A2} \mid A)$. If, in addition

$$p(\mathbf{I} \mid \mathbf{y}_o, \mathbf{y}_m, \boldsymbol{\theta}_{A2}, A) = p(\mathbf{I} \mid \mathbf{y}_o, \boldsymbol{\theta}_{A2}, A) \tag{2.37}$$

then the missing observables $\mathbf{y}_m$ are said to be missing at random. From (2.36)

$$\begin{aligned} p(\mathbf{y}_o, \mathbf{I} \mid \boldsymbol{\theta}_A, A) &= \int p(\mathbf{y}_o, \mathbf{y}_m \mid \boldsymbol{\theta}_{A1}, A) p(\mathbf{I} \mid \mathbf{y}_o, \mathbf{y}_m, \boldsymbol{\theta}_{A2}, A)\, d\nu(\mathbf{y}_m) \\ &= p(\mathbf{I} \mid \mathbf{y}_o, \boldsymbol{\theta}_{A2}, A) \int p(\mathbf{y}_o, \mathbf{y}_m \mid \boldsymbol{\theta}_{A1}, A)\, d\nu(\mathbf{y}_m) \\ &= p(\mathbf{I} \mid \mathbf{y}_o, \boldsymbol{\theta}_{A_2}, A) p(\mathbf{y}_o \mid \boldsymbol{\theta}_{A_1}, A). \end{aligned} \tag{2.38}$$

[If $p(\mathbf{I} \mid \mathbf{y}, \boldsymbol{\theta}_{A2}, A) = p(\mathbf{I} \mid \boldsymbol{\theta}_{A2}, A)$, then the missing observables $\mathbf{y}_m$ are said to be missing completely at random, which of course implies (2.37). For further discussion, see Gelman et al. (1995) or Little and Rubin (2002).] Comparison of (2.38) with (2.33) and (2.34) shows that $\mathbf{y}_o$ is an ancillary statistic with respect to $\boldsymbol{\theta}_{A2}$. If the vector of interest $\boldsymbol{\omega}$ does not depend on $\boldsymbol{\theta}_{A2}$, as is generally the case, then the investigator need be concerned only with $p(\boldsymbol{\theta}_{A1} \mid A)$ and $\int p(\mathbf{y}_o, \mathbf{y}_m \mid \boldsymbol{\theta}_{A1}, A)\, d\nu(\mathbf{y}_m)$. The last expression is cumbersome, in principle, but can often be managed easily using simulation methods; see Example 5.2.1 and Exercises 5.3.3, 6.4.3, and 7.1.1.

2.2.3 Nuisance Parameters

If $\boldsymbol{\theta}'_A = (\boldsymbol{\theta}'_{A1}, \boldsymbol{\theta}'_{A2})$ and $p(\boldsymbol{\omega} \mid \mathbf{y}, \boldsymbol{\theta}_A, A) = p(\boldsymbol{\omega} \mid \mathbf{y}, \boldsymbol{\theta}_{A2}, A)$ but there is no ancillary statistic with respect to $\boldsymbol{\theta}_{A2}$, then $\boldsymbol{\theta}_{A1}$ is a vector of nuisance parameters. In non-Bayesian econometrics nuisance parameters can be troublesome, because test and related statistics pertaining to $E[h(\boldsymbol{\omega}) \mid \mathbf{y}, \boldsymbol{\theta}_A, A]$ depend on the value of the unknown parameter vector $\boldsymbol{\theta}_{A2}$. Nuisance parameters never present any fundamental difficulty in Bayesian inference, because they are marginalized in the posterior distribution of $\boldsymbol{\omega}$:

$$\begin{aligned} p(\boldsymbol{\omega} \mid \mathbf{y}^o, A) &= \int_{\Theta_A} p(\boldsymbol{\omega} \mid \mathbf{y}^o, \boldsymbol{\theta}_A, A) p(\boldsymbol{\theta}_A \mid \mathbf{y}^o, A)\, d\nu(\boldsymbol{\theta}_A) \\ &= \int_{\Theta_{A2}} p(\boldsymbol{\omega} \mid \mathbf{y}^o, \boldsymbol{\theta}_{A2}, A) p(\boldsymbol{\theta}_{A2} \mid \mathbf{y}^o, A) \\ &\quad \times \left[\int_{\Theta_{A1}} p(\boldsymbol{\theta}_{A1} \mid \boldsymbol{\theta}_{A2}, \mathbf{y}^o, A)\, d\nu(\boldsymbol{\theta}_{A1}) \right] d\nu(\boldsymbol{\theta}_{A2}) \\ &= \int_{\Theta_{A2}} p(\boldsymbol{\omega} \mid \mathbf{y}^o, \boldsymbol{\theta}_{A2}, A) p(\boldsymbol{\theta}_{A2} \mid \mathbf{y}^o, A)\, d\nu(\boldsymbol{\theta}_{A2}). \end{aligned}$$

In a posterior simulator, if $\boldsymbol{\theta}_A^{(m)} \sim p(\boldsymbol{\theta}_A \mid \mathbf{y}^o, A)$ then $\boldsymbol{\theta}_{A1}^{(m)}$ can be ignored and $\boldsymbol{\omega}^{(m)} \sim p(\boldsymbol{\omega} \mid \mathbf{y}^o, \boldsymbol{\theta}_{A2}^{(m)}, A)$.

Example 2.2.4 Precision as a Nuisance Parameter in the Normal Linear Regression Model It is often the case that $p(\boldsymbol{\omega} \mid \boldsymbol{\beta}, h, A) = p(\boldsymbol{\omega} \mid \boldsymbol{\beta}, A)$ in this model. In the context of Example 2.1.2 the precision h is then a nuisance parameter. At a formal level, we may therefore work directly with the posterior distribution $\boldsymbol{\beta} \mid (\mathbf{y}^o, \mathbf{X}, A)$ instead of $(\boldsymbol{\beta}, h) \mid (\mathbf{y}^o, \mathbf{X}, A)$ by integrating h from the joint distribution whose kernel is (2.18c)–(2.18d). At a practical level this is challenging since there is no closed-form solution except in a few limiting cases like the one presented in Example 3.2.1. Numerical procedures provide a ready solution to the practical problem for the situation of Example 2.1.2; see Example 4.3.1.

Exercise 2.2.1 Models for Positive Observables The observable y_t is strictly positive. In model A, the distribution of y_t is exponential:

$$y_t \mid (\theta, A) \sim \exp(\theta^{-1}), \quad p(y_t \mid \theta, A) = \theta \exp(-\theta y_t) I_{(0,\infty)}(y_t).$$

In model B, the distribution of y_t is half-normal:

$$\begin{aligned} y_t \mid (h, B) \sim HN(0, h^{-1}), \quad & p(y_t \mid h, B) \\ &= (\pi/2)^{-1/2} h^{1/2} \exp(-h y_t^2/2) I_{(0,\infty)}(y_t). \end{aligned}$$

The observables $y_1, \ldots, y_T$ are independently and identically distributed. Indicate a vector of sufficient statistics in model A, and one in model B.

(For continuation, see Exercise 2.3.3.)

Exercise 2.2.2 A Complete Uniform Distribution Model Suppose that $y_1, \ldots, y_T$ are independently distributed, each with a uniform distribution on the interval $[0, \theta]$.

(a) What is $p(y_1, \ldots, y_T \mid \theta, A)$?

(b) Find a 2×1 sufficient statistic vector for θ.

(c) Find the maximum likelihood estimator of θ. What important regularity condition underlying the conventional asymptotic distribution theory of maximum likelihood estimators is violated in this case?

(d) Suppose that the model is completed with the prior density

$$p(\theta \mid A) = \lambda \exp(-\lambda\theta) I_{(0,\infty)}(\theta),$$

where λ is a specified positive constant. Find a kernel of the posterior density for θ.

(e) Suppose that the model is completed with the prior density $p(\theta \mid A) = c^{-1} I_{(0,c)}(\theta)$, where c is a specified positive constant. Find the posterior density (not a kernel) and the moments $E(\theta \mid \mathbf{y}^o, A)$ and $var(\theta \mid \mathbf{y}^o, A)$.

(For continuation, see Exercise 2.3.2.)

Exercise 2.2.3 Sufficiency and Ancillarity for the Uniform Distribution Suppose that $(y_t, t = 1, \ldots, T)$ are independently distributed, each with a uniform distribution on the interval $[\theta, \theta + 1]$. Define

$$y_{\min} = \min_{t=1,\ldots,T}(y_t);\ y_{\max} = \max_{t=1,\ldots,T}(y_t);\ y^* = (y_{\min} + y_{\max})/2;\ r = y_{\max} - y_{\min}.$$

(a) Show that (y^*, r) is a sufficient statistic.

(b) Show that y^* is not a sufficient statistic.

(c) Show that the distribution of y_t is location-invariant: $p(y_t \mid \theta + a) = p(y_t - a \mid \theta)$.

(d) Show that the distribution of r does not depend on θ.

(e) Show that r is ancillary with respect to θ.

Exercise 2.2.4 A Truncated Normal Distribution Suppose that $y_t \overset{\text{i.i.d.}}{\sim} N(\mu, h^{-1})$ but that y_t is also truncated below at a:

$$p(y_t \mid \mu, h, a, A) \propto \exp[-h(y_t - \mu)^2/2] I_{[a,\infty)}(y_t).$$

(a) Write the probability density $p(y_1, \ldots, y_T \mid \mu, h, a, A)$.

(b) Find a nontrivial vector of sufficient statistics $\mathbf{s}$ in a model with this observables density. (The trivial vector of sufficient statistics is the entire data set.)

(c) Are any of the sufficient statistics in (b) ancillary with respect to (μ, h)?

Exercise 2.2.5 Conditioning in the Normal Linear Regression Model Assume the observables distribution

$$y_{1t} = \alpha y_{2t} + \boldsymbol{\beta}_1' \mathbf{x}_{1t} + \boldsymbol{\beta}_2' \mathbf{x}_{2t} + \varepsilon_{1t} \tag{2.39}$$

$$y_{2t} = \boldsymbol{\gamma}_1' \mathbf{x}_{1t} + \boldsymbol{\gamma}_2' \mathbf{x}_{3t} + \varepsilon_{2t} \tag{2.40}$$

$$\begin{pmatrix} \varepsilon_{1t} \\ \varepsilon_{2t} \end{pmatrix} \overset{\text{i.i.d.}}{\sim} N\left\{ \begin{pmatrix} 0 \\ 0 \end{pmatrix}, \begin{bmatrix} h_1^{-1} & 0 \\ 0 & h_2^{-1} \end{bmatrix} \right\}$$

for $t = 1, \ldots, T$. The covariates $\mathbf{x}_{1t}$, $\mathbf{x}_{2t}$, and $\mathbf{x}_{3t}$ are all fixed. This is a simple example of what is called a "recursive simultaneous equations system" in econometrics. Note that y_{2t} is determined in (2.40); conditional on y_{2t}, y_{1t} is then determined independently in (2.39).

(a) Indicate a vector of sufficient statistics. (There is more than one right answer. But a shorter vector is better than a longer vector.)

(b) Suppose that the vector of interest $\boldsymbol{\omega}$ is a function of only α and $\boldsymbol{\beta}_1$. What are the ancillary statistics, if any? What are the nuisance parameters, if any?

(c) In the same situation as (b), can you reformulate the model and thereby reduce the vector of sufficient statistics? If so, indicate any ancillary statistics and nuisance parameters in this new model.

(d) Suppose that what is ultimately of interest is $P(y_{1,T+1} > c \mid \mathbf{x}_{T+1} = \mathbf{x}_{T+1})$, where it is assumed that the same model will apply in period $T + 1$ and

$$\mathbf{x}'_{T+1} = (\mathbf{x}'_{1,T+1}, \mathbf{x}'_{2,T+1}, \mathbf{x}'_{3,T+1}).$$

What are the ancillary statistics, if any? What are the nuisance parameters, if any?

2.3 CONJUGATE PRIOR DISTRIBUTIONS

The densities $p(\boldsymbol{\theta}_A \mid A)$ and $p(\mathbf{Y}_T \mid \boldsymbol{\theta}_A, A)$ together represent a belief regarding the observables $\mathbf{Y}_T$. In selecting the distribution of unobservables, or the conditional distribution of observables, the richer the class of functional forms from which to choose, the more adequate the representation of prior beliefs possible. On the other hand, the choice is constrained by the tractability of the posterior density $p(\boldsymbol{\theta}_A \mid \mathbf{Y}_T^o, A) \propto p(\boldsymbol{\theta}_A \mid A)p(\mathbf{Y}_T^o \mid \boldsymbol{\theta}_A, A)$, which is jointly determined by the choice of functional forms for the data density and prior density. The search for rich tractable classes of prior distributions may be formalized by considering classes of prior densities, $p(\boldsymbol{\theta}_A \mid \boldsymbol{\gamma}_A, A)$. In this approach, $\boldsymbol{\gamma}_A$ is a vector of constants that indexes prior beliefs. In fact, we have already considered this approach in Example 2.1.2, in which the parameters indexing prior beliefs were $\underline{\boldsymbol{\beta}}$, $\underline{\mathbf{H}}$, $\underline{s}^2$, and $\underline{\nu}$.

Definition 2.3.1 Suppose that the observables density $p(\mathbf{Y}_T \mid \boldsymbol{\theta}_A, A)$ has the $r \times 1$ sufficient statistic vector $\mathbf{s}_T = \mathbf{s}_T(\mathbf{Y}_T; A)$, that r is fixed as T varies, and $(\mathbf{s}_T)_1 = T$. Denote the corresponding likelihood function

$$L(\boldsymbol{\theta}_A; \mathbf{s}_T^o, A) = L(\boldsymbol{\theta}_A; \mathbf{s}_T(\mathbf{Y}_T^o, A), A) = L(\boldsymbol{\theta}_A; \mathbf{Y}_T^o, A) \propto p(\mathbf{Y}_T^o \mid \boldsymbol{\theta}_A, A).$$

Then the *conjugate family of prior densities with respect to* $p(\mathbf{Y}_T \mid \boldsymbol{\theta}_A, A)$ is $\{p(\boldsymbol{\theta}_A \mid \boldsymbol{\gamma}_A, A), \boldsymbol{\gamma}_A \in \Gamma_A\}$, where

$$p(\boldsymbol{\theta}_A \mid \boldsymbol{\gamma}_A, A) \propto L(\boldsymbol{\theta}_A; \boldsymbol{\gamma}_A, A) = k(\boldsymbol{\theta}_A \mid \boldsymbol{\gamma}_A, A) \tag{2.41}$$

and

$$\Gamma_A = \left\{\boldsymbol{\gamma}_A : \int_{\Theta_A} k(\boldsymbol{\theta}_A \mid \boldsymbol{\gamma}_A, A)\, d\nu(\boldsymbol{\theta}_A) < \infty\right\}.$$

The kernel of any conjugate prior density may be interpreted as a likelihood function corresponding to a *notional data set* $\mathbf{Z}_{(\boldsymbol{\gamma}_A)_1}$ with sample size $(\boldsymbol{\gamma}_A)_1$ and sufficient statistic $\mathbf{s}_{(\boldsymbol{\gamma}_A)_1}(\mathbf{Z}_{(\boldsymbol{\gamma}_A)_1})$. To the extent that we can represent prior beliefs arising from notional data with the same probability density functional form as the likelihood function, a conjugate prior distribution will provide a good representation of belief. Because of (2.41), the prior density and the likelihood function have exactly the same functional form in $\boldsymbol{\theta}_A$.

Example 2.3.1 The Conjugate Prior Density in a Simplified Normal Linear Model Suppose that in the normal linear regression model the precision is known, $h = h_0$. From the proof of the Gauss–Markov theorem, recall

$$(\mathbf{y}^o - \mathbf{X}\boldsymbol{\beta})'(\mathbf{y}^o - \mathbf{X}\boldsymbol{\beta}) = s^2 + (\boldsymbol{\beta} - \mathbf{b})'\mathbf{X}'\mathbf{X}(\boldsymbol{\beta} - \mathbf{b}), \tag{2.42}$$

where $s^2 = (\mathbf{y}^o - \mathbf{X}\mathbf{b})'(\mathbf{y}^o - \mathbf{X}\mathbf{b})$. Hence from (2.17), we obtain

$$p(\mathbf{y} \mid \boldsymbol{\beta}, \mathbf{X}, A) \propto \exp[-h_0(\boldsymbol{\beta} - \mathbf{b})\mathbf{X}'\mathbf{X}(\boldsymbol{\beta} - \mathbf{b})/2]. \tag{2.43}$$

Thus $\mathbf{b}$ and $\mathbf{X}'\mathbf{X}$ are sufficient statistics. Since (2.43) is the kernel of a normal density in $\boldsymbol{\beta}$, $\boldsymbol{\beta} \mid A \sim N(\underline{\boldsymbol{\beta}}, \underline{\mathbf{H}}^{-1})$ is the conjugate prior distribution of $\boldsymbol{\beta}$, with $\boldsymbol{\gamma}_A = \{\underline{\boldsymbol{\beta}}, \underline{\mathbf{H}}\}$. This prior distribution corresponds to a notional data set in which the least-squares coefficient vector ("$\mathbf{b}$") is $\underline{\boldsymbol{\beta}}$ and the moment matrix of the covariates ("$\mathbf{X}'\mathbf{X}$") is $h_0^{-1}\underline{\mathbf{H}}$.

A special instance is $y_t \overset{\text{i.i.d.}}{\sim} N(\mu, 1)$. The conjugate prior distribution is $\mu \sim N(\underline{\mu}, \underline{h}^{-1})$. The notional data set corresponds to a sample of size $\underline{h}$ with sample mean $\underline{\mu}$.

More generally, the conjugate prior distribution can be expressed in terms of notional data in the form

$$\underset{q \times k}{\mathbf{R}}\,\boldsymbol{\beta} \sim N(\mathbf{r}, \mathbf{V}). \tag{2.44}$$

Often $\mathbf{V}$ is a diagonal matrix, and then (2.44) may be interpreted as the combination of q independent components of information about $\boldsymbol{\beta}$, in the same way that the covariates $\mathbf{X}$ provide T such independent components in the normal linear model. From (2.44) the pdf of the random vector $\mathbf{z} = \mathbf{R}\boldsymbol{\beta}$ is

$$p(\mathbf{z}) = (2\pi)^{-q/2}\,|\mathbf{V}|^{-1/2}\exp[-(\mathbf{r} - \mathbf{R}\boldsymbol{\beta})'\mathbf{V}^{-1}(\mathbf{r} - \mathbf{R}\boldsymbol{\beta})/2].$$

Hence if $\text{rank}(\mathbf{R}) = k$, then (2.44) implies $\boldsymbol{\beta} \mid A \sim N(\underline{\boldsymbol{\beta}}, \underline{\mathbf{H}}^{-1})$, with $\underline{\mathbf{H}} = \mathbf{R}'\mathbf{V}^{-1}\mathbf{R}$ and $\underline{\boldsymbol{\beta}} = (\mathbf{R}'\mathbf{V}^{-1}\mathbf{R})^{-1}\mathbf{R}'\mathbf{V}^{-1}\mathbf{r}$. This representation of prior information was introduced by Theil and Goldberger (1961); recall the similar development for actual data in Exercise 2.1.3.

The following extension of the idea of a conjugate family of prior densities will prove useful in subsequent work.

Definition 2.3.2 In the data density $p(\mathbf{Y}_T \mid \boldsymbol{\theta}_A, A)$ let $\boldsymbol{\theta}'_A = (\boldsymbol{\theta}'_{A1}, \boldsymbol{\theta}'_{A2})$ and fix $\boldsymbol{\theta}_{A2} = \boldsymbol{\theta}^0_{A2}$. Suppose that the data density $p(\mathbf{Y}_T \mid \boldsymbol{\theta}_{A1}, \boldsymbol{\theta}^0_{A2}, A)$ has the $r^* \times 1$

sufficient statistic vector $\mathbf{s}_T^* = \mathbf{s}_T^*(\mathbf{Y}_T; \boldsymbol{\theta}_{A2}^0, A)$, that r^* is fixed as T varies, and $(\mathbf{s}_T^*)_1 = T$. Denote the corresponding partial likelihood function

$$L(\boldsymbol{\theta}_{A1}; \boldsymbol{\theta}_{A2}^0, \mathbf{s}_T^{*o}, A) = L(\boldsymbol{\theta}_{A1}; \boldsymbol{\theta}_{A2}^0, \mathbf{Y}_T^o, A) \propto p(\mathbf{Y}_T^o \mid \boldsymbol{\theta}_{A1}, \boldsymbol{\theta}_{A2}^0, A).$$

Then the *conditionally conjugate family of prior densities with respect to* $p(\mathbf{Y}_T \mid \boldsymbol{\theta}_{A1}, \boldsymbol{\theta}_{A2}^0, A)$ is $\{p(\boldsymbol{\theta}_{A1} \mid \boldsymbol{\gamma}_A^*, \boldsymbol{\theta}_{A2}^0, A), \boldsymbol{\gamma}_A^* \in \Gamma_A^*\}$, where

$$p(\boldsymbol{\theta}_{A1} \mid \boldsymbol{\gamma}_A^*, \boldsymbol{\theta}_{A2}^0, A) \propto L(\boldsymbol{\theta}_{A1}; \boldsymbol{\theta}_{A2}^0, \boldsymbol{\gamma}_A^*, A) = k(\boldsymbol{\theta}_{A1} \mid \boldsymbol{\gamma}_A^*, \boldsymbol{\theta}_{A2}^0, A)$$

and

$$\Gamma_A^* = \left\{\boldsymbol{\gamma}_A^* : \int_{\Theta_{A1}} k(\boldsymbol{\theta}_{A1} \mid \boldsymbol{\gamma}_A^*, \boldsymbol{\theta}_{A2}^0, A)\, d\nu(\boldsymbol{\theta}_{A1}) < \infty\right\}.$$

Example 2.3.2 Conditionally Conjugate Prior Distributions in the Normal Linear Model In Example 2.1.2 the prior distribution (2.11)–(2.12) for the parameter vector $\boldsymbol{\theta}_A' = (\boldsymbol{\beta}', h)$ was indexed by $\boldsymbol{\gamma}_A = \{\underline{\boldsymbol{\beta}}, \underline{\mathbf{H}}, \underline{s}^2, \underline{\nu}\}$. In Example 2.2.1 it was seen that $\mathbf{s}_T = (T, \mathbf{b}, s^2, \mathbf{X}'\mathbf{X})$ is a sufficient statistic because

$$p(\mathbf{y}^o \mid \boldsymbol{\beta}, h, \mathbf{X}, A) \propto h^{T/2} \exp\{-h[s^2 + (\boldsymbol{\beta}-\mathbf{b})'\mathbf{X}'\mathbf{X}(\boldsymbol{\beta}-\mathbf{b})]/2\} \tag{2.45}$$

where $s^2 = (\mathbf{y}^o - \mathbf{X}\mathbf{b})'(\mathbf{y}^o - \mathbf{X}\mathbf{b})$. For $h = h_0$, Example 2.3.1 provides the conditionally conjugate prior density. Conditioning on $\boldsymbol{\beta} = \boldsymbol{\beta}_0$ and employing (2.42), we obtain

$$p(\mathbf{y}^o \mid h, \boldsymbol{\beta} = \boldsymbol{\beta}_0, \mathbf{X}, A) \propto h^{T/2} \exp(-\bar{s}^2 h/2)$$

where $\bar{s}^2 = s^2 + (\boldsymbol{\beta}_0 - \mathbf{b})'\mathbf{X}'\mathbf{X}(\boldsymbol{\beta}_0 - \mathbf{b})$. Hence the prior density (2.15) is conditionally conjugate. It corresponds to a notional sample of $\underline{\nu} - 2$ observations of $\varepsilon_t \overset{\text{i.i.d.}}{\sim} N(0, h^{-1})$ in which $\sum_{t=1}^{\underline{\nu}-2} \varepsilon_t^2 = \underline{s}^2$.

Example 2.3.3 The Conjugate Prior Distribution in the Normal Linear Regression Model The prior density (2.11)–(2.12) is conditionally conjugate but not conjugate. To obtain a conjugate family of prior densities, regard (2.45) as a density kernel in h and $\boldsymbol{\beta}$, and seek to determine its form. Observe that

$$\int_{\mathbb{R}^k} p(\mathbf{y}^o \mid \boldsymbol{\beta}, h, \mathbf{X}, A)\, d\boldsymbol{\beta} \propto h^{T/2} \exp(-hs^2/2)$$
$$\cdot \int_{\mathbb{R}^k} \exp[-h(\boldsymbol{\beta}-\mathbf{b})'\mathbf{X}'\mathbf{X}(\boldsymbol{\beta}-\mathbf{b})/2]\, d\boldsymbol{\beta}$$
$$= h^{T/2} \exp(-hs^2/2)(2\pi)^{k/2} h^{-k/2} \left|\mathbf{X}'\mathbf{X}\right|^{-1/2} \propto h^{(T-k)/2} \exp(-hs^2/2), \tag{2.46}$$

and that the kernel of (2.45) in $\boldsymbol{\beta}$ is

$$\exp[-h(\boldsymbol{\beta}-\mathbf{b})'\mathbf{X}'\mathbf{X}(\boldsymbol{\beta}-\mathbf{b})/2]. \tag{2.47}$$

Since (2.46) is the kernel of a chi-square density in s^2h and (2.47) is the kernel of a multivariate normal density in $\boldsymbol{\beta}$, it follows that the conjugate prior distribution can be represented

$$\underline{s}^2 h \mid A \sim \chi^2(\underline{\nu}), \tag{2.48}$$

$$\boldsymbol{\beta} \mid (h, A) \sim N(\underline{\boldsymbol{\beta}}, h^{-1}\underline{\mathbf{H}}^{-1}). \tag{2.49}$$

The combination (2.48)–(2.49) is a specific instance of a *normal-gamma prior*, so called because the distribution (2.49) is normal, and because any gamma distribution for h can be written in the form (2.48).

The corresponding posterior density kernel is

$$\begin{aligned} &\left|h\underline{\mathbf{H}}\right|^{1/2} h^{(T+\underline{\nu}-2)/2} \exp(-\underline{s}^2 h/2) \\ &\cdot \exp\{-h[(\boldsymbol{\beta} - \underline{\boldsymbol{\beta}})'\underline{\mathbf{H}}(\boldsymbol{\beta} - \underline{\boldsymbol{\beta}}) + (\mathbf{y}^o - \mathbf{X}\boldsymbol{\beta})'(\mathbf{y}^o - \mathbf{X}\boldsymbol{\beta})]/2\}. \end{aligned} \tag{2.50}$$

From (2.42) the term in brackets in this expression is

$$s^2 + (\boldsymbol{\beta} - \overline{\boldsymbol{\beta}})'\overline{\mathbf{H}}(\boldsymbol{\beta} - \overline{\boldsymbol{\beta}}) + \underline{\boldsymbol{\beta}}'\underline{\mathbf{H}}\underline{\boldsymbol{\beta}} + \mathbf{b}'\mathbf{X}'\mathbf{X}\mathbf{b} - \overline{\boldsymbol{\beta}}'\overline{\mathbf{H}}\overline{\boldsymbol{\beta}}, \tag{2.51}$$

where

$$\overline{\mathbf{H}} = \underline{\mathbf{H}} + \mathbf{X}'\mathbf{X} \quad \text{and} \quad \overline{\boldsymbol{\beta}} = \overline{\mathbf{H}}^{-1}(\underline{\mathbf{H}}\underline{\boldsymbol{\beta}} + \mathbf{X}'\mathbf{X}\mathbf{b}). \tag{2.52}$$

[Note that the definition of $\overline{\mathbf{H}}$ in (2.52) is not the same as $\overline{\mathbf{H}}$ in the model with a conditionally conjugate prior distribution (2.19).] The posterior density kernel in $\boldsymbol{\beta}$ alone is $\exp[-h(\boldsymbol{\beta} - \overline{\boldsymbol{\beta}})'\overline{\mathbf{H}}(\boldsymbol{\beta} - \overline{\boldsymbol{\beta}})/2]$, whence $\boldsymbol{\beta} \mid (h, \mathbf{y}^o, \mathbf{X}, A) \sim N[\overline{\boldsymbol{\beta}}, (h\overline{\mathbf{H}})^{-1}]$. Let

$$Q^* = s^2 + \underline{\boldsymbol{\beta}}'\underline{\mathbf{H}}\underline{\boldsymbol{\beta}} + \mathbf{b}'\mathbf{X}'\mathbf{X}\mathbf{b} - \overline{\boldsymbol{\beta}}'\overline{\mathbf{H}}\overline{\boldsymbol{\beta}}, \tag{2.53}$$

and then substitute (2.53) in (2.51) and (2.51) in (2.50):

$$\begin{aligned} p(\boldsymbol{\beta}, h \mid \mathbf{y}^o, \mathbf{X}, A) &\propto h^{(T+k+\underline{\nu}-2)/2} \\ &\cdot \exp\{-h[\underline{s}^2 + (\boldsymbol{\beta} - \overline{\boldsymbol{\beta}})'\overline{\mathbf{H}}(\boldsymbol{\beta} - \overline{\boldsymbol{\beta}}) + Q^*]/2\}. \end{aligned} \tag{2.54}$$

Integrating this expression with respect to $\boldsymbol{\beta}$, we obtain

$$\begin{aligned} p(h \mid \mathbf{y}^o, \mathbf{X}, A) &\propto h^{(T+k+\underline{\nu}-2)/2}(2\pi)^{k/2}\left|h\overline{\mathbf{H}}\right|^{-1/2} \exp[-h(\underline{s}^2 + Q^*)/2] \\ &\propto h^{(T+\underline{\nu}-2)/2} \exp[-h(\underline{s}^2 + Q^*)/2], \end{aligned}$$

and so

$$\overline{s}^2 h \mid (\mathbf{y}^o, \mathbf{X}, A) \sim \chi^2(\overline{\nu}). \tag{2.55}$$

where

$$\overline{s}^2 = \underline{s}^2 + Q^* \text{ and } \overline{\nu} = \underline{\nu} + T. \tag{2.56}$$

To get some insight into Q^*, a little manipulation of (2.53) yields

$$Q^* = s^2 + (\overline{\boldsymbol{\beta}} - \mathbf{b})'\mathbf{X}'\mathbf{X}(\overline{\boldsymbol{\beta}} - \mathbf{b}) + (\overline{\boldsymbol{\beta}} - \underline{\boldsymbol{\beta}})'\underline{\mathbf{H}}(\overline{\boldsymbol{\beta}} - \underline{\boldsymbol{\beta}}). \tag{2.57}$$

The first term in this expression is the sum of squared residuals, which clearly is informative for h—in fact, T/s^2 is the maximum likelihood estimate of h. The last two terms in (2.57) measure the distance between the prior mean $\underline{\boldsymbol{\beta}}$ and the least-squares vector $\mathbf{b}$. Note that if $\underline{\boldsymbol{\beta}} = \mathbf{b}$, then $\overline{\boldsymbol{\beta}} = \mathbf{b}$ as well and these terms vanish. The second term in (2.57) is the distance between $\mathbf{b}$ and $\overline{\boldsymbol{\beta}}$ using a metric proportional to data precision $h\mathbf{X}'\mathbf{X}$, and the third term is the distance between $\underline{\boldsymbol{\beta}}$ and $\overline{\boldsymbol{\beta}}$ using a metric proportional to prior precision $h\underline{\mathbf{H}}$. If these distances are large, then in the context of the model the explanation is that h is small. A larger value of $(\overline{\boldsymbol{\beta}} - \mathbf{b})'\mathbf{X}'\mathbf{X}(\overline{\boldsymbol{\beta}} - \mathbf{b})$ or $(\overline{\boldsymbol{\beta}} - \underline{\boldsymbol{\beta}})'\underline{\mathbf{H}}(\overline{\boldsymbol{\beta}} - \underline{\boldsymbol{\beta}})$ contributes to a larger value of Q^* and hence a smaller value of h by means of (2.55).

Note that we may also integrate (2.54) with respect to h to obtain $p(\boldsymbol{\beta} \mid \mathbf{y}^o, \mathbf{X}, A)$. The kernel of (2.54) in h is that of the distribution

$$[\underline{s}^2 + (\boldsymbol{\beta} - \overline{\boldsymbol{\beta}})'\overline{\mathbf{H}}(\boldsymbol{\beta} - \overline{\boldsymbol{\beta}}) + Q^*]h \sim \chi^2(T + k + \underline{\nu}).$$

Referring to the constant of integration for this distribution [e.g., see (2.15)] and integrating (2.54) with respect to h, we obtain

$$p(\boldsymbol{\beta} \mid \mathbf{y}^o, \mathbf{X}, A) \propto [\underline{s}^2 + (\boldsymbol{\beta} - \overline{\boldsymbol{\beta}})'\overline{\mathbf{H}}(\boldsymbol{\beta} - \overline{\boldsymbol{\beta}}) + Q^*]^{-(T+k+\underline{\nu})/2}.$$

The kernel of this expression is that of a *multivariate Student-tdistribution* (Johnson and Kotz 1972, Chapter 37; Zellner 1971, Appendix B.2) with location vector $\overline{\boldsymbol{\beta}}$, scale matrix $\overline{s}^2\overline{\mathbf{H}}^{-1}$, and $T + \underline{\nu}$ degrees of freedom:

$$\boldsymbol{\beta} \mid (\mathbf{y}^o, \mathbf{X}, A) \sim t[\overline{\boldsymbol{\beta}}, (T + \underline{\nu})^{-1}\overline{s}^2\overline{\mathbf{H}}^{-1}; T + \underline{\nu}]. \tag{2.58}$$

In Example 2.3.3 the potential for an analytically tractable posterior density inherent in conjugate prior densities was realized. The prior density, likelihood kernel, and posterior density were all members of the normal-gamma family. This commonality of distribution families generalizes to a much wider class of observables distributions.

Definition 2.3.3 The *exponential family of distributions* consists of all distributions with the observables density

$$p(\mathbf{Y}_T \mid \boldsymbol{\theta}_A, A) = [g(\boldsymbol{\theta}_A)]^T \left[\prod_{t=1}^{T} f(\mathbf{y}_t)\right] \exp\left\{\sum_{i=1}^{r} c_i \phi_i(\boldsymbol{\theta}_A)\left[\sum_{t=1}^{T} h_i(\mathbf{y}_t)\right]\right\},$$

where $\Theta_A \supset \boldsymbol{\theta}_A$, $f(\cdot)$ and $\{c_i, \phi_i(\cdot), h_i(\cdot)\}_{i=1}^r$ are all specified.

Examples within the exponential family include the normal, Bernoulli, Poisson, and exponential distributions. A sufficient statistic is

$$\left[T, \sum_{t=1}^{T} h_i(\mathbf{y}_t^o)(i = 1, \ldots, r)\right].$$

The conjugate family of prior densities is

$$p(\boldsymbol{\theta}_A \mid \boldsymbol{\gamma}_A, A) \propto [g(\boldsymbol{\theta}_A)]^{\gamma_1} \exp\left[\sum_{i=1}^{r} c_i \phi_i(\boldsymbol{\theta}_A)\gamma_{i+1}\right], \tag{2.59}$$

$$\boldsymbol{\gamma}_A \in \Gamma_A = \left\{\boldsymbol{\gamma}_A : \int_{\Theta_A} [g(\boldsymbol{\theta}_A)]^{\gamma_1} \exp\left[\sum_{i=1}^{r} c_i \phi_i(\boldsymbol{\theta}_A)\gamma_{i+1}\right] d\boldsymbol{\theta}_A < \infty\right\}.$$

The posterior density of $\boldsymbol{\theta}_A$ has kernel

$$[g(\boldsymbol{\theta}_A)]^{\gamma_1+T} \exp\left\{\sum_{i=1}^{r} c_i \phi_i(\boldsymbol{\theta}_A)\left[\gamma_{i+1} + \sum_{t=1}^{T} h_i(\mathbf{y}_t)\right]\right\}. \tag{2.60}$$

Thus the conjugate prior may be interpreted as a set of γ_1 observations, with sufficient statistics $\gamma_i (i = 2, \ldots, r+1)$. The multiplicative interaction between the prior and the likelihood function is of precisely the same form as the multiplicative interaction between successive observations. Consequently it is a simple matter to combine prior information of this form as well. Specifically, given n independent experts with prior density kernels

$$[g(\boldsymbol{\theta}_A)]^{\gamma_1^j} \exp\left[\sum_{i=1}^{r} c_i \phi_i(\boldsymbol{\theta}_A)\gamma_{i+1}^j\right] (j = 1, \ldots, n),$$

the joint prior density is of the form (2.59) with $\gamma_i = \sum_{j=1}^{n} \gamma_i^j (i = 1, \ldots, r+1)$.

Exercise 2.3.1 Some Conjugate Prior Distributions In each case find the conjugate prior distribution corresponding to the observables distribution, and provide a "notional data" interpretation for the prior:

(a) y_t $(t = 1, \ldots, T)$ is i.i.d. uniform on (θ_1, θ_2).

(b) $y_t \overset{\text{i.i.d.}}{\sim} N(0, \sigma^2)$ $(t = 1, \ldots, T)$.

(c) $y_t \overset{\text{i.i.d.}}{\sim} HN(\mu, 1)$ $(t = 1, \ldots, T)$, where HN is the half-normal distribution

$$p(y_t \mid \mu) = (\pi/2)^{-1/2} \exp[-(y_t - \mu)^2/2] \cdot I_{(\mu,\infty)}(y_t).$$

(d) $y_t \overset{\text{i.i.d.}}{\sim}$ Poisson(θ); that is, $P(y_t = j \mid \theta) = \exp(-\theta)\theta^j/j!$ $(j = 0, 1, 2, \ldots)$.

Exercise 2.3.2 A Complete Uniform Distribution Model (This is a continuation of Exercise 2.2.2.) The observables $y_1, \ldots, y_T$ are independently distributed, each with a uniform distribution on the interval $[0, \theta]$.

(a) Show that $p(\theta \mid A) \propto \theta^{-\gamma_1} I_{[\gamma_2,\infty)}(\theta)$ is the kernel of the family of conjugate prior densities for θ. Express the properly normalized prior density function. What is the set Γ_A of permissible values of $(\gamma_1, \gamma_2)'$?

(b) Express the posterior density (not merely the kernel) corresponding to the likelihood function [derived in Exercise 2.2.2(a)] and the prior density in (a).

(For continuation, see Exercise 2.5.1.)

Exercise 2.3.3 Models for Positive Observables (This is a continuation of Exercise 2.2.1.) The observables y_t are i.i.d. and strictly positive. In model A

$$p(y_t \mid \theta, A) = \theta \exp(-\theta y_t) I_{(0,\infty)}(y_t),$$

while in model B

$$p(y_t \mid h, B) = (\pi/2)^{-1/2} h^{1/2} \exp(-h y_t^2/2) I_{(0,\infty)}(y_t).$$

(a) Derive the conjugate prior densities $p(\theta \mid A)$ and $p(h \mid B)$. Make sure that the densities are properly normalized—that is, they should integrate to one over the relevant range (but you do not need to demonstrate that fact). In each case, if the prior distribution is from a common parametric distribution family (e.g., normal, uniform, ...), name the family and indicate the parameter(s).

(b) Express the posterior density kernel for each model. If either posterior distribution is from a common parametric distribution family, name the family and indicate the parameter(s). If possible, indicate the posterior mean and variance of the single unobservable in each case.

(For continuation, see Exercise 2.4.6.)

Exercise 2.3.4 Completing the Argument Derive (2.57) from (2.53).

Exercise 2.3.5 Uniform Distribution on the Centered Disk Suppose that the 2×1 random vectors $\mathbf{y}_t$ are independently and uniformly distributed on a disk centered at $(0, 0)$ with radius r:

$$p(\mathbf{y}_t \mid r, A) = \pi^{-1} r^{-2} I_{S(r)}(\mathbf{y}_t) \ (t = 1, \ldots, T)$$

where $S(r) = \{(y_1, y_2) : y_1^2 + y_2^2 \leq r^2\}$.

(a) Find a sufficient statistic in a model with this observables density.

(b) What is the maximum likelihood estimate of r?

(c) Find the family of conjugate prior distributions for r.

(d) Let $q = r^{-1}$ and suppose that the prior distribution of q is $\underline{s}^2 q \sim \chi^2(\underline{\nu})$. What is the posterior distribution of q (or r)? Be as specific as possible.

Exercise 2.3.6 Interval Data Suppose that $\widetilde{y}_1, \ldots, \widetilde{y}_T \overset{\text{i.i.d.}}{\sim} N(\mu, h^{-1})$ but none of the $\widetilde{y}_t$ are observable. Instead, we observe

$$y_t = \begin{cases} 1 & \text{if} \quad \widetilde{y}_t \leq c_1 \\ 2 & \text{if} \quad c_1 < \widetilde{y}_t \leq c_2 \\ 3 & \text{if} \quad \widetilde{y}_t > c_2 \end{cases}.$$

The constants c_1 and c_2 are known. (This kind of problem can arise in survey data when respondents are asked to provide intervals. For example, respondents are generally more willing to indicate that their income is within a bracket than they are to provide actual income.)

(a) What is the vector of unobservables in this model?

(b) Write the likelihood function for μ and h. Provide a 3×1 sufficient statistic vector.

(c) What is the conjugate prior distribution for μ and h?

(Exercise 6.2.2 is related, and extends this exercise.)

Exercise 2.3.7 Spells of Employment and the Exponential Distribution An economic consultant for a fast-food chain has been given a random sample of the chain's service workers. In her model, A, the length of time y between the time the worker is hired and the time a worker quits has an exponential distribution with parameter θ^{-1}, $p(y \mid \theta) = \theta \exp(-\theta y)$. For the purposes of this problem, assume that no one is ever laid off or fired. The only way of leaving employment is by quitting. You can also assume that no one ever has more than one spell of employment with the company—if they quit, they never come back.

In this problem we consider the case in which the consultant's data consist entirely of "complete spells;" that is, for each individual t in the sample the consultant observes the length of time, y_t, between hiring and quitting.

(a) Express the joint density of the observables and find a sufficient statistic vector for θ.

(b) Show that the conjugate prior distribution of θ has the form $\underline{s}^2\theta \sim \chi^2(\underline{\nu})$, and provide an "artificial data" interpretation of $(\underline{s}^2, \underline{\nu})$.

(c) Using the prior density in (b), express the kernel of the posterior density. Show that the posterior distribution of θ has a gamma distribution of the form $\overline{s}^2\theta \sim \chi^2(\overline{\nu})$. Express $\overline{s}^2$ and $\overline{\nu}$ in terms of the sufficient statistics from (a) and (γ_1, γ_2) from (b).

Exercise 2.3.8 Censored Spells in the Exponential Model In the same situation as Exercise 2.3.7, suppose instead that the consultant's data do not consist entirely of complete spells. Instead, the consultant collects data by gathering a random sample of the records of everyone hired in the first quarter of 1998. Because of the limitations of her budget, she can gather records only through the end of 2000. T_1 of the individuals in her sample quit before the end of 2000, and for these individuals she knows the length of the spell of employment y_t. T_2 of the individuals in her sample were still working for the fast-food chain at the end of 2000, and for these individuals she has only a lower bound z_t on the spell of employment.

(a) Express the joint density of the observables in this situation.

(b) What is a vector of sufficient statistics for θ? (Do better than the trivial answer $y_1^o, \ldots, y_{T_1}^o, z_1^o, \ldots z_{T_2}^o$.)

(c) Is the prior distribution from Exercise 2.3.7(b) still conjugate? If it is, provide an "artificial data" interpretation of the distribution. If it is not, find the conjugate prior distribution in this situation.

[This exercise, and Exercise 2.3.7, are examples of Bayesian survival analysis for which there is a substantial literature. The simulation methods developed in Chapter 4 were first applied to survival analysis by Dellaportas and Smith (1993). For applications in economics, see DeJong (1993) and Campolieti (2000, 2001).]

2.4 BAYESIAN DECISION THEORY AND POINT ESTIMATION

The elements of Bayesian decision theory are isomorphic to those of behavior under uncertainty in economics. The connection was first developed in a series of papers by Friedman and Savage (1948, 1952) and Savage (1951), and classic expositions of this and related work are Berger (1985) and Pratt et al. (1995). Both Bayesian decisionmakers and economic agents associate a cardinal measure with all possible combinations of random elements in their environment that they cannot control, and those elements that they do control. The latter are called "actions" in Bayesian decision theory and "choices" in economics. The mapping to a cardinal measure is a loss function in the former and a utility function in the latter, but except for a change in sign they serve the same purpose. The decisionmaker takes the Bayes action that minimizes the expected value of his loss function; the economic agent makes the choice that maximizes the expected value of her utility function. The formal setup for the Bayesian decisionmaker is as follows.

Definition 2.4.1 The elements of a *Bayesian decision problem* are an *action* $\mathbf{a} \in A \subseteq \mathbb{R}^m$ controlled by the decisionmaker, a *loss function* $L(\mathbf{a}, \boldsymbol{\omega})$ depending on the action and a vector of interest $\boldsymbol{\omega} \in \Omega \subseteq \mathbb{R}^q$, and a distribution function P for $\boldsymbol{\omega}$. The objective of the decisionmaker is to minimize the *Bayes risk function*

$$R(\mathbf{a}) = E[L(\mathbf{a}, \boldsymbol{\omega})] = \int_{\Omega} L(\mathbf{a}, \boldsymbol{\omega}) p(\boldsymbol{\omega})\, d\nu(\boldsymbol{\omega}).$$

The decision problem originates in the need to choose $\mathbf{a}$. The loss function is the criterion for choosing $\mathbf{a}$. It specifies the vector of interest $\boldsymbol{\omega}$, which in turn suggests the data and models the decisionmaker may wish to use. These data and models dictate the form of P, which need not be specified at this level of generality.

Definition 2.4.2 If the Bayes risk function $R(\mathbf{a})$ exists for all $\mathbf{a} \in A$, then any solution $\widehat{\mathbf{a}} = \arg\min_{\mathbf{a} \in A} R(\mathbf{a})$ of the Bayesian decision problem is a *Bayes action*. The associated *Bayes risk* is $R(\widehat{\mathbf{a}})$.

In application the relevant conditioning set is the information available at the time the action $\mathbf{a}$ is taken. This might be when no data are available (as when an experiment is designed), when some data are available (as when deciding which experiment, if any, to conduct next), or when data are being observed regularly (as in forecasting situations). The term *Bayes risk* is sometimes confined to the case of a prior distribution [see, e.g., Berger (1985), Section 1.3]. The broader definition is adopted here because of its utility in applied Bayesian decisionmaking.

The random vector $\boldsymbol{\omega}$ in Definitions 2.4.1 and 2.4.2 is the vector of interest identified at the end of Section 1.2. In the class size example introduced in Section 1.1.1, the vector of interest ω is the average test score in the school district. If T is the number of teachers in the school district, S is the number of students, c is the cost of each teacher, and the school district places the value d on each test point for each student each year, then the loss function might be $L(T, \omega) = cT - dS\omega$. The relationship between class size and test scores creates a link between T and ω. Uncertainty about the link is reflected in the distribution of ω conditional on T. Such problems can be solved routinely using the simulation methods set forth in Chapter 4, and we return to this problem with such methods in hand in Example 5.1.2.

Nevertheless, simple loss functions teach a great deal about the structure of Bayesian decision problems. In particular, three solutions are often applied by investigators who do not expressly articulate the loss function that supports the solution. The critical client is then well served by examining whether her loss function is well approximated by the one that the investigator has assumed implicitly.

Definition 2.4.3 The loss function $L(\mathbf{a}, \boldsymbol{\omega})$ is a *quadratic loss function* if

$$L(\mathbf{a}, \boldsymbol{\omega}) = (\mathbf{a} - \boldsymbol{\omega})'\mathbf{Q}(\mathbf{a} - \boldsymbol{\omega}), \tag{2.61}$$

where $\mathbf{Q}$ is a positive definite matrix.

This definition is more general than it might first seem; any second-order polynomial in $\mathbf{a}$ and $\boldsymbol{\omega}$ can be brought into the form $(\mathbf{a} - \boldsymbol{\omega})'\mathbf{Q}(\mathbf{a} - \boldsymbol{\omega})$ plus a random term that is unaffected by $\mathbf{a}$. If $\mathbf{Q}$ is positive definite, then the loss function is quadratic. Note the symmetry in the loss function of Definition 2.4.3; for a given action $\mathbf{a}$ the realized loss is the same if the outcome is $\boldsymbol{\omega} = \mathbf{a} + \boldsymbol{\delta}$ or $\boldsymbol{\omega} = \mathbf{a} - \boldsymbol{\delta}$. An actual application may or may not be well served by this characteristic, and this point should be examined before proceeding with a quadratic loss function, which leads to the following simple implication.

Theorem 2.4.1 Bayes Action with a Quadratic Loss Function If the loss function is (2.61) and $A = \mathbb{R}^q$, then the Bayes action is $\widehat{\mathbf{a}} = E(\boldsymbol{\omega})$ and the Bayes risk is $R(\widehat{\mathbf{a}}) = \text{tr}[\mathbf{Q}\ \text{var}(\boldsymbol{\omega})]$.

Proof: Note that

$$R(\mathbf{a}) = E[L(\mathbf{a}, \boldsymbol{\omega})] = \int_{\Omega} (\mathbf{a} - \boldsymbol{\omega})'\mathbf{Q}(\mathbf{a} - \boldsymbol{\omega}) p(\boldsymbol{\omega})\, d\nu(\boldsymbol{\omega})$$

is twice differentiable:

$$\partial R(\mathbf{a})/\partial \mathbf{a} = 2 \int_{\Omega} \mathbf{Q}(\mathbf{a} - \boldsymbol{\omega}) p(\boldsymbol{\omega})\, d\nu(\boldsymbol{\omega}),$$
$$\partial^2 R(\mathbf{a})/\partial \mathbf{a} \partial \mathbf{a}' = 2\mathbf{Q}.$$

These conditions imply that $\widehat{\mathbf{a}} = E(\boldsymbol{\omega})$ is the unique Bayes action. The Bayes risk is

$$E(\widehat{\mathbf{a}}-\boldsymbol{\omega})'\mathbf{Q}(\widehat{\mathbf{a}}-\boldsymbol{\omega}) = \text{tr}E[\mathbf{Q}(\widehat{\mathbf{a}}-\boldsymbol{\omega})(\widehat{\mathbf{a}}-\boldsymbol{\omega})']$$
$$= \text{tr}E\{\mathbf{Q}[\boldsymbol{\omega}-\mathbf{E}(\boldsymbol{\omega})][\boldsymbol{\omega}-\mathbf{E}(\boldsymbol{\omega})]'\} = \text{tr}[\mathbf{Q}\ \text{var}(\boldsymbol{\omega})]. \quad \blacksquare$$

This result is strong and perhaps surprising in two dimensions: (1) the action is the same no matter what the positive definite matrix $\mathbf{Q}$—a change in $\mathbf{Q}$ will affect Bayes risk but will leave the Bayes action unchanged; and (2) it is only the mean of $\boldsymbol{\omega}$ that matters for the Bayes action. Other properties of the distribution are irrelevant beyond the fact that variance must exist for the problem to be well defined. Second moments matter for Bayes risk, but not for the choice of $\mathbf{a}$.

Note that if $\boldsymbol{\omega}$ were replaced by its mean, which we can denote $\overline{\boldsymbol{\omega}}$, in the quadratic loss function, then $L(\mathbf{a}, \overline{\boldsymbol{\omega}}) = (\mathbf{a}-\overline{\boldsymbol{\omega}})'\mathbf{Q}(\mathbf{a}-\overline{\boldsymbol{\omega}})$. The loss-minimizing action is $\widehat{\mathbf{a}} = \overline{\boldsymbol{\omega}}$, which is also the solution of the actual Bayesian decision problem. This fact is sometimes referred to as the *certainty equivalence principle*, after Simon (1956), who introduced it in a more general context. A corollary is that if only means are reported by an investigator, then the application of the investigator's findings is effectively restricted to decisions well characterized by quadratic loss. This may or may not be a reasonable limitation. It depends on the application at hand.

Because of their symmetry, quadratic loss functions are especially inappropriate if the consequences of an action being "too high" are quite different in severity, as will be the case if an action is similarly "too low." This asymmetry is characteristic of many situations that are described as "risky" in the colloquial use of that word. The following loss function explicitly incorporates asymmetry, for the case of an action with a single dimension.

Definition 2.4.4 If $a \in A \subseteq \mathbb{R}$ and $\omega \in \Omega \subseteq A$, the loss function $L(a, \omega)$ is a *linear–linear loss function* if

$$L(a, \omega) = (1 - q)(a - \omega) I_{(-\infty,a)}(\omega) + q(\omega - a) I_{(a,\infty)}(\omega) \tag{2.62}$$

where $q \in (0, 1)$.

To take a familiar example, consider an individual with modest savings who is about to retire. Let a represent the individual's postretirement economic standard of living and ω the return on her savings. If $\omega > a$, she will find her wealth accumulating, whereas if $\omega < a$, she will eventually become destitute. In an application of a linear–linear (or "lin-lin") loss function, it would be the case that $q < \frac{1}{2}$.

Theorem 2.4.2 Bayes Action with a Linear–Linear Loss Function If the loss function $L(a, \omega)$ is (2.62), the random variable ω is absolutely continuous and $A = \mathbb{R}$, then the Bayes action is

$$\widehat{a} = \{a : P(\omega \leq a) = q\} \tag{2.63}$$

and the Bayes risk is

$$R(\widehat{a}) = q(1-q) \cdot [E(\widehat{a} - \omega \mid \omega \leq \widehat{a}) + E(\omega - \widehat{a} \mid \omega > \widehat{a})].$$

Proof: To verify the solution note that

$$R(a) = E[L(a, \omega)] = (1-q)\int_{-\infty}^{a} (a-\omega)p(\omega)\,d\omega + q\int_{a}^{\infty} (\omega - a)p(\omega)\,d\omega.$$

This function is twice differentiable:

$$\begin{aligned} dR(a)/da &= (1-q)P(\omega \leq a) - qP(\omega > a), \\ d^2R(a)/da^2 &= p(a). \end{aligned}$$

The first-order condition implies

$$(1-q)P(\omega \leq a) = q[1 - P(\omega \leq a)] \Leftrightarrow P(\omega \leq a) = q. \tag{2.64}$$

If $p(\widehat{a}) > 0$ for $\widehat{a}$ satisfying (2.64), there is a unique Bayes action, and if not, then $\widehat{a}$ is set valued. Substituting $a = \widehat{a}$ in (2.62) and taking the expectation, we obtain

$$\begin{aligned} R(\widehat{a}) &= (1-q)E[(\widehat{a} - \omega) \mid \omega \leq \widehat{a}]P(\omega \leq \widehat{a}) \\ &\quad + qE[(\omega - \widehat{a}) \mid \omega > \widehat{a}]P(\omega > \widehat{a}) \\ &= q(1-q)\{E[(\widehat{a} - \omega) \mid \omega \leq \widehat{a}] + E[(\omega - \widehat{a}) \mid \omega > \widehat{a}]\}. \end{aligned}$$ ■

The Bayes action $\widehat{a}$ is the qth quantile of the distribution of ω. If q is small (large), then the loss if ω is larger (smaller) than a is relatively small (large), and so the Bayes action is small (large) relative to typical realizations of ω. Note the implication of Theorem 2.4.2 for our retiree: because $q < \frac{1}{2}$, she will choose a standard of living, a, less than the median of ω (which would be the Bayes action if $q = \frac{1}{2}$). The Bayes action for the linear–linear function provides some insight into the structure of the value at risk problem introduced in Section 1.1.2. In this problem ω is the value of the portfolio in the future period t^*. For any quadratic loss function $L(a, \omega) = (a - \omega)^2 q$, the solution is $a = E(\omega)$. Such a loss function is inappropriate if the decisionmaker is concerned about the risk undertaken by institutions with fiduciary responsibilities to preserve capital. A loss function of the form (2.62) would be more appropriate, and if we take $q = .05$, then the Bayes action is the 5% quantile of the distribution of ω.

It is common to see reports of "most probable" or "most likely" values of random variables. One of the attractions of these solutions is that they are easy to compute. Unlike the quadratic and linear–linear loss functions, no integration of probability densities is required. The formal rationale for this action rests on some rather strong requirements.

Definition 2.4.5 If the distribution of $\boldsymbol{\omega}$ is absolutely continuous, the loss function $L(\mathbf{a}, \boldsymbol{\omega}; \varepsilon)$ is a *zero–one loss function* if $\mathbf{a} \in \Omega$ and

$$L(\mathbf{a}, \boldsymbol{\omega}; \varepsilon) = 1 - I_{N_\varepsilon(\mathbf{a})}(\boldsymbol{\omega}),$$

where $N_\varepsilon(\mathbf{a})$ is an open ε neighborhood of $\mathbf{a}$.

Theorem 2.4.3 Bayes Action with a Zero–One Loss Function Suppose that $p(\boldsymbol{\omega})$ is a continuous function with a unique mode at $\boldsymbol{\omega} = \widehat{\mathbf{a}}$, and $A = \mathbb{R}^q$. Let $\widehat{\mathbf{a}}(\varepsilon)$ be the Bayes action for a zero–one loss function. Then $\lim_{\varepsilon \to 0} \widehat{\mathbf{a}}(\varepsilon) = \widehat{\mathbf{a}}$.

Proof: Given any $\delta > 0$, let $\Omega_\delta = \{\boldsymbol{\omega} : p(\boldsymbol{\omega}) > p(\widehat{\mathbf{a}}) - \delta\}$. For δ sufficiently small, Ω_δ is an open neighborhood of the mode $\widehat{\mathbf{a}}$. There exists $\varepsilon^* > 0$ such that if $\varepsilon < \varepsilon^*$, then $N_\varepsilon(\widehat{\mathbf{a}}) \subseteq \Omega_\delta$. Hence $\forall \varepsilon < \varepsilon^*$, $N_\varepsilon(\widehat{\mathbf{a}}(\varepsilon)) \cap \Omega_\delta \neq \{\varnothing\}$. (Why?) Since δ can be arbitrarily small, $\widehat{\mathbf{a}}$ must be a limit point of $N_\varepsilon(\widehat{\mathbf{a}}(\varepsilon))$. ■

Of course, if the mean and the mode of $\boldsymbol{\omega}$ are the same, then quadratic loss and zero–one loss lead to the same Bayes action. This action will also result if in addition the distribution of $\boldsymbol{\omega}$ is symmetric about its mean, and linear–linear loss applies element by element with $q = \frac{1}{2}$. It is rare to find modes reported without at least an implicit appeal to unimodality, symmetry, or both. The solution of the zero–one loss Bayesian decision problem is appealing because of its computational simplicity rather then its approximation of common actual loss functions.

As a practical matter, attention need not be confined to these or other loss functions because they lead to analytically simple solutions. Example 5.1.2 illustrates the use of a realistic but analytically intractable loss function, together with a posterior simulator, to find the Bayes action. For an application in marketing, see Rossi et al. (1996).

Bayesian point estimation is a Bayes action corresponding to a loss function in which the vector of interest $\boldsymbol{\omega}$ is the vector of unobservables $\boldsymbol{\theta}_A$. In the usual setup in which the relevant distribution is the posterior, the Bayes action is the point estimate

$$\widehat{\boldsymbol{\theta}}_A = \arg\min_{\widetilde{\boldsymbol{\theta}}_A} \int_{\Theta_A} L(\widetilde{\boldsymbol{\theta}}_A, \boldsymbol{\theta}_A) p(\boldsymbol{\theta}_A \mid \mathbf{y}^o, A)\, d\nu(\boldsymbol{\theta}_A).$$

Bayesian point estimation differs fundamentally from non-Bayesian approaches to estimation that seek a rule, or estimator $\boldsymbol{\theta}_A^*(\mathbf{y})$, to minimize

$$\int_{\Psi} L[\boldsymbol{\theta}_A^*(\mathbf{y}), \boldsymbol{\theta}_A] p(\mathbf{y} \mid \boldsymbol{\theta}_A, A)\, d\nu(\mathbf{y}).$$

In the solution of this problem $\widehat{\boldsymbol{\theta}}_A$ nearly always depends on $\boldsymbol{\theta}_A$ as well as on $\mathbf{y}$, and consequently $\widehat{\boldsymbol{\theta}}_A$ is almost never feasible. There is no single principle for eliminating the dependence of $\widehat{\boldsymbol{\theta}}_A$ on $\boldsymbol{\theta}_A$, and since this can usually be done in a number of reasonable ways there is often a proliferation of non-Bayesian estimates in any particular application. By contrast, given a complete model and a loss function, the Bayesian point estimate is well defined as long as $E[L(\widehat{\boldsymbol{\theta}}_A, \boldsymbol{\theta}_A) \mid \mathbf{y}^o, A]$ is well defined and finite for at least some $\widehat{\boldsymbol{\theta}}_A$.

The solution of Bayesian decision problems in the three specific cases just considered carries through directly to the specific case of point estimation of the vector of unobservables $\boldsymbol{\theta}_A$. A quadratic loss function $L(\widehat{\boldsymbol{\theta}}_A, \boldsymbol{\theta}_A)$ leads to the posterior mean $\widehat{\boldsymbol{\theta}}_A = E(\boldsymbol{\theta}_A \mid \mathbf{y}^o, A)$. The linear–linear loss function for a parameter selects the posterior qth quantile. The limiting zero–one loss function leads to the mode of $p(\boldsymbol{\theta}_A \mid \mathbf{y}^o, A)$ when applied to the entire vector $\boldsymbol{\theta}_A$.

Point estimation of parameters is heavily emphasized in statistics—perhaps more so in the non-Bayesian than the Bayesian literature, but the latter also positions point estimation prominently. This nearly always reflects the evolution of technology rather than the underlying decision problem. The parameter vector $\boldsymbol{\theta}_A$, in the context of a complete model, provides expression of

$$p(\boldsymbol{\omega} \mid A) = \int_{\Theta_A} p(\boldsymbol{\omega} \mid \boldsymbol{\theta}_A, A) p(\boldsymbol{\theta}_A \mid A)\, d\nu(\boldsymbol{\theta}_A)$$

in a way that facilitates Bayesian updating with data:

$$p(\boldsymbol{\omega} \mid \mathbf{y}^o, A) = \int_{\Theta_A} p(\boldsymbol{\omega} \mid \boldsymbol{\theta}_A, A) p(\boldsymbol{\theta}_A \mid \mathbf{y}^o, A)\, d\nu(\boldsymbol{\theta}_A). \tag{2.65}$$

Even technically oriented decisionmakers have little use for parameters, which are intermediate devices of interest to investigators. Occasionally some components of $\boldsymbol{\omega}$ correspond to certain components of $\boldsymbol{\theta}_A$, but this emphasizes that it is the vector of interest $\boldsymbol{\omega}$ and not the parameter vector $\boldsymbol{\theta}_A$ that matters in the decision. It is never the case that

$$p(\boldsymbol{\omega} \mid \mathbf{y}^o, A) = p(\boldsymbol{\omega} \mid \widehat{\boldsymbol{\theta}}_A, A), \tag{2.66}$$

an assumption commonly made in non-Bayesian statistics. There may, of course, be cases in which the difference between (2.65) and (2.66) is negligible, but this again enforces the point that the governing rule is (2.65) rather than (2.66). The burden is to demonstrate the closeness of (2.66) to (2.65) if we are to use (2.66).

To illustrate the techniques involved in analytic approaches to Bayesian point estimation, and provide some more insight into the normal linear regression model, consider the tasks of estimating $\boldsymbol{\beta}$ and h, respectively.

Example 2.4.1 Estimation of $\boldsymbol{\beta}$ in the Normal Linear Regression Model In the normal linear regression model (2.10)–(2.12)

$$\boldsymbol{\beta} \mid (h, \mathbf{y}^o, \mathbf{X}, A) \sim N(\overline{\boldsymbol{\beta}}, \overline{\mathbf{H}}^{-1}).$$

(Throughout this example we condition on h as well as the data.) Given any quadratic loss function, $\widehat{\boldsymbol{\beta}} = E[\boldsymbol{\beta} \mid h, \mathbf{y}^o, \mathbf{X}, A] = \overline{\boldsymbol{\beta}}$. Given a zero–one loss function, $\widehat{\beta} = \overline{\beta}$, as well. Given the linear–linear loss function

$$L(\widehat{\beta}_j, \beta_j) = (1-q)(\widehat{\beta}_j - \beta_j) I_{(-\infty, \widehat{\beta}_j)}(\beta_j) + q(\widehat{\beta}_j - \beta_j) I_{(\widehat{\beta}_j, \infty)}(\beta_j),$$

$$\widehat{\beta}_j = \overline{\beta}_j + (\overline{h}^{jj})^{1/2} \Phi^{-1}(q),$$

where $\overline{h}^{jj}$ denotes the (j, j)th entry of $\overline{\mathbf{H}}^{-1}$ and Φ^{-1} is the inverse cdf of the standard normal distribution. If $q = \frac{1}{2}$, then $\widehat{\beta}_j = \overline{\beta}_j$. Clearly the equivalence of the three estimates is driven by the unimodality and symmetry of the posterior density of $\boldsymbol{\beta}$, and will emerge whenever the posterior pdf of the parameter estimated is unimodal and symmetric.

Example 2.4.2 Estimation of h, $\sigma^2 = \mathbf{h}^{-1}$, and $\sigma = \mathbf{h}^{-1/2}$ in the Normal Linear Regression Model In the normal linear regression model (2.10)–(2.12)

$$\overline{s}^2 h \mid (\boldsymbol{\beta}, \mathbf{y}^o, \mathbf{X}, A) \sim \chi^2(\overline{\nu}),$$

where $\overline{s}^2$ and $\overline{\nu}$ are as defined in (2.26). Since the mean of a chi-square random variable is its degrees of freedom, $\widehat{h} \mid (\boldsymbol{\beta}, \mathbf{y}^o, \mathbf{X}, A) = \overline{\nu}/\overline{s}^2$ if the loss function is quadratic. From (2.14) the mode of the chi square pdf is its degrees of freedom less two (or else zero, whichever is larger), and so the zero–one loss estimate is $\widehat{h} \mid (\boldsymbol{\beta}, \mathbf{y}^o, \mathbf{X}, A) = (\overline{\nu} - 2)/\overline{s}^2$ if $\overline{\nu} \geq 2$.

Through the same change of variable to $\sigma^2 = h^{-1}$ undertaken in the prior leading to the pdf (2.16), we have

$$p[\sigma^2 \mid \boldsymbol{\beta}, \mathbf{y}^o, \mathbf{X}, A] = [2^{\overline{\nu}/2}\Gamma(\overline{\nu}/2)]^{-1}(\overline{s}^2)^{\overline{\nu}/2}(\sigma^2)^{-(\overline{\nu}+2)/2}\exp(-\overline{s}^2/2\sigma^2). \quad (2.67)$$

$E[\sigma^2 \mid (\boldsymbol{\beta}, \mathbf{y}^o, \mathbf{X}, A)] = \overline{s}^2/(\overline{\nu} - 2)$ if $\overline{\nu} > 2$, but estimation of σ^2 under quadratic loss has no solution if $\overline{\nu} \leq 2$. The mode of $p[\sigma^2 \mid (\boldsymbol{\beta}, \mathbf{y}^o, \mathbf{X}, A)]$ occurs at $\overline{s}^2/(\overline{\nu} + 2)$, which is therefore the zero–one loss estimate of σ^2 in this model.

If we transform instead to the standard deviation $\sigma = h^{-1/2}$, by the usual change of variable

$$p[\sigma \mid \boldsymbol{\beta}, \mathbf{y}^o, \mathbf{X}, A] = [2^{(\overline{\nu}-2)/2}\Gamma(\overline{\nu}/2)]^{-1}(\overline{s}^2)^{\overline{\nu}/2}\sigma^{-(\overline{\nu}+1)}\exp(-\overline{s}^2/2\sigma^2). \qquad (2.68)$$

The corresponding posterior mean and Bayes estimate under a quadratic loss function are

$$\widehat{\sigma} = \frac{2^{-1/2}\Gamma[(\overline{\nu}-1)/2]}{\Gamma(\overline{\nu}/2)}\overline{s} = \frac{(\overline{\nu}/2)^{1/2}\Gamma[(\overline{\nu}-1)/2]}{\Gamma(\overline{\nu}/2)}(\overline{s}^2/\overline{\nu})^{1/2}. \qquad (2.69)$$

With a zero–one loss function the Bayes estimate of σ is the posterior mode $\overline{s}/(\overline{\nu}+1)^{1/2}$.

In all three cases there is no simple closed-form expression for the median of the posterior distribution, which is in turn the estimate under a symmetric linear–linear loss function. But it is easy to compute the estimates, given $\overline{s}^2$ and $\overline{\nu}$, using standard software for the inverse of the cdf of a chi square random variable. Here are some values in the case of h:

$\overline{s}^2$	$\overline{\nu}$	Mean	Mode	Median
5	5	1.000	0.60	.8702
20	20	1.000	0.90	.9669
100	100	1.000	0.98	.9934

Note that for a fixed value of $\overline{s}^2/\overline{\nu}$, the estimates converge as $\overline{\nu}$ increases. This is due to the fact that the distribution of a normalized chi square random variable approaches the standard normal distribution as $\overline{\nu}$ increases, and the normal pdf is symmetric about its mean. Also note that the mode is always below the median and the mean is always above, and that the distance between the mode and the median is about twice that between the mean and the median.

Exercise 2.4.1 Generalizing the Quadratic Loss Function Consider the weighted squared-error loss function

$$L(\widetilde{\boldsymbol{\omega}}, \boldsymbol{\omega}) = w(\boldsymbol{\omega})(\widetilde{\boldsymbol{\omega}} - \boldsymbol{\omega})'\mathbf{Q}(\widetilde{\boldsymbol{\omega}} - \boldsymbol{\omega})$$

where $\mathbf{Q}$ is a positive definite matrix and $w(\boldsymbol{\omega}) > 0 \; \forall \; \boldsymbol{\omega} \in \Omega$. Show that the corresponding Bayes estimate of $\boldsymbol{\omega}$ is

$$\widehat{\boldsymbol{\omega}} = \frac{E[\boldsymbol{\omega} w(\boldsymbol{\omega}) \mid \mathbf{y}^o, A]}{E[w(\boldsymbol{\omega}) \mid \mathbf{y}^o, A]}.$$

Exercise 2.4.2 Generalizing the Linear–Linear Loss Function In this exercise the distribution of ω need not be absolutely continuous. This implies that $\widehat{a}$ defined in (2.63) need not exist. However, if (2.63) is replaced by

$$\widehat{a} = \{a : P(\omega \leq a) \geq q, P(\omega \geq a) \geq 1 - q\} \tag{2.70}$$

then $\widehat{a}$ will always exist, although it still need not be unique.

(a) Under what conditions will the loss functions

$$L_1(a, \omega) = (1 - q_1)(a - \omega)I_{(-\infty,a)}(\omega) + q_1(\omega - a)I_{(a,\infty)}(\omega)$$

and

$$L_2(a, \omega) = (1 - q_2)(a - \omega)I_{(-\infty,a)}(\omega) + q_2(\omega - a)I_{(a,\infty)}(\omega),$$

where $q_1 \in (0, 1)$ and $q_2 \in (0, 1)$ but $q_1 \neq q_2$, lead to the same Bayes action $\widehat{a}$?

(b) The risk function can be defined with respect to the probability measure P of the random variable ω as

$$R(a) = (1 - q) \int_{-\infty}^{a} (a - \omega)\, dP(\omega)\, d\omega + q \int_{a}^{\infty} (\omega - a)\, dP(\omega)\, d\omega.$$

Show that the right derivative of $R(\cdot)$ is

$$dR/da^+ = (1 - q)P(\omega \leq a) - qP(\omega > a)$$

and the left derivative is

$$dR/da^- = (1 - q)P(\omega < a) - qP(\omega \geq a).$$

(c) Show that if $P(\omega \leq a) < q$, then $dR/da^+ < 0$, and if $P(\omega \geq a) < 1 - q$, then $dR/da^- > 0$. Conclude that $\widehat{a}$ defined in (2.70) is the set of Bayes actions.

(d) Under what conditions is $\widehat{a}$ unique?

Exercise 2.4.3 Linex Loss Function Zellner (1986a) proposed the linear–exponential (or "linex") loss function

$$L(a, \omega) = \exp[r(a - \omega)] - r(a - \omega) - 1,$$

where $r \neq 0$.

(a) Show that the Bayes action is $\widehat{a} = -r^{-1} \log\{E[\exp(-r\omega)]\}$.

(b) Show that if, in addition, $\omega \sim N(\mu, h^{-1})$, then $\widehat{a} = \mu - (r/2h)$.

Exercise 2.4.4 Properties of Estimates Let $\boldsymbol{\theta}'_A = (\boldsymbol{\theta}'_{A1}, \boldsymbol{\theta}'_{A2})$.

(a) Consider two cases. In both cases, the loss function is quadratic. In the first case it is of the form $L(\widehat{\boldsymbol{\theta}}_A, \boldsymbol{\theta}_A)$, but in the second case it is of the form $L(\widehat{\boldsymbol{\theta}}_{A1}, \boldsymbol{\theta}_{A1})$, but is still quadratic. Is the Bayes estimate of $\boldsymbol{\theta}_{A1}$ the same in the two cases?

(b) Consider two cases. In both cases, the loss function is zero–one. In the first case it is of the form $L(\widehat{\boldsymbol{\theta}}_A, \boldsymbol{\theta}_A)$, but in the second case it is of the form $L(\widehat{\boldsymbol{\theta}}_{A1}, \boldsymbol{\theta}_{A1})$, but is still zero–one. Is the Bayes estimate of $\boldsymbol{\theta}_{A1}$ the same in the two cases?

Exercise 2.4.5 Point Estimation for the Lognormal Distribution Suppose $\log(y_t) \mid (\mu, h, A) \overset{\text{i.i.d.}}{\sim} N(\mu, h^{-1})$ $(t = 1, \ldots, T)$.

(a) Beginning with the pdf of the univariate normal distribution, derive and express the pdf of y in terms of μ and h. Then derive $E(y_t \mid \mu, h, A)$. In doing this you may find it convenient to use the moment generating function for $z \sim N(\mu, h^{-1})$, which is $E[\exp(tz)] = \exp(t\mu + t^2/2h)$

(b) Here, and for the rest of this exercise, suppose that h is known but μ is not, and $\mu \mid A \sim N(\underline{\mu}, \underline{h}^{-1})$. Derive the posterior distribution for μ.

(c) Find the Bayes estimate $\widehat{\omega}$ of $\omega = E(y_t \mid \mu, h, A)$, given a quadratic loss function.

(d) Find the Bayes estimate $\widehat{\omega}$ of $\omega = E(y_t \mid \mu, h, A)$, given the loss function $L(\widehat{\omega}, \omega) = |\omega - \widehat{\omega}|$.

(e) Find the Bayes estimate $\widehat{\omega}$ of $\omega = E(y_t \mid \mu, h, A)$, given a zero–one loss function.

Exercise 2.4.6 Models for Positive Observables Recall the exponential model in Exercise 2.3.3. The observables $y_1, \ldots, y_T$ are i.i.d., $p(y_t \mid \theta, A) = \theta \exp\,(-\theta y_t) I_{(0,\infty)}(y_t)$. Suppose $\omega = y_{T+1}$, which is (as yet) unobserved, independent of the observed $y_1 = y_1^o, \ldots, y_T = y_T^o$, and has the same distribution as each of $y_1, \ldots, y_T$. *Find* the estimate $\widehat{\omega}$ of ω implied by the loss function

$$(\widehat{\omega} - \omega) I_{(-\infty,\widehat{\omega})}(\omega) + 3(\omega - \widehat{\omega}) I_{(\widehat{\omega},\infty)}(\omega).$$

Make your answer compact, and use conventional notation to the extent you can. (For continuation, see Exercise 2.6.2.)

Exercise 2.4.7 Decisionmaking under Uncertainty In deciding whether to permit a merger of two large firms selling the same product, a government regulatory body (GRB) considers the change ω in the price of the product that will occur following the merger. The GRB can take the action $a = 1$ (permit the merger) or $a = 0$ (forbid the merger). Its loss function is $L(a, \omega)$ with $L(0, \omega) = 0$. The GRB is uncertain about ω, but given all available information, $\omega \sim N(\overline{\omega}, \overline{h}_\omega^{-1})$.

(a) Suppose that the GRB's loss function is $L(1, \omega) = (\omega - \omega^*)^3 + g(\omega - \omega^*)^2$. In this loss function, $\omega^* > 0$ and $g > 0$. What does the loss function express about the GRB's attitude toward price changes following the merger?

(b) Continuing to assume the same loss function as in (a), express the GRB's decision rule in terms of $\overline{\omega}$, $\overline{h}_\omega$, ω^*, and g.

(c) Now suppose instead that the GRB's loss function is $L(1, \omega) = \exp(f\omega) - b$. In this loss function $f > 0$ and $b > 0$. Does this loss function express attitudes toward price change similar to those for the loss function in (a)?

(d) Continuing to assume the loss function in (c), express the GRB's decision rule in terms of $\overline{\omega}$, $\overline{h}_\omega$, f,and b.

(e) Suppose that the GRB's staff reports $\overline{\omega}$ but not $\overline{h}_\omega$. Would this matter in part (b)? In part (d)?

2.5 CREDIBLE SETS

A credible set conveys one aspect of uncertainty about a vector of interest $\boldsymbol{\omega} \in \Omega$ with pdf $p(\boldsymbol{\omega})$. It is a mapping from the distribution of $\boldsymbol{\omega}$ to a subset of Ω containing $\boldsymbol{\omega}$ with given probability.

Definition 2.5.1 A set $C \subseteq \Omega$ such that

$$P(\boldsymbol{\omega} \in C) = \int_C p(\boldsymbol{\omega})\, d\nu(\boldsymbol{\omega}) = 1 - \alpha \tag{2.71}$$

is a $100(1 - \alpha)\%$ *credible set for* $\boldsymbol{\omega}$ *with respect to* $p(\boldsymbol{\omega})$.

If the distribution of $\boldsymbol{\omega}$ is absolutely continuous, so that $d\nu(\boldsymbol{\omega}) = d\boldsymbol{\omega}$, then C must exist, and except possibly in the case $\alpha = 0$, C will not be unique. On the other hand, if $\boldsymbol{\omega}$ is a discrete random variable then C will exist only for certain values of α. In what follows in this section, for any $S \subseteq \Omega$, $\overline{S} = \Omega - S$, and $\nu(S) = \int_S d\nu(\boldsymbol{\omega})$.

A posterior credible set is defined using (2.71) and $p(\boldsymbol{\omega}) = p(\boldsymbol{\omega} \mid \mathbf{y}^o, A)$. It differs fundamentally from a non-Bayesian confidence region. The latter is a set $R(\mathbf{y}) \subseteq \Omega$ such that

$$\begin{aligned} &P[\boldsymbol{\omega} \in R(\mathbf{y}) \mid \boldsymbol{\theta}_A, A] \\ &\quad = \int_\Psi \left[\int_{R(\mathbf{y})} p(\boldsymbol{\omega} \mid \mathbf{y}, \boldsymbol{\theta}_A, A)\, d\nu(\boldsymbol{\omega}) \right] p(\mathbf{y} \mid \boldsymbol{\theta}_A, A)\, d\nu(\mathbf{y}) = 1 - \alpha. \end{aligned}$$

With the exception of a few elementary cases, it is generally impossible to find an expression for $P[\boldsymbol{\omega} \in R(\mathbf{Y}_T) \mid \boldsymbol{\theta}_A, A]$ that does not involve the unobservables $\boldsymbol{\theta}_A$. The problem is essentially the same as that arising for non-Bayesian point estimates. Even when these problems can be solved, non-Bayesian confidence regions easily can lead to awkward results.

Example 2.5.1 A Simple Confidence Interval Suppose that $\{y_t\}_{t=1}^{T}$ are independently and uniformly distributed on the interval $[\theta - .5, \theta + .5]$, and $T = 25$. A 95% confidence interval for θ is $[\widehat{\theta} - .056, \widehat{\theta} + .056]$, where

$$\widehat{\theta} = [\min(y_t) + \max(y_t)]/2.$$

Consider case 1:

$$\min(y_t) = 3.10, \ \max(y_t) = 3.20, \ R = [3.094, 3.206],$$

and case 2:

$$\min(y_t) = 3.00, \ \max(y_t) = 3.96, \ R = [3.424, 3.536].$$

Both results defy common sense. In case 1, θ could be as small as 2.7 and as large as 3.6, and in case 2, it is certain that θ must be in the interval [3.46, 3.50]. In both cases the difficulty is that it is intuitive to condition on the data actually observed, whereas this non-Bayesian procedure provides intervals that include θ with probability .75 in hypothetical repetitions of the experiment of collecting samples of size $T = 25$.

To construct credible sets, introduce the prior distribution $\theta \sim N(\underline{\theta}, \underline{h}^{-1})$. Then 100% credible sets are [2.7, 3.6] in case 1 and [3.46, 3.50] in case 2. As $\underline{h} \to 0$, 95% credible sets are [2.7225, 3.5775] in case 1, and [3.461, 3.499] in case 2. In general, credible sets are not hard to determine for any given values of $\underline{\theta}$ and $\underline{h}$.

Credible sets are invariant under transformation. If

$$P(\boldsymbol{\omega} \in C) = \int_C p(\boldsymbol{\omega})\, d\nu(\boldsymbol{\omega}) = 1 - \alpha,$$

$f : \Omega \to \widetilde{\Omega}$ is one-to-one, $\widetilde{\boldsymbol{\omega}} = f(\boldsymbol{\omega})$, and $\widetilde{C}$ is the image of C under f, then $P(\widetilde{\boldsymbol{\omega}} \in \widetilde{C}) = 1 - \alpha$. Clearly credible sets are not unique, in general. For example, if the posterior distribution of $\boldsymbol{\omega}$ is absolutely continuous and $\alpha > 0$, then there are uncountably many solutions of (2.71) for C. Nor, in general, need a credible set of a specified size exist if the posterior distribution is not absolutely continuous. An interesting subset of all credible sets is the set of highest probability density (HPD) credible sets.

Definition 2.5.2 A 100(1-α)% *highest probability density* (HPD) *credible set* for $\boldsymbol{\omega}$ with respect to $p(\boldsymbol{\omega})$ is a 100(1-α)% credible set C for $\boldsymbol{\omega}$ with the property that if $\boldsymbol{\omega}_1 \in C$ and $\boldsymbol{\omega}_2 \in \overline{C}$ then $p(\boldsymbol{\omega}_1) \geq p(\boldsymbol{\omega}_2) \forall \boldsymbol{\omega}_1 \in C$ and all $\boldsymbol{\omega}_2 \in \overline{C}$.

The elements of the set of HPD credible sets are often unique up to sets of ν-measure 0. This is always the case if there exists a function $c : (0, 1) \to \mathbb{R}^+$ with the property that $P[\boldsymbol{\omega} : p(\boldsymbol{\omega}) \geq c(\alpha)] = (1 - \alpha)$. For example, HPD credible sets for multivariate normal distributions are unique and consist of ellipses and their interiors, like those shown in Figure 2.1. On the other hand, HPD credible sets for a uniform distribution are not unique.

It is natural to cast the choice of a particular credible set from among all possible credible sets as a Bayesian decision problem. The HPD credible sets correspond to the set of solutions of one such problem.

Theorem 2.5.1 Optimality of Highest-Density Regions Suppose that the distribution of ω is absolutely continuous and $p(\omega)$ is the pdf of ω. For all $C \subseteq \Omega$, let $\nu(C)$ be the Lebesgue (ordinary) measure of C. Let $C^*(\alpha) = \{C : P(\omega \in C) = 1 - \alpha\}$. Given the loss function $L(C, \omega) = k\nu(C) - I_C(\omega)$, with $k > 0$ and defined on $[C^*(\alpha) \times \Omega] \to \mathbb{R}$, C is a solution of

$$\min_{\widetilde{C}} E[L(\widetilde{C}, \omega)] \tag{2.72}$$

if and only if for all $\omega_1 \in C$ and $\omega_2 \in \overline{C}$, $p(\omega_1) \geq p(\omega_2)$, except possibly for a collection of (ω_1, ω_2) with probability zero.

Proof: For any $C \in C^*(\alpha)$, $E[L(C, \omega)] = k\nu(C) - (1 - \alpha)$, and consequently the solutions of (2.72) and the problem $\min_{C \in C^*(\alpha)} \nu(C)$ are the same.

Suppose that for all $\omega_1 \in C$ and $\omega_2 \in \overline{C}$, $p(\omega_1) \geq p(\omega_2)$, except possibly for a collection of (ω_1, ω_2) with posterior probability zero. For any other $D \in C^*(\alpha)$, we obtain

$$P(\omega \in C) = P(\omega \in D) \Rightarrow P(\omega \in C \cap \overline{D}) = P(\omega \in \overline{C} \cap D),$$

and consequently

$$\begin{aligned}\inf_{\omega \in C \cap \overline{D}} p(\omega)\nu(C \cap \overline{D}) &\leq \int_{C \cap \overline{D}} p(\omega)\, d\nu(\omega) \\ &= \int_{\overline{C} \cap D} p(\omega)\, d\nu(\omega) \leq \sup_{\overline{C} \cap D} p(\omega)\nu(\overline{C} \cap D).\end{aligned}$$

By assumption $\sup_{\overline{C} \cap D} p(\omega) \leq \inf_{\omega \in C \cap \overline{D}} p(\omega)$, and so $\nu(C \cap \overline{D}) \leq \nu(\overline{C} \cap D)$, whence $\nu(C) \leq \nu(D)$.

Now suppose the contrary—that there exist $E \subseteq C$ and $B \subseteq \overline{C}$ such that $\omega_1 \in E, \omega_2 \in B \Rightarrow p(\omega_2) \geq p(\omega_1)$, and $P(\omega \in E) = P(\omega \in B) > 0$. Let $D = (C \cap \overline{E}) \cup B$. Then $D \in C^*(\alpha)$, and by the argument in the previous paragraph $\nu(D) < \nu(C)$. ■

HPD credible sets are not invariant under transformation, as illustrated in Figure 2.2. Figure 2.2a illustrates $p(\omega)$, with the solid line indicating the unique 80% HPD interval. The transformation $\omega^* = f(\omega) = \omega^{1/2}$, illustrated in Figure 2.2b, leads to the pdf for ω^* shown in Figure 2.2d. The solid line in that figure indicates the image of the HPD interval from Figure 2.2a. Clearly this is not the 80% HPD interval for ω^*, which is indicated by the solid line in Figure 2.2c.

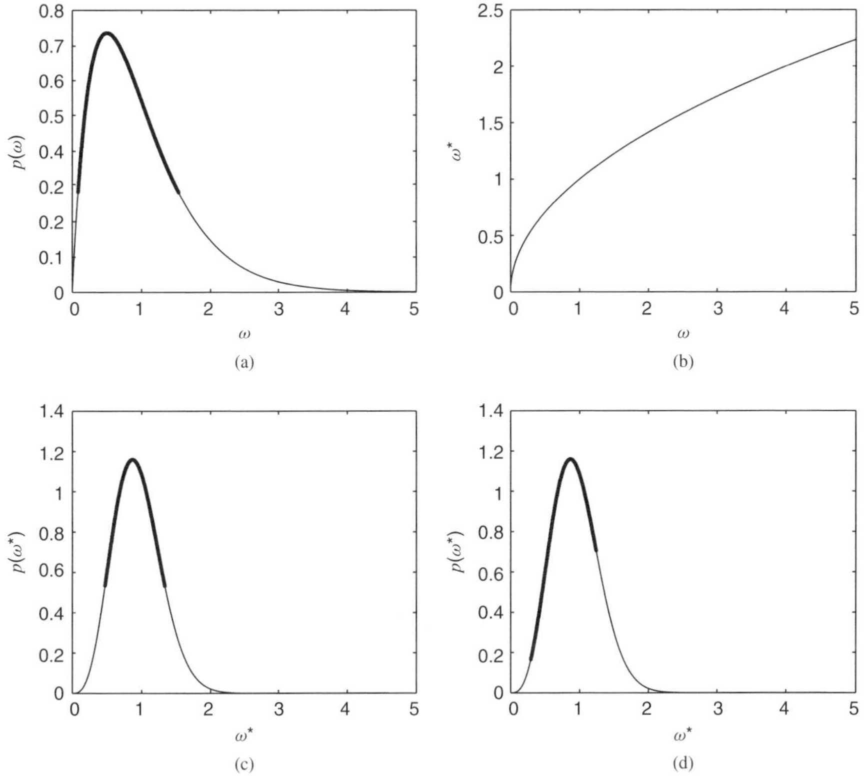

Figure 2.2. Some credible sets and projections under transformation: (a) HPD region for ω; (b) mapping into ω^*; (c) HPD region for ω^*; (d) ω HPD region mapped into ω^*.

More generally, if $f : \Omega \rightarrow \Omega^*$ is a one-to-one function, then $p(\boldsymbol{\omega}^*) = p[f^{-1}(\boldsymbol{\omega}^*)]J[f^{-1}(\boldsymbol{\omega}^*)]$ where $J(\boldsymbol{\omega}) = |\partial f/\partial \boldsymbol{\omega}|^{-1}$ is the Jacobian of transformation. If $p(\boldsymbol{\omega}_1) \geq p(\boldsymbol{\omega}_2)$ but $p(\boldsymbol{\omega}_1)J(\boldsymbol{\omega}_1) < p(\boldsymbol{\omega}_2)J(\boldsymbol{\omega}_2)$, then $p(\boldsymbol{\omega}_1^*) < p(\boldsymbol{\omega}_2^*)$ for $\boldsymbol{\omega}_1^* = f(\boldsymbol{\omega}_1)$, $\boldsymbol{\omega}_2^* = f(\boldsymbol{\omega}_2)$.

Example 2.5.2 Highest Conditional Posterior Density Regions in the Normal Linear Regression Model In the model (2.10)–(2.12) the conditional posterior distribution of $\boldsymbol{\beta}$ is (2.23). From the corresponding density kernel (2.22), a highest posterior density region is of the form

$$\{\boldsymbol{\beta} : (\boldsymbol{\beta} - \overline{\boldsymbol{\beta}})'\overline{\mathbf{H}}(\boldsymbol{\beta} - \overline{\boldsymbol{\beta}}) \leq c\}.$$

Because $(\boldsymbol{\beta} - \overline{\boldsymbol{\beta}})'\overline{\mathbf{H}}(\boldsymbol{\beta} - \overline{\boldsymbol{\beta}}) \mid (h, \mathbf{y}^o, \mathbf{X}, A) \sim \chi^2(k)$, a $100(1-\alpha)\%$ highest conditional posterior density region for $\boldsymbol{\beta}$ is

$$C = \{\boldsymbol{\beta} : (\boldsymbol{\beta} - \overline{\boldsymbol{\beta}})'\overline{\mathbf{H}}(\boldsymbol{\beta} - \overline{\boldsymbol{\beta}}) \leq \chi^2_{\alpha}(k)\}.$$

In the same model, the conditional posterior distribution of h is (2.25) and the corresponding conditional posterior density kernel (2.24) is unimodal. If $\overline{\nu} = T + \underline{\nu} > 2$, then $p(h \mid \boldsymbol{\beta}, \mathbf{y}^o, \mathbf{X}, A)$ is monotone increasing for $h < (\overline{\nu} - 2)/\overline{s}^2$ and monotone decreasing for $h > (\overline{\nu} - 2)/\overline{s}^2$. Moreover, $\lim_{h \to 0} p(h|\boldsymbol{\beta}, \mathbf{y}^o, \mathbf{X}, A) = \lim_{h \to \infty} p(h|\boldsymbol{\beta}, \mathbf{y}^o, \mathbf{X}, A) = 0$. Therefore a highest conditional, posterior density region is of the form (c_1, c_2), with

$$p(h = c_1 \mid \boldsymbol{\beta}, \mathbf{y}^o, \mathbf{X}, A) = p(h = c_2 \mid \boldsymbol{\beta}, \mathbf{y}^o, \mathbf{X}, A)$$

and

$$\int_{c_1}^{c_2} p(h \mid \boldsymbol{\beta}, \mathbf{y}^o, \mathbf{X}, A)\, dh = 1 - \alpha.$$

If $T + \underline{\nu} \leq 2$, then $p(h \mid \boldsymbol{\beta}, \mathbf{y}^o, \mathbf{X}, A)$ is monotone decreasing and the region is of the form $(0, c)$ with $c = \chi^2_\alpha(T + \underline{\nu})/\overline{s}^2$. A similar analysis may be applied to obtain highest conditional posterior density regions for σ^2, beginning from (2.67), and for σ, beginning from (2.68).

Exercise 2.5.1 A Complete Uniform Distribution Model (This is a continuation of Exercise 2.3.2.) The observables $y_1, \ldots, y_T$ are independently distributed, each with a uniform distribution on the interval $[0, \theta]$. The posterior distribution is that corresponding to the conjugate prior distribution, found in Exercise 2.3.2(b).

(a) Find the 90% highest posterior density interval for θ.

(b) Find the Bayes estimate $\widehat{\theta}$ of θ given, alternatively

- **(i)** A quadratic loss function
- **(ii)** A linear–linear loss function

$$L(\widehat{\theta}, \theta) = (1 - q)(\widehat{\theta} - \theta) I_{(-\infty, \widehat{\theta})}(\theta) + q(\theta - \widehat{\theta}) I_{[\widehat{\theta}, \infty)}(\theta);$$

- **(iii)** A zero–one loss function $L(\widehat{\theta}, \theta; \varepsilon) = 1 - I_{N_\varepsilon(\widehat{\theta})}(\theta)$.

Exercise 2.5.2 Credible Sets under Transformation In a complete model suppose that the likelihood function is $L(\boldsymbol{\theta}_A; \mathbf{y}^o, A)$ and the prior density is $p(\boldsymbol{\theta}_A \mid A)$. Let $\boldsymbol{\gamma}_A = f(\boldsymbol{\theta}_A)$, where f is one-to-one. Let $L^*(\boldsymbol{\gamma}_A; \mathbf{y}^o, A)$ be the corresponding likelihood function for $\boldsymbol{\gamma}_A$, and let $p^*(\boldsymbol{\gamma}_A \mid A)$ be the corresponding prior density for $\boldsymbol{\gamma}_A$. Provide either a proof or a specific counterexample for each of the following statements:

(a) If $\widehat{\boldsymbol{\theta}}_A$ is the mode of $L(\boldsymbol{\theta}_A; \mathbf{y}^o, A)$, then $\widehat{\boldsymbol{\gamma}}_A = f(\widehat{\boldsymbol{\theta}}_A)$ is the mode of $L^*(\boldsymbol{\gamma}_A; \mathbf{y}^o, A)$.

(b) If $\widehat{\boldsymbol{\theta}}_A$ is the mode of $p(\boldsymbol{\theta}_A \mid \mathbf{y}^o, A)$, then $\widehat{\boldsymbol{\gamma}}_A = f(\widehat{\boldsymbol{\theta}}_A)$ is the mode of $p(\boldsymbol{\gamma}_A \mid \mathbf{y}^o, A)$.

(c) If R is a 95% highest posterior density region for $\boldsymbol{\theta}_A$, and S is the image of R under f, then S is a 95% highest posterior density region for $\boldsymbol{\gamma}_A$.

Exercise 2.5.3 Point Estimates and Credible Sets In a complete model $y_t \mid (\mu, A) \overset{\text{i.i.d.}}{\sim} N(\mu, 1)$ $(t = 1, \ldots, T)$ and the prior distribution is $\mu \sim N(0, 4)$ truncated to $\mu \geq 0$: $p(\mu \mid A) = (2\pi)^{-1/2} \exp(-\mu^2/8) I_{[0,\infty)}(\mu)$. In answering each of the following questions, express your answer in terms of the sample size T, the observed sample mean $\overline{y}_T^o$, and the pdf and cdf ϕ and Φ (respectively) of the standard normal distribution.

(a) What is the Bayes estimate of μ given
  (i) A quadratic loss function?
  (ii) A zero–one loss function?
  (iii) A linear–linear loss function with $q = .75$?
(b) What is a 90% highest posterior density region for μ?

[*Hints*: **(i)** You may need to consider more than one case in each situation, depending on the values of $\overline{y}_T^o$ and T; **(ii)** a standard result in distribution theory states that if $x \sim N(0, 1)$ but is truncated to $x \geq c$, then $E(x) = \phi(c)/[1 - \Phi(c)]$.]

2.6 MODEL COMPARISON

Often we must reach a conclusion or make a decision based on several models rather than one, and there is a large literature on model selection. The complete probability structure introduced in Section 1.5 suggests that model averaging, not model choice, is the essence of the problem. This insight goes back at least to Jeffreys (1939). Its importance in statistics was recognized by Roberts (1965) and in econometrics by Zellner (1971) and Leamer (1978). More recently Draper (1995), Chatfield (1995), Kass and Raftery (1995), and Hoeting et al. (1999) have reviewed theoretical and practical aspects of Bayesian model averaging.

For specificity denote the models $j = 1, \ldots, J$. Model j has unobservables vector $\boldsymbol{\theta}_{A_j}$, unobservables prior density $p(\boldsymbol{\theta}_{A_j} \mid A_j)$, observables density $p(\mathbf{y} \mid \boldsymbol{\theta}_{A_j}, A_j)$, and vector of interest density $p(\boldsymbol{\omega} \mid \mathbf{y}, \boldsymbol{\theta}_{A_j}, A_j)$. The J models are related by their predictions for a common set of observables $\mathbf{y}$ and a common vector of interest $\boldsymbol{\omega}$. The numbers of unobservables in the models may or may not be the same and various models may or may not nest one another. The vector of interest is substantively the same for all models, although its distribution is model-specific. The specification of the collection of J models is completed with the prior probabilities $p(A_j)$ $(j = 1, \ldots, J)$ for the respective models, and $\sum_{j=1}^{J} p(A_j) = 1$. There is no essential conceptual distinction between model and prior—we could just as well regard the entire collection as a single model, with $\{p(A_j), p(\boldsymbol{\theta}_{A_j} \mid A_j)\, d\nu_j(\boldsymbol{\theta}_{A_j})\}_{j=1}^{J}$ providing the prior distribution of unobservables. At an operational level the distinction is usually clarified by the fact that we may undertake the essential computations one model at a time.

2.6.1 Marginal Likelihoods

The density $p(\boldsymbol{\omega} \mid \mathbf{y}^o)$ is ultimately of interest. The formal solution is

$$p(\boldsymbol{\omega} \mid \mathbf{y}^o) = \sum_{j=1}^{J} p(\boldsymbol{\omega} \mid \mathbf{y}^o, A_j) p(A_j \mid \mathbf{y}^o), \tag{2.73}$$

known as *model averaging*. In expression (2.73), we obtain

$$p(A_j \mid \mathbf{y}^o) = p(\mathbf{y}^o \mid A_j) p(A_j) / p(\mathbf{y}^o) \propto p(\mathbf{y}^o \mid A_j) p(A_j). \tag{2.74}$$

Thus the weight $p(A_j \mid \mathbf{y}^o)$ in (2.73) is the product of the model prior probability $p(A_j)$, and the marginal likelihood $p(\mathbf{y}^o \mid A_j)$ of Definition 2.1.1. As noted there

$$p(\mathbf{y}^o \mid A_j) = \int_{\Theta_{A_j}} p(\mathbf{y}^o \mid \boldsymbol{\theta}_{A_j}, A_j) p(\boldsymbol{\theta}_{A_j} \mid A_j) \, d\nu_j(\boldsymbol{\theta}_{A_j}). \tag{2.75}$$

It is important to recognize that if $p(\mathbf{y}^o \mid \boldsymbol{\theta}_{A_j}, A_j)$ is replaced by the corresponding likelihood function $L(\boldsymbol{\theta}_{A_j}; \mathbf{y}^o, A_j) \propto p(\mathbf{y}^o \mid \boldsymbol{\theta}_{A_j}, A_j)$ in (2.75), then, unless the constants of proportionality are the same across all models, (2.73) will no longer be true. Ignoring this fact can be extremely misleading because omitted constants in likelihood functions can vary by many orders of magnitude from one model to the next with the same data set.

The development in (2.73)–(2.75) indicates that the marginal likelihood $p(\mathbf{y}^o \mid A_j)$ is the key additional component, beyond the analysis of the individual models A_j, that model averaging requires.

Example 2.6.1 Marginal Likelihood in the Normal Linear Regression Model with Fixed Precision If $h = h_0$ and $\boldsymbol{\beta} \mid A \sim N(\underline{\boldsymbol{\beta}}, \underline{\mathbf{H}}^{-1})$, then

$$\begin{aligned} &p(\boldsymbol{\beta} \mid A) p(\mathbf{y}^o \mid \boldsymbol{\beta}, h_0, \mathbf{X}, A) \\ &\quad = (2\pi)^{-(T+k)/2} h_0^{T/2} \left|\underline{\mathbf{H}}\right|^{1/2} \\ &\quad \cdot \exp\{-[h_0(\mathbf{y}^o - \mathbf{X}\boldsymbol{\beta})'(\mathbf{y}^o - \mathbf{X}\boldsymbol{\beta}) + (\boldsymbol{\beta} - \underline{\boldsymbol{\beta}})'\underline{\mathbf{H}}(\boldsymbol{\beta} - \underline{\boldsymbol{\beta}})]/2\}. \end{aligned} \tag{2.76}$$

Completing the square as in Example 2.1.2, the term in brackets is

$$(\boldsymbol{\beta} - \overline{\boldsymbol{\beta}})'\overline{\mathbf{H}}(\boldsymbol{\beta} - \overline{\boldsymbol{\beta}}) + Q, \tag{2.77}$$

with $\overline{\mathbf{H}} = \underline{\mathbf{H}} + h_0\mathbf{X}'\mathbf{X}$, $\overline{\boldsymbol{\beta}} = \overline{\mathbf{H}}^{-1}(\underline{\mathbf{H}}\underline{\boldsymbol{\beta}} + h_0\mathbf{X}'\mathbf{X}\mathbf{b})$, and

$$Q = h_0 \mathbf{y}^{o\prime}\mathbf{y}^o + \underline{\boldsymbol{\beta}}'\underline{H}\underline{\boldsymbol{\beta}} - \overline{\boldsymbol{\beta}}'\overline{H}\overline{\boldsymbol{\beta}} \tag{2.78}$$

$$= h_0 s^2 + (\mathbf{b} - \overline{\boldsymbol{\beta}})' h_0 \mathbf{X}'\mathbf{X}(\mathbf{b} - \overline{\boldsymbol{\beta}}) + (\underline{\boldsymbol{\beta}} - \overline{\boldsymbol{\beta}})'\underline{\mathbf{H}}(\underline{\boldsymbol{\beta}} - \overline{\boldsymbol{\beta}}), \tag{2.79}$$

where $s^2 = (\mathbf{y}^o - \mathbf{X}\mathbf{b})'(\mathbf{y}^o - \mathbf{X}\mathbf{b})$. [Expression (2.79) follows from (2.78) in the same way as (2.57) from (2.53).] Substituting (2.79) in (2.77) and then (2.77) in (2.76), the marginal likelihood is

$$\begin{aligned}
&\int_{\mathbb{R}^k} p(\boldsymbol{\beta} \mid A) p(\mathbf{y}^o \mid \boldsymbol{\beta}, h_0, \mathbf{X}, A)\, d\boldsymbol{\beta} \\
&= (2\pi)^{-(T+k)/2} h_0^{T/2} \left|\underline{\mathbf{H}}\right|^{1/2} \int_{\mathbb{R}^k} \exp\{-[(\boldsymbol{\beta}-\overline{\boldsymbol{\beta}})'\overline{\mathbf{H}}(\boldsymbol{\beta}-\overline{\boldsymbol{\beta}}) + Q]/2\}\, d\boldsymbol{\beta} \\
&= (2\pi)^{-T/2} h_0^{T/2} \left|\underline{\mathbf{H}}\right|^{1/2} \left|\overline{\mathbf{H}}\right|^{-1/2} \exp(-Q/2) \qquad (2.80) \\
&= (2\pi)^{-T/2} h_0^{T/2} (\left|\underline{\mathbf{H}}\right| / \left|\overline{\mathbf{H}}\right|)^{1/2} \\
&\quad \cdot \exp\{-[h_0 s^2 + (\mathbf{b}-\overline{\boldsymbol{\beta}})' h_0 \mathbf{X}'\mathbf{X}(\mathbf{b}-\overline{\boldsymbol{\beta}}) + (\underline{\boldsymbol{\beta}} - \overline{\boldsymbol{\beta}})'\underline{\mathbf{H}}(\underline{\boldsymbol{\beta}} - \overline{\boldsymbol{\beta}})]/2\}.
\end{aligned}$$

This expression indicates those features of a model that contribute to a higher marginal likelihood, and thus a greater weight in model averaging.

Example 2.6.2 Marginal Likelihood in the Normal Linear Regression Model with Conjugate Prior Example 2.3.3 showed that the conjugate prior distribution in the normal linear regression model (2.10) is given by (2.48) and (2.49). The posterior density kernel in standard form is

$$\begin{aligned}
&p(h \mid A) p(\boldsymbol{\beta} \mid h, A) p(\mathbf{y}^o \mid \mathbf{X}, \boldsymbol{\beta}, h, A) \\
&= [2^{\underline{\nu}/2}\Gamma(\underline{\nu}/2)]^{-1} (2\pi)^{-(T+k)/2} (\underline{s}^2)^{\underline{\nu}/2} \left|\underline{\mathbf{H}}\right|^{1/2} h^{(T+k+\underline{\nu}-2)/2} \exp(-\underline{s}^2 h/2) \\
&\quad \cdot \exp\{-h[(\boldsymbol{\beta} - \underline{\boldsymbol{\beta}})'\underline{\mathbf{H}}(\boldsymbol{\beta} - \underline{\boldsymbol{\beta}}) + (\mathbf{y}^o - \mathbf{X}\boldsymbol{\beta})'(\mathbf{y}^o - \mathbf{X}\boldsymbol{\beta})]/2\}.
\end{aligned}$$

The last term in brackets can be expressed $(\boldsymbol{\beta}-\overline{\boldsymbol{\beta}})'\overline{\mathbf{H}}(\boldsymbol{\beta}-\overline{\boldsymbol{\beta}}) + Q^*$. The posterior parameters $\overline{\boldsymbol{\beta}}$ and $\overline{\mathbf{H}}$ are defined in (2.52) and Q^* is given in (2.57). Then

$$\begin{aligned}
&\int_{\mathbb{R}^k} p(h \mid A) p(\boldsymbol{\beta} \mid h, A) p(\mathbf{y}^o \mid \boldsymbol{\beta}, h, \mathbf{X}, A)\, d\boldsymbol{\beta} \\
&= [2^{\underline{\nu}/2}\Gamma(\underline{\nu}/2)]^{-1} (2\pi)^{-T/2} (\underline{s}^2)^{\underline{\nu}/2} (\left|\underline{\mathbf{H}}\right| / \left|\overline{\mathbf{H}}\right|)^{1/2} \\
&\quad \cdot h^{(T+\underline{\nu}-2)/2} \exp[-(\underline{s}^2 + Q^*)h/2]. \qquad (2.81)
\end{aligned}$$

The last line in this expression is a kernel of the density corresponding to the distribution $(\underline{s}^2 + Q^*)h \sim \chi^2(T + \underline{\nu})$, from which [recall (2.15)]

$$\int_0^\infty h^{(T+\underline{\nu}-2)/2} \exp[-(\underline{s}^2 + Q^*)h/2]\, dh \qquad (2.82)$$

$$= 2^{(T+\underline{\nu})/2}\Gamma[(T + \underline{\nu})/2](\underline{s}^2 + Q^*)^{-(T+\underline{\nu})/2}. \qquad (2.83)$$

Substituting for Q^* from (2.57) and then placing (2.83) in (2.81), the marginal likelihood is

$$\int_{\mathbb{R}^k}\int_0^\infty p(h \mid A)p(\boldsymbol{\beta} \mid h, A)p(\mathbf{y}^o \mid \boldsymbol{\beta}, h, \mathbf{X}, A)\, dh\, d\boldsymbol{\beta}$$
$$= \pi^{-T/2}\{\Gamma[(T+\underline{\nu})/2]/\,\Gamma(\underline{\nu}/2)\}(|\underline{\mathbf{H}}|\,/\,|\overline{\mathbf{H}}|)^{1/2}(\underline{s}^2)^{\underline{\nu}/2}$$
$$\cdot\,[\underline{s}^2 + s^2 + (\mathbf{b}-\overline{\boldsymbol{\beta}})'\mathbf{X}'\mathbf{X}(\mathbf{b}-\overline{\boldsymbol{\beta}}) + (\underline{\boldsymbol{\beta}} - \overline{\boldsymbol{\beta}})'\underline{\mathbf{H}}(\underline{\boldsymbol{\beta}} - \overline{\boldsymbol{\beta}})]^{-(T+\underline{\nu})/2}. \quad (2.84)$$

From (2.74), the ratio of posterior probabilities of two models is

$$\frac{P(A_j \mid \mathbf{y}^o)}{P(A_k \mid \mathbf{y}^o)} = \frac{P(A_j)}{P(A_k)} \cdot \frac{p(\mathbf{y}^o \mid A_j)}{p(\mathbf{y}^o \mid A_k)}. \quad (2.85)$$

This ratio is central in comparing models.

Definition 2.6.1 In favor of the model A_j versus the model A_k, the *prior odds ratio* is $P(A_j)/P(A_k)$; the *Bayes factor* is $p(\mathbf{y}^o \mid A_j)/p(\mathbf{y}^o \mid A_k)$; and the *posterior odds ratio* is $P(A_j \mid \mathbf{y}^o)/P(A_k \mid \mathbf{y}^o)$.

In (2.85) the posterior odds ratio is expressed as the product of the prior odds ratio and the Bayes factor. The Bayes factor, in turn, is the ratio of marginal likelihoods. If a Bayesian investigator reports the marginal likelihood of a model, then others can use this to make comparisons with other models and include the model in the process of model averaging. In special cases there are analytical expressions for Bayes factors, and these reveal those properties of models and data that are important in the posterior odds ratio.

Example 2.6.3 Bayes Factor for Two Normal Linear Regression Models with Conjugate Priors Suppose that there are two models

$$\mathbf{y} \sim N(\mathbf{X}_j\boldsymbol{\beta}_j, h_j^{-1}\mathbf{I}_T),$$
$$\underline{s}_j^2 h_j \mid A_j \sim \chi^2(\underline{\nu}_j), \quad \boldsymbol{\beta}_j \mid (h_j, A_j) \sim N(\underline{\boldsymbol{\beta}}_j, \underline{\mathbf{H}}_j^{-1}) \quad (j = 1, 2).$$

Note that the vector $\mathbf{y}$ is the same for the two models. They may therefore be used in model averaging, and can be regarded as competing specifications for the observable $\mathbf{y}$. From (2.84) and Definition 2.6.1 the Bayes factor in favor of model A_1 versus model A_2 is

$$\frac{p(\mathbf{y}^o \mid \mathbf{X}_1, A_1)}{p(\mathbf{y}^o \mid \mathbf{X}_2, A_2)} = \frac{\Gamma[(T+\underline{\nu}_1)/2]\Gamma(\underline{\nu}_2/2)}{\Gamma[(T+\underline{\nu}_2)/2]\Gamma(\underline{\nu}_1/2)} \cdot \left(\frac{|\underline{\mathbf{H}}_1|\,|\overline{\mathbf{H}}_2|}{|\underline{\mathbf{H}}_2|\,|\overline{\mathbf{H}}_1|}\right)^{1/2} \cdot \frac{(\underline{s}_1^2)^{\underline{\nu}_1/2}}{(\underline{s}_2^2)^{\underline{\nu}_2/2}}$$
$$\cdot\,\frac{[\underline{s}_1^2 + s_1^2 + (\mathbf{b}_1-\overline{\boldsymbol{\beta}}_1)'\mathbf{X}_1'\mathbf{X}_1(\mathbf{b}_1-\overline{\boldsymbol{\beta}}_1) + (\underline{\boldsymbol{\beta}}_1 - \overline{\boldsymbol{\beta}}_1)'\underline{\mathbf{H}}_1(\underline{\boldsymbol{\beta}}_1 - \overline{\boldsymbol{\beta}}_1)]^{-(T+\underline{\nu}_1)/2}}{[\underline{s}_2^2 + s_2^2 + (\mathbf{b}_2-\overline{\boldsymbol{\beta}}_2)'\mathbf{X}_2'\mathbf{X}_2(\mathbf{b}_2-\overline{\boldsymbol{\beta}}_2) + (\underline{\boldsymbol{\beta}}_2 - \overline{\boldsymbol{\beta}}_2)'\underline{\mathbf{H}}_2(\underline{\boldsymbol{\beta}}_2 - \overline{\boldsymbol{\beta}}_2)]^{-(T+\underline{\nu}_2)/2}}. \quad (2.86)$$

In the same situation the likelihood ratio statistic is $(s_1^2/s_2^2)^{-T/2}$. All of the modifications of this statistic in (2.86) come about because of the prior distributions in the two models. Conventional non-Bayesian hypothesis tests comparing the two models are based on the fact that if one of the models nests the other—that is, if either $\mathbf{X}_1 = \mathbf{X}_2\mathbf{A}_{21}$ or $\mathbf{X}_2 = \mathbf{X}_1\mathbf{A}_{12}$—then $(s_1^2/s_2^2)^{-T/2}$ has a convenient sampling distribution. There is no such nesting required for the use of Bayes factors and posterior odds ratios. Furthermore, while $\mathbf{X}_1 = \mathbf{X}_2\mathbf{A}_{21}$ implies that the likelihood ratio $(s_1^2/s_2^2)^{-T/2}$ cannot exceed 1, the Bayes factor can be any positive number.

Inspection of (2.86) indicates those aspects of model and observed data that lead to a higher Bayes factor in favor of the first model. If $\underline{\nu}_1 = \underline{\nu}_2 = \underline{\nu}$, $\underline{s}_1^2 = \underline{s}_2^2 = \underline{s}^2$, $\underline{\boldsymbol{\beta}}_1 = \mathbf{b}_1$ and $\underline{\boldsymbol{\beta}}_2 = \mathbf{b}_2$, then the second line of (2.86) reduces to $[(\underline{s}^2 + s_1^2)/(\underline{s}^2 + s_2^2)]^{-(T+\underline{\nu})/2}$. If $\underline{s}^2$ is small relative to s_1^2 and s_2^2, and $\underline{\nu}$ is small relative to T, this is a minor modification of the likelihood ratio. The first line of (2.86) favors the model in which the prior precision is a relatively more important component of the posterior precision of the coefficient vector. In a situation in which prior means agreed with the least-squares fit, the model that concentrates its prior probability more intensely on this point is favored in the first line.

In the model averaging process (2.73) it is the relative posterior probabilities, or equivalently the posterior odds ratios, of the models under consideration that matter. In general, there is no sense in which we are forced to choose among models. In some cases, however, choice of models is the essence of the decision problem; see, for example, Bajari and Lee (2003), who use alternative models and Bayes factors in deciding whether there has been collusion at an auction. With no real loss of generality, assume that there are only two models in the choice set. Treating model choice as a Bayes action, let $L(A_i, A_j)$ denote the loss incurred in choosing model A_i conditional on model A_j being true. Suppose further that $L(A_i, A_i) = 0$ and $L(A_i, A_j) > 0$ if $i \neq j$. Given the data $\mathbf{y}^o$, the expected loss from choosing model A_i is $P(A_j \mid \mathbf{y}^o)L(A_i, A_j)$ $(j \neq i)$, and so the Bayes action is to choose model 1 if $P(A_2 \mid \mathbf{y}^o)L(A_1, A_2) < P(A_1 \mid \mathbf{y}^o)L(A_2, A_1)$, that is, if

$$\frac{P(A_1 \mid \mathbf{y}^o)}{P(A_2 \mid \mathbf{y}^o)} = \frac{P(A_1)p(\mathbf{y}^o \mid A_1)}{P(A_2)p(\mathbf{y}^o \mid A_2)} > \frac{L(A_1, A_2)}{L(A_2, A_1)}.$$

Definition 2.6.2 In choosing between two models, the ratio of loss functions $L(A_1, A_2)/L(A_2, A_1)$ is the *Bayes critical value.*

We choose model 1 if the posterior odds ratio in favor of it exceeds the Bayes critical value. For reasons of economy an investigator may therefore report only the marginal likelihood, leaving it to his or her clients—the users of the investigator's research—to provide their own prior model probabilities and loss functions. The steps of reporting marginal likelihoods and Bayes factors are sometimes called *hypothesis testing* as well.

2.6.2 Predictive Densities

The marginal likelihood of model A_j, $p(\mathbf{Y}_T^o \mid A_j)$, is the measure of how well model A_j predicted the data $\mathbf{Y}_T^o$ that are relevant for the comparison of model j with other models. In fact, there is a more formal link between the marginal likelihood of a model and the adequacy of the model's predictions that underscores the predictive interpretation of $p(\mathbf{Y}_T^o \mid A_j)$. To establish this link, first consider the distribution of $\mathbf{y}_{T+1}, \ldots, \mathbf{y}_F$ conditional on $\mathbf{Y}_T^o$ and model A.

Definition 2.6.3 The *predictive density* of $\mathbf{y}_{T+1}, \ldots, \mathbf{y}_F$ conditional on $\mathbf{Y}_T^o$ and model A is

$$p(\mathbf{y}_{T+1}, \ldots, \mathbf{y}_F \mid \mathbf{Y}_T^o, A). \tag{2.87}$$

The predictive density is relevant after formulation of model A and observing $\mathbf{Y}_T = \mathbf{Y}_T^o$, but before observing $\mathbf{y}_{T+1}, \ldots, \mathbf{y}_F$. Once $\mathbf{y}_{T+1}, \ldots, \mathbf{y}_F$ are known, we can evaluate (2.87) at the observed values.

Definition 2.6.4 The *predictive likelihood* of $\mathbf{y}_{T+1}^o, \ldots, \mathbf{y}_F^o$ conditional on $\mathbf{Y}_T^o$ and the model A is the real number $p(\mathbf{y}_{T+1}^o, \ldots, \mathbf{y}_F^o \mid \mathbf{Y}_T^o, A)$.

It is natural to compare how well alternative models predict the same set of observations, by comparing their predictive likelihoods.

Definition 2.6.5 The *predictive Bayes factor* in favor of model A_j, versus model A_k, is $p(\mathbf{y}_{T+1}^o, \ldots, \mathbf{y}_F^o \mid \mathbf{Y}_T^o, A_j)/p(\mathbf{y}_{T+1}^o, \ldots, \mathbf{y}_F^o \mid \mathbf{Y}_T^o, A_k)$.

There is a formal link between predictive likelihood and marginal likelihood that is illuminating and useful, dating at least to Geisel (1975).

Theorem 2.6.1 Representation of Predictive Likelihood The predictive likelihood is a ratio of marginal likelihoods:

$$p(\mathbf{y}_{T+1}^o, \ldots, \mathbf{y}_F^o \mid \mathbf{Y}_T^o, A) = p(\mathbf{Y}_F^o \mid A)/p(\mathbf{Y}_T^o \mid A).$$

Proof:

$$\begin{aligned} p(\mathbf{Y}_F \mid A) &= p(\mathbf{Y}_F \mid \mathbf{Y}_T, A)p(\mathbf{Y}_T \mid A) \\ &= p(\mathbf{y}_{T+1}, \ldots, \mathbf{y}_F \mid \mathbf{Y}_T, A)p(\mathbf{Y}_T \mid A). \end{aligned}$$ ■

Theorem 2.6.1 shows that the predictive likelihood is the multiplicative updating factor applied to the marginal likelihood $p(\mathbf{Y}_T^o \mid A)$, after the observations $\mathbf{y}_{T+1}^o, \ldots, \mathbf{y}_F^o$ become available, that produces the new marginal likelihood $p(\mathbf{Y}_F^o \mid A)$. This updating relationship is quite general.

Corollary 2.6.1 Decomposition of the Marginal Likelihood For any strictly increasing sequence of integers $\{s_j\}_{j=0}^q$ with $s_0 = 0$ and $s_q = T$, the marginal likelihood may be decomposed

$$p(\mathbf{Y}_T^o \mid A) = \prod_{\tau=1}^{q} p(\mathbf{y}_{s_{\tau-1}+1}^o, \ldots, \mathbf{y}_{s_\tau}^o \mid \mathbf{Y}_{s_{\tau-1}}^o, A). \tag{2.88}$$

This result immediately implies that the Bayes factor in favor of model A_j versus model A_k can be decomposed in terms of predictive Bayes factors:

$$\frac{p(\mathbf{Y}_T^o \mid A_j)}{p(\mathbf{Y}_T^o \mid A_k)} = \prod_{\tau=1}^{q} \left[\frac{p(\mathbf{y}_{s_{\tau-1}+1}^o, \ldots, \mathbf{y}_{s_\tau}^o \mid \mathbf{Y}_{s_{\tau-1}}^o, A_j)}{p(\mathbf{y}_{s_{\tau-1}+1}^o, \ldots, \mathbf{y}_{s_\tau}^o \mid \mathbf{Y}_{s_{\tau-1}}^o, A_k)} \right].$$

The decomposition in Corollary 2.6.1 summarizes the "out of sample prediction record" of the model as expressed in the predictive likelihoods. In the sense made precise by (2.88) and the use of $p(\mathbf{Y}_T^o \mid A)$ in model averaging [(2.73) and (2.74)], there is no distinction between a model's adequacy and its out of sample prediction record. The decomposition (2.88) may be interpreted as a formal expression of Milton Friedman's well-known identification of a model's evaluation with its predictive performance: "Theory is to be judged by its predictive power ... The only relevant test of the *validity* of a hypothesis is comparison of its predictions with experience" [see Friedman (1953), pp. 8–9; emphasis in original]. There are striking similarities between Friedman (1953) and Jeffreys (1939, 1961). The third edition (Jeffreys 1961) contains, in Chapter 1, essentially the results presented here for the very special case of deterministic dichotomous outcomes.

Example 2.6.4 Predictive Densities in the Normal Linear Regression Model with Fixed Precision Suppose that the specification of the normal linear regression model (2.10)–(2.11) applies to F observations. Precision is fixed at $h = h_0$. The covariate matrix $\mathbf{X}$ and outcome vector $\mathbf{y}^o$ for the first T observations are known. For the last $f = F - T$ observations the covariate matrix $\widetilde{\mathbf{X}}$ is known but the corresponding outcome vector $\widetilde{\mathbf{y}}^o$ is not. Thus

$$\boldsymbol{\beta} \mid (\mathbf{y}^o, \mathbf{X}, \widetilde{\mathbf{X}}, A) \sim N(\overline{\boldsymbol{\beta}}, \overline{\mathbf{H}}^{-1}),$$

with $\overline{\mathbf{H}} = \underline{\mathbf{H}} + h_0\mathbf{X}'\mathbf{X}$ and $\overline{\boldsymbol{\beta}} = \overline{\mathbf{H}}^{-1}(\underline{\mathbf{H}}\underline{\boldsymbol{\beta}} + h_0\mathbf{X}'\mathbf{y}^o)$. Since

$$\widetilde{\mathbf{y}} \mid (\boldsymbol{\beta}, \mathbf{y}^o, \mathbf{X}, \widetilde{\mathbf{X}}, A) \sim N(\widetilde{\mathbf{X}}\boldsymbol{\beta}, h_0^{-1}\mathbf{I}_f)$$

it follows that

$$\widetilde{\mathbf{y}} \mid (\mathbf{y}^o, \mathbf{X}, \widetilde{\mathbf{X}}, A) \sim N(\widetilde{\mathbf{X}}\overline{\boldsymbol{\beta}}, \widetilde{\mathbf{X}}\overline{\mathbf{H}}^{-1}\widetilde{\mathbf{X}}' + h_0^{-1}\mathbf{I}_f).$$

When $\tilde{\mathbf{y}} = \tilde{\mathbf{y}}^o$ is observed, the predictive likelihood for the last f observations is

$$p(\tilde{\mathbf{y}}^o \mid \mathbf{y}^o, \mathbf{X}, \tilde{\mathbf{X}}, A) = (2\pi)^{-f/2} \left|\tilde{\mathbf{X}}\overline{\mathbf{H}}^{-1}\tilde{\mathbf{X}}' + h_0^{-1}\mathbf{I}_f\right|^{-1/2} \cdot \exp[-(\tilde{\mathbf{y}}^o - \tilde{\mathbf{X}}\overline{\boldsymbol{\beta}})'(\tilde{\mathbf{X}}\overline{\mathbf{H}}^{-1}\tilde{\mathbf{X}}' + h_0^{-1}\mathbf{I}_f)^{-1}(\tilde{\mathbf{y}}^o - \tilde{\mathbf{X}}\overline{\boldsymbol{\beta}})/2]. \tag{2.89}$$

From (2.78) and (2.80) the marginal likelihood for the first T observations is

$$p(\mathbf{y}^o \mid \mathbf{X}, A) = (2\pi)^{-T/2} h_0^{T/2} \left|\underline{\mathbf{H}}\right|^{1/2} \left|\overline{\mathbf{H}}\right|^{-1/2} \cdot \exp[-(h_0\mathbf{y}^{o\prime}\mathbf{y}^o + \underline{\boldsymbol{\beta}}'\underline{\mathbf{H}}\underline{\boldsymbol{\beta}} - \overline{\boldsymbol{\beta}}'\overline{\mathbf{H}}\overline{\boldsymbol{\beta}})/2] \tag{2.90}$$

and the marginal likelihood for all F observations $\mathbf{y}_F^{o\prime} = (\mathbf{y}^{o\prime}, \tilde{\mathbf{y}}^{o\prime})$ is

$$p(\mathbf{y}_F^o \mid \mathbf{X}, \tilde{\mathbf{X}}, A) = (2\pi)^{-F/2} h_0^{F/2} \left|\underline{\mathbf{H}}\right|^{1/2} \left|\overline{\overline{\mathbf{H}}}\right|^{-1/2} \cdot \exp[-(h_0\mathbf{y}_F^{o\prime}\mathbf{y}_F^o + \underline{\boldsymbol{\beta}}'\underline{\mathbf{H}}\underline{\boldsymbol{\beta}} - \overline{\overline{\boldsymbol{\beta}}}'\overline{\overline{\mathbf{H}}}\overline{\overline{\boldsymbol{\beta}}})/2]. \tag{2.91}$$

In (2.91) $\overline{\overline{\mathbf{H}}} = \overline{\mathbf{H}} + h_0\tilde{\mathbf{X}}'\tilde{\mathbf{X}}$, and $\overline{\overline{\boldsymbol{\beta}}} = \overline{\overline{\mathbf{H}}}^{-1}(\underline{\mathbf{H}}\underline{\boldsymbol{\beta}} + h_0\mathbf{X}'\mathbf{y}^o + h_0\tilde{\mathbf{X}}'\tilde{\mathbf{y}}^o)$.

Theorem 2.6.1 states that (2.89) is the ratio of (2.91) to (2.90). This follows directly once we establish two facts:

$$h_0^{f/2}|\overline{\mathbf{H}} + h_0\tilde{\mathbf{X}}'\tilde{\mathbf{X}}|^{-1/2}/|\overline{\mathbf{H}}|^{-1/2} = |\tilde{\mathbf{X}}\overline{\mathbf{H}}^{-1}\tilde{\mathbf{X}}' + h_0^{-1}\mathbf{I}_f|^{-1/2} \tag{2.92}$$

and

$$(\tilde{\mathbf{y}}^o - \tilde{\mathbf{X}}\overline{\boldsymbol{\beta}})'(\tilde{\mathbf{X}}\overline{\mathbf{H}}^{-1}\tilde{\mathbf{X}}' + h_0^{-1}\mathbf{I}_f)^{-1}(\tilde{\mathbf{y}}^o - \tilde{\mathbf{X}}\overline{\boldsymbol{\beta}}) = h_0\tilde{\mathbf{y}}^{o\prime}\tilde{\mathbf{y}}^o + \overline{\boldsymbol{\beta}}'\overline{\mathbf{H}}\overline{\boldsymbol{\beta}} - \overline{\overline{\boldsymbol{\beta}}}'\overline{\overline{\mathbf{H}}}\overline{\overline{\boldsymbol{\beta}}}. \tag{2.93}$$

Restate (2.92) as

$$\left|\overline{\mathbf{H}} + h_0\tilde{\mathbf{X}}'\tilde{\mathbf{X}}\right| = \left|\overline{\mathbf{H}}\right| \cdot \left|\tilde{\mathbf{X}}\overline{\mathbf{H}}^{-1}\tilde{\mathbf{X}}' + h_0^{-1}\mathbf{I}_f\right| h_0^f,$$

and exploit the fact that if

$$\mathbf{E} = \begin{bmatrix} \mathbf{A} & \mathbf{C} \\ \mathbf{B} & \mathbf{D} \end{bmatrix}$$

and $\mathbf{A}$ and $\mathbf{D}$ are nonsingular then

$$|\mathbf{E}| = |\mathbf{A}| \cdot \left|\mathbf{D} - \mathbf{B}\mathbf{A}^{-1}\mathbf{C}\right| = |\mathbf{D}| \cdot \left|\mathbf{A} - \mathbf{C}\mathbf{D}^{-1}\mathbf{B}\right| \tag{2.94}$$

[Rao (1965), Problem 1b.2.4]. Letting

$$\mathbf{E} = \begin{bmatrix} -\overline{\mathbf{H}} & h_0^{1/2}\tilde{\mathbf{X}}' \\ h_0^{1/2}\tilde{\mathbf{X}} & \mathbf{I}_f \end{bmatrix},$$

from (2.94) we have

$$\begin{aligned}|-\overline{\mathbf{H}}| \cdot \left|\mathbf{I}_f + h_0\widetilde{\mathbf{X}}\overline{\mathbf{H}}^{-1}\widetilde{\mathbf{X}}'\right| &= |-\overline{\mathbf{H}} - h_0\widetilde{\mathbf{X}}'\widetilde{\mathbf{X}}| \Longrightarrow |\overline{\mathbf{H}} + h_0\widetilde{\mathbf{X}}'\widetilde{\mathbf{X}}| \\ &= |\overline{\mathbf{H}}| \cdot \left|\mathbf{I}_f + h_0\widetilde{\mathbf{X}}\overline{\mathbf{H}}^{-1}\widetilde{\mathbf{X}}'\right| = |\overline{\mathbf{H}}| \cdot \left|\widetilde{\mathbf{X}}\overline{\mathbf{H}}^{-1}\widetilde{\mathbf{X}}' + h_0^{-1}\mathbf{I}_f\right| \cdot h_0^f.\end{aligned}$$

Turning to (2.93), we obtain the equation

$$\overline{\overline{\boldsymbol{\beta}}}'\overline{\overline{\mathbf{H}}}\overline{\overline{\boldsymbol{\beta}}} = \overline{\boldsymbol{\beta}}'\overline{\mathbf{H}}\overline{\boldsymbol{\beta}} - \overline{\boldsymbol{\beta}}'\widetilde{\mathbf{X}}'\mathbf{V}\widetilde{\mathbf{X}}\overline{\boldsymbol{\beta}} + 2\overline{\boldsymbol{\beta}}'\widetilde{\mathbf{X}}'\mathbf{V}\widetilde{\mathbf{y}}^o + h_0\widetilde{\mathbf{y}}^{o\prime}\widetilde{\mathbf{y}}^o - \widetilde{\mathbf{y}}^{o\prime}\mathbf{V}\widetilde{\mathbf{y}}^o \tag{2.95}$$

where $\mathbf{V} = (\widetilde{\mathbf{X}}\overline{\mathbf{H}}^{-1}\widetilde{\mathbf{X}}' + h_0^{-1}\mathbf{I}_f)^{-1}$. Note that

$$\overline{\overline{\mathbf{H}}}^{-1} = (\overline{\mathbf{H}} + h_0\widetilde{\mathbf{X}}'\widetilde{\mathbf{X}})^{-1} = \overline{\mathbf{H}}^{-1} - \overline{\mathbf{H}}^{-1}\widetilde{\mathbf{X}}'\mathbf{V}\widetilde{\mathbf{X}}\overline{\mathbf{H}}^{-1}.$$

Hence

$$\begin{aligned}\overline{\overline{\boldsymbol{\beta}}} &= \overline{\overline{\mathbf{H}}}^{-1}(\overline{\mathbf{H}}\overline{\boldsymbol{\beta}} + h_0\widetilde{\mathbf{X}}'\widetilde{\mathbf{y}}^o) \\ &= (\overline{\mathbf{H}}^{-1} - \overline{\mathbf{H}}^{-1}\widetilde{\mathbf{X}}'\mathbf{V}\widetilde{\mathbf{X}}\overline{\mathbf{H}}^{-1})(\overline{\mathbf{H}}\overline{\boldsymbol{\beta}} + h_0\widetilde{\mathbf{X}}'\widetilde{\mathbf{y}}^o) \\ &= \overline{\boldsymbol{\beta}} + h_0\overline{\mathbf{H}}^{-1}\widetilde{\mathbf{X}}'\widetilde{\mathbf{y}}^o - \overline{\mathbf{H}}^{-1}\widetilde{\mathbf{X}}'\mathbf{V}\widetilde{\mathbf{X}}\overline{\boldsymbol{\beta}} - h_0\overline{\mathbf{H}}^{-1}\widetilde{\mathbf{X}}'\mathbf{V}\widetilde{\mathbf{X}}\overline{\mathbf{H}}^{-1}\widetilde{\mathbf{X}}'\widetilde{\mathbf{y}}^o.\end{aligned} \tag{2.96}$$

Note also that

$$\overline{\overline{\boldsymbol{\beta}}}'\overline{\overline{\mathbf{H}}} = \overline{\boldsymbol{\beta}}'\overline{\mathbf{H}} + h_0\widetilde{\mathbf{y}}^{o\prime}\widetilde{\mathbf{X}}. \tag{2.97}$$

The left side of (2.95) is the product of (2.97) and (2.96). Expanding this product, we have

$$\overline{\overline{\boldsymbol{\beta}}}'\overline{\overline{\mathbf{H}}}\overline{\overline{\boldsymbol{\beta}}} = \overline{\boldsymbol{\beta}}'\overline{\mathbf{H}}\overline{\boldsymbol{\beta}} \tag{2.98a}$$

$$+ h_0\overline{\boldsymbol{\beta}}'\widetilde{\mathbf{X}}'\widetilde{\mathbf{y}}^o \tag{2.98b}$$

$$- \overline{\boldsymbol{\beta}}'\widetilde{\mathbf{X}}'\mathbf{V}\widetilde{\mathbf{X}}\overline{\boldsymbol{\beta}} \tag{2.98c}$$

$$- h_0\overline{\boldsymbol{\beta}}'\widetilde{\mathbf{X}}'\mathbf{V}\widetilde{\mathbf{X}}\overline{\mathbf{H}}^{-1}\widetilde{\mathbf{X}}'\widetilde{\mathbf{y}}^o \tag{2.98d}$$

$$+ h_0\widetilde{\mathbf{y}}^{o\prime}\widetilde{\mathbf{X}}\overline{\boldsymbol{\beta}} \tag{2.98e}$$

$$+ h_0^2\widetilde{\mathbf{y}}^{o\prime}\widetilde{\mathbf{X}}\overline{\mathbf{H}}^{-1}\widetilde{\mathbf{X}}'\widetilde{\mathbf{y}}^o \tag{2.98f}$$

$$- h_0\widetilde{\mathbf{y}}^{o\prime}\widetilde{\mathbf{X}}\overline{\mathbf{H}}^{-1}\widetilde{\mathbf{X}}'\mathbf{V}\widetilde{\mathbf{X}}\overline{\boldsymbol{\beta}} \tag{2.98g}$$

$$- h_0^2\widetilde{\mathbf{y}}^{o\prime}\widetilde{\mathbf{X}}\overline{\mathbf{H}}^{-1}\widetilde{\mathbf{X}}'\mathbf{V}\widetilde{\mathbf{X}}\overline{\mathbf{H}}^{-1}\widetilde{\mathbf{X}}'\widetilde{\mathbf{y}}^o. \tag{2.98h}$$

Expressions (2.98a) and (2.98c) are the first two terms on the right side of (2.95). Expression (2.98b) is the same as (2.98e), and (2.98d) is the same as (2.98g). Twice the sum of (2.98b) and (2.98d) is

$$\begin{aligned}2(h_0\overline{\boldsymbol{\beta}}'\widetilde{\mathbf{X}}'\widetilde{\mathbf{y}}^o - h_0\overline{\boldsymbol{\beta}}'\widetilde{\mathbf{X}}'\mathbf{V}\widetilde{\mathbf{X}}\overline{\mathbf{H}}^{-1}\widetilde{\mathbf{X}}'\widetilde{\mathbf{y}}^o) &= 2h_0\overline{\boldsymbol{\beta}}'\widetilde{\mathbf{X}}'(\mathbf{I}_f - \mathbf{V}\widetilde{\mathbf{X}}\overline{\mathbf{H}}^{-1}\widetilde{\mathbf{X}}')\widetilde{\mathbf{y}}^o \\ &= 2h_0\overline{\boldsymbol{\beta}}'\widetilde{\mathbf{X}}'[\mathbf{I}_f - \mathbf{V}(\mathbf{V}^{-1} - h_0^{-1}\mathbf{I}_f)]\widetilde{\mathbf{y}}^o \\ &= 2\overline{\boldsymbol{\beta}}'\widetilde{\mathbf{X}}'\mathbf{V}\widetilde{\mathbf{y}}^o.\end{aligned}$$

This is the third term on the right side of (2.95). Finally, the sum of (2.98f) and (2.98h) is

$$h_0^2\widetilde{\mathbf{y}}^{o\prime}(\widetilde{\mathbf{X}}\overline{\mathbf{H}}^{-1}\widetilde{\mathbf{X}}' - \widetilde{\mathbf{X}}\overline{\mathbf{H}}^{-1}\widetilde{\mathbf{X}}'\mathbf{V}\widetilde{\mathbf{X}}\overline{\mathbf{H}}^{-1}\widetilde{\mathbf{X}}')\widetilde{\mathbf{y}}^o.$$

Employing the relationships $\mathbf{V}\widetilde{\mathbf{X}}\overline{\mathbf{H}}^{-1}\widetilde{\mathbf{X}}' = \widetilde{\mathbf{X}}\overline{\mathbf{H}}^{-1}\widetilde{\mathbf{X}}'\mathbf{V} = \mathbf{I}_f - h_0^{-1}\mathbf{V}$, this expression is

$$\begin{aligned}h_0^2\widetilde{\mathbf{y}}^{o\prime}[\widetilde{\mathbf{X}}\overline{\mathbf{H}}^{-1}\widetilde{\mathbf{X}}' - \widetilde{\mathbf{X}}\overline{\mathbf{H}}^{-1}\widetilde{\mathbf{X}}'(\mathbf{I}_f - h_0^{-1}\mathbf{V})]\widetilde{\mathbf{y}}^o &= h_0\widetilde{\mathbf{y}}^{o\prime}\widetilde{\mathbf{X}}\overline{\mathbf{H}}^{-1}\widetilde{\mathbf{X}}'\mathbf{V}\widetilde{\mathbf{y}}^o \\ &= h_0\widetilde{\mathbf{y}}^{o\prime}(\mathbf{I}_f - h_0^{-1}\mathbf{V})\widetilde{\mathbf{y}}^o \\ &= h_0\widetilde{\mathbf{y}}^{o\prime}\widetilde{\mathbf{y}}^o - \widetilde{\mathbf{y}}^{o\prime}\mathbf{V}\widetilde{\mathbf{y}}^o.\end{aligned}$$

These are the last two terms on the right side of (2.95).

Exercise 2.6.1 Comparison of Simple Normal Models Suppose $y_t \overset{\text{i.i.d.}}{\sim} N(\mu, 1)$. The sample size is $T = 10$; $\overline{y} = -0.2$ and $\sum_{t=1}^{10} y_t^2/10 = 1$. For each of the following prior distributions, compute the numerical value of the log marginal likelihood. Explain the ordering of the values that you obtain.

(a) $\mu = 0$.

(b) $\mu \sim N(0, 1)$. (This is a special case of Example 2.6.1.)

(c) $\mu \sim N(0, 1)$ truncated to $\mu > 0$:

$$p(\mu) = (\pi/2)^{-1/2}\exp(-\mu^2/2)I_{(0,\infty)}(\mu).$$

(This is a variant on Example 2.6.1.)

Exercise 2.6.2 Models for Positive Observables (This is a continuation of Exercise 2.4.6.) The observables $y_1, \ldots, y_T$ are i.i.d. and strictly positive. In model A

$$p(y_t \mid \theta, A) = \theta\exp(-\theta y_t)I_{(0,\infty)}(y_t)$$

while in model B

$$p(y_t \mid h, B) = (\pi/2)^{-1/2}h^{1/2}\exp(-hy_t^2/2)I_{(0,\infty)}(y_t).$$

Derive the Bayes factor in favor of model A.

Exercise 2.6.3 Model Combination Suppose that there are three models (A, B, C) for the observable $\mathbf{y}$. Each model completely specifies the distribution of $\mathbf{y}$—there are no unobservables in any of the models. The complete specification is

Model	Model Prior Probability	$\mathbf{y}$ Density	ω Density
A	$p(A)$	$p(\mathbf{y} \mid A)$	$p(\omega \mid \mathbf{y}, A)$
B	$p(B)$	$p(\mathbf{y} \mid B)$	$p(\omega \mid \mathbf{y}, B)$
C	$p(C)$	$p(\mathbf{y} \mid C)$	$p(\omega \mid \mathbf{y}, C)$

(a) Suppose that the investigator's problem is to choose one of these models subject to the following loss function:

Choice ↓ Truth⟶	A	B	C
A	0	$L(A, B)$	$L(A, C)$
B	$L(B, A)$	0	$L(B, C)$
C	$L(C, A)$	$L(C, B)$	0

Formulate an explicit rule for model choice. Be as specific as you can.

(b) Now suppose that the investigator's problem is to estimate ω, subject to a quadratic loss function. What is her estimate? Be as specific as possible.

(c) Finally, suppose that the investigator's problem is to form an estimate $\widehat{\omega}$, using the loss function $|\widehat{\omega} - \omega|$. What is her estimate? Be as specific as possible.

CHAPTER 3

Topics in Bayesian Inference

This chapter continues the development of principles of Bayesian inference. While the topics treated here are not essential to the specific models taken up in Chapters 5–7, they provide a greater depth of understanding that often yields dividends in Bayesian investigation. Much of the chapter addresses the prior distribution of unobservables. The development of hierarchical priors (Section 3.1) illustrates how models can be enriched with large numbers of unobservables as long as prior information provides sufficient structure. The treatment of improper prior distributions (Section 3.2) emphasizes their interpretation as limits of proper priors. Section 3.3 provides one approach to the common situation in which the investigator does not know the client's prior distribution or even the client. The chapter treats two other topics, as well. Asymptotic analysis (Section 3.4) derives conditions under which posterior distributions collapse to points as sample size increases, and further conditions that imply that the posterior distribution approaches the normal distribution. The chapter concludes with a discussion of the likelihood principle, which states that data-based information is conveyed entirely through the likelihood function. Bayesian inference is always consistent with the likelihood principle, and Section 3.5 illustrates how violations of this principle can lead to unreasonable decisions.

3.1 HIERARCHICAL PRIORS AND LATENT VARIABLES

The use of unobservable, or latent, variables has a history of more than a half-century in econometrics and the social sciences; see Goldberger (1974) or Griliches (1977) for a recounting of the origins of modeling with latent variables. In Bayesian statistics, the concept of the hierarchical prior distribution was introduced by Lindley and Smith (1972) and Smith (1973). In both cases the techniques have substantially increased the flexibility and applicability of inference and are widely used. While developed independently, the two principles are identical from a Bayesian perspective. Moreover there is a natural congruence between these

Contemporary Bayesian Econometrics and Statistics, by John Geweke

methods and Markov chain Monte Carlo posterior simulation methods discussed in Section 4.3. The following example illustrates the main ideas in a simple setting.

Example 3.1.1 Prior Distributions in a Model for Many Means Outcomes y_{it} are observed for individuals $i = 1, \ldots, n$ and time periods (or trials) $t = 1, \ldots, T$. Suppose that a complete model A_1 includes

$$y_{it} = \mu_i + \varepsilon_{it}, \quad \varepsilon_{it} \overset{\text{i.i.d.}}{\sim} N(0, h^{-1}), \tag{3.1}$$

$$\boldsymbol{\mu} \mid A_1 \sim N(\underline{\boldsymbol{\mu}}, \underline{\mathbf{H}}^{-1}), \tag{3.2}$$

where $\boldsymbol{\mu}' = (\mu_1, \ldots, \mu_n)'$, and an independent prior $p(h \mid A_1)$ that need not be further specified for the purposes of this example. The prior distribution of $\boldsymbol{\mu}$ incorporates the idea that there is substantial uncertainty about the means μ_i, but that relative to this uncertainty the means are likely to be similar, although not identical. This idea could be expressed through $E(\boldsymbol{\mu} \mid A) = \underline{\boldsymbol{\mu}} = \boldsymbol{\iota}_n \underline{\mu}$, where $\boldsymbol{\iota}_n$ is an $n \times 1$ vector of ones, $\text{var}(\mu_i \mid A) = \underline{\mathbf{H}}^{ii} = \underline{h}^{-1}$, and $\text{cov}(\mu_i, \mu_j \mid A) = \underline{\mathbf{H}}^{ij} = \underline{\rho}\,\underline{h}^{-1}$. The investigator provides the numerical values of $\underline{\mu}$, $\underline{h} > 0$, and $\underline{\rho} \in (0, 1)$. The closer $\underline{\rho}$ is to one, the more similar are the means μ_i in the prior specification of the model.

An alternative, but equivalent, complete model A_2 retains (3.1) but in place of (3.2) introduces a hierarchical prior distribution. This distribution begins with a new unobservable

$$\mu \mid A_2 \sim N(\underline{\mu}, \underline{\rho}\,\underline{h}^{-1}). \tag{3.3}$$

Then the means $\mu_1, \ldots, \mu_n$ are conditionally independent, with

$$\mu_i \mid (\mu, A_2) \sim N[\mu, (1 - \underline{\rho})\underline{h}^{-1}]. \tag{3.4}$$

Taken together, (3.3) and (3.4) are equivalent to (3.2).

Yet a third complete model A_3 substitutes

$$y_{it} = \widetilde{z}_i + \varepsilon_{it}$$

for (3.1). The random variables $\widetilde{z}_i$ are latent—that is, they are never observed. If A_3 specifies the distribution of latent variables

$$\widetilde{z}_i \mid A_3 \overset{\text{i.i.d.}}{\sim} N[\mu, (1 - \underline{\rho})\underline{h}^{-1}]$$

and the prior distribution $\mu \mid A_3 \sim N(\underline{\mu}, \underline{\rho}\,\underline{h}^{-1})$, then A_3 is equivalent to A_2; in fact, $\widetilde{z}_i = \mu_i$.

More generally, a complete model A with a two-tier hierarchical prior distribution specifies a conditional observables density

$$p(\mathbf{y} \mid \boldsymbol{\lambda}_A, \boldsymbol{\psi}_A, A). \tag{3.5}$$

In the first tier, the prior distribution of $\boldsymbol{\lambda}_A$ is expressed conditional on a vector of unobservable *hyperparameters* $\boldsymbol{\phi}_A$:

$$p(\boldsymbol{\lambda}_A \mid \boldsymbol{\phi}_A, A). \tag{3.6}$$

The term "hyperparameter" denotes the fact that $\boldsymbol{\phi}_A$ is not a parameter of the observables density (3.5). Rather, it is a convenient construct for expressing uncertainty about $\boldsymbol{\lambda}_A$, by means of the prior density

$$p(\boldsymbol{\psi}_A, \boldsymbol{\phi}_A \mid A), \tag{3.7}$$

which completes the model. The prior density of all the unobservables is

$$p(\boldsymbol{\lambda}_A, \boldsymbol{\psi}_A, \boldsymbol{\phi}_A \mid A) = p(\boldsymbol{\lambda}_A \mid \boldsymbol{\phi}_A, A) p(\boldsymbol{\psi}_A, \boldsymbol{\phi}_A \mid A). \tag{3.8}$$

A simple complete latent variable model B includes a vector of unobserved (latent) variables $\widetilde{\mathbf{z}}$ and a conditional observables density

$$p(\mathbf{y} \mid \widetilde{\mathbf{z}}, \boldsymbol{\psi}_B, B), \tag{3.9}$$

a model for the latent variables

$$p(\widetilde{\mathbf{z}} \mid \boldsymbol{\phi}_B, B), \tag{3.10}$$

and a prior density for the unobservables $\boldsymbol{\psi}_B$ and $\boldsymbol{\phi}_B$:

$$p(\boldsymbol{\psi}_B, \boldsymbol{\phi}_B \mid B). \tag{3.11}$$

Then the prior density of the full vector of unobservables $\boldsymbol{\theta}'_B = (\widetilde{\mathbf{z}}', \boldsymbol{\psi}'_B, \boldsymbol{\phi}'_B)$ is

$$p(\widetilde{\mathbf{z}}, \boldsymbol{\psi}_B, \boldsymbol{\phi}_B \mid B) = p(\widetilde{\mathbf{z}} \mid \boldsymbol{\phi}_B, B) p(\boldsymbol{\psi}_B, \boldsymbol{\phi}_B \mid B). \tag{3.12}$$

Comparing (3.5)–(3.8) with (3.9)–(3.12), it is apparent that the complete model with a two-tier hierarchical prior distribution and the simple latent variable model with a conventional prior distribution are formally identical. Since

$$p(\mathbf{y} \mid \boldsymbol{\phi}_A, \boldsymbol{\psi}_A, A) = \int_{\Lambda_A} p(\boldsymbol{\lambda}_A \mid \boldsymbol{\phi}_A, A) p(\mathbf{y} \mid \boldsymbol{\lambda}_A, \boldsymbol{\psi}_A, A)\, d\nu(\boldsymbol{\lambda}_A),$$

the unobservables vector $\boldsymbol{\lambda}_A$ is formally redundant. The reason for introducing $\boldsymbol{\lambda}_A$ is that it facilitates expression of the prior distribution, makes analysis of the posterior distribution easier, or both. Likewise the latent variable model implies

$$p(\mathbf{y} \mid \boldsymbol{\phi}_B, \boldsymbol{\psi}_B, B) = \int_Z p(\widetilde{\mathbf{z}} \mid \boldsymbol{\phi}_B, B) p(\mathbf{y} \mid \widetilde{\mathbf{z}}, \boldsymbol{\psi}_B, B)\, d\nu(\widetilde{\mathbf{z}}).$$

An advantage of the latent variable formulation in a Bayesian context is that it obviates the need to evaluate the likelihood function

$$p(\mathbf{y}^o \mid \boldsymbol{\phi}_B, \boldsymbol{\psi}_B, B) = \int_Z p(\widetilde{\mathbf{z}} \mid \boldsymbol{\phi}_B, B) p(\mathbf{y}^o \mid \widetilde{\mathbf{z}}, \boldsymbol{\psi}_B, B)\, d\nu(\widetilde{\mathbf{z}})$$

analytically, a task that is impossible in some applications, for example, the multinomial probit model [see Geweke et al. (1994) or McCulloch and Rossi (1994)].

A further advantage of hierarchical prior distributions is that they are often the natural vocabulary for generalizing a model, and they facilitate the expression of conditional posterior distributions that are central in the Markov chain Monte Carlo posterior simulators introduced in Section 4.3.

Example 3.1.2 Posterior Distributions in a Model for Many Means In the complete model A_2 of Example 3.1.1 consisting of (3.1), (3.3), and (3.4), the conditional posterior distributions of the means μ_i are independent:

$$\mu_i \mid (\mu, h, \mathbf{y}^o, A_2) \sim N(\overline{\mu}_i, \overline{h}_i^{-1}) \ (i = 1, \ldots, n) \tag{3.13}$$

with

$$\overline{h}_i = (1-\underline{\rho})^{-1}\underline{h} + Th,\ \overline{\mu}_i = \overline{h}_i^{-1}\left[(1-\underline{\rho})^{-1}\underline{h}\mu + h\sum_{t=1}^{T} y_{it}\right]. \tag{3.14}$$

Note that $\underline{\mu}$ does not appear in (3.14). The conditional posterior distribution of μ is

$$\mu \mid (\mu_1, \ldots, \mu_n, h, \mathbf{y}^o, A_2) \sim N(\overline{\mu}, \overline{h}^{-1}) \tag{3.15}$$

with

$$\overline{h} = \underline{\rho}^{-1}\underline{h} + n(1-\underline{\rho})^{-1}\underline{h},\ \overline{\mu} = \overline{h}^{-1}\left[\underline{\rho}^{-1}\underline{h}\underline{\mu} + (1-\underline{\rho})^{-1}\underline{h}\sum_{i=1}^{n} \mu_i\right]. \tag{3.16}$$

Note that the data $\mathbf{y}^o$ do not appear in the latter distribution. The conditional distributions (3.13) and (3.15) are the basis for a Markov chain Monte Carlo posterior simulator.

In Example 3.1.1 the investigator specified numerical values for $\underline{\rho}$ and $\underline{h}$ in the prior distribution. Suppose instead that the investigator regards $\underline{\rho}$ and $\underline{h}$ as unobservables, and reflecting this fact replaces them with the symbols ρ^* and h^*, assigning them the independent prior distributions $\underline{s}^{*2}h^* \mid A_2 \sim \chi^2(\underline{\nu}^*)$ and $\rho^* \mid A_2 \sim \text{uniform}(\underline{\rho}_1^*, \underline{\rho}_2^*)$, where $\underline{s}^{*2}$, $\underline{\nu}^*$, $\underline{\rho}_1^*$, and $\underline{\rho}_2^*$, represent positive real numbers with $\underline{\rho}_1^* < \underline{\rho}_2^* < 1$. Note that the conditional distributions of the means μ_i and the hyperparameter μ remain as they are in (3.13)–(3.16)—in particular, $\underline{s}^{*2}$, $\underline{\nu}^*$, $\underline{\rho}_1^*$, and $\underline{\rho}_2^*$ do not appear in these expressions. Moreover, the conditional posterior distributions for ρ^* and h^* do not depend on $\mathbf{y}^o$ or h.

This example indicates how a hierarchy of prior distributions may be extended. The fact that the conditional posterior distribution of μ does not depend on the data in (3.15) is the manifestation of a universal characteristic of the vector of hyperparameters in a model with a two-tier hierarchical prior. From (3.5)–(3.7), we obtain

$$p(\boldsymbol{\phi}_A \mid \boldsymbol{\lambda}_A, \boldsymbol{\psi}_A, \mathbf{y}^o, A) \propto p(\boldsymbol{\psi}_A, \boldsymbol{\phi}_A \mid A) p(\boldsymbol{\lambda}_A \mid \boldsymbol{\psi}_A, \boldsymbol{\phi}_A, A) p(\mathbf{y}^o \mid \boldsymbol{\lambda}_A, \boldsymbol{\psi}_A, A)$$
$$\propto p(\boldsymbol{\psi}_A, \boldsymbol{\phi}_A \mid A) p(\boldsymbol{\lambda}_A \mid \boldsymbol{\psi}_A, \boldsymbol{\phi}_A, A).$$

See Exercise 3.1.2 for an extension of this idea.

Exercise 3.1.1 Completing the Argument Derive (3.13)–(3.16).

Exercise 3.1.2 Multitier Prior Distributions Consider a model with an $(n-1)$-tier hierarchical prior distribution of the unobservables. The conditional pdf of the observables is $p(\mathbf{y} \mid \boldsymbol{\theta}_{A1}, A)$, and the prior density is

$$p(\boldsymbol{\theta}_A \mid A) = p(\boldsymbol{\theta}_{An} \mid A) \prod_{i=1}^{n-1} p(\boldsymbol{\theta}_{Ai} \mid \boldsymbol{\theta}_{Ai+1}, A),$$

where $\boldsymbol{\theta}'_A = (\boldsymbol{\theta}'_{A1}, \ldots, \boldsymbol{\theta}'_{An})$. Show that in the conditional posterior densities $p[\boldsymbol{\theta}_{Aj} \mid \boldsymbol{\theta}_{Ai}(i \neq j), \mathbf{y}^o, A]$, the vectors $\boldsymbol{\theta}_{Ai}$ do not actually appear unless $i = j-1$ or $i = j+1$, and the data vector $\mathbf{y}^o$ does not appear unless $j = 1$.

Exercise 3.1.3 Hierarchical Prior Distributions In model A, the distribution of $\mathbf{y}' = (y_1, \ldots, y_T)$ conditional on $\mathbf{x}' = (x_{1-p}, \ldots, x_{T-1})$ is

$$y_t = \beta_0 + \sum_{s=1}^{p} \beta_s x_{t-s} + \varepsilon_t, \quad \boldsymbol{\varepsilon} \mid (h, \mathbf{x}, A) \sim N(\mathbf{0}, h^{-1}\mathbf{I}_T),$$

where $\boldsymbol{\varepsilon} = (\varepsilon_1, \ldots, \varepsilon_T)'$. The investigator would like to complete the model with the independent prior distributions

$$\boldsymbol{\beta} = (\beta_0, \ldots, \beta_p)' \sim N(\underline{\boldsymbol{\beta}}, \underline{\mathbf{H}}^{-1}), \ \underline{s}^2 h \sim \chi^2(\underline{\nu}).$$

Her beliefs are well represented by $\underline{\boldsymbol{\beta}} = \mathbf{0}$. In choosing $\underline{\mathbf{H}}$ she wishes to express the idea that for $s = 1, \ldots, p$ the coefficients β_s are likely to be smaller in absolute value, the greater is s, but she is not sure how quickly they become small.

(a) Set up a hierarchical prior distribution that could represent the investigator's beliefs about $\beta_1, \ldots, \beta_p$; that is, set $\underline{\mathbf{H}} = \mathbf{H}(\phi)$, and then choose a prior distribution $p(\phi \mid \lambda)$ in which the investigator will fix the value of λ.

(b) Corresponding to the prior distribution you chose in (a), express:

(i) $p(\boldsymbol{\beta} \mid \mathbf{y}, \mathbf{x}, h, \underline{\boldsymbol{\beta}}, \underline{s}^2, \underline{\nu}, \phi, \lambda, A)$; besides $\boldsymbol{\beta}$, this expression should involve only $\mathbf{y}$, $\mathbf{x}$, h, $\underline{\boldsymbol{\beta}}$, and ϕ.

(ii) $p(h \mid \mathbf{y}, \mathbf{x}, \boldsymbol{\beta}, \underline{\boldsymbol{\beta}}, \underline{s}^2, \underline{\nu}, \phi, \lambda, A)$; besides h, this expression should involve only $\mathbf{y}$, $\mathbf{x}$, $\boldsymbol{\beta}$, $\underline{s}^2$, and $\underline{\nu}$.

(iii) $p(\phi \mid \mathbf{y}, \mathbf{x}, h, \boldsymbol{\beta}, \underline{\boldsymbol{\beta}}, \underline{s}^2, \underline{\nu}, \lambda, A)$; besides ϕ, this expression should involve only $\boldsymbol{\beta}$, $\underline{\boldsymbol{\beta}}$, and λ.

3.2 IMPROPER PRIOR DISTRIBUTIONS

Bayesian investigators often report results using prior distributions that are widely dispersed, so that their densities are nearly flat, at least over the range of the parameter space Θ_A in which the likelihood function is concentrated. As we shall see in Section 8.4, there can be sound technical reasons for doing this. But this procedure also looks appealing on the grounds that a nearly "flat" prior density seems to convey very little information, and is therefore appropriate in communicating results to a diverse group of people who may have very different priors. This rationale is misleading, and this can be seen by considering the effects of reparameterization of a model. Suppose that from the model A we create the model B by taking $\boldsymbol{\theta}_B = f(\boldsymbol{\theta}_A)$, and $f(\cdot)$ is one-to-one. Then we can write $\boldsymbol{\theta}_A = h(\boldsymbol{\theta}_B)$ with $h = f^{-1}$ and $\boldsymbol{\theta}_B \in \Theta_B$, with Θ_B the image of Θ_A under $f(\cdot)$. The new observables density is

$$p(\mathbf{y} \mid \boldsymbol{\theta}_B, B) = p[\mathbf{y} \mid \boldsymbol{\theta}_A = h(\boldsymbol{\theta}_B), A]. \tag{3.17}$$

The new prior density is

$$p(\boldsymbol{\theta}_B \mid B) = p[\boldsymbol{\theta}_A = h(\boldsymbol{\theta}_B) \mid A] \left|[\partial h(\boldsymbol{\theta}_B)/\partial \boldsymbol{\theta}'_B]\right|. \tag{3.18}$$

Note that because of the Jacobian term in (3.18), $p(\boldsymbol{\theta}_B \mid B)$ can be made nearly "flat" when $p(\boldsymbol{\theta}_A \mid A)$ is not, by appropriate choice of f, and vice versa. For the vector of interest $\boldsymbol{\omega}$, we have

$$p(\boldsymbol{\omega} \mid \mathbf{y}, \boldsymbol{\theta}_B, B) = p[\boldsymbol{\omega} \mid y, \boldsymbol{\theta}_A = h(\boldsymbol{\theta}_B), A]. \tag{3.19}$$

For purposes of learning about $\boldsymbol{\omega}$ it does not matter which model is used because

$$\begin{aligned} p(\boldsymbol{\omega} \mid \mathbf{y}, B) &\propto \int_{\Theta_B} p(\boldsymbol{\omega} \mid \mathbf{y}, \boldsymbol{\theta}_B, B) p(\mathbf{y} \mid \boldsymbol{\theta}_B, B) p(\boldsymbol{\theta}_B \mid B)\, d\nu(\boldsymbol{\theta}_B) \\ &= \int_{\Theta_B} p[\boldsymbol{\omega} \mid \mathbf{y}, \boldsymbol{\theta}_A = h(\boldsymbol{\theta}_B), A] p[\mathbf{y} \mid \boldsymbol{\theta}_A = h(\boldsymbol{\theta}_B), A] \\ &\quad \cdot p[\boldsymbol{\theta}_A = h(\boldsymbol{\theta}_B) \mid A] \left|[\partial h(\boldsymbol{\theta}_B)/\partial \boldsymbol{\theta}'_B]\right| d\nu(\boldsymbol{\theta}_B) \end{aligned}$$

$$= \int_{\Theta_A} p(\boldsymbol{\omega} \mid \mathbf{y}, \boldsymbol{\theta}_A, A) p(\mathbf{y} \mid \boldsymbol{\theta}_A, A) p(\boldsymbol{\theta}_A \mid A)$$

$$\cdot \left|[\partial f(\boldsymbol{\theta}_A)/\partial \boldsymbol{\theta}_A']\right|^{-1} \left|[\partial f(\boldsymbol{\theta}_A)/\partial \boldsymbol{\theta}_A']\right| d\nu(\boldsymbol{\theta}_A)$$

$$= \int_{\Theta_A} p(\boldsymbol{\omega} \mid \mathbf{y}, \boldsymbol{\theta}_A, A) p(\mathbf{y} \mid \boldsymbol{\theta}_A, A) p(\boldsymbol{\theta}_A \mid A)\, d\nu(\boldsymbol{\theta}_A)$$

$$\propto p(\boldsymbol{\omega} \mid \mathbf{y}, A).$$

[The first equality simply substitutes from (3.19), (3.17) and (3.18). The second equality is a conventional change of variable from $\boldsymbol{\theta}_B$ to $\boldsymbol{\theta}_A$.] Evidence about $\boldsymbol{\omega}$ will be the same in models A and B, and yet the prior in B can be manipulated to be nearly "flat" in the parameter space Θ_B. Thus the shape of the prior alone is no indication of how much information it conveys.

To develop prior distributions that may nonetheless prove useful in communicating results, consider a sequence of models $A = \{A_j\}_{j=1}^{\infty}$, each with the same parameter vector $\boldsymbol{\theta}_A$, data density $p(\mathbf{y} \mid \boldsymbol{\theta}_A, A)$, and vector of interest $\boldsymbol{\omega}$, but with different prior densities $p(\boldsymbol{\theta}_A \mid A_j)$. Let $k(\boldsymbol{\theta}_A \mid A_j)$ be a sequence of kernels corresponding to the sequence of prior densities, $k(\boldsymbol{\theta}_A \mid A_j) \propto p(\boldsymbol{\theta}_A \mid A_j)$. Then the corresponding sequence of posterior densities for $\boldsymbol{\theta}_A$ is

$$p(\boldsymbol{\theta}_A \mid \mathbf{y}^o, A_j) \propto p(\mathbf{y}^o \mid \boldsymbol{\theta}_A, A) k(\boldsymbol{\theta}_A \mid A_j).$$

It may turn out that $k(\boldsymbol{\theta}_A \mid A_j)$ has a pointwise limit $k(\boldsymbol{\theta}_A \mid A)$ that is not finitely integrable—that is, it is not the kernel of any pdf. At the same time, it may be the case that $\lim_{j\to\infty} p(\boldsymbol{\theta}_A \mid \mathbf{y}, A_j) \propto k(\boldsymbol{\theta}_A \mid A) p(\mathbf{y} \mid \boldsymbol{\theta}_A, A)$ is finitely integrable and therefore is a well-defined posterior density kernel. For many purposes, analysis may be carried out using $k(\boldsymbol{\theta}_A \mid A)$, ignoring the fact that it cannot be the kernel of a prior density. The following definition, theorem, and three corollaries develop these ideas more carefully.

Definition 3.2.1 Let $k(\boldsymbol{\theta}_A \mid A_j)$ be a sequence of prior density kernels for which $k(\boldsymbol{\theta}_A \mid A) = \lim_{j\to\infty} k(\boldsymbol{\theta}_A \mid A_j) \forall \boldsymbol{\theta}_A \in \Theta_A$ exists but is not finitely integrable. If $\lim_{j\to\infty} p(\boldsymbol{\theta}_A \mid \mathbf{y}^o, A_j) \propto k(\boldsymbol{\theta}_A \mid A) p(\mathbf{y}^o \mid \boldsymbol{\theta}_A, A)$ exists and is finitely integrable, then $k(\boldsymbol{\theta}_A \mid A)$ is an *improper prior density kernel for* $\boldsymbol{\theta}_A$ in the model A with data $\mathbf{y}^o$.

An attraction of using an improper prior distribution is that it can reflect some limiting properties of the sequence of distributions $\boldsymbol{\omega} \mid (\mathbf{Y}_T^o, A_j)$ and moments $E[h(\boldsymbol{\omega})|(\mathbf{y}^o, A_j)]$. It is important to establish the conditions under which this will happen, and to see exactly which limiting properties are reflected in the posterior distribution that employs the improper prior distribution.

Theorem 3.2.1 Convergence of Posterior Densities Given a Sequence of Prior Densities Let $p(\boldsymbol{\theta}_A \mid A_j)$ be a sequence of prior densities with corresponding kernels $k(\boldsymbol{\theta}_A \mid A_j)$. Suppose that for all $\boldsymbol{\theta}_A \in \Theta_A$, $k(\boldsymbol{\theta}_A \mid A_j)$ is monotone nondecreasing with

$$\lim_{j\to\infty} k(\boldsymbol{\theta}_A \mid A_j) = k(\boldsymbol{\theta}_A \mid A),$$

where $k(\boldsymbol{\theta}_A|A)$ is an improper prior density kernel for $\boldsymbol{\theta}_A$. Suppose further that

$$c_j = \int_{\Theta_A} p(\mathbf{y}^o \mid \boldsymbol{\theta}_A, A) k(\boldsymbol{\theta}_A \mid A_j)\, d\nu(\boldsymbol{\theta}_A) \tag{3.20}$$

has finite limit c. Then $\boldsymbol{\theta}_A|(\mathbf{y}^o, A_j) \xrightarrow{d} \boldsymbol{\theta}_A|(\mathbf{y}^o, A)$ and a kernel of the limiting posterior distribution $\lim_{j\to\infty} p(\boldsymbol{\theta}_A \mid \mathbf{y}^o, A_j)$ is $p(\mathbf{y}^o \mid \boldsymbol{\theta}_A, A)k(\boldsymbol{\theta}_A \mid A)$.

Proof: Clearly

$$\lim_{j\to\infty} p(\mathbf{y}^o \mid \boldsymbol{\theta}_A, A_j)k(\boldsymbol{\theta}_A \mid A_j) = p(\mathbf{y}^o \mid \boldsymbol{\theta}_A, A)k(\boldsymbol{\theta}_A \mid A) \forall \boldsymbol{\theta}_A \in \Theta_A.$$

By the monotone convergence theorem (Royden 1968, Section 4.2)

$$c = \int_{\Theta_A} p(\mathbf{y}^o \mid \boldsymbol{\theta}_A, A)k(\boldsymbol{\theta}_A \mid A)\, d\nu(\boldsymbol{\theta}_A).$$

Consequently

$$\lim_{j\to\infty} p(\mathbf{y}^o \mid \boldsymbol{\theta}_A, A_j)k(\boldsymbol{\theta}_A \mid A_j)/c_j = p(\mathbf{y}^o \mid \boldsymbol{\theta}_A, A)k(\boldsymbol{\theta}_A \mid A)/c \;\forall\; \boldsymbol{\theta}_A \in \Theta_A,$$

which is equivalent to $\lim_{j\to\infty} p(\boldsymbol{\theta}_A \mid \mathbf{y}^o, A_j) = p(\boldsymbol{\theta}_A \mid \mathbf{y}^o, A) \forall \boldsymbol{\theta}_A \in \Theta_A$. By Scheffe's theorem (Rao 1965, Theorem 2c.4.xv), $\boldsymbol{\theta}_A|(\mathbf{y}^o, A_j) \xrightarrow{d} \boldsymbol{\theta}_A|(\mathbf{y}^o, A)$. ■

When the conditions of Theorem 3.2.1 are satisfied, reports of posterior densities of parameters using the improper prior can be interpreted as limits of sequences of posterior densities employing priors whose kernels converge to the improper prior kernel. These conditions imply that there exist convergent sequences of credible sets, as well—that is, $P(\boldsymbol{\theta}_A \in C \mid \mathbf{y}^o, A) = 1 - \alpha \Rightarrow \lim_{j\to\infty} P(\boldsymbol{\theta}_A \in C \mid \mathbf{y}^o, A_j) = 1 - \alpha$. Under further weak conditions, the improper prior also provides limits of moments of $\boldsymbol{\theta}_A$ and functions of $\boldsymbol{\theta}_A$.

Corollary 3.2.1 Convergence of Posterior Moments Given a Sequence of Prior Densities Suppose $\boldsymbol{\theta}_A|(\mathbf{y}^o, A_j) \xrightarrow{d} \boldsymbol{\theta}_A|(\mathbf{y}^o, A)$ and $g : \Theta_A \to \mathbb{R}$ is continuous. If $\lim_{j\to\infty} E[g(\boldsymbol{\theta}_A)| \mathbf{y}^o, A_j] = g^*$, then $g^* = E[g(\boldsymbol{\theta}_A)| \mathbf{y}^o, A]$.

Proof: The conditions imply $g(\boldsymbol{\theta}_A)|(\mathbf{y}^o, A_j) \xrightarrow{d} g(\boldsymbol{\theta}_A)|(\mathbf{y}^o, A)$ (Rao 1965, Theorem 2c.4.xii). The result follows from Rao (1965), Theorem 2c.4.viii. ■

More generally, it is usually the case that $\boldsymbol{\omega} \mid (\mathbf{y}^o, A_j) \xrightarrow{d} \boldsymbol{\omega} \mid (\mathbf{y}^o, A)$, but some conditions on $p(\boldsymbol{\omega} \mid y^o, \boldsymbol{\theta}_A, A_j)$ are needed.

Corollary 3.2.2 Convergence of the Posterior Distribution of a Vector of Interest Given a Sequence of Prior Densities Suppose $\boldsymbol{\theta}_A|(\mathbf{y}^o, A_j) \xrightarrow{d} \boldsymbol{\theta}_A|(\mathbf{y}^o, A)$, and for all $\boldsymbol{\omega} \in \Omega$, $g(\boldsymbol{\theta}_A; \boldsymbol{\omega}) = p(\boldsymbol{\omega} \mid \mathbf{y}^o, \boldsymbol{\theta}_A, A_j)$ is a continuous function of $\boldsymbol{\theta}_A$. Then $\boldsymbol{\omega} \mid (\mathbf{y}^o, A_j) \xrightarrow{d} \boldsymbol{\omega} \mid (\mathbf{y}^o, A)$.

Proof: The conditions imply

$$\liminf_{j\to\infty} \int_{\Theta_A} g(\boldsymbol{\theta}_A; \boldsymbol{\omega}) p(\boldsymbol{\theta}_A \mid \mathbf{y}^o, A_j)\, d\nu(\boldsymbol{\theta}_A)$$
$$\geq \int_{\Theta_A} g(\boldsymbol{\theta}_A; \boldsymbol{\omega}) p(\boldsymbol{\theta}_A \mid \mathbf{y}^o, A)\, d\nu(\boldsymbol{\theta}_A),$$

(Rao 1965, Theorem 2c.4.vii), which is equivalent to

$$\liminf_{j\to\infty} p(\boldsymbol{\omega} \mid \mathbf{y}^o, A_j) \geq p(\boldsymbol{\omega} \mid \mathbf{y}^o, A).$$

But

$$\int_{\Omega} p(\boldsymbol{\omega} \mid \mathbf{y}^o, A_j)\, d\nu(\boldsymbol{\omega}) = \int_{\Omega} p(\boldsymbol{\omega} \mid \mathbf{y}^o, A)\, d\nu(\boldsymbol{\omega}) = 1\ \forall j,$$

and hence $\lim_{j\to\infty} p(\boldsymbol{\omega} \mid \mathbf{y}^o, A_j) = p(\boldsymbol{\omega} \mid \mathbf{y}^o, A)$ except possibly on a set of ν-measure zero. Thus

$$\lim_{j\to\infty} \int_{\Omega} \left| p(\boldsymbol{\omega} \mid \mathbf{y}^o, A_j) - p(\boldsymbol{\omega} \mid \mathbf{y}^o, A) \right| d\nu(\boldsymbol{\omega}) = 0,$$

and the result follows from Scheffe's theorem. ■

Finally, posterior moments $E[h(\boldsymbol{\omega}) \mid \mathbf{y}^o, A_j]$ converge if in addition $h(\cdot)$ is continuous.

Corollary 3.2.3 Convergence of Posterior Moments of a Vector of Interest Given a Sequence of Prior Densities Suppose $\boldsymbol{\omega} \mid (\mathbf{y}^o, A_j) \xrightarrow{d} \boldsymbol{\omega} \mid (\mathbf{y}^o, A)$ and $h : \Omega \to \mathbb{R}$ is continuous. If $E[h(\boldsymbol{\omega}) \mid \mathbf{y}^o, A_j] \to h^*$, then $h^* = E[h(\boldsymbol{\omega}) \mid \mathbf{y}^o, A]$.

Proof: Identical to that of Corollary 3.2.2. ■

Example 3.2.1 A Sequence of Diffuse Priors for $\boldsymbol{\beta}$ in the Normal Linear Regression Model In the normal linear regression model (2.10) fix the precision at $h = h_0$ and consider the sequence of prior distributions

$$\boldsymbol{\beta} \mid A_j \sim N[\underline{\boldsymbol{\beta}}, (\underline{a}_j \underline{\mathbf{H}})^{-1}] (j = 1, 2, \ldots).$$

The sequence $\{\underline{a}_j\}$ is monotone decreasing with $\lim_{j\to\infty}\underline{a}_j = 0$. A corresponding sequence of prior density kernels for $\boldsymbol{\beta}$ is

$$k(\boldsymbol{\beta} \mid A_j) = \exp[-\underline{a}_j(\boldsymbol{\beta} - \underline{\boldsymbol{\beta}})'\underline{\mathbf{H}}(\boldsymbol{\beta} - \underline{\boldsymbol{\beta}})/2] \propto p(\boldsymbol{\beta} \mid A_j).$$

The function $k(\boldsymbol{\beta} \mid A_j)$ is monotone increasing to 1 except at $\boldsymbol{\beta} = \underline{\boldsymbol{\beta}}$ where $k(\boldsymbol{\beta} \mid A_j) = 1$. Hence it satisfies the condition on kernels in Theorem 3.2.1 with $k(\boldsymbol{\beta} \mid A) = 1 \forall \boldsymbol{\beta} \in \mathbb{R}^k$. Proceeding as in (2.18a)–(2.21), the corresponding sequence of posterior density kernels is

$$p(\mathbf{y}^o \mid \boldsymbol{\beta}, \mathbf{X}, A_j)k(\boldsymbol{\beta} \mid A_j) = (2\pi)^{-T/2}h_0^{T/2} \cdot \exp\{-[(\boldsymbol{\beta} - \overline{\boldsymbol{\beta}}_j)'\overline{\mathbf{H}}_j(\boldsymbol{\beta} - \overline{\boldsymbol{\beta}}_j) + Q_j]/2\}, \quad (3.21)$$

where

$$\overline{\mathbf{H}}_j = \underline{a}_j\underline{\mathbf{H}} + h_0\mathbf{X}'\mathbf{X}, \quad \overline{\boldsymbol{\beta}}_j = \overline{\mathbf{H}}_j^{-1}(\underline{a}_j\underline{\mathbf{H}}\underline{\boldsymbol{\beta}} + h_0\mathbf{X}'\mathbf{X}\mathbf{b}),$$

and

$$Q_j = h_0\mathbf{y}^{o\prime}\mathbf{y}^o + \underline{a}_j\underline{\boldsymbol{\beta}}'\underline{H}\underline{\boldsymbol{\beta}} - \overline{\boldsymbol{\beta}}_j'\overline{H}\,\overline{\boldsymbol{\beta}}_j.$$

Thus

$$\int_{\mathbb{R}^k} p(\mathbf{y}^o \mid \boldsymbol{\beta}, \mathbf{X}, A_j)k(\boldsymbol{\beta} \mid A_j)\,d\boldsymbol{\beta} = (2\pi)^{(k-T)/2}h_0^{T/2}\exp(-Q_j/2)\left|\overline{\mathbf{H}}_j\right|^{-1/2} < \infty,$$

which converges to

$$(2\pi)^{(k-T)/2}h_0^{(T-k)/2}\exp[-h_0(\mathbf{y}^o - \mathbf{X}\mathbf{b})'(\mathbf{y}^o - \mathbf{X}\mathbf{b})/2]\left|\mathbf{X}'\mathbf{X}\right|^{-1/2}.$$

Hence from (3.21) and Theorem 3.2.1 $\boldsymbol{\beta} \mid (\mathbf{y}^o, \mathbf{X}, A_j) \xrightarrow{d} \boldsymbol{\beta} \mid (\mathbf{y}^o, \mathbf{X}, A)$, with the kernel density of the latter distribution given by

$$p(\mathbf{y}^o \mid \boldsymbol{\beta}, \mathbf{X}, A_j)k(\boldsymbol{\beta} \mid A) \propto \exp[-h_0(\boldsymbol{\beta} - \mathbf{b})'\mathbf{X}'\mathbf{X}(\boldsymbol{\beta} - \mathbf{b})/2]$$

which shows that $\boldsymbol{\beta} \mid (\mathbf{y}^o, \mathbf{X}, A) \sim N[\mathbf{b}, (h_0\mathbf{X}'\mathbf{X})^{-1}]$. From Corollary 3.2.1, we have

$$E[\boldsymbol{\beta} \mid (\mathbf{y}^o, \mathbf{X}, A_j)] \to \mathbf{b}, \ \mathrm{var}[\boldsymbol{\beta} \mid (\mathbf{y}^o, \mathbf{X}, A_j)] \to (h_0\mathbf{X}'\mathbf{X})^{-1}.$$

Suppose that the function of interest is the $r \times 1$ vector $\boldsymbol{\omega} = y_* = \mathbf{X}_*\boldsymbol{\beta} + \varepsilon_*$, where $\boldsymbol{\varepsilon}_* \sim N(\mathbf{0}, h_0^{-1}\mathbf{I}_r)$ is independent of $\boldsymbol{\varepsilon}$. (This is the conventional "prediction problem" discussed in many basic econometrics texts.) Then

$$p(\boldsymbol{\omega} \mid \boldsymbol{\beta}, \mathbf{X}_*, A) \propto \exp[-h_0(\boldsymbol{\omega} - \mathbf{X}_*\boldsymbol{\beta})'(\boldsymbol{\omega} - \mathbf{X}_*\boldsymbol{\beta})/2],$$

which is continuous in $\boldsymbol{\beta}$. From Corollary 3.2.2, we obtain

$$\boldsymbol{\omega} \mid (\mathbf{y}^o, \mathbf{X}, A_j) \xrightarrow{d} \boldsymbol{\omega} \mid (\mathbf{y}^o, \mathbf{X}, A) \sim N\{\mathbf{X}_*\mathbf{b},\ h_0^{-1}[(\mathbf{X}'\mathbf{X})^{-1} + \mathbf{I}_r]\}.$$

Since the elements of $\boldsymbol{\omega}$ and $\boldsymbol{\omega}\boldsymbol{\omega}'$ are continuous functions of $\boldsymbol{\omega}$, it is also the case (Corollary 3.2.3) that

$$E[\boldsymbol{\omega} \mid (\mathbf{y}^o, \mathbf{X}, A_j)] \to E[\boldsymbol{\omega} \mid (\mathbf{y}^o, \mathbf{X}, A)] = \mathbf{X}^*\mathbf{b}$$

and

$$\text{var}[\boldsymbol{\omega} \mid (\mathbf{y}^o, \mathbf{X}, A_j)] \to \text{var}[\boldsymbol{\omega} \mid (\mathbf{y}^o, \mathbf{X}, A)] = h_0^{-1}[(\mathbf{X}'\mathbf{X})^{-1} + \mathbf{I}_r].$$

An important limitation of improper priors is that they lead to models whose marginal likelihood is zero.

Theorem 3.2.2 Marginal Likelihood Given an Improper Prior The conditions of Theorem 3.2.1 imply

$$\lim_{j\to\infty} \int_{\Theta_A} p(\boldsymbol{\theta}_A \mid A_j) p(\mathbf{y}^o \mid \boldsymbol{\theta}_A, A)\, d\nu(\boldsymbol{\theta}_A) = 0.$$

Proof: Let $d_j = \int_{\Theta_A} k(\boldsymbol{\theta}_A \mid A_j)\, d\nu(\boldsymbol{\theta}_A)$. Then

$$\int_{\Theta_A} p(\boldsymbol{\theta}_A \mid A_j) p(\mathbf{y}^o \mid \boldsymbol{\theta}_A, A)\, d\nu(\boldsymbol{\theta}_A) = c_j / d_j,$$

where c_j is as defined in (3.20). Since $d_j \to \infty$ and $c_j \to c < \infty$, the result follows. ∎

As consequences, a model A with an improper prior distribution has no weight in model averaging (2.73)–(2.74) and will never be selected in a model choice decision problem. The latter result is widely known as "Lindley's paradox," after Lindley (1957) and Bartlett (1957).

Example 3.2.2 Marginal Likelihood for a Sequence of Diffuse Priors for $\boldsymbol{\beta}$ in the Normal Linear Regression Model Continuing with Example 3.2.1, from (2.80) the marginal likelihood of the jth model in the sequence is

$$(2\pi)^{-T/2} h_0^{T/2} \left|a_j \underline{\mathbf{H}}\right|^{1/2} \left|a_j \underline{\mathbf{H}} + h_0 \mathbf{X}'\mathbf{X}\right|^{-1/2} \exp(-Q_j/2). \qquad (3.22)$$

Since $\lim_{j\to\infty} Q_j = -h_0(\mathbf{y}^o - \mathbf{X}\mathbf{b})'(\mathbf{y}^o - \mathbf{X}\mathbf{b})$, the limit of (3.22) is zero.

Theorem 3.2.1, Corollaries 3.2.1–3.2.3, and Example 3.2.1 show that there are reasonable conditions under which the use of an improper prior can be interpreted as a limiting case of the use of a sequence of prior distributions. This sequence in turn may be interpreted as "increasingly less concentrated" in the sense that $\lim_{j\to\infty} p(\boldsymbol{\theta}_A \mid A_j) = 0 \forall \boldsymbol{\theta}_A \in \Theta_A$. But the argument made at the start of this section shows that interpreting this sequence as an approach to "uninformative" priors is treacherous because it need not be invariant under transformation. In view of this difficulty, Jeffreys (1961) proposed a particular prior density that is unique under transformation.

Definition 3.2.2 If $\{\mathbf{y}_t\}$ is i.i.d. with pdf $p(\mathbf{y}_t \mid \boldsymbol{\theta}_A, A)$, and $p(\mathbf{y}_t \mid \boldsymbol{\theta}_A, A)$ is differentiable with respect to $\boldsymbol{\theta}_A \forall \boldsymbol{\theta}_A \in \Theta_A$, then the *Jeffreys invariant prior density kernel* is

$$k(\boldsymbol{\theta}_A \mid A) \propto \left| E\left[\frac{\partial \log p(\mathbf{y}_t \mid \boldsymbol{\theta}_A, A)}{\partial \boldsymbol{\theta}_A} \cdot \frac{\partial \log p(\mathbf{y}_t \mid \boldsymbol{\theta}_A, A)}{\partial \boldsymbol{\theta}'_A} \mid \boldsymbol{\theta}_A, A\right]\right|^{1/2} \tag{3.23}$$

$$= \left| -E\left[\frac{\partial^2 \log p(\mathbf{y}_t \mid \boldsymbol{\theta}_A, A)}{\partial \boldsymbol{\theta}_A \partial \boldsymbol{\theta}'_A}\right] \mid \boldsymbol{\theta}_A, A\right|^{1/2}. \tag{3.24}$$

Note that expectation is with respect to the random vector $\mathbf{y}_t$ and not the constant vector $\boldsymbol{\theta}_A$ in (3.23). The equality in (3.24) is a property of probability densities widely used in non-Bayesian statistics; for example, see Poirier (1995), Theorem 6.5.1.

Theorem 3.2.3 Invariance of the Jeffreys Prior The Jeffreys invariant prior density is invariant under one-to-one reparameterization.

Proof: Construct the model B by taking $\boldsymbol{\theta}_B = f(\boldsymbol{\theta}_A)$, $\boldsymbol{\theta}_A = h(\boldsymbol{\theta}_B)$ and $p(\mathbf{y}_t \mid \boldsymbol{\theta}_B, B) = p[\mathbf{y}_t \mid \boldsymbol{\theta}_A = h(\boldsymbol{\theta}_B), A]$. Applying (3.23), we obtain

$$\begin{aligned}
k(\boldsymbol{\theta}_B \mid B) &= k[h(\boldsymbol{\theta}_B) \mid A] \cdot \left|[\partial h(\boldsymbol{\theta}_B)/\partial \boldsymbol{\theta}'_B]\right| \\
&= \left| E\left\{\frac{\partial \log p[\mathbf{y}_t \mid h(\boldsymbol{\theta}_B), A]}{\partial h(\boldsymbol{\theta}_B)} \cdot \frac{\partial \log p[\mathbf{y}_t \mid h(\boldsymbol{\theta}_B), A]}{\partial h(\boldsymbol{\theta}_B)'}\right\}\right|^{1/2} \\
&\quad \times \left| \left[\frac{\partial h(\boldsymbol{\theta}_B)}{\partial \boldsymbol{\theta}'_B}\right] \mid \boldsymbol{\theta}_A, A\right| \\
&= \left| E\left\{\left[\frac{\partial h(\boldsymbol{\theta}_B)}{\partial \boldsymbol{\theta}'_B}\right] \frac{\partial \log p[\mathbf{y}_t \mid h(\boldsymbol{\theta}_B), A]}{\partial h(\boldsymbol{\theta}_B)} \frac{\partial \log p[\mathbf{y}_t \mid h(\boldsymbol{\theta}_B), A]}{\partial h(\boldsymbol{\theta}_B)'}\right.\right. \\
&\quad \times \left.\left.\left[\frac{\partial h(\boldsymbol{\theta}_B)}{\partial \boldsymbol{\theta}'_B}\right]'\right\} \mid \boldsymbol{\theta}_A, A\right|^{1/2} \\
&= \left| E\left\{\frac{\partial \log p[\mathbf{y}_t \mid \boldsymbol{\theta}_B, B]}{\partial \boldsymbol{\theta}_B} \frac{\partial \log p[\mathbf{y}_t \mid \boldsymbol{\theta}_B, B]}{\partial \boldsymbol{\theta}'_B}\right\} \mid \boldsymbol{\theta}_B, B\right|^{1/2}. \qquad \blacksquare
\end{aligned}$$

Example 3.2.3 Jeffreys Invariant Prior for the Bernoulli Distribution For a sequence of i.i.d. Bernoulli trials with outcomes $y_t = 0$ or $y_t = 1$, $p(y_t \mid \theta) = \theta^{y_t}(1-\theta)^{(1-y_t)}$. Then $\log p(y_t \mid \theta) = y_t \log(\theta) + (1-y_t)\log(1-\theta)$, and

$d \log p(y_t \mid \theta)/d\theta = \theta^{-1} y_t + (1-\theta)^{-1}(y_t - 1)$. Since

$$E[d \log p(y_t \mid \theta)/d\theta]^2 = E[\theta^{-1} y_t + (1-\theta)^{-1}(y_t - 1)]^2$$
$$= \theta(\theta^{-1})^2 + (1-\theta)(1-\theta)^{-2} = \theta^{-1}(1-\theta)^{-1},$$

the Jeffreys invariant prior density kernel is $k(\theta) \propto \theta^{-1/2}(1-\theta)^{-1/2}$.

Exercise 3.2.1 Improper Prior Distributions in the Normal Linear Regression Model In Example 3.2.1 the precision h was fixed. Suppose instead that there is a sequence of prior distributions for h, $\underline{s}_j^2 h \mid A_j \sim \chi^2(\underline{\nu}_j)$.

(a) Let $\underline{s}_j^2 = q\underline{\nu}_j$ where $q > 0$, and suppose $\lim_{j\to\infty} \underline{\nu}_j = 0$. Find a corresponding sequence of kernels $k(h \mid A_j)$ satisfying the conditions of Theorem 3.2.1 and for which $\lim_{j\to\infty} k(h \mid A_j) = k(h \mid A) = h^{-1} I_{(0,\infty)}(h)$.

(b) Suppose that in the normal linear regression model of Example 2.1.2

$$q\underline{\nu}_j h \mid A_j \sim \chi^2(\underline{\nu}_j) \quad \text{and} \quad \boldsymbol{\beta} \mid A_j \sim N[\underline{\boldsymbol{\beta}}, (a_j \underline{\mathbf{H}})^{-1}]$$

are the independent prior distributions for h and $\boldsymbol{\beta}$. Also suppose that $\text{rank}(\mathbf{X}) = k$. For $\lim_{j\to\infty} \underline{\nu}_j = \lim_{j\to\infty} a_j = 0$, write the limiting posterior density kernel $k(\boldsymbol{\beta}, h| \mathbf{y}^o, \mathbf{X}, A)$. Show that this is a density kernel of the distribution

$$s^2 h \mid (\mathbf{y}^o, \mathbf{X}, A) \sim \chi^2(T-k),$$
$$\boldsymbol{\beta} | (h, \mathbf{y}^o, \mathbf{X}, A) \sim N[\mathbf{b}, (h\mathbf{X}'\mathbf{X})^{-1}],$$

where $\mathbf{b} = (\mathbf{X}'\mathbf{X})^{-1}\mathbf{X}'\mathbf{y}^o$ and $s^2 = (\mathbf{y} - \mathbf{X}\mathbf{b})'(\mathbf{y} - \mathbf{X}\mathbf{b})$. Thus $\boldsymbol{\beta}$ and h have a normal-gamma posterior distribution. It follows (recall Example 2.3.3) that

$$\boldsymbol{\beta} \mid (\mathbf{y}^o, \mathbf{X}, A) \sim t[\mathbf{b}, (T-k)^{-1} s^2 (\mathbf{X}'\mathbf{X})^{-1}; T-k].$$

(c) Suppose that in the normal linear regression model with conjugate prior distribution of Example 2.3.3

$$q\underline{\nu}_j h \mid A_j \sim \chi^2(\underline{\nu}_j) \text{ and } \boldsymbol{\beta} \mid (h, A_j) \sim N[\underline{\boldsymbol{\beta}}, (a_j h \underline{\mathbf{H}})^{-1}].$$

Also suppose that $\text{rank}(\mathbf{X}) = k$. For $\lim_{j\to\infty} \underline{\nu}_j = \lim_{j\to\infty} a_j = 0$, write the limiting posterior density kernel $k(\boldsymbol{\beta}, h \mid \mathbf{y}^o, \mathbf{X}, A)$. Show that this is a density kernel of the distribution

$$s^2 h \mid (\mathbf{y}^o, \mathbf{X}, A) \sim \chi^2(T),$$
$$\boldsymbol{\beta} | (h, \mathbf{y}^o, \mathbf{X}, A) \sim N[\mathbf{b}, h(\mathbf{X}'\mathbf{X})^{-1}].$$

It follows that $\boldsymbol{\beta} \mid (\mathbf{y}^o, \mathbf{X}, A) \sim t[\mathbf{b}, s^2(\mathbf{X}'\mathbf{X})^{-1}; T]$.

(d) The conventional non-Bayesian treatment of the normal linear model is derived as follows:

$$\begin{aligned} hs^2 \mid (\boldsymbol{\beta}, h, \mathbf{X}, A) &\sim \chi^2(T-k), \\ \mathbf{b} \mid (\boldsymbol{\beta}, h, \mathbf{X}, A) &\sim N[\boldsymbol{\beta}, (h\mathbf{X}'\mathbf{X})^{-1}], \\ \mathbf{b} \mid (\boldsymbol{\beta}, \mathbf{X}, A) &\sim t[\boldsymbol{\beta}, (T-k)^{-1}s^2(\mathbf{X}'\mathbf{X})^{-1}; T-k]. \end{aligned}$$

It is common to give an informal Bayesian interpretation of these results in statements such as "The probability that β_2 is negative is...." Using the results in parts (b) and (c), provide a formal Bayesian interpretation of such statements.

Exercise 3.2.2 An Invariant Prior Distribution Suppose that the observables are independently and uniformly distributed on the interval $(0, \theta)$.

(a) What is the Jeffreys invariant prior distribution for θ? Is this prior conjugate?

(b) Consistent with Theorem 3.2.1, can you find a sequence of proper prior densities $p(\theta \mid A_j)$ with kernels $k(\theta \mid A_j) \geq k(\theta \mid A_{j-1})$, that has as its pointwise limit the prior distribution you found in (a)?

Exercise 3.2.3 Jeffreys Prior for the Exponential Distribution Sometimes the pdf of the exponential distribution is written $p(y \mid \theta, A) = \theta \exp(-\theta y)$, and sometimes it is written $p(y \mid \lambda, A) = \lambda^{-1} \exp(-y/\lambda)$.

(a) Derive the Jeffreys prior for θ and the Jeffreys prior for λ. Show that these priors are improper.

(b) Derive the corresponding posterior densities for an i.i.d. sample $\mathbf{y}^o$ and show that for any finite interval S of the real line, $P(\theta \in S \mid \mathbf{y}^o, A) = P(\lambda^{-1} \in S \mid \mathbf{y}^o, A)$.

(c) Suppose that instead of the Jeffreys prior for θ, we used the improper "flat" prior $p(\theta) \propto I_{(0,\infty)}(\theta)$. Given a sample of size $T = 1$ with single observation $y_1^o = 1$, compute the posterior probability that $\theta < 1$.

(d) Suppose now that we used the same improper "flat" prior for λ. Try to find the posterior probability that $\lambda > 1$, given the same single observation $y_1^o = 1$.

Exercise 3.2.4 Jeffreys Prior for the Poisson Distribution Find the Jeffreys prior for the parameter θ of a Poisson distribution [see Exercise 2.3.1(d)], assuming i.i.d. sampling. Is the prior conjugate?

Exercise 3.2.5 Lindley's Paradox Suppose $y_t \overset{\text{i.i.d.}}{\sim} N(\mu, h^{-1})$ where h is known. Here are four alternative prior distributions for μ:

- $\mu = 0$
- $p(\mu) \propto I_{(0,c)}(\mu)$

- $p(\mu) \propto I_{(0,\infty)}(\mu)$
- $p(\mu) \propto I_{(-\infty,\infty)}(\mu)$

Given a sample of size T, there are six distinct Bayes factors for pairs of these hypotheses that could be formed.

(a) Which Bayes factors will be trivially zero or infinite, and why?
(b) For the nontrivial pairs, express the Bayes factors using standard notation.

Exercise 3.2.6 Predictive Densities and Improper Priors Suppose that in the normal linear model (Example 2.1.2) h is fixed at $h = h_0$. Partition $\mathbf{X}$ and $\mathbf{y}$:

$$\mathbf{X} = \begin{bmatrix} \underset{T_1 \times k}{\mathbf{X}_1} \\ \underset{T_2 \times k}{\mathbf{X}_2} \end{bmatrix}, \quad \mathbf{y} = \begin{pmatrix} \underset{T_1 \times 1}{\mathbf{y}_1} \\ \underset{T_2 \times 1}{\mathbf{y}_2} \end{pmatrix}.$$

In model A, $\boldsymbol{\beta} \sim N(\underline{\boldsymbol{\beta}}, \underline{\mathbf{H}}^{-1})$. There is a sequence of models $\{B_j\}$ that differ from A and from each other only with respect to the prior distribution for $\boldsymbol{\beta} : \boldsymbol{\beta} \mid B_j \sim N(\underline{\boldsymbol{\beta}}, (j+1)\underline{\mathbf{H}}^{-1})$. There are data $\mathbf{y}_1^o$ for the observable $\mathbf{y}_1$, but $\mathbf{y}_2$ has not been observed. The covariate matrix $\mathbf{X}$ is known.

(a) Find the limiting distribution $\lim_{j\to\infty} \boldsymbol{\beta} \mid (\mathbf{X}_1, \mathbf{y}_1^o, B_j)$.
(b) Show that $\lim_{j\to\infty} p(\mathbf{y}_1^o \mid \mathbf{X}_1, B_j)/p(\mathbf{y}_1^o \mid \mathbf{X}_1, A) = 0$.
(c) Find the distribution $\mathbf{y}_2 \mid (\mathbf{X}_1, \mathbf{X}_2, \mathbf{y}_1^o, A)$ and the limiting distribution $\lim_{j\to\infty} \mathbf{y}_2 \mid (\mathbf{X}_1, \mathbf{X}_2, \mathbf{y}_1^o, B_j)$.
(d) Now suppose that we obtain the data $\mathbf{y}_2^o$. Is it the case that

$$\lim_{j\to\infty} \frac{p(\mathbf{y}_2^o \mid \mathbf{X}_1, \mathbf{X}_2, \mathbf{y}_1^o, B_j)}{p(\mathbf{y}_2^o \mid \mathbf{X}_1, \mathbf{X}_2, \mathbf{y}_1^o, A)} = 0?$$

3.3 PRIOR ROBUSTNESS AND THE DENSITY RATIO CLASS

In many instances Bayesian investigators do not know their clients' priors, or even the identity of their clients. For example, the investigator may be an academic economist and the clients, the readers of an article published by the economist. A number of approaches can be taken in this situation. One is to report posterior moments corresponding to alternative priors, but such an enumeration can become tiresome long before reasonable possibilities for priors are exhausted. Another is to provide clients with the ability to modify priors simply and directly and examine the impact on posterior moments. Section 8.4 discusses this approach. Another possibility is to report a range along with each posterior moment, corresponding to all possible prior distributions within a specified class of distributions. Several interesting classes of prior distributions have been proposed, reviewed in Berger (1994) and Wasserman (1992).

This section takes up the density ratio class of prior distributions. This class consists of all prior distributions with a probability density kernel $k(\boldsymbol{\theta}_A \mid A)$ that satisfies the inequalities $a(\boldsymbol{\theta}_A) \leq k(\boldsymbol{\theta}_A \mid A) \leq b(\boldsymbol{\theta}_A)$, where $a(\boldsymbol{\theta}_A)$ and $b(\boldsymbol{\theta}_A)$ are kernels of prior densities that yield proper posterior densities. A case of particular interest is $b(\boldsymbol{\theta}_A) = r \cdot a(\boldsymbol{\theta}_A)$. The density ratio class then permits ratios of prior probabilities of subsets of Θ_A to vary by a factor of up to r from the corresponding ratios implied by the prior density kernel $k(\boldsymbol{\theta}_A \mid A)$. If we interpret improper prior density kernels as assigning relative probabilities to $\Theta_A^* \subseteq \Theta_A$ for which $\int_{\Theta_A^*} k(\boldsymbol{\theta}_A \mid A)\, d\boldsymbol{\theta}_A < \infty$, then the density ratio class can be extended to improper prior distributions with the same interpretation.

The development here was first presented in Geweke and Petrella (1998). It builds on work in Wasserman and Kadane (1992) and Lavine (1991a, 1991b), and is the basis for the routine and efficient computation of bounds of posterior moments $E(\boldsymbol{\omega} \mid y^o, A)$ approximated by posterior simulators developed in Section 8.5.

In this section, let $g(\boldsymbol{\theta}_A) = \int_\Omega h(\boldsymbol{\omega}) p(\boldsymbol{\omega} \mid \mathbf{y}^o, \boldsymbol{\theta}_A, A)\, d\nu(\boldsymbol{\omega})$. For any prior kernel $k(\boldsymbol{\theta}_A \mid A)$, proper or improper, we obtain

$$E[g(\boldsymbol{\theta}_A) \mid \mathbf{y}^o, A] = \frac{\displaystyle\int_{\Theta_A} k(\boldsymbol{\theta}_A \mid A) p(\mathbf{y}^o \mid \boldsymbol{\theta}_A, A) g(\boldsymbol{\theta}_A)\, d\nu(\theta_A)}{\displaystyle\int_{\Theta_A} k(\boldsymbol{\theta}_A \mid A) p(\mathbf{y}^o \mid \boldsymbol{\theta}_A, A)\, d\nu(\theta_A)}. \tag{3.25}$$

Let $a(\boldsymbol{\theta}_A)$ and $b(\boldsymbol{\theta}_A)$ be given functions for which $0 \leq a(\boldsymbol{\theta}_A) \leq k(\boldsymbol{\theta}_A \mid A) \leq b(\boldsymbol{\theta}_A)$ $\forall \boldsymbol{\theta}_A \in \Theta_A$, and $b(\boldsymbol{\theta}_A) p(\mathbf{y}^o \mid \boldsymbol{\theta}_A, A)$ is finitely integrable on Θ_A. The formal problem is to determine the range of values of (3.25) over the set S of all prior density kernels $k(\boldsymbol{\theta}_A \mid A)$ satisfying $0 \leq a(\boldsymbol{\theta}_A) \leq k(\boldsymbol{\theta}_A \mid A) \leq b(\boldsymbol{\theta}_A)$, that is, to determine

$$\underline{E}[g(\boldsymbol{\theta}_A) \mid \mathbf{y}^o, A] = \inf_{k \in S} E[g(\boldsymbol{\theta}_A) \mid \mathbf{y}^o, A]$$

and

$$\overline{E}[g(\boldsymbol{\theta}_A) \mid \mathbf{y}^o, A] = \sup_{k \in S} E[g(\boldsymbol{\theta}_A) \mid \mathbf{y}^o, A].$$

Because $\underline{E}[g(\boldsymbol{\theta}_A) \mid \mathbf{y}^o, A] = -\overline{E}[-g(\boldsymbol{\theta}_A) \mid \mathbf{y}^o, A]$, only the maximization problem need be considered. The following result was shown in DeRobertis and Hartigan (1981), Proposition 4.1; Lavine (1991b), Claim 3; Wasserman and Kadane (1992), Theorem 4(b); and Wasserman et al. (1993), p. 308. A proof is included here because it parallels a similar result based on posterior simulators presented in Section 8.5.

Theorem 3.3.1 Bounding a Posterior Moment over the Density Ratio Class of Priors Let $b(\boldsymbol{\theta}_A)$ be a prior density kernel for which the posterior density kernel $b(\boldsymbol{\theta}_A) p(\mathbf{y}^o \mid \boldsymbol{\theta}_A, A)$ is finitely integrable on Θ_A, and let $a(\boldsymbol{\theta}_A) \leq b(\boldsymbol{\theta}_A)$ be a second prior density kernel. Let S be the set of all prior density kernels k satisfying

$a(\boldsymbol{\theta}_A) \leq k(\boldsymbol{\theta}_A) \leq b(\boldsymbol{\theta}_A) \forall \boldsymbol{\theta}_A \in \Theta_A$, and suppose that (3.25) is bounded above for $k \in S$. Let

$$k(\boldsymbol{\theta}_A; c) = \begin{cases} a(\boldsymbol{\theta}_A) & \text{if} \quad g(\boldsymbol{\theta}_A) \leq c \\ b(\boldsymbol{\theta}_A) & \text{if} \quad g(\boldsymbol{\theta}_A) > c \end{cases}.$$

Then the unique solution of

$$f(c) = \int_{\Theta_A} [g(\boldsymbol{\theta}_A) - c] p(\mathbf{y}^o \mid \boldsymbol{\theta}_A, A) k(\boldsymbol{\theta}_A; c)\, d\nu(\boldsymbol{\theta}_A) = 0 \tag{3.26}$$

is $\widehat{c} = \overline{E}[g(\boldsymbol{\theta}_A) \mid \mathbf{y}^o, A] = \sup_{k \in S} E[g(\boldsymbol{\theta}_A) \mid \mathbf{y}^o, A]$.

Proof: Since (3.25) is bounded above for $k \in S$, $f(c)$ is finite for all real c. Moreover $f(c)$ is differentiable and

$$f'(c) \leq -\int_{\Theta_A} p(\mathbf{y}^o \mid \boldsymbol{\theta}_A, A) a(\boldsymbol{\theta}_A)\, d\nu(\boldsymbol{\theta}_A) \tag{3.27}$$

for all c. Hence (3.26) has exactly one solution.

For all $k \in S$, $k(\boldsymbol{\theta}_A) \leq k(\boldsymbol{\theta}_A; \widehat{c})$ if $g(\boldsymbol{\theta}_A) - \widehat{c} > 0$ and $k(\boldsymbol{\theta}_A) \geq k(\boldsymbol{\theta}_A; \widehat{c})$ if $g(\boldsymbol{\theta}_A) - \widehat{c} < 0$. Hence

$$\int_{\Theta_A} [g(\boldsymbol{\theta}_A) - c] p(\mathbf{y}^o \mid \boldsymbol{\theta}_A, A) k(\boldsymbol{\theta}_A)\, d\nu(\boldsymbol{\theta}_A) \leq 0,$$

and

$$\frac{\int_{\Theta_A} k(\boldsymbol{\theta}_A) p(\mathbf{y}^o \mid \boldsymbol{\theta}_A) g(\boldsymbol{\theta}_A)\, d\nu(\boldsymbol{\theta}_A)}{\int_{\Theta_A} k(\boldsymbol{\theta}_A) p(\mathbf{y}^o \mid \boldsymbol{\theta}_A)\, d\nu(\boldsymbol{\theta}_A)} \leq \frac{\int_{\Theta_A} k(\boldsymbol{\theta}_A; \widehat{c}) p(\mathbf{y}^o \mid \boldsymbol{\theta}_A) g(\boldsymbol{\theta}_A)\, d\nu(\boldsymbol{\theta}_A)}{\int_{\Theta_A} k(\boldsymbol{\theta}_A; \widehat{c}) p(\mathbf{y}^o \mid \boldsymbol{\theta}_A)\, d\nu(\boldsymbol{\theta}_A)} = c.$$

■

Because Theorem 3.3.1 remains valid with the formal substitution $p(\mathbf{y}^o \mid \boldsymbol{\theta}_A, A) = 1$, it provides bounds on prior expectations in a density ratio class of prior distributions as well.

Example 3.3.1 Density Ratio Bounds for the Normal Density Suppose that there is a single unobservable θ_A, $p(\mathbf{y}^o \mid \theta_A, A) a(\theta_A)$ is the kernel of the standard normal distribution, and $b(\theta_A)/a(\theta_A) = r > 1$. For $g(\theta_A) = \theta_A$, (3.26) then becomes

$$\int_{-\infty}^{c} (\theta_A - c)\phi(\theta_A)\, d\theta_A + r \int_{c}^{\infty} (\theta_A - c)\phi(\theta_A)\, d\theta_A = 0, \tag{3.28}$$

where $\phi(\cdot)$ is the pdf of the standard normal distribution. Denoting the cdf of the standard normal distribution by $\Phi(\cdot)$ and using the relation $\int_{-\infty}^{c} z\phi(z)\,dz = -\phi(c)$ [Johnson et al. (1994), (13.134)], it follows from (3.28) that

$$(r-1)[\phi(c) + c\Phi(c)] - rc = 0.$$

The unique solution of this equation, $c = \gamma(r)$, is displayed in the solid line in Figure 3.1.

The function $\gamma(r)$ provides some guidance in choosing r in this and similar density ratio classes. For example, $r = 10$ permits the prior mean of any one parameter to shift up or down by about 0.9 prior standard deviation. To allow a shift of 1.5 prior standard deviations in a prior mean requires $r = 52.3$. Larger shifts in the prior mean require very large values of r because the tails of the normal distribution decline rapidly.

Example 3.3.2 Density Ratio Bounds for Student-*t* Densities In the same situation as Example 3.3.1 suppose instead that $p(\mathbf{y}^o \mid \theta_A, A)a(\theta_A)$ is the kernel of the standard Student-t distribution with $\lambda > 1$ degrees of freedom. Then c is the

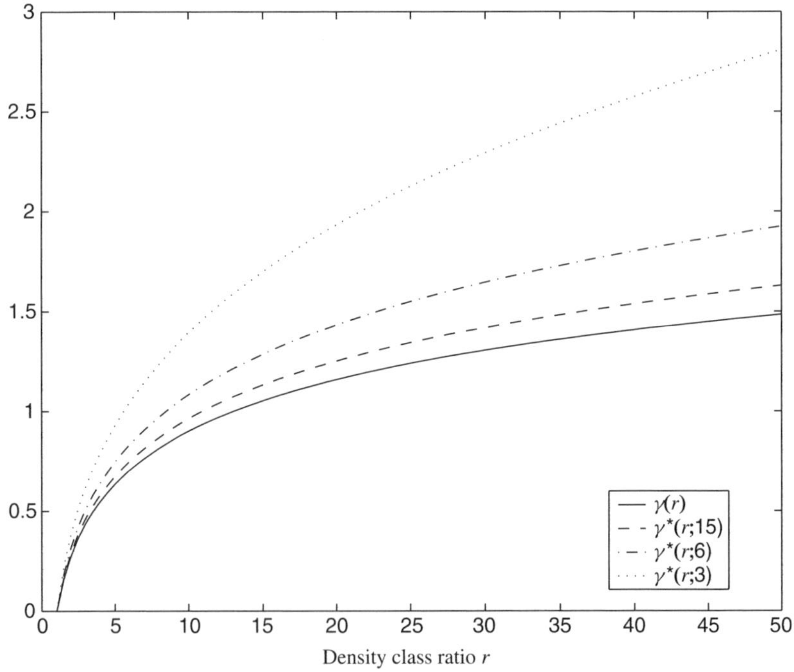

Figure 3.1. Gamma functions for normal, $t(15)$, $t(6)$, and $t(3)$ distributions.

root of

$$\int_{-\infty}^{c} (\theta_A - c) t(\theta_A; \lambda)\, d\theta_A + r \int_{c}^{\infty} (\theta_A - c) t(\theta_A; \lambda)\, d\theta_A = 0,$$

where $t(\cdot; \lambda)$ is the pdf of the standard Student-t distribution with λ degrees of freedom. The unique solution of this equation, $c = \gamma^*(r; \lambda)$, is also shown in Figure 3.1 for several values of λ. Note that, for small λ, a given value of r permits a much larger shift for this set of prior distributions than does the normal case. This is because the tails of the Student-t density with low degrees of freedom are much thicker than those of the normal density.

Exercise 3.3.1 Completing the Argument Derive (3.27).

Exercise 3.3.2 Extending Examples 3.3.1 and 3.3.2 In this exercise, the random variable x represents any unobservable, for example, $x = g(\boldsymbol{\theta}_A)$. Its distribution could be the posterior, the prior, or some other distribution.

(a) Let $a(x)$ be any density kernel of the $N(\mu, h^{-1})$ distribution, and $b(x) = r \cdot a(x)$, where $r > 1$. Show that if the random variable x has probability density kernel $k(x)$ and $a(x) \leq k(x) \leq b(x)$, then $E(x) \leq \mu + h^{-1/2}\gamma(r)$.

(b) In the same situation as (a), suppose instead that $a(x)$ is any density kernel of the Student-t distribution with $\lambda > 1$ degrees of freedom. Show that $E(x) \leq \mu + h^{-1/2}\gamma^*(r; \lambda)$.

3.4 ASYMPTOTIC ANALYSIS

Asymptotic analysis addresses properties of the limiting behavior of a posterior density $p(\boldsymbol{\theta}_A \mid \mathbf{Y}_T^o, A)$ as sample size $T \to \infty$. These properties depend on the behavior of the sequence $\mathbf{Y}_T^o = \{\mathbf{y}_1^o, \ldots, \mathbf{y}_T^o\}$. We shall assume *in this section only* that $\mathbf{Y}_T^o$ is the observed value of a random vector $\mathbf{Y}_T$ with probability density $p(\mathbf{Y}_T \mid D)$, where D is the *data-generating process*. The vectors $\boldsymbol{\theta}_A$ and $\mathbf{Y}_T$ are random, with density $p(\mathbf{Y}_T \mid D) p(\boldsymbol{\theta}_A \mid \mathbf{Y}_T, A)$. We do not assume that there necessarily exists any $\boldsymbol{\theta}_A \in \Theta_A$ such that $p(\mathbf{Y}_T \mid \boldsymbol{\theta}_A, A) = p(\mathbf{Y}_T \mid D)$. In asymptotic analysis, $\boldsymbol{\theta}_A$ and $\mathbf{Y}_T$ appear repeatedly in circumstances where one or the other can be either a random vector or the argument of a function. For clarity we shall adopt the convention, *in this section only*, of using a tilde to distinguish a random vector from the argument of a function. Thus, for example, $P(\widetilde{\mathbf{Y}}_T \in C \mid D) = \int_C p(\mathbf{Y}_T \mid D)\, d\nu(\mathbf{Y}_T)$. Finally, we shall assume *in this section only* that the dimension of the $k_A \times 1$ vector $\boldsymbol{\theta}_A$ is fixed and does not depend on sample size.

The case in which $\boldsymbol{\theta}_A$ is discrete provides one of the most important conclusions of asymptotic analysis without the technical conditions required in the case of continuous $\boldsymbol{\theta}_A$.

Theorem 3.4.1 Asymptotic Concentration of the Posterior Distribution for a Discrete Parameter Vector Suppose that

1. The vector of unobservables $\widetilde{\boldsymbol{\theta}}_A$ has a discrete prior distribution over a finite set of n points $\boldsymbol{\theta}_{Ai}$, and $p(\boldsymbol{\theta}_{Ai} \mid A) > 0\, (i = 1, \ldots, n)$.
2. $T^{-1} \log p(\widetilde{\mathbf{Y}}_T \mid \boldsymbol{\theta}_{Ai}, A) \overset{\text{a.s.}}{\to} \ell(\boldsymbol{\theta}_{Ai}; A)\, (i = 1, \ldots, n)$.
3. For one $j \in \{1, \ldots, n\}$, $\ell(\boldsymbol{\theta}_{Aj}; A) > \ell(\boldsymbol{\theta}_{Ai}; A)$ for all $i = 1, \ldots, n, i \neq j$.

Then

$$\lim_{T \to \infty} P(\widetilde{\boldsymbol{\theta}}_A = \boldsymbol{\theta}_{Aj} \mid \widetilde{\mathbf{Y}}_T, A) = 1. \tag{3.29}$$

Proof: For all $i \neq j$

$$T^{-1} \log[p(\boldsymbol{\theta}_{Ai} \mid \widetilde{\mathbf{Y}}_T, A)/p(\boldsymbol{\theta}_{Aj} \mid \widetilde{\mathbf{Y}}_T, A)] = T^{-1} \log[p(\boldsymbol{\theta}_{Ai} \mid A)/p(\boldsymbol{\theta}_{Aj} \mid A)]$$
$$+ T^{-1} \log[p(\widetilde{\mathbf{Y}}_T \mid \boldsymbol{\theta}_{Ai}, A)/p(\widetilde{\mathbf{Y}}_T \mid \boldsymbol{\theta}_{Aj}, A)] \overset{\text{a.s.}}{\to} \ell(\boldsymbol{\theta}_{Ai}; A) - \ell(\boldsymbol{\theta}_{Aj}; A). \qquad \blacksquare$$

This result does not require that $\widetilde{\mathbf{y}}_t$ be i.i.d., either in $p(\widetilde{\mathbf{Y}}_T \mid D)$ or in any of $p(\widetilde{\mathbf{Y}}_T \mid \boldsymbol{\theta}_{Ai}, A)$ $(i = 1, \ldots, n)$. However, if $\widetilde{\mathbf{Y}}_T$ is i.i.d. in both D and A, then condition 2 of Theorem 3.4.1 may be restated in terms of Kullback–Leibler information. [On the wider significance of Kullback–Leibler information, see Mittelhammer et al. (2000), Section 13.1.1.]

Definition 3.4.1 Given two densities $p(\mathbf{y} \mid A)$ and $p(\mathbf{y} \mid B)$ for the same observable $\widetilde{\mathbf{y}}$ and defined with respect to the same measure ν, the *Kullback–Leibler information criterion* (KLIC) *distance* from $p(\mathbf{y} \mid A)$ to $p(\mathbf{y} \mid B)$ is

$$K[p(\mathbf{y} \mid A), p(\mathbf{y} \mid B)] = \int_{\Psi} \log[p(\mathbf{y} \mid A)/p(\mathbf{y} \mid B)] p(\mathbf{y} \mid A)\, d\nu(\mathbf{y})$$
$$= E\{\log[p(\widetilde{\mathbf{y}}|A)/p(\widetilde{\mathbf{y}}|B)] \mid A\}.$$

Note that the KLIC distance is directed

$$K[p(\mathbf{y} \mid A), p(\mathbf{y} \mid B)] \neq K[p(\mathbf{y} \mid B), p(\mathbf{y} \mid A)],$$

and one can be finite while the other is infinite. Clearly $K[p(\mathbf{y} \mid A), p(\mathbf{y} \mid A)] = 0$, and by Jensen's inequality for a convex function

$$K[p(\mathbf{y} \mid A), p(\mathbf{y} \mid B)] = -E\{\log[p(\widetilde{\mathbf{y}}|B)/p(\widetilde{\mathbf{y}}|A)] \mid A\}$$
$$\geq -\log\{E[p(\widetilde{\mathbf{y}}|B)/p(\widetilde{\mathbf{y}}|A)] \mid A\} = -\log(1) = 0.$$

Condition 3 of Theorem 3.4.1 may now be restated, for the case of i.i.d. distributions, as

$$E[\log p(\widetilde{\mathbf{y}}| \boldsymbol{\theta}_{Aj}, A) \mid D] > E[\log p(\widetilde{\mathbf{y}}| \boldsymbol{\theta}_{Ai}, A) \mid D]$$
$$\Leftrightarrow \int_{\Psi} \log p(\mathbf{y} \mid \boldsymbol{\theta}_{Aj}, A) p(\mathbf{y} \mid D)\, d\nu(\mathbf{y}) > \int_{\Psi} \log p(\mathbf{y} \mid \boldsymbol{\theta}_{Ai}, A) p(\mathbf{y} \mid D)\, d\nu(\mathbf{y})$$
$$\Leftrightarrow K[p(\mathbf{y} \mid D), p(\mathbf{y} \mid \theta_{Aj}, A)] < K[p(\mathbf{y} \mid D), p(\mathbf{y} \mid \theta_{Ai}, A)]$$

for each $i \neq j$. More succinctly, as sample size increases, the posterior distribution places all probability on the parameter vector $\boldsymbol{\theta}_{Aj}$ that provides the smallest KLIC distance from the data-generating density $p(\mathbf{y} \mid D)$ to the model density $p(\mathbf{y} \mid \boldsymbol{\theta}_{Ai}, A)$ $(i = 1, \ldots, n)$. Conclusion (3.29) of Theorem 3.4.1 is often summarized by saying that $\boldsymbol{\theta}_{Aj}$ is the *pseudotrue value* of $\widetilde{\boldsymbol{\theta}}_A$.

Example 3.4.1 Asymptotic Concentration in the Bernoulli Model with Discrete Parameter Suppose that in model A, $\widetilde{\mathbf{y}}_t$ is an i.i.d. Bernoulli random variable with $P(\widetilde{y}_t = 1 \mid p, A) = p$. The prior distribution places positive probability on only the three points $p = p_1 = \frac{1}{4}$, $p = p_2 = \frac{1}{2}$, and $p = p_3 = \frac{3}{4}$. Suppose that in the data-generating process D, $\widetilde{y}_t$ is an i.i.d. Bernoulli random variable with $P(\widetilde{y}_t = 1 \mid D) = p^*$, and $p^* \in (0, 1)$. Then

$$E[\log p(\widetilde{y}_t \mid p_j, A) \mid D] = p^* \log p_j + (1 - p^*) \log(1 - p_j).$$

One can show that $P(p = 1/2 \mid \widetilde{\mathbf{Y}}_T, A) \overset{\text{a.s.}}{\to} 1$ if and only if $p^* \in (0.36907, 0.63093)$.

When the unobservables $\widetilde{\boldsymbol{\theta}}_A$ are continuously distributed, the posterior probability attached to any single point is always zero, for each $\boldsymbol{\theta}_A \in \Theta_A$ and for all T, and consequently this is true for each $\boldsymbol{\theta}_A \in \Theta_A$ in the limit as well. In this case it is useful to frame asymptotic analysis in terms of limiting probabilities of a neighborhood of a point $\boldsymbol{\theta}_A^*$ with the distinguishing features indicated in the following result.

Theorem 3.4.2 Asymptotic Concentration of the Posterior Distribution for a Continuous Parameter Vector Suppose that

1. The prior distribution of $\widetilde{\boldsymbol{\theta}}_A$ is absolutely continuous and $P(\widetilde{\boldsymbol{\theta}}_A \in C \mid A) > 0$ for all $C \subseteq \Theta_A$ for which $\int_C d\boldsymbol{\theta}_A > 0$.
2. $T^{-1} \log p(\widetilde{\mathbf{Y}}_T \mid \boldsymbol{\theta}_A, A) \overset{\text{a.s.}}{\to} \ell(\boldsymbol{\theta}_A; A)$ uniformly for all $\boldsymbol{\theta}_A \in \Theta_A$.
3. $\ell(\boldsymbol{\theta}_A; A)$ is a continuous function of $\boldsymbol{\theta}_A$ with a unique global mode at $\boldsymbol{\theta}_A = \boldsymbol{\theta}_A^*$, and there exist $\underline{\ell}$ and $\overline{\ell}$ for which $M(\boldsymbol{\theta}_A^*) = \{\boldsymbol{\theta}_A : \underline{\ell} < \ell(\boldsymbol{\theta}_A) < \overline{\ell}\}$ is a bounded open neighborhood of $\boldsymbol{\theta}_A^*$.

Then for any open neighborhood $N(\boldsymbol{\theta}_A^*)$ of $\boldsymbol{\theta}_A^*$,

$$\lim_{T \to \infty} P[\widetilde{\boldsymbol{\theta}}_A \in N(\boldsymbol{\theta}_A^*) \mid \widetilde{\mathbf{Y}}_T, A] = 1. \tag{3.30}$$

Proof: Define $G(\boldsymbol{\theta}_A^*) = N(\boldsymbol{\theta}_A^*) \cap M(\boldsymbol{\theta}_A^*)$. Let $\overline{G}(\boldsymbol{\theta}_{A*}) = \Theta_A - G(\boldsymbol{\theta}_A^*)$, $\ell_0 = \ell(\boldsymbol{\theta}_A^*; A)$, and $\ell_1 = \sup_{\overline{G}(\boldsymbol{\theta}_A^*)} \ell(\boldsymbol{\theta}_A; A)$. By virtue of condition 3, $\ell_0 > \ell_1$. Let $\ell_2 = (\ell_0 + \ell_1)/2$ and define

$$H(\boldsymbol{\theta}_A^*) = G(\boldsymbol{\theta}_A^*) \cap \{\boldsymbol{\theta}_A : \ell(\boldsymbol{\theta}_A) > \ell_2\}.$$

Then

$$\frac{P[\widetilde{\boldsymbol{\theta}}_A \in \overline{N}(\boldsymbol{\theta}_A^*) \mid \widetilde{\mathbf{Y}}_T, A]}{P[\widetilde{\boldsymbol{\theta}}_A \in N(\boldsymbol{\theta}_A^*) \mid \widetilde{\mathbf{Y}}_T, A]} < \frac{P[\widetilde{\boldsymbol{\theta}}_A \in \overline{G}(\boldsymbol{\theta}_A^*) \mid \widetilde{\mathbf{Y}}_T, A]}{P[\widetilde{\boldsymbol{\theta}}_A \in H(\boldsymbol{\theta}_A^*) \mid \widetilde{\mathbf{Y}}_T, A]}$$

$$< \frac{P[\widetilde{\boldsymbol{\theta}}_A \in \overline{G}(\boldsymbol{\theta}_A^*) \mid A]}{P[\widetilde{\boldsymbol{\theta}}_A \in H(\boldsymbol{\theta}_A^*) \mid A]} \cdot \frac{\sup_{\boldsymbol{\theta}_A \in \overline{G}(\boldsymbol{\theta}_A^*)} p(\widetilde{\mathbf{Y}}_T \mid \boldsymbol{\theta}_A, A)}{\inf_{\boldsymbol{\theta}_A \in H(\boldsymbol{\theta}_A^*)} p(\widetilde{\mathbf{Y}}_T \mid \boldsymbol{\theta}_A, A)}. \tag{3.31}$$

From the continuity of ℓ and condition 1 $P[\widetilde{\boldsymbol{\theta}}_A \in H(\boldsymbol{\theta}_A^*) \mid A] > 0$. Then, from the almost sure uniform convergence of $T^{-1} \log p(\widetilde{\mathbf{Y}}_T \mid \boldsymbol{\theta}_A, A)$

$$T^{-1} \log \left[\frac{\sup_{\boldsymbol{\theta}_A \in \overline{G}(\boldsymbol{\theta}_A^*)} p(\widetilde{\mathbf{Y}}_T \mid \boldsymbol{\theta}_A, A)}{\inf_{\boldsymbol{\theta}_A \in H(\boldsymbol{\theta}_A^*)} p(\widetilde{\mathbf{Y}}_T \mid \boldsymbol{\theta}_A, A)} \right] \overset{\text{a.s.}}{\to} \ell_1 - \ell_2 < 0.$$

Consequently the almost sure limit of (3.31) is 0. ■

Condition (3.30) of this theorem is often summarized by saying that $\boldsymbol{\theta}_A^*$ is the pseudotrue value of $\boldsymbol{\theta}_A$. If $\widetilde{\mathbf{y}}_t$ is i.i.d. in both D and A, then

$$K[p(\mathbf{y} \mid D), p(\mathbf{y} \mid \boldsymbol{\theta}_A^*, A)] < K[p(\mathbf{y} \mid D), p(\mathbf{y} \mid \boldsymbol{\theta}_A, A)]$$

for all $\boldsymbol{\theta}_A \in \Theta_A$ except $\boldsymbol{\theta}_A^*$. Condition 2 is key in applying Theorem 3.4.2. This condition is seldom satisfied for a natural parameter space Θ_A. Instead, the parameter space must be further restricted to a closed and bounded subset Θ_A^* of Θ_A. Then, showing that $T^{-1} \log p(\widetilde{\mathbf{Y}}_T \mid \boldsymbol{\theta}_A, A) \overset{\text{a.s.}}{\to} \ell(\boldsymbol{\theta}_A; A)$ for all $\boldsymbol{\theta}_A \in \Theta_A^*$ is sufficient for condition 2. (Condition 3 requires that $\boldsymbol{\theta}_A^*$ be an interior point of Θ_A^*.) Non-Bayesian approaches to consistency of point estimates encounter a similar technical complication; see, for example, Amemiya (1985) Theorems 4.2.1 and 4.2.2.

Example 3.4.2 Asymptotic Concentration in the Bernoulli Model with Continuous Parameter In the i.i.d. Bernoulli setting of Example 3.4.1, but with a continuous prior distribution, we have

$$T^{-1} \log p(\widetilde{\mathbf{Y}}_T \mid p, A) \overset{\text{a.s.}}{\to} p^* \log(p) + (1 - p^*) \log(1 - p),$$

where p^* is the value of p in the i.i.d. Bernoulli data-generating process D. If the prior distribution has support (p_1, p_2) with $0 < p_1 < p^* < p_2 < 1$, then the conditions of Theorem 3.4.2 are satisfied and $P[p \in (p^* - \varepsilon, p^* + \varepsilon) \mid \widetilde{\mathbf{Y}}_T, A] \overset{\text{a.s.}}{\to} 1$ for all $\varepsilon > 0$.

Example 3.4.3 Asymptotic Concentration in the Normal Linear Regression Model Suppose that, as in Example 2.1.2, the conditional distribution of observables specified by model A is $\widetilde{\mathbf{y}}|(h, \boldsymbol{\beta}, \widetilde{\mathbf{X}}, A) \sim N(\widetilde{\mathbf{X}}\boldsymbol{\beta}, h^{-1}\mathbf{I}_T)$. Suppose that in the data-generating process D

$$T^{-1}\begin{bmatrix}\widetilde{\mathbf{y}}'\widetilde{\mathbf{y}} & \widetilde{\mathbf{y}}'\widetilde{\mathbf{X}}\\ \widetilde{\mathbf{X}}'\widetilde{\mathbf{y}} & \widetilde{\mathbf{X}}'\widetilde{\mathbf{X}}\end{bmatrix} \overset{\text{a.s.}}{\rightarrow} \begin{bmatrix}\sigma_{yy} & \boldsymbol{\sigma}_{y\mathbf{x}}\\ \boldsymbol{\sigma}_{\mathbf{x}y} & \Sigma_{\mathbf{xx}}\end{bmatrix}, \tag{3.32}$$

a positive definite matrix. D is otherwise left unspecified, and it is not necessarily the case that the distribution of $\widetilde{\mathbf{y}}|(\widetilde{\mathbf{X}}, D)$ is normal, or that the observables $\widetilde{\mathbf{y}}_t$ are independent conditional on $\widetilde{\mathbf{X}}$. Then

$$\log p(\mathbf{y} \mid \boldsymbol{\beta}, h, \mathbf{X}, A) = [-T\log(2\pi) + T\log h - h(\mathbf{y}'\mathbf{y} - 2\mathbf{y}'\mathbf{X}\boldsymbol{\beta} + \boldsymbol{\beta}'\mathbf{X}'\mathbf{X}\boldsymbol{\beta})]/2$$

and

$$T^{-1}\log p(\widetilde{\mathbf{y}} \mid \boldsymbol{\beta}, h, \widetilde{\mathbf{X}}, A) \overset{\text{a.s.}}{\rightarrow} [-\log(2\pi) + \log h - h(\sigma_{yy} - 2\boldsymbol{\sigma}_{y\mathbf{x}}\boldsymbol{\beta} + \boldsymbol{\beta}'\Sigma_{\mathbf{xx}}\boldsymbol{\beta})]/2.$$

The unique global maximum of the latter function occurs at the point $\boldsymbol{\beta}^* = \Sigma_{\mathbf{xx}}^{-1}\boldsymbol{\sigma}_{\mathbf{x}y}$, $h^* = (\sigma_{yy} - \sigma_{y\mathbf{x}}\Sigma_{\mathbf{xx}}^{-1}\boldsymbol{\sigma}_{\mathbf{x}y})^{-1}$. Condition 1 of Theorem 3.4.2 is satisfied if the prior distribution of $\boldsymbol{\beta}$ and h is absolutely continuous, condition 2 is satisfied if the support of $p(\boldsymbol{\beta}, h \mid A)$ bounds h and $\boldsymbol{\beta}'\boldsymbol{\beta}$ from above, and condition 3 is satisfied if this support includes the point $(\boldsymbol{\beta}^*, h^*)$.

Given further regularity conditions, the posterior distribution, appropriately scaled, converges in distribution to a normal distribution as sample size T increases. The regularity conditions in the following result, due to Chen (1985), are similar to typical conditions for the asymptotic sampling-theoretic distribution of the maximum likelihood estimator; for example, see Amemiya (1985), Theorem 4.2.4. In fact, these regularity conditions can be weakened, to include cases involving nonstationary time series in which the limiting sampling theoretic distributions of maximum likelihood estimators are not normal; see Sweeting and Adekola (1987) for one such development and Sims and Uhlig (1991) for a significant application in time series econometrics.

Theorem 3.4.3 Asymptotic Posterior Distribution for a Continuous Parameter Vector Suppose that for all T, $\log p(\boldsymbol{\theta}_A \mid \mathbf{Y}_T, A)$ is twice differentiable for all $\boldsymbol{\theta}_A \in \Theta_A$ and $\mathbf{Y}_T \in \Psi_T$. Denote $L_T(\boldsymbol{\theta}_A) = \log p(\boldsymbol{\theta}_A \mid \widetilde{\mathbf{Y}}_T, A)$, $L_T'(\boldsymbol{\theta}_A) = \partial L_T(\boldsymbol{\theta}_A)/\partial\boldsymbol{\theta}_A$ and $L_T''(\boldsymbol{\theta}_A) = \partial^2 L_T(\boldsymbol{\theta}_A)/\partial\boldsymbol{\theta}_A\,\partial\boldsymbol{\theta}_A'$. Suppose that with probability 1 there exists finite T^* such that

1. For all $T > T^*$, $L_T(\boldsymbol{\theta}_A)$ has a strict local maximum at $\boldsymbol{\theta}_{AT} = \boldsymbol{\theta}_{AT}^*$, at which point $L'(\boldsymbol{\theta}_{AT}^*) = \mathbf{0}$ and $L_T''(\boldsymbol{\theta}_{AT}^*)$ is a negative definite matrix.
2. For all $T > T^*$, the largest eigenvalue σ_T^2 of the positive definite matrix $\Sigma_T = -[L_T''(\boldsymbol{\theta}_{AT}^*)]^{-1}$ satisfies the condition $\lim_{T\to\infty}\sigma_T^2 = 0$.
3. Given any $\varepsilon > 0$ there exists $T(\varepsilon)$ and $\delta(\varepsilon) > 0$ such that for all $T > T(\varepsilon)$ and $\boldsymbol{\theta}_A \in \{\boldsymbol{\theta}_A : (\boldsymbol{\theta}_A - \boldsymbol{\theta}_{AT}^*)'(\boldsymbol{\theta}_A - \boldsymbol{\theta}_{AT}^*) < \delta(\varepsilon)\}$,

$$\mathbf{I}_k - \mathbf{B}(\varepsilon) \le L_T''(\boldsymbol{\theta}_A)[L_T''(\boldsymbol{\theta}_A^*)]^{-1} \le \mathbf{I}_k + \mathbf{B}(\varepsilon)$$

where $\mathbf{B}(\varepsilon)$ is a positive semidefinite matrix whose largest eigenvalue tends to zero as $\varepsilon \to 0$.

4. There exists a point $\boldsymbol{\theta}_A^* \in \Theta_A$ such that for any open neighborhood $N(\boldsymbol{\theta}_A^*)$ of $\boldsymbol{\theta}_A^*$

$$\lim_{T\to\infty} P[\widetilde{\boldsymbol{\theta}}_A \in N(\boldsymbol{\theta}_A^*) \mid \widetilde{\mathbf{Y}}_T, A] = 1.$$

For all $T > T^*$, let $\Sigma_T^{-1/2}$ be any $k_A \times k_A$ matrix such that $(\Sigma_T^{-1/2})'\Sigma_T^{-1/2} = \Sigma_T^{-1}$, let $\widetilde{\boldsymbol{\theta}}_{AT}$ be the random vector corresponding to the pdf $p(\boldsymbol{\theta}_A \mid \widetilde{\mathbf{Y}}_T, A)$, and let $\widetilde{\mathbf{z}}_T = \Sigma_T^{-1/2}(\widetilde{\boldsymbol{\theta}}_{AT} - \boldsymbol{\theta}_A^*)$. Then $\widetilde{\mathbf{z}}_T \xrightarrow{d} N(\mathbf{0}, \mathbf{I}_{k_A})$.

Proof: See Chen (1985). ■

Condition 4 of this theorem is the conclusion of Theorem 3.4.2. Conditions 1, 2, and 3 are more primitive and are often easier to verify; see Exercises 3.4.1(c) and 3.4.2(c). Theorem 3.4.3 does not play as vital a role in Bayesian analysis as do central limit theorems in non-Bayesian approaches. In non-Bayesian approaches central limit theorems provide the basis for approximate inference when, as is nearly always the case, the relevant exact sampling-theoretic distributions are unknown. In Bayesian analysis, the exact posterior distribution is known, in principle, in any complete model. Chapter 4 shows how posterior simulators can be used to reveal the posterior distribution. Theorem 3.4.3 provides the conditions under which this distribution will be approximately normal, and Theorem 3.4.2 provides an interpretation of the unobservables $\widetilde{\boldsymbol{\theta}}_A$ on which the support of the posterior distribution is concentrated.

Exercise 3.4.1 Asymptotic Analysis of the Exponential Observables Distribution In an investigator's model A, $\widetilde{y}_t$ is i.i.d. and $p(y_t \mid \theta, A) = \theta \exp(-\theta y) I_{(0,\infty)}(y)$. In the data-generating process D, $\widetilde{y}_t$ is i.i.d., $P(\widetilde{y}_t > 0 \mid D) = 1$, and $E(\widetilde{y}_t \mid D) = \mu < \infty$. The investigator undertakes Bayesian inference using an i.i.d sample $\widetilde{\mathbf{Y}}_T = \{\widetilde{y}_t\ (t = 1, \ldots, T)\}$ drawn from a population with density $p(\mathbf{y} \mid D)$.

(a) Show that $T^{-1} \log p(\widetilde{\mathbf{Y}}_T \mid \theta, A) \xrightarrow{\text{a.s.}} \ell(\theta; A)$, and $\theta^* = \arg\max_\theta \ell(\theta; A) = \mu^{-1}$.

(b) Provide further conditions sufficient for θ^* to be the pseudotrue value of θ.

(c) Show that given the conditions in (b), conditions 1, 2, and 3 of Theorem 3.4.3 are also satisfied.

Exercise 3.4.2 Asymptotic Analysis of Omitted Covariates An investigator's model A for observables is the normal linear regression model

$$\widetilde{y}_t = \boldsymbol{\beta}'\widetilde{\mathbf{x}}_t + \widetilde{\varepsilon}_t,\ \widetilde{\varepsilon}_t \overset{\text{i.i.d.}}{\sim} N(0, h^{-1}).$$

In the data-generating process D, $\widetilde{y}_t = \boldsymbol{\gamma}'\widetilde{\mathbf{x}}_t + \boldsymbol{\delta}'\widetilde{\mathbf{z}}_t + \eta_t$, $\eta_t \overset{\text{i.i.d.}}{\sim} N(0, j^{-1})$ and

$$T^{-1}\begin{bmatrix}\widetilde{\mathbf{X}}'\widetilde{\mathbf{X}} & \widetilde{\mathbf{X}}'\widetilde{\mathbf{Z}} \\ \widetilde{\mathbf{Z}}'\widetilde{\mathbf{X}} & \widetilde{\mathbf{Z}}'\widetilde{\mathbf{Z}}\end{bmatrix} \overset{\text{a.s.}}{\rightarrow} \begin{bmatrix}\Sigma_{XX} & \Sigma_{XZ} \\ \Sigma_{ZX} & \Sigma_{ZZ}\end{bmatrix} = \mathbf{Q}^*$$

where $\mathbf{Q}^*$ is positive definite.

(a) Show that

$$T^{-1}\log p(\widetilde{\mathbf{Y}}_T \mid \boldsymbol{\beta}, h, \widetilde{\mathbf{X}}, A) \overset{\text{a.s.}}{\rightarrow} \ell(\boldsymbol{\beta}, h; \boldsymbol{\gamma}, \boldsymbol{\delta}, j, \mathbf{Q}^*, A).$$

Find $\arg\max_{\boldsymbol{\beta},h} \ell(\boldsymbol{\beta}, h; \boldsymbol{\gamma}, \boldsymbol{\delta}, j, \mathbf{Q}^*, A)$.

(b) Provide further conditions sufficient for $(\boldsymbol{\beta}^*, h^*)$ to be the pseudotrue value of $(\boldsymbol{\beta}, h)$.

(c) Show that given the conditions in (b), conditions 1, 2, and 3 of Theorem 3.4.3 are also satisfied.

3.5 THE LIKELIHOOD PRINCIPLE

Suppose $\boldsymbol{\omega} = g(\boldsymbol{\theta}_A)$. Then the posterior moment $E[g(\boldsymbol{\theta}_A) \mid \mathbf{y}^o, A]$ can be expressed

$$E[\boldsymbol{\omega} \mid \mathbf{y}^o, A] = E[g(\boldsymbol{\theta}_A) \mid \mathbf{y}^o, A] = \int_{\Theta_A} g(\boldsymbol{\theta}_A)p(\boldsymbol{\theta}_A \mid \mathbf{y}^o, A)\, d\nu(\boldsymbol{\theta}_A)$$

$$= \frac{\int_{\Theta_A} g(\boldsymbol{\theta}_A)L(\boldsymbol{\theta}_A; \mathbf{y}^o, A)p(\boldsymbol{\theta}_A \mid A)\, d\nu(\boldsymbol{\theta}_A)}{\int_{\Theta_A} L(\boldsymbol{\theta}_A; \mathbf{y}^o, A)p(\boldsymbol{\theta}_A \mid A)\, d\nu(\boldsymbol{\theta}_A)}.$$

Consequently, in forming posterior moments, we never have recourse to the data beyond $L(\boldsymbol{\theta}_A; \mathbf{y}^o, A)$. All information in the data about $g(\boldsymbol{\theta}_A)$ is conveyed through the likelihood function. This result is a consequence of posterior conditioning. It can also be obtained from a different set of first principles, developed by Barnard (1949) and Fisher (1956) and fully exposited by Birnbaum (1962). Berger and Wolpert (1988) provide a thorough exposition of the topic and Poirier (1988) provides an introduction written specifically for economists.

The likelihood principle, formally defined below, states that if two data-generating mechanisms lead to the same likelihood function, then they convey exactly the same evidence about the unknown parameters. This is not a self-evident assertion, especially in the context of non-Bayesian statistics. The following example, which is generalized at the end of this section, illustrates this point.

Example 3.5.1 Stopping Rules from Bernoulli Trials. Consider two coin-flipping experiments involving independent flips of a coin for which $P(\text{heads}) = \theta$.

Let $y_t = 1$ if a head occurs in trial t and $y_t = 0$ if a tail occurs. In experiment 1 the coin is flipped T times. The number of heads, h_T, is a sufficient statistic because

$$p(y_1, \ldots, y_T \mid \theta) = \prod_{t=1}^{T} \theta^{y_t}(1-\theta)^{(1-y_t)} = \theta^{h_T}(1-\theta)^{(T-h_T)}.$$

(The support of this distribution is all possible $\{y_t\}_{t=1}^{T}$.) The distribution of h_T itself is binomial:

$$p(h_T \mid \theta) = \binom{T}{h_T}\theta^{h_T}(1-\theta)^{(T-h_T)}.$$

In experiment 2 the coin is flipped until m heads have been observed. The total number of flips, T_m, is a sufficient statistic because

$$p(y_1, \ldots, y_{T_m} \mid \theta) = \prod_{t=1}^{T_m} \theta^{y_t}(1-\theta)^{(1-y_t)} = \theta^{m}(1-\theta)^{(T_m-m)}.$$

(The support of this distribution is all possible $\{y_t\}_{t=1}^{s}$ for which $\sum_{t=1}^{s} y_t = m$ and $y_s = 1$.) The distribution of T_m itself is negative binomial:

$$p(T_m \mid \theta) = \binom{T_m - 1}{m - 1}\theta^{m}(1-\theta)^{(T_m-m)}.$$

Suppose that the number of heads in the two experiments turns out to be the same ($h_T = m$) and the number of tails also turns out to be the same ($T - h_T = T_m - m$). The likelihood principle asserts that in this case the conclusions about θ must be the same in the two experiments.

The formal development here will assert two basic principles, the weak sufficiency principle and the weak conditionality principle, and then show that the two together are equivalent to the likelihood principle. We may accept or reject either of the two basic principles, but if we accept them both, then we also accept the likelihood principle. If we reject the likelihood principle, then we also reject the weak sufficiency principle, the weak conditionality principle, or both. The formal development begins with two definitions.

Definition 3.5.1 An *experiment* is characterized by the triplet

$$E = \{\mathbf{y}, \boldsymbol{\theta}_A, p(\mathbf{y} \mid \boldsymbol{\theta}_A, A)\}.$$

Definition 3.5.2 The *evidence* about $\boldsymbol{\theta}_A$ arising from E and the realization $\mathbf{y} = \mathbf{y}^o$, denoted $Ev(E, \mathbf{y}^o)$, is any inference, conclusion, or report concerning $\boldsymbol{\theta}_A$ based on E and $\mathbf{y} = \mathbf{y}^o$.

Examples of evidence include point estimates, hypothesis tests, posterior distributions, and general discussions at the ends of scientific papers using data.

Definition 3.5.3 Weak Sufficiency Principle Suppose that $\mathbf{s}(\mathbf{y}; A)$ is a sufficient statistic in the experiment $E = \{\mathbf{y}, \boldsymbol{\theta}_A, p(\mathbf{y} \mid \boldsymbol{\theta}_A, A)\}$. Let two runs of the experiment result in realizations $\mathbf{y}_1^o$ and $\mathbf{y}_2^o$, respectively, and suppose $\mathbf{s}(\mathbf{y}_1^o; A) = \mathbf{s}(\mathbf{y}_2^o; A)$. Then $Ev(E, \mathbf{y}_1^o) = Ev(E, \mathbf{y}_2^o)$.

The reasonableness of the weak sufficiency principle is inherent in the ex post formulation of sufficiency (2.29), which, we have seen, is logically equivalent to the ex ante definition (2.28). The development of the weak conditionality principle is based on the concept of a mixed experiment.

Definition 3.5.4 Given two experiments

$$E_j = \{\mathbf{y}_j, \boldsymbol{\theta}_A, p(\mathbf{y}_j \mid \boldsymbol{\theta}_A, A_j)\}(j = 1, 2)$$

involving the same parameter vector $\boldsymbol{\theta}_A$, a *mixed experiment*, based on E_1 and E_2, is $E_* = [\mathbf{y}_*, \boldsymbol{\theta}_A, p(j, \mathbf{y}_j \mid \boldsymbol{\theta}_A, A)]$ where the random vector $\mathbf{y}_* = (j, \mathbf{y}_j)$, $A = (A_1, A_2)$, and $p(j, \mathbf{y}_j \mid \boldsymbol{\theta}_A, A) = .5p(\mathbf{y}_j \mid \boldsymbol{\theta}_A, A_j)$.

Thus in a mixed experiment $p(j \mid \boldsymbol{\theta}_A, A) = .5$. Which experiment is actually conducted does not depend on $\boldsymbol{\theta}_A$.

Definition 3.5.5 Weak Conditionality Principle. Consider two experiments, $E_j = \{\mathbf{y}_j, \boldsymbol{\theta}_A, p(\mathbf{y}_j \mid \boldsymbol{\theta}_A, A_j)\}(j = 1, 2)$, as well as the mixed experiment $E_* = [\mathbf{y}_*, \boldsymbol{\theta}_A, p(j, \mathbf{y}_j \mid \boldsymbol{\theta}_A, A)]$. Then $Ev[E_*, (j, \mathbf{y}_j)] = Ev(E_j, \mathbf{y}_j)$.

Example 3.5.2 Assessing the Quality of a Scientific Paper The editor of a scientific journal wishes to learn about the quality, $\boldsymbol{\theta}_A$, of a scientific paper. He can send the paper to either referee A or referee B, each of whom has expertise in different areas relevant to the paper. The editor may decide to send the paper to A, to send it to B, or to flip a coin and send it to A if the coin is heads and to B if it is tails. The weak conditionality principle asserts that once the report from the known referee is in hand, the editor's findings about the quality of the scientific paper will be the same whether he chose the referee deliberately or flipped a coin.

Definition 3.5.6 Likelihood Principle. Consider the two experiments $E_j = [\mathbf{y}_j, \boldsymbol{\theta}_A, p(\mathbf{y}_j \mid \boldsymbol{\theta}_A, A_j)](j = 1, 2)$. Suppose that for the particular realizations $\mathbf{y}_1^o$ and $\mathbf{y}_2^o$, the respective likelihood functions satisfy

$$L(\boldsymbol{\theta}_A; \mathbf{y}_1^o, A_1) \propto L(\boldsymbol{\theta}_A; \mathbf{y}_2^o, A_2).$$

Then $Ev(E_1, \mathbf{y}_1^o) = Ev(E_2, \mathbf{y}_2^o)$.

Theorem 3.5.1 Equivalence of the Likelihood Principle, and the Weak Sufficiency and Weak Conditionality Principles The weak sufficiency principle and the weak conditionality principle are together equivalent to the likelihood principle.

Proof: Assume the likelihood principle. The antecedents of the weak sufficiency principle are that $s(\mathbf{y}; A)$ is a sufficient statistic in the experiment $E = [\mathbf{y}, \boldsymbol{\theta}_A, p(\mathbf{y} \mid \boldsymbol{\theta}_A, A)]$ and that $\mathbf{s}(\mathbf{y}_1^o; A) = \mathbf{s}(\mathbf{y}_2^o; A)$. By the factorization criterion

$$L(\boldsymbol{\theta}_A; \mathbf{y}_j^o, A) = p[\mathbf{s}(\mathbf{y}_j^o, A) \mid \boldsymbol{\theta}_A, A] r(\mathbf{y}_j^o; A) \quad (j = 1, 2).$$

Hence $L(\boldsymbol{\theta}_A; \mathbf{y}_1^o, A) \propto L(\boldsymbol{\theta}_A; \mathbf{y}_2^o, A)$, and by the likelihood principle $Ev(E, \mathbf{y}_1^o) = Ev(E, \mathbf{y}_2^o)$, thus establishing the weak sufficiency principle.

The mixed experiment $E_* = [(j, \mathbf{y}_j), \boldsymbol{\theta}_A, p(j, \mathbf{y}_j \mid \boldsymbol{\theta}_A, A)]$ has likelihood function

$$\begin{aligned} L(\boldsymbol{\theta}_A; j^o, \mathbf{y}_{j^o}^o, A) &\propto p(j^o, \mathbf{y}_{j^o}^o \mid \boldsymbol{\theta}_A, A) \\ &= .5 p(\mathbf{y}_{j^o}^o \mid \boldsymbol{\theta}_A, A_{j^o}) \propto L(\boldsymbol{\theta}_A; \mathbf{y}_{j^o}^o, A_{j^o}). \end{aligned}$$

Hence $Ev[E_*, (j, \mathbf{y}_j)] = Ev(E_j, \mathbf{y}_j)$, thus establishing the weak conditionality principle.

Taking up the converse, suppose that for the particular realizations $\mathbf{y}_1^o$ and $\mathbf{y}_2^o$, the likelihood functions from the two experiments satisfy $L(\boldsymbol{\theta}_A; \mathbf{y}_1^o, A_1) \propto L(\boldsymbol{\theta}_A; \mathbf{y}_2^o, A_2)$. The proof proceeds by creating identical sufficient statistics based on a mixed experiment employing the weak conditionality principle; the weak sufficiency principle then gives the result:

1. Define the mixed experiment E_* as in the weak conditionality principle, from which we know $Ev[E_*, (j, \mathbf{y}_j)] = Ev[E_j, \mathbf{y}_j]$. Let $\mathbf{y}_1^o$ and $\mathbf{y}_2^o$ be the two realizations in the antecedent of the likelihood principle.
2. For the mixed experiment E_* with random outcomes $(j, \mathbf{y}_j)$, define the statistic

$$Q(j, \mathbf{y}_j) = \begin{cases} (1, \mathbf{y}_1^o) & \text{if} \quad j = 2 \text{ and } \mathbf{y}_2 = \mathbf{y}_2^o \\ (j, \mathbf{y}_j) & \text{otherwise} \end{cases}.$$

   Thus the outcomes $(1, \mathbf{y}_1^o)$ and $(2, \mathbf{y}_2^o)$ result in the same values of Q. [The motivation in this construction is to deliberately blur the outcomes $(1, \mathbf{y}_1^o)$ and $(2, \mathbf{y}_2^o)$ and then show that we lose nothing.]
3. Note that Q is sufficient for $\boldsymbol{\theta}_A$ from the definition (2.28):

$$p[(j, \mathbf{y}_j) \mid Q \neq (1, \mathbf{y}_1^o)] = \begin{cases} 1 & \text{if } (j, \mathbf{y}_j) = Q \\ 0 & \text{otherwise} \end{cases};$$

$$p[(1, \mathbf{y}_1^o) \mid Q = (1, \mathbf{y}_1)] = \frac{.5p(\mathbf{y}_1^o \mid \boldsymbol{\theta}_A, A_1)}{.5\sum_{j=1}^{2} p(\mathbf{y}_j^o \mid \boldsymbol{\theta}_A, A_j)} = \frac{c}{c+1};$$

$$p[(2, \mathbf{y}_2^o) \mid Q = (1, \mathbf{y}_1^o)] = \frac{1}{c+1}.$$

(In the last two equations $c = p(\mathbf{y}_1^o \mid \boldsymbol{\theta}_A, A_1)/p(\mathbf{y}_2^o \mid \boldsymbol{\theta}_A, A_2)$. The ratio does not involve $\boldsymbol{\theta}_A$, from the antecedent of the likelihood principle.)

4. Since the sufficient statistic Q is the same regardless of whether the outcome is $\mathbf{y}_1^o$ or $\mathbf{y}_2^o$, the weak sufficiency principle implies $Ev(E_1, \mathbf{y}_1^o) = Ev(E_2, \mathbf{y}_2^0)$. ■

The likelihood principle extends the conclusion of Example 3.5.1, to sequential experiments and stopping rules generally.

Definition 3.5.7 A *sequential experiment* is an experiment in which the *stopping time* T is random.

Definition 3.5.8 A *stopping rule* in a sequential experiment is a sequence of probabilities $\boldsymbol{\tau} = \{\tau_t\}_{t=0}^{\infty}$ in which τ_0 is constant and

$$\tau_m = P(T = m \mid \mathbf{Y}_m, \boldsymbol{\theta}_A, A) = P(T = m \mid \mathbf{Y}_m, A) = \tau_m(\mathbf{Y}_m; A).$$

Note that the stopping probability may depend on the observables $\mathbf{Y}_m$, but not on the unobservables $\boldsymbol{\theta}_A$.

Corollary 3.5.1 Stopping Rule Corollary of the Likelihood Principle In a sequential experiment $Ev[E, (T, \mathbf{Y}_T)]$ depends on $(T, \mathbf{Y}_T)$ only through $L(\boldsymbol{\theta}_A; \mathbf{Y}_T, A)$. The likelihood principle implies that the stopping rule $\boldsymbol{\tau}$ is irrelevant.

Proof

$$p(T, \mathbf{Y}_T \mid \boldsymbol{\theta}_A, A) = (1 - \tau_0) \prod_{t=1}^{T-1} [1 - \tau_t(\mathbf{Y}_t; A)]\tau_T(\mathbf{Y}_T; A)p(\mathbf{Y}_T \mid \boldsymbol{\theta}_A, A).$$

Hence $L(\boldsymbol{\theta}_A; T, \mathbf{Y}_T, A) \propto p(\mathbf{Y}_T \mid \boldsymbol{\theta}_A, A)$. ■

Exercise 3.5.1 The Likelihood Principle, Conditioning, and Non-Bayesian Statistics (I) [From Berger and Wolpert (1988), Example 10, p 21.] Either of two experiments, E_1 or E_2, can be undertaken to learn about a parameter θ. In each experiment a single random outcome y will be observed. The distribution of y depends on θ in each experiment, as follows:

	Experiment A		
	$P(y=1 \mid \theta)$	$P(y=2 \mid \theta)$	$P(y=3 \mid \theta)$
$\theta = 0$	.900	.050	.050
$\theta = 1$	.090	.055	.855

	Experiment B		
	$P(y=1 \mid \theta)$	$P(y=2 \mid \theta)$	$P(y=3 \mid \theta)$
$\theta = 0$	.260	.730	.010
$\theta = 1$	.026	.803	.171

(a) If you were able to choose which experiment to carry out, which one would you choose? Provide as much formal justification for your answer as you can.

(b) Take the likelihood principle as given, and show that the outcome $y = 1$ conveys the same evidence about θ regardless of which experiment is performed. The same is true for $y = 2$, and the same is true for $y = 3$.

(c) Is there a logical conflict between your answers to (a) and (b)? Why or why not?

(d) Consider a non-Bayesian (classical) test that accepts $\theta = 0$ when $y = 1$ and decides $\theta = 1$ otherwise. Show that this is a most powerful test, with error probabilities (of type I and type II, respectively) .10 and .09 in experiment A and .74 and .026 in experiment B. (Recall what "most powerful" means in this context; any other test has either a higher probability of type I error, or a higher probability of type II error, or both.)

Exercise 3.5.2 The Likelihood Principle, Conditioning, and Non-Bayesian Statistics (II) [From Berger and Wolpert (1988), Example 1, p 5.] Suppose that the random variables $x_t (t = 1, \ldots, T)$ are independent, and $P(x_t = \theta - 1 \mid \theta) = P(x_t = \theta + 1 \mid \theta) = \frac{1}{2}$.

(a) Experiment 1 consists of collecting a sample of size $T = 2$. Show that a 75% confidence interval of smallest size for θ is

$$C(x_1, x_2) = \begin{cases} \text{the point } (x_1 + x_2)/2 & \text{if} \quad x_1 \neq x_2 \\ \text{the point } x_1 - 1 & \text{if} \quad x_1 = x_2 \end{cases}$$

[Thus, if used repeatedly, $C(x_1, x_2)$ would contain θ with probability .75.] The evidence in experiment 1 is that $C(x_1, x_2)$ constitutes a 75% confidence interval for θ.

(b) In the context of (a), suppose that you observe $x_1 = 2$ and $x_2 = 4$. Then the 75% confidence interval in (a) is the point 3. Is this consistent with common sense about the reliability of the conclusion that $\theta = 3$?

(c) In experiment 2 x_t is drawn repeatedly, and the experiment concludes with the first occurrence of $x_t \neq x_1$; thus the size, T, of the collected sample is random. The evidence in experiment 2 is that $\theta = (x_1 + x_T)/2$ is a 100% confidence interval for θ. Is the evidence from experiments 1 and 2 consistent with the likelihood principle? Provide a formal answer.

CHAPTER 4

Posterior Simulation

Bayesian inference requires that we be able to access the posterior distribution of the vector of interest $\boldsymbol{\omega}$ in one or more models. In all except simple illustrative cases this cannot be done analytically. This chapter describes algorithms for simulating a sequence $\{\boldsymbol{\omega}^{(m)}\}$ whose distribution is closely related to the distribution $\boldsymbol{\omega} \mid (\mathbf{y}^o, A)$. The sequence $\{\boldsymbol{\omega}^{(1)}, \ldots, \boldsymbol{\omega}^{(M)}\}$ can be used to approximate posterior moments of the form $E[h(\boldsymbol{\omega}) \mid \mathbf{y}^o, A]$ and Bayes actions of the form $\widehat{\mathbf{a}} = \arg\min_{\mathbf{a} \in A} E[L(\mathbf{a}, \boldsymbol{\omega}) \mid \mathbf{y}^o, A]$ arbitrarily well: the larger is M, the better is the approximation. Taken together, these algorithms are known generically as *posterior simulation methods*. The simplest possible relation between $\{\boldsymbol{\omega}^{(m)}\}$ and $p(\boldsymbol{\omega} \mid \mathbf{y}^o, A)$ is $\boldsymbol{\omega}^{(m)} \overset{\text{i.i.d.}}{\sim} p(\boldsymbol{\omega} \mid \mathbf{y}^o, A)$. In this case it is possible to learn a great deal about the posterior distribution of $\boldsymbol{\omega}$, as detailed in the next section. In most models it is not known how to construct such a sequence. Fortunately there are more sophisticated posterior simulators that typically can be used. One approach is to find a distribution that is similar to the posterior distribution, and from which i.i.d. drawings can be made. It is possible to correct for the difference in the simulation and posterior distributions, in such a way that posterior moments can be approximated arbitrarily well. Section 4.2 makes clear the sense in which the simulation and posterior distributions must be similar, and details two kinds of corrections that can be made. Section 4.4 takes up some variants on these methods that can greatly increase the amount of information about the posterior distribution in a given number of simulations.

As the dimension of the space in which simulations are carried out becomes large, it is often increasingly difficult to find a single distribution that is sufficiently similar to the posterior distribution that this approach is practical. A different class of simulation methods, known as Markov chain Monte Carlo (MCMC), constructs sequences that are neither independent nor identically distributed, but converge in distribution to the posterior distribution. These simulators are more sophisticated, and they make the posterior distribution accessible in a very large set of econometric and statistical models. Section 4.3 provides an informal introduction to these

Contemporary Bayesian Econometrics and Statistics, by John Geweke

methods, with a treatment in greater depth in Section 4.5. Section 4.6 takes up some combinations of these methods that widen the range of problems that can be attacked successfully by MCMC methods. Section 4.7 discusses the evaluation of the numerical accuracy of MCMC simulators.

While our primary interest is in simulating the posterior distribution of a function of interest $h(\boldsymbol{\omega})$, the methods developed in this chapter can be used for any distribution. For example, they can be used to explore the implications of a prior distribution. To reflect this generality in the notation, this chapter takes the canonical simulation problem to be that of learning about the distribution of ω, where $\boldsymbol{\omega} \sim p(\boldsymbol{\omega} \mid \boldsymbol{\theta}, I)$ and $\boldsymbol{\theta} \sim p(\boldsymbol{\theta} \mid I)$, where I denotes the distribution of interest. In this formulation $\boldsymbol{\theta} \in \Theta$, $\boldsymbol{\omega} \in \Omega$, $p(\boldsymbol{\theta} \mid I)$ is any density with respect to a measure $d\nu(\boldsymbol{\theta})$, and $p(\boldsymbol{\omega} \mid \boldsymbol{\theta}, I)$ is any conditional density with respect to a measure $d\nu(\boldsymbol{\omega})$.

4.1 DIRECT SAMPLING

Suppose that from the probability density $p(\boldsymbol{\theta}, \boldsymbol{\omega} \mid I)$ it is possible to simulate pairs of independent identically distributed (i.i.d.) drawings $\boldsymbol{\theta}^{(m)} \sim p(\boldsymbol{\theta} \mid I)$ and $\boldsymbol{\omega}^{(m)} \sim p(\boldsymbol{\omega} \mid \boldsymbol{\theta}^{(m)}, I)$. An example of such a density is the posterior density in Example 2.3.3, the normal linear regression model with conjugate prior distribution. In that example the corresponding posterior distribution is represented as

$$\overline{s}^2 h \mid (\mathbf{y}^o, \mathbf{X}, A) \sim \chi^2(\overline{\nu}), \qquad \boldsymbol{\beta} \mid (h, \mathbf{y}^o, \mathbf{X}, A) \sim N(\overline{\boldsymbol{\beta}}, h^{-1}\overline{\mathbf{H}}^{-1}).$$

The following result shows that it is possible to use the sequence $\{\boldsymbol{\theta}^{(m)}\}$ to approximate several aspects of the distribution of $\boldsymbol{\omega}$, including moments.

Theorem 4.1.1 Approximation of Distributions and Moments by Direct Sampling Suppose that the sequence $\{\boldsymbol{\theta}^{(m)}, \boldsymbol{\omega}^{(m)}\}$ is i.i.d., with $\boldsymbol{\theta}^{(m)} \in \Theta$, $\boldsymbol{\omega}^{(m)} \in \Omega$, $\boldsymbol{\theta}^{(m)} \sim p(\boldsymbol{\theta} \mid I)$, and $\boldsymbol{\omega}^{(m)} \mid \boldsymbol{\theta}^{(m)} \sim p(\boldsymbol{\omega} \mid \boldsymbol{\theta}^{(m)}, I)$. Let $h : \Omega \to \mathbb{R}^1$ and consider several additional conditions:

1. $E[h(\boldsymbol{\omega}) \mid I] = \overline{h}$.
2. $\text{var}[h(\boldsymbol{\omega}) \mid I] = \sigma^2$.
3. For given $p \in (0, 1)$, there is a unique h_p such that the statements

$$P[h(\boldsymbol{\omega}) \leq h_p \mid I] \geq p \text{ and } P[h(\boldsymbol{\omega}) \geq h_p \mid I] \geq 1 - p$$

are both true.
4. The pdf $p[h(\boldsymbol{\omega})|I]$ is continuous, and for the unique h_p corresponding to p, $p[h(\boldsymbol{\omega}) = h_p \mid I] > 0$.

Then

(a) $\boldsymbol{\omega}^{(m)} \overset{\text{i.i.d.}}{\sim} p(\boldsymbol{\omega} \mid I)$, regardless of whether any of conditions 1–4 are true.

(b) Given condition 1, $\overline{h}^{(M)} = M^{-1}\sum_{m=1}^{M} h(\boldsymbol{\omega}^{(m)}) \overset{\text{a.s.}}{\to} \overline{h}$.

(c) Given conditions 1 and 2, $M^{1/2}(\overline{h}^{(M)} - \overline{h}) \overset{d}{\to} N(0, \sigma^2)$ and

$$\widehat{\sigma}^{2(M)} = M^{-1}\sum_{m=1}^{M}[h(\boldsymbol{\omega}^{(m)}) - \overline{h}^{(M)}]^2 \overset{\text{a.s.}}{\to} \sigma^2.$$

Let $\widehat{h}_p^{(M)}$ be any real number such that

$$M^{-1}\sum_{m=1}^{M} I_{(-\infty,\widehat{h}_p^{(M)}]}[h(\boldsymbol{\omega}^{(m)})] \geq p \quad \text{and}$$

$$M^{-1}\sum_{m=1}^{M} I_{[\widehat{h}_p^{(M)},\infty)}[h(\boldsymbol{\omega}^{(m)})] \geq 1 - p.$$

Then

(d) Given condition 3, $\widehat{h}_p^{(M)} \overset{\text{a.s.}}{\to} h_p$.

(e) Given conditions 3 and 4

$$M^{1/2}(\widehat{h}_p^{(M)} - h_p) \overset{d}{\to} N\{0, p(1-p)/p[h(\boldsymbol{\omega}) = h_p \mid I]^2\}.$$

Proof: Conclusion (a) is just a consequence of the definitions of conditional, joint, and marginal probability. Conclusion (b) follows from the strong law of large numbers [see Casella and Berger (2002), Theorem 5.5.9] and (c), from the Lindeberg–Lévy central limit theorem (Casella and Berger 2002, Theorem 5.5.15). Conclusion (d) is immediate from 6f.2(i) in Rao (1965) and (e), from 6f.2(ii) in Rao (1965). ■

In the approximation of any quantity, some assessment of the error is essential. For the approximation of moments by direct sampling, the relevant assessment is given by conclusion (c) in Theorem 4.1.1. Given conditions 1 and 2, the approximation $\overline{h}^{(M)} \dot{\sim} N(\overline{h}, \widehat{\sigma}^{2(M)}/M)$ is valid for large M. In this case $(\widehat{\sigma}^{2(M)}/M)^{1/2}$ is known as the *numerical standard error* (NSE) of $\overline{h}^{(M)}$.

Example 4.1.1 Direct Sampling in the Normal Linear Regression Model with Conjugate Prior Distribution In the context of Example 2.3.3 suppose that the vector of interest ω is the unobserved outcome y corresponding to a new experiment in which $\mathbf{x} = \mathbf{x}^*$; thus $\omega \mid (\mathbf{x}^*, \boldsymbol{\beta}, h, A) \sim N(\boldsymbol{\beta}'\mathbf{x}^*, h^{-1})$. If

$$\overline{s}^2 h^{(m)} \overset{\text{i.i.d.}}{\sim} \chi^2(\overline{\nu}),$$

$$\boldsymbol{\beta}^{(m)} \mid h^{(m)} \sim N[\overline{\boldsymbol{\beta}}, (h^{(m)}\overline{\mathbf{H}})^{-1}],$$

$$\omega^{(m)} \mid (\boldsymbol{\beta}^{(m)}, h^{(m)}) \sim N(\boldsymbol{\beta}^{(m)\prime}\mathbf{x}^*, h^{(m)-1}),$$

then $\omega^{(m)} \overset{\text{i.i.d.}}{\sim} p(\omega \mid \mathbf{x}^*, \mathbf{y}^o, \mathbf{X}, A)$, where $\overline{\boldsymbol{\beta}}$ and $\overline{\mathbf{H}}$ are as defined in (2.52) and $\overline{s}^2$ and $\overline{\nu}$ are as defined in (2.56).

From the sample $\{\omega^{(1)}, \ldots, \omega^{(M)}\}$ we may approximate

$$p = P(\omega \leq a \mid \mathbf{x}^*, \mathbf{y}^o, \mathbf{X}, A) = E[I_{(-\infty,a]}(\omega) \mid \mathbf{x}^*, \mathbf{y}^o, \mathbf{X}, A] \tag{4.1}$$

by $p^{(M)} = M^{-1} \sum_{m=1}^{M} I_{(-\infty,a]}(\omega^{(m)})$. The approximate variance of this numerical approximation to p is $p(1-p)/M$, and this variance may in turn be approximated by $p^{(M)}(1-p^{(M)})/M$. (For an improvement on this approximation, see Exercise 4.4.1.)

Suppose we were to estimate ω. If the loss function is quadratic, then $\widehat{\omega} = E(\omega \mid \mathbf{x}^*, \mathbf{y}^o, \mathbf{X}, A)$ can be approximated by $M^{-1} \sum_{m=1}^{M} \omega^{(m)}$. If the loss function is $L(\widetilde{\omega}, \omega) = (1-q)(\widetilde{\omega} - \omega) I_{(-\infty,\widetilde{\omega}]}(\omega) + q(\omega - \widetilde{\omega}) I_{(\widetilde{\omega},\infty)}(\omega)$, then $\widehat{\omega}$ is quantile q of the $\omega \mid (\mathbf{x}^*, \mathbf{y}^o, \mathbf{X}, A)$ distribution: $P(\omega \leq \widehat{\omega} \mid \mathbf{x}^*, \mathbf{y}^o, \mathbf{X}, A) = q$. The estimate $\widehat{\omega}$ can be approximated by the corresponding quantile of $\{\omega^{(m)}\}_{m=1}^{M}$: the values $\omega^{(m)}$ will in general all be different, and we choose that $\omega^{(m^*)}$ such that the fraction of $\omega^{(m)}$ less than or equal to $\omega^{(m^*)}$ is at least q and the fraction of $\omega^{(m)}$ greater than or equal to $\omega^{(m^*)}$ is at least $1-q$.

Theorem 4.1.1 can be used to solve Bayesian decision and estimation problems for specific loss functions. If the loss function is quadratic (Definition 2.4.3) or linear–exponential (Exercise 2.4.3), the Bayes action is a posterior moment and conclusions (b) and (c) are relevant. If the loss function is linear–linear (Definition 2.4.4), then conclusions (d) and (e) are relevant. In general, however, Bayesian decision problems need not reduce to posterior moments or quantiles. The following result applies when the loss function $L(\mathbf{a}, \boldsymbol{\omega})$ is a smooth function of the Bayes action $\mathbf{a}$.

Theorem 4.1.2 Approximation of Bayes Actions by Direct Sampling Suppose that the sequence $\{\boldsymbol{\theta}^{(m)}, \boldsymbol{\omega}^{(m)}\}$ is i.i.d., with $\boldsymbol{\theta}^{(m)} \sim p(\boldsymbol{\theta} \mid I)$ and $\boldsymbol{\omega}^{(m)} \mid (\boldsymbol{\theta}^{(m)}, I) \sim p(\boldsymbol{\omega} \mid \boldsymbol{\theta}^{(m)}, I)$. Let $L(\mathbf{a}, \boldsymbol{\omega}) \geq 0$ be a loss function defined on $A \times \Omega$, where A is an open subset of $\mathbb{R}^m$. Suppose that the risk function

$$R(\mathbf{a}) = \int_{\Omega} \int_{\Theta} L(\mathbf{a}, \boldsymbol{\omega}) p(\boldsymbol{\theta} \mid I) p(\boldsymbol{\omega} \mid \boldsymbol{\theta}, I)\, d\boldsymbol{\theta}\, d\boldsymbol{\omega}$$

has a strict global minimum at $\widehat{\mathbf{a}} \in A \subseteq \mathbb{R}^m$. Consider several additional conditions, for a suitably defined open neighborhood of $\widehat{\mathbf{a}}$, $N(\widehat{\mathbf{a}})$:

1. $M^{-1} \sum_{m=1}^{M} L(\mathbf{a}, \boldsymbol{\omega}^{(m)})$ converges uniformly to $R(\mathbf{a})$ for all $\mathbf{a} \in N(\widehat{\mathbf{a}})$, almost surely.
2. $\partial L(\mathbf{a}, \boldsymbol{\omega})/\partial \mathbf{a}$ exists and is a continuous function of $\mathbf{a}$, for all $\boldsymbol{\omega} \in \Omega$ and all $\mathbf{a} \in N(\widehat{\mathbf{a}})$.
3. $\partial^2 L(\mathbf{a}, \boldsymbol{\omega})/\partial \mathbf{a}\, \partial \mathbf{a}'$ exists and is a continuous function of $\mathbf{a}$, for all $\boldsymbol{\omega} \in \Omega$ and all $\mathbf{a} \in N(\widehat{\mathbf{a}})$.

4. $\mathbf{B} = \text{var}[\partial L(\mathbf{a}, \boldsymbol{\omega})/\partial \mathbf{a}|_{\mathbf{a}=\widehat{\mathbf{a}}} \mid I]$ exists and is finite.

5. $\mathbf{H} = E[\partial^2 L(\mathbf{a}, \boldsymbol{\omega})/\partial \mathbf{a}\, \partial \mathbf{a}'|_{\mathbf{a}=\widehat{\mathbf{a}}} \mid I]$ exists and is finite and nonsingular.

6. For any $\varepsilon > 0$, there exists M_ε such that

$$P[\sup_{\mathbf{a} \in N(\widehat{\mathbf{a}})} \left|\partial^3 L(\mathbf{a}, \boldsymbol{\omega})/\partial a_i \partial a_j \partial a_k\right| < M_\varepsilon \mid I] \geq 1 - \varepsilon$$

for all $i, j, k = 1, \ldots, m$.

Let A_M be the set of all roots of $M^{-1} \sum_{m=1}^{M} \partial L(\mathbf{a}, \boldsymbol{\omega}^{(m)})/\partial \mathbf{a} = \mathbf{0}$. Then

(a) Given conditions 1 and 2, for any $\varepsilon > 0$

$$\lim_{M \to \infty} P[\inf_{\mathbf{a} \in A_M} (\mathbf{a} - \widehat{\mathbf{a}})'(\mathbf{a} - \widehat{\mathbf{a}}) > \varepsilon \mid I] = 0. \quad (4.2)$$

(b) Given conditions 1–6, if $\widehat{\mathbf{a}}_M$ is any element of A_M such that $\widehat{\mathbf{a}}_M \xrightarrow{p} \widehat{\mathbf{a}}$, then

(i) $M^{1/2}(\widehat{\mathbf{a}}_M - \widehat{\mathbf{a}}) \xrightarrow{d} N(\mathbf{0}, \mathbf{H}^{-1}\mathbf{B}\mathbf{H}^{-1})$.

(ii) $M^{-1} \sum_{m=1}^{M} \partial L(\mathbf{a}, \boldsymbol{\omega}^{(m)})/\partial \mathbf{a}|_{\mathbf{a}=\widehat{\mathbf{a}}_M} \cdot \partial L(\mathbf{a}, \boldsymbol{\omega}^{(m)})/\partial \mathbf{a}'|_{\mathbf{a}=\widehat{\mathbf{a}}_M} \xrightarrow{p} \mathbf{B}$.

(iii) $M^{-1} \sum_{m=1}^{M} \partial^2 L(\mathbf{a}, \boldsymbol{\omega}^{(m)})/\partial \mathbf{a}\, \partial \mathbf{a}'|_{\mathbf{a}=\widehat{\mathbf{a}}_M} \xrightarrow{p} \mathbf{H}$.

Proof: Result (a) follows from Amemiya (1985), Theorem 4.1.2. Result (b) follows from Amemiya (1985), Theorems 4.1.3 and 4.1.4. ■

Theorem 4.1.2 is widely and readily applicable. The conditions can be verified directly in most cases. Beyond the posterior simulator the computations require coding of the loss function and its first two derivatives. Once this is accomplished, conventional and widely available optimization software can be applied directly to the function $R_M(\mathbf{a}) = M^{-1} \sum_{m=1}^{M} L(\mathbf{a}, \boldsymbol{\omega}^{(m)})$ of $\mathbf{a}$. We can show that $R_M(\widehat{\mathbf{a}}_M) \xrightarrow{p} R_M(\widehat{\mathbf{a}})$, and in fact [see Shao (1989)] $R_M(\widehat{\mathbf{a}}_M) - R_M(\widehat{\mathbf{a}}) = O(M^{-1} \log - \log M)$. The caveats generally applicable to numerical optimization are relevant here. Unless $L(\mathbf{a}, \boldsymbol{\omega})$ is known to be concave, a local minimum need not be a global one. The result is the posterior simulation approximation $\widehat{\mathbf{a}}_M$ of the Bayes action $\widehat{\mathbf{a}}$. The usual advice, to iterate to a minimum from alternative starting values, applies here, but in posterior simulation we also have the alternative of solving the problem with a larger sample size M. Result (b) provides numerical standard errors for assessing the accuracy of the approximation that are valid for large M.

Exercise 4.1.1 The Inverse CDF Method This method applies, in principle, to any random variable for which it is possible to compute the inverse of the cumulative distribution. It is generally limited to univariate distributions. Its efficiency, relative to other methods described in the next section, depends on the time required to compute the inverse cdf.

(a) Suppose that a random variable has cdf $F(x)$ and $F(x) = p$ if and only if $x = F^{-1}(p)$. Show that if u is uniformly distributed on $[0, 1]$, then $x = F^{-1}(u)$ has cdf $F(x)$.

(b) How would you use the inverse cdf method to simulate an exponentially distributed random variable with mean μ?

Exercise 4.1.2 Simulation from the Bivariate Normal Distribution Suppose that

$$f(x, y) = \begin{cases} (x^2 + y^2)^{1/2} \text{ if } x^2 + y^2 < 1; \\ 1 \text{ if } x^2 + y^2 \geq 1. \end{cases}$$

Furthermore, $(x, y)' \sim N(\boldsymbol{\mu}, \Sigma)$; $\boldsymbol{\mu}$ and Σ are known. In what follows

$$\boldsymbol{\mu} = \begin{pmatrix} \mu_1 \\ \mu_2 \end{pmatrix}, \quad \Sigma = \begin{bmatrix} \sigma_{11} & \sigma_{12} \\ \sigma_{12} & \sigma_{22} \end{bmatrix},$$

and

$$\varepsilon_j^{(m)} (j = 1, 2;\ m = 1, 2, \ldots)$$

are mutually independent, standard normal random variables.

(a) Show that $E[f(x, y)]$ and $\text{var}[f(x, y)]$ are finite.

(b) Consider the simulation

$$x^{(m)} = \mu_1 + \sigma_{11}^{1/2} \varepsilon_1^{(m)},$$
$$y^{(m)} = \mu_2 + (\sigma_{12}/\sigma_{11})(x^{(m)} - \mu_1) + (\sigma_{22} - \sigma_{12}^2/\sigma_{11})^{1/2} \varepsilon_2^{(m)}.$$

Show that

$$M^{-1} \sum_{m=1}^{M} f(x^{(m)}, y^{(m)}) \overset{\text{a.s.}}{\rightarrow} E[f(x, y)].$$

(For continuation, see Exercise 4.5.1.)

4.2 ACCEPTANCE AND IMPORTANCE SAMPLING

Suppose that we cannot derive a method for drawing i.i.d. random vectors $\boldsymbol{\theta}^{(m)}$ directly from the density $p(\boldsymbol{\theta} \mid I)$ but we can simulate i.i.d. drawings from a density $p(\boldsymbol{\theta} \mid S)$ that is similar to $p(\boldsymbol{\theta} \mid I)$. Then it may be possible to learn about many aspects of the vector of interest $\boldsymbol{\omega}$, but the sense in which $p(\boldsymbol{\theta} \mid S)$ is similar to $p(\boldsymbol{\theta} \mid I)$ is critical. We consider two approaches here: acceptance sampling and importance sampling. Section 4.3.2 discusses a closely related third approach, the independence Metropolis chain.

4.2.1 Acceptance Sampling

Acceptance sampling may be used to learn about any distribution of $\boldsymbol{\theta}$ with generic density $p(\boldsymbol{\theta} \mid I)$, given a *source density* $p(\boldsymbol{\theta} \mid S)$ from which i.i.d. random variables can be drawn. For the methods discussed in this section, it is necessary that we be able to evaluate an arbitrary kernel $k(\boldsymbol{\theta} \mid S)$ of $p(\boldsymbol{\theta} \mid S)$, and an arbitrary kernel $k(\boldsymbol{\theta} \mid I)$ of $p(\boldsymbol{\theta} \mid I)$.

Figure 4.1 provides the intuition of acceptance sampling. The heavy solid curve represents the density of interest $p(\theta \mid I)$ and the lighter solid curve, the source density $p(\theta \mid S)$. The ratio $p(\theta \mid I)/p(\theta \mid S)$ is bounded above by a constant a. In Figure 4.1, $p(1.16 \mid I)/p(1.16 \mid S) = a = 1.86$, and the dotted curve is $a \cdot p(\theta \mid S)$. The idea is to draw θ^* from $p(\theta \mid S)$, and accept the draw with probability $p(\theta^* \mid I)/[a \cdot p(\theta^* \mid S)]$. For example, if $\theta^* = 0$, then the draw is accepted with probability .269, whereas if $\theta^* = 1.16$, then the draw is accepted with probability 1. The accepted values in fact simulate i.i.d. drawings from the density of interest $p(\theta \mid I)$.

Theorem 4.2.1 Acceptance Sampling Let $k(\boldsymbol{\theta} \mid I) = c_I \cdot p(\boldsymbol{\theta} \mid I)$ be a kernel of the density of interest $p(\boldsymbol{\theta} \mid I)$, and let $k(\boldsymbol{\theta} \mid S) = c_S \cdot p(\boldsymbol{\theta} \mid S)$ be a kernel of the source density $p(\boldsymbol{\theta} \mid S)$. Let $r = \sup_{\boldsymbol{\theta} \in \Theta} k(\boldsymbol{\theta} \mid I)/k(\boldsymbol{\theta} \mid S) < \infty$. Suppose that $\boldsymbol{\theta}^{(m)}$ is drawn as follows:

1. Draw u uniform on [0, 1].
2. Draw $\boldsymbol{\theta}^* \sim p(\boldsymbol{\theta} \mid S)$.

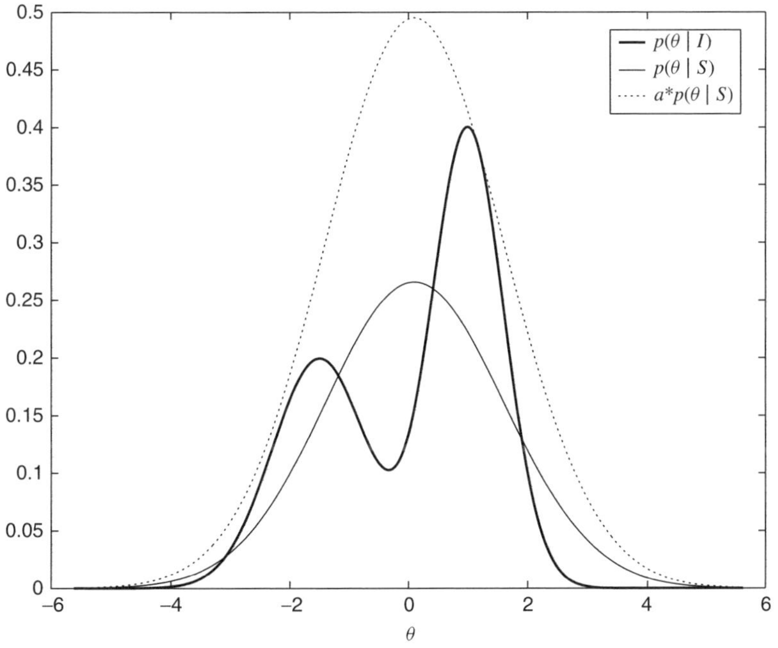

Figure 4.1. Acceptance sampling.

3. If $u > k(\boldsymbol{\theta}^* \mid I)/[r \cdot k(\boldsymbol{\theta}^* \mid S)]$, then return to step 1.
4. Set $\boldsymbol{\theta}^{(m)} = \boldsymbol{\theta}^*$.

If the draws in steps 1 and 2 are independent, then $\boldsymbol{\theta}^{(m)} \sim p(\boldsymbol{\theta} \mid I)$.

Proof: Let Θ^* denote the support of $p(\boldsymbol{\theta} \mid S)$; $r < \infty$ implies $\Theta \subseteq \Theta^*$. The unconditional probability of proceeding from step 3 to step 4 is

$$\int_{\Theta^*} \{k(\boldsymbol{\theta} \mid I)/[rk(\boldsymbol{\theta} \mid S)]\} p(\boldsymbol{\theta} \mid S)\, d\nu(\boldsymbol{\theta}) = c_I/rc_S. \tag{4.3}$$

Let A be any subset of Θ. The unconditional probability of proceeding from step 3 to step 4 with $\boldsymbol{\theta} \in A$ is

$$\int_A \{k(\boldsymbol{\theta} \mid I)/[rk(\boldsymbol{\theta} \mid S)]\} p(\boldsymbol{\theta} \mid S)\, d\nu(\boldsymbol{\theta}) = \int_A k(\boldsymbol{\theta} \mid I)\, d\nu(\boldsymbol{\theta})/rc_S. \tag{4.4}$$

The probability that $\boldsymbol{\theta} \in A$, conditional on proceeding from step 3 to step 4, is the ratio of (4.4) to (4.3), which is $\int_A k(\boldsymbol{\theta} \mid I)\, d\nu(\boldsymbol{\theta})/c_I = \int_A p(\boldsymbol{\theta} \mid I)\, d\nu(\boldsymbol{\theta}) = P(\boldsymbol{\theta} \in A \mid I)$. ■

The proof of Theorem 4.2.1 provides the key to the efficiency of this algorithm. Regardless of the choices of kernels, the unconditional probability in (4.3) is $c_I/rc_S = \inf_{\boldsymbol{\theta} \in \Theta} p(\boldsymbol{\theta} \mid S)/p(\boldsymbol{\theta} \mid I) = a^{-1}$. If we wish to generate M draws of $\boldsymbol{\theta}$ using acceptance sampling, the expected number of times we will have to draw u, draw $\boldsymbol{\theta}^*$, and compute $k(\boldsymbol{\theta}^* \mid I)/[r \cdot k(\boldsymbol{\theta}^* \mid S)]$ is $M \cdot \sup_{\boldsymbol{\theta} \in \Theta} p(\boldsymbol{\theta} \mid I)/p(\boldsymbol{\theta} \mid S) = M \cdot a$. The computational efficiency of the algorithm is driven by those $\boldsymbol{\theta}$ for which $p(\boldsymbol{\theta} \mid S)$ has the most severe undersampling relative to $p(\boldsymbol{\theta} \mid I)$. In most applications the time-consuming part of the algorithm is the evaluation of the kernels $k(\boldsymbol{\theta} \mid S)$ and $k(\boldsymbol{\theta} \mid I)$, especially the latter. [If $p(\boldsymbol{\theta} \mid I)$ is a posterior density, then evaluation of $k(\boldsymbol{\theta} \mid I)$ entails computing the likelihood function.] In such cases $a = \sup_{\boldsymbol{\theta} \in \Theta} k(\boldsymbol{\theta} \mid I)/k(\boldsymbol{\theta} \mid S)$ is indeed the relevant measure of inefficiency.

The retained values of $\boldsymbol{\theta}$ that constitute $\{\boldsymbol{\theta}^{(m)}\}$ are independent, and all are drawn from the distribution with density $p(\boldsymbol{\theta} \mid I)$. If we also draw $\boldsymbol{\omega}^{(m)} \sim p(\boldsymbol{\omega} \mid \boldsymbol{\theta}^{(m)}, I)$, then Theorem 4.1.1 applies directly to the sequence $\{\boldsymbol{\theta}^{(m)}, \boldsymbol{\omega}^{(m)}\}$.

Example 4.2.1 Restricted Normal Linear Regression Model Suppose that the coefficient vector $\boldsymbol{\beta}$ in the normal linear regression model is restricted to a set $S \subseteq \mathbb{R}^k$. For example, the signs of coefficients might be restricted, or functional form restrictions might limit the range of $\boldsymbol{\beta}$. To consider a simple case, suppose that precision h is fixed and

$$p(\boldsymbol{\beta} \mid A) \propto \exp[-(\boldsymbol{\beta} - \underline{\boldsymbol{\beta}})'\underline{\mathbf{H}}(\boldsymbol{\beta} - \underline{\boldsymbol{\beta}})/2] I_S(\boldsymbol{\beta}).$$

Then

$$p(\boldsymbol{\beta} \mid \mathbf{y}^o, \mathbf{X}, A) \propto \exp[-(\boldsymbol{\beta} - \overline{\boldsymbol{\beta}})'\,\overline{\mathbf{H}}(\boldsymbol{\beta} - \overline{\boldsymbol{\beta}})/2] I_S(\boldsymbol{\beta})$$

with $\overline{\mathbf{H}}$ as defined in (2.19) and $\overline{\boldsymbol{\beta}}$ in (2.20). If the source distribution for acceptance sampling is $N(\overline{\boldsymbol{\beta}}, \overline{\mathbf{H}})$, then a draw is accepted if $\boldsymbol{\beta} \in S$ and rejected if $\boldsymbol{\beta} \notin S$. The fraction accepted is the posterior probability that $\boldsymbol{\beta} \in S$ when the prior distribution for $\boldsymbol{\beta}$ is $N(\underline{\boldsymbol{\beta}}, \underline{\mathbf{H}}^{-1})$, and there are no restrictions. For more details and applications, see Geweke (1986), which develops this method in the context of the improper prior distribution of Exercise 3.2.1.

The lower the acceptance probability, the lower is the computational efficiency of the method. In fact, the acceptance probability can be so low that the algorithm effectively halts computations. This is most likely to be a problem when k is large and acceptance sampling is incorporated in the algorithms described in Section 4.3. In general it is never wise simply to assume that inequality constraints can be handled effectively by acceptance sampling. In specific cases superior alternatives are available. Once such alternative is developed in Section 5.3, building on the algorithm in the next example.

Example 4.2.2 Acceptance Sampling for the Truncated Univariate Normal Distribution Tailoring the source density to the density of interest can be critical to the effectiveness of acceptance sampling. A problem that arises frequently is that of sampling from a standard normal distribution truncated to the interval (a, b). If the source density is normal, then the acceptance probability is $\Phi(b) - \Phi(a)$, which can be quite small. The inverse cdf method [see Exercise 4.1.1(a)] involves finding the root of a nonlinear equation that itself requires evaluation of an integral. Moreover, for sufficiently large yet finite values of a, any numerical integration routine returns $\Phi^{-1}(a) = 1$, thus making it impossible to draw from the normal distribution truncated to (a, ∞). An efficient algorithm described in Geweke (1991) applies acceptance sampling with the source distribution as follows:

Characteristics of the Truncation			Source Distribution
$a < 0 < b < \infty$	$a \geq -t_1$ and $b \leq t_1$		Uniform(a, b)
	$a < -t_1$ or $b > t_1$		$N(0, 1)$
$0 \leq a < b < \infty$	$f(a)/f(b) \leq t_2$		Uniform(a, b)
	$f(a)/f(b) > t_2$	$a \leq t_3$	$\lvert N(0, 1)\rvert$
		$a > t_3$	$a + \exp(a^{-1})$
$b = \infty$	$a \leq t_4$		$N(0, 1)$
	$a > t_4$		$a + \exp(a^{-1})$

where $f(x) = \exp(-x^2/2)$, and the constants t_j are part of the design of the algorithm. The cases $-\infty < a < b < 0$ and $a = -\infty$ require only changes in sign. Geweke (1991) suggests $t_1 = .375$, $t_2 = 2.18$, $t_3 = .725$, and $t_4 = .45$ and reports that the method requires from one-sixth to one-half the time of the inverse cdf method, depending on the configuration of a and b.

4.2.2 Importance Sampling

Rather than accept only a fraction of the draws from the source density, it is possible to retain all of them, and consistently approximate $E[h(\boldsymbol{\omega}) \mid I]$ by appropriately weighting the draws. The probability density function of the source distribution is then called the *importance sampling density*, a term due to Hammersly and Handscomb (1964), who were among the first to propose the method. It appears to have been introduced to the econometrics literature by Kloek and van Dijk (1978). As with acceptance sampling, denote the source density by $p(\boldsymbol{\theta} \mid S)$ and an arbitrary kernel of the source density by $k(\boldsymbol{\theta} \mid S) = c_S \cdot p(\boldsymbol{\theta} \mid S)$ for any $c_S > 0$. Denote an arbitrary kernel of the density of interest by $k(\boldsymbol{\theta} \mid I) = c_I \cdot p(\boldsymbol{\theta} \mid I)$ for any $c_I > 0$. The following result is similar to Geweke (1989a), Theorem 2.

Theorem 4.2.2 Approximation of Moments by Importance Sampling Suppose that the sequence $\{\boldsymbol{\theta}^{(m)}, \boldsymbol{\omega}^{(m)}\}$ is independent and identically distributed, with $\boldsymbol{\theta}^{(m)} \sim p(\boldsymbol{\theta} \mid S)$ and $\boldsymbol{\omega}^{(m)} \sim p(\boldsymbol{\omega} \mid \boldsymbol{\theta}^{(m)}, I)$. Define the weighting function $w(\boldsymbol{\theta}) = k(\boldsymbol{\theta} \mid I)/k(\boldsymbol{\theta} \mid S)$, let $h : \Omega \to \mathbb{R}^1$, and consider several additional conditions:

1. $E[h(\boldsymbol{\omega}) \mid I] = \overline{h}$ exists.
2. $\operatorname{var}[h(\boldsymbol{\omega}) \mid I] = \sigma^2$ exists.
3. The support of $p(\boldsymbol{\theta} \mid S)$ includes Θ.
4. $w(\boldsymbol{\theta})$ is bounded above on Θ.

Then

(a) Given conditions 1 and 3

$$\overline{h}^{(M)} = \frac{\sum_{m=1}^{M} w(\boldsymbol{\theta}^{(m)}) h(\boldsymbol{\omega}^{(m)})}{\sum_{m=1}^{M} w(\boldsymbol{\theta}^{(m)})} \xrightarrow{\text{a.s.}} \overline{h}. \tag{4.5}$$

(b) Given conditions 1–4

$$M^{1/2}(\overline{h}^{(M)} - \overline{h}) \xrightarrow{d} N(0, \tau^2)$$

and

$$\widehat{\tau}^{2(M)} = \frac{M \sum_{m=1}^{M} [h(\boldsymbol{\omega}^{(m)}) - \overline{h}^{(M)}]^2 w(\boldsymbol{\theta}^{(m)})^2}{\left[\sum_{m=1}^{M} w(\boldsymbol{\theta}^{(m)})\right]^2} \xrightarrow{\text{a.s.}} \tau^2. \tag{4.6}$$

Proof: The sequence $\{\boldsymbol{\omega}^{(m)}\}$ is i.i.d., and from conditions 1 and 3,

$$E[w(\boldsymbol{\theta}) \mid S] = \int_{\Theta} \frac{k(\boldsymbol{\theta} \mid I)}{k(\boldsymbol{\theta} \mid S)} p(\boldsymbol{\theta} \mid S)\, d\nu(\boldsymbol{\theta}) = \frac{c_I}{c_S} = \overline{w}.$$

By the strong law of large numbers, we obtain

$$\overline{w}^{(M)} = M^{-1} \sum_{m=1}^{M} w(\boldsymbol{\theta}^{(m)}) \overset{\text{a.s.}}{\rightarrow} \overline{w}. \tag{4.7}$$

The sequence $\{w(\boldsymbol{\theta}^{(m)}),\ h(\boldsymbol{\omega}^{(m)})\}$ is also i.i.d., and

$$\begin{aligned} E[w(\boldsymbol{\theta})h(\boldsymbol{\omega}) \mid S] &= \int_{\Theta} w(\boldsymbol{\theta}) \left[\int_{\Omega} h(\boldsymbol{\omega}) p(\boldsymbol{\omega} \mid \boldsymbol{\theta}, I)\, d\nu(\boldsymbol{\omega}) \right] p(\boldsymbol{\theta} \mid S)\, d\nu(\boldsymbol{\theta}) \\ &= (c_I/c_S) \int_{\Theta} \int_{\Omega} h(\boldsymbol{\omega}) p(\boldsymbol{\omega} \mid \boldsymbol{\theta}, I) p(\boldsymbol{\theta} \mid I)\, d\nu(\boldsymbol{\omega})\, d\nu(\boldsymbol{\theta}) \\ &= (c_I/c_S) E[h(\boldsymbol{\omega}) \mid I] = \overline{w} \cdot \overline{h}. \end{aligned}$$

By the strong law of large numbers

$$\overline{wh}^{(M)} = M^{-1} \sum_{m=1}^{M} w(\boldsymbol{\theta}^{(m)}) h(\boldsymbol{\omega}^{(m)}) \overset{\text{a.s.}}{\rightarrow} \overline{w} \cdot \overline{h}. \tag{4.8}$$

Since the fraction in (4.5) is the ratio of the left side of (4.8) to the left side of (4.7), (a) is established.

Turning to (b), first note that

$$\begin{aligned} E[w(\boldsymbol{\theta})^2 h(\boldsymbol{\omega})^2 \mid S] &= \int_{\Theta} w(\boldsymbol{\theta})^2 \left[\int_{\Omega} h(\boldsymbol{\omega})^2 p(\boldsymbol{\omega} \mid \boldsymbol{\theta}, I)\, d\nu(\boldsymbol{\omega}) \right] p(\boldsymbol{\theta} \mid S)\, d\nu(\boldsymbol{\theta}) \\ &= \int_{\Theta} w(\boldsymbol{\theta}) \frac{k(\boldsymbol{\theta} \mid I)}{k(\boldsymbol{\theta} \mid S)} \left[\int_{\Omega} h(\boldsymbol{\omega})^2 p(\boldsymbol{\omega} \mid \boldsymbol{\theta}, I)\, d\nu(\boldsymbol{\omega}) \right] p(\boldsymbol{\theta} \mid S)\, d\nu(\boldsymbol{\theta}) \\ &= (c_I/c_S) \int_{\Theta} w(\boldsymbol{\theta}) \left[\int_{\Omega} h(\boldsymbol{\omega})^2 p(\boldsymbol{\omega} \mid \boldsymbol{\theta}, I)\, d\nu(\boldsymbol{\omega}) \right] p(\boldsymbol{\theta} \mid I)\, d\nu(\boldsymbol{\theta}). \end{aligned} \tag{4.9}$$

Condition 4 bounds (4.9) by

$$(c_I/c_S) E[h(\boldsymbol{\omega})^2 \mid I] \sup_{\boldsymbol{\theta} \in \Theta} w(\boldsymbol{\theta}) \tag{4.10}$$

and from condition 2 (4.10) is finite. Taking the specific case $h(\boldsymbol{\omega}) = 1$, it follows that $E[w(\boldsymbol{\theta})^2 \mid S] < \infty$ as well. Hence

$$\mathbf{V} = \text{var}\left[\begin{pmatrix} w(\boldsymbol{\theta}) \\ w(\boldsymbol{\theta}) h(\boldsymbol{\omega}) \end{pmatrix} \mid S \right]$$

is finite, and excepting the trivial case in which $h(\boldsymbol{\omega})$ is almost surely constant, $\mathbf{V}$ is nonsingular. By the Lindeberg–Lévy central limit theorem

$$M^{1/2}\left[\begin{pmatrix}\overline{w}^{(M)} \\ \overline{wh}^{(M)}\end{pmatrix} - \begin{pmatrix}\overline{w} \\ \overline{wh}\end{pmatrix}\right] \xrightarrow{d} N(\mathbf{0}, \mathbf{V}).$$

Utilizing the asymptotic (in M) expansion known as the *delta method* (Casella and Berger 2002, Section 5.5.4; Greene 2003, Section D.2.7), since

$$\frac{\overline{wh}^{(M)}}{\overline{w}^{(M)}} = \frac{\overline{wh}}{\overline{w}} + \frac{\overline{wh}^{(M)} - \overline{wh}}{\overline{w}} - \frac{\overline{wh} \cdot (\overline{w}^{(M)} - \overline{w})}{\overline{w}^2} + o_p(M^{-1/2}),$$

we have

$$M^{1/2}\left(\frac{\overline{wh}^{(M)}}{\overline{w}^{(M)}} - \overline{h}\right) \xrightarrow{d} N(0, \tau^2),$$

where

$$\tau^2 = \overline{w}^{-2}\{\mathrm{var}[w(\boldsymbol{\theta})h(\boldsymbol{\omega}) \mid S] - 2\overline{h}\mathrm{cov}[w(\boldsymbol{\theta})h(\boldsymbol{\omega}), w(\boldsymbol{\theta}) \mid S]$$
$$+ \overline{h}^2\, \mathrm{var}[w(\boldsymbol{\theta}) \mid S]\} = \overline{w}^{-2}\, \mathrm{var}\{[w(\boldsymbol{\theta})h(\boldsymbol{\omega}) - \overline{h}w(\boldsymbol{\theta})] \mid S\}.$$

This is consistently approximated by

$$\frac{M^{-1}\sum_{m=1}^{M}[w(\boldsymbol{\theta}^{(m)})h(\boldsymbol{\omega}^{(m)}) - \overline{h}^{(M)}w(\boldsymbol{\theta}^{(m)})]^2}{\left[M^{-1}\sum_{m=1}^{M} w(\boldsymbol{\theta}^{(m)})\right]^2},$$

which is equivalent to (4.6). ■

An apparent attraction of importance sampling, relative to acceptance sampling, is that it is formally easier to apply in approximating moments. It is necessary to establish condition 1 of Theorem 4.2.2 regardless of the method of evaluation, and whether condition 3 holds should be immediately apparent. By contrast, acceptance sampling requires that the upper bound of the weight function $w(\boldsymbol{\theta})$ be determined before a simulation consistent approximation to a moment can be constructed. The analog of this condition in Theorem 4.2.2 is condition 4, but note that this condition is needed only for the evaluation of numerical accuracy. Furthermore, if $w(\boldsymbol{\theta})$ has a known bound, then condition 4 holds, but for condition 4 to hold it is not necessary to know the bound. As a practical matter, however, if condition 4 does not hold, then convergence is typically quite slow, and the inability to evaluate numerical accuracy renders numerical approximations unreliable. Thus a practical advantage of importance sampling, as opposed to acceptance sampling, is that the former is practical if we establish the existence of an upper bound for $w(\boldsymbol{\theta})$, whereas in the latter we must evaluate this bound.

Example 4.2.3 Reweighting to a Different Prior Distribution Suppose that we have available an i.i.d. sample from a posterior density $p(\boldsymbol{\theta}_A \mid \mathbf{y}^o, A_1)$, corresponding to model A_1 with prior density $p(\boldsymbol{\theta}_A \mid A_1)$. Now suppose that we wish to investigate a second model A_2 with the same observables density but prior density $p(\boldsymbol{\theta}_A \mid A_2)$. If $p(\boldsymbol{\theta}_A \mid A_2)/p(\boldsymbol{\theta}_A \mid A_1)$ is bounded above on Θ_A, then $p(\boldsymbol{\theta}_A \mid \mathbf{y}^o, A_1)$ is an importance sampling density for $p(\boldsymbol{\theta}_A \mid \mathbf{y}^o, A_2)$ with conditions 3 and 4 of Theorem 4.2.2 satisfied. The weight function is $p(\boldsymbol{\theta}_A \mid A_2)/p(\boldsymbol{\theta}_A \mid A_1)$. This procedure is sometimes called "reweighting of the posterior simulation sample." See Section 8.4 for further discussion.

Example 4.2.4 A Hybrid Acceptance and Importance Sampling Algorithm Given density of interest $p(\boldsymbol{\theta} \mid I)$ and source density $p(\boldsymbol{\theta} \mid S)$, suppose that it is known that $p(\boldsymbol{\theta} \mid I)/p(\boldsymbol{\theta} \mid S)$ is bounded on Θ but the bound is unknown. Define the importance sampling density $p(\boldsymbol{\theta} \mid a, S)$ with kernel

$$k(\boldsymbol{\theta} \mid a, S) = \begin{cases} p(\boldsymbol{\theta} \mid I) \text{ if } p(\boldsymbol{\theta} \mid I)/p(\boldsymbol{\theta} \mid S) \leq a \\ p(\boldsymbol{\theta} \mid S) \text{ if } p(\boldsymbol{\theta} \mid I)/p(\boldsymbol{\theta} \mid S) > a \end{cases}.$$

Applying Theorem 4.2.1 to the density of interest kernel $k(\boldsymbol{\theta} \mid a, S)$ and source density $p(\boldsymbol{\theta} \mid S)$, we see that $\sup_{\boldsymbol{\theta} \in \Theta}[k(\boldsymbol{\theta} \mid a, S)/p(\boldsymbol{\theta} \mid S)] \leq \max(a, 1)$, and therefore i.i.d. draws from $p(\boldsymbol{\theta} \mid a, S)$ are possible. Importance sampling (Theorem 4.2.2) applies to the density of interest $p(\boldsymbol{\theta} \mid I)$ and importance sampling density $p(\boldsymbol{\theta} \mid a, S)$. Conditions 1–4 of Theorem 4.2.2 apply to $p(\boldsymbol{\theta} \mid a, S)$ to the extent that they apply to $p(\boldsymbol{\theta} \mid S)$. This strategy can be useful if generating the vector $\boldsymbol{\omega}$ is expensive relative to drawing $\boldsymbol{\theta}$ from $p(\boldsymbol{\theta} \mid S)$, and deciding to accept, reject, or weight the draws. We reject many draws $\boldsymbol{\theta}$ (without drawing $\boldsymbol{\omega}$), which would have had small weights with the importance sampling density $p(\boldsymbol{\theta} \mid S)$ (but nonetheless would have required drawing $\boldsymbol{\omega}$).

The ratio σ^2/τ^2 of the variance of a moment estimate based on hypothetical i.i.d. draws to the limiting variance of the estimate based on the importance sample is known as the *relative numerical efficiency* (RNE) of the simulator. If we can sample directly from the density $p(\boldsymbol{\theta} \mid I)$—equivalently, if $k(\boldsymbol{\theta} \mid S) \propto k(\boldsymbol{\theta} \mid I)$, or the weight function $w(\boldsymbol{\theta})$ is constant—then the RNE is 1.0. It is generally lower for importance sampling. The RNE is inversely proportional to the number of iterations of the posterior simulator required to achieve a given NSE. The NSE of the approximation is roughly proportional to $\sigma \cdot (M \cdot \text{RNE})^{-1/2}$.

Theorem 4.2.3 Approximation of Bayes Actions by Importance Sampling Suppose that the sequence $\{\boldsymbol{\theta}^{(m)}, \boldsymbol{\omega}^{(m)}\}$ is independent and identically distributed, with $\boldsymbol{\theta}^{(m)} \sim p(\boldsymbol{\theta} \mid S)$ and $\boldsymbol{\omega}^{(m)} \mid \boldsymbol{\theta}^{(m)} \sim p(\boldsymbol{\omega} \mid \boldsymbol{\theta}^{(m)}, I)$. Suppose that the support of $p(\boldsymbol{\theta} \mid S)$ includes Θ, and the weighting function $w(\boldsymbol{\theta}) = k(\boldsymbol{\theta} \mid I)/k(\boldsymbol{\theta} \mid S)$ is bounded above. Let $L(\mathbf{a}, \boldsymbol{\omega}) \geq 0$ be a loss function defined on $A \times \Omega$, where A is an open subset of $\mathbb{R}^m$. Suppose that the risk function

$$R(\mathbf{a}) = \int_\Omega \int_\Theta L(\mathbf{a}, \boldsymbol{\omega}) p(\boldsymbol{\theta} \mid I) p(\boldsymbol{\omega} \mid \boldsymbol{\theta}, I)\, d\boldsymbol{\theta}\, d\boldsymbol{\omega}$$

has a strict global minimum at $\widehat{\mathbf{a}} \in A \subseteq \mathbb{R}^m$. Consider several additional conditions, for a suitably defined open neighborhood of $\widehat{\mathbf{a}}$, $N(\widehat{\mathbf{a}})$:

1. $M^{-1} \sum_{m=1}^{M} L(\mathbf{a}, \boldsymbol{\omega}^{(m)})$ converges uniformly to $R(\mathbf{a})$ for all $\mathbf{a} \in N(\widehat{\mathbf{a}})$, almost surely.
2. $\partial L(\mathbf{a}, \boldsymbol{\omega})/\partial \mathbf{a}$ exists and is a continuous function of $\mathbf{a}$, for all $\boldsymbol{\omega} \in \Omega$ and all $\mathbf{a} \in N(\widehat{\mathbf{a}})$.
3. $\partial^2 L(\mathbf{a}, \boldsymbol{\omega})/\partial \mathbf{a}\, \partial \mathbf{a}'$ exists and is a continuous function of $\mathbf{a}$, for all $\boldsymbol{\omega} \in \Omega$ and all $\mathbf{a} \in N(\widehat{\mathbf{a}})$.
4. $\mathbf{B} = \text{var}[\partial L(\mathbf{a}, \boldsymbol{\omega})/\partial \mathbf{a}|_{\mathbf{a}=\widehat{\mathbf{a}}} w(\boldsymbol{\theta})^{1/2} \mid S]$ exists and is finite.
5. $\mathbf{H} = E[\partial^2 L(\mathbf{a}, \boldsymbol{\omega})/\partial \mathbf{a}\, \partial \mathbf{a}'|_{\mathbf{a}=\widehat{\mathbf{a}}} \mid S]$ exists and is finite and nonsingular.
6. For any $\varepsilon > 0$, there exists M_ε such that

$$P[\sup_{\mathbf{a} \in N(\widehat{\mathbf{a}})} \left|\partial^3 L(\mathbf{a}, \boldsymbol{\omega})/\partial a_i\, \partial a_j\, \partial a_k\right| < M_\varepsilon \mid S] \geq 1 - \varepsilon$$

   for all $i, j, k = 1, \ldots, m$.

Let A_M be the set of all roots of $M^{-1} \sum_{m=1}^{M} [\partial L(\mathbf{a}, \boldsymbol{\omega}^{(m)})/\partial \mathbf{a}] w(\boldsymbol{\theta}^{(m)}) = 0$. Then

(a) Given conditions 1 and 2, for any $\varepsilon > 0$, we obtain

$$\lim_{M \to \infty} P[\inf_{\mathbf{a} \in A_M} (\mathbf{a} - \widehat{\mathbf{a}})'(\mathbf{a} - \widehat{\mathbf{a}}) > \varepsilon \mid S] = 0.$$

(b) Given conditions 1–6, if $\widehat{\mathbf{a}}_M$ is any element of A_M such that $\widehat{\mathbf{a}}_M \xrightarrow{p} \widehat{\mathbf{a}}$, then

(i) $M^{1/2}(\widehat{\mathbf{a}}_M - \widehat{\mathbf{a}}) \xrightarrow{d} N(\mathbf{0}, \mathbf{H}^{-1}\mathbf{B}\mathbf{H}^{-1})$.

(ii) $M^{-1} \sum_{m=1}^{M} w(\boldsymbol{\theta}^{(m)})^2 \partial L(\mathbf{a}, \boldsymbol{\omega}^{(m)})/\partial \mathbf{a}|_{\mathbf{a}=\widehat{\mathbf{a}}_M} \cdot \partial L(\mathbf{a}, \boldsymbol{\omega}^{(m)})/\partial \mathbf{a}'|_{\mathbf{a}=\widehat{\mathbf{a}}_M} \xrightarrow{p} \mathbf{B}$.

(iii) $M^{-1} \sum_{m=1}^{M} w(\boldsymbol{\theta}^{(m)}) \partial^2 L(\mathbf{a}, \boldsymbol{\omega}^{(m)})/\partial \mathbf{a}\, \partial \mathbf{a}'|_{\mathbf{a}=\widehat{\mathbf{a}}_M} \xrightarrow{p} \mathbf{H}$.

Proof: Consider the auxiliary problem in which the pdf of $\boldsymbol{\theta}$ and $\boldsymbol{\omega}$ is $p(\boldsymbol{\theta} \mid S) p(\boldsymbol{\omega} \mid \boldsymbol{\theta}, I)$ and the loss function is $L(\mathbf{a}, \boldsymbol{\omega}) w(\boldsymbol{\theta})$. Because the support of $p(\boldsymbol{\theta} \mid S)$ includes Θ, the unique Bayes action in the auxiliary problem is also $\widehat{\mathbf{a}}$. Then apply Theorem 4.1.2 directly to the auxiliary problem. ■

Note that if $w(\boldsymbol{\theta})$ is bounded above on Θ, which is condition 4 of Theorem 4.2.2, then $\text{var}[\partial L(\mathbf{a}, \boldsymbol{\omega})/\partial \mathbf{a}|_{\mathbf{a}=\widehat{\mathbf{a}}} w(\boldsymbol{\theta})^{1/2} \mid S]$ will exist and be finite if the same is true of $\text{var}[\partial L(\mathbf{a}, \boldsymbol{\omega})/\partial \mathbf{a}|_{\mathbf{a}=\widehat{\mathbf{a}}} \mid I]$. The same optimization algorithms may be applied to compute $\widehat{\mathbf{a}}_M$ here as in the case of direct sampling, except that they are used with $R_M^*(\mathbf{a}) = M^{-1}\sum_{m=1}^{M} L(\mathbf{a}, \boldsymbol{\omega}^{(m)})w(\boldsymbol{\theta}^{(m)})$ rather than with $M^{-1}\sum_{m=1}^{M} L(\mathbf{a}, \boldsymbol{\omega}^{(m)})$. Note, however, that $R_M^*(\widehat{\mathbf{a}}_M)/M^{-1}\sum_{m=1}^{M} w(\boldsymbol{\theta}^{(m)}) \xrightarrow{p} R(\widehat{\mathbf{a}})$. The results of Theorem 4.2.3 may be found in Shao (1989). Importance sampling can be especially attractive when the distribution of $\boldsymbol{\omega}$ depends on the action $\mathbf{a}$, a case we do not consider here; see Geyer (1996).

Exercise 4.2.1 Sampling from the Tail of the Normal Distribution Suppose that the density of interest is the univariate standard normal, truncated to (a, ∞), where $a > 0$. The source density is a translated exponential density with pdf $p(x \mid S) = \theta^{-1}\exp[-\theta(x-a)]I_{(a,\infty)}(x)$. Show that the optimal choice of θ is $\theta = [a + (4+a^2)^{1/2}]/2$. Note that as $a \to \infty$, $\theta/a \to 1$. [Geweke (1991) reports that in the context of the algorithm described in Example 4.2.2 the gain in efficiency from using the optimal value of θ, relative to the simpler choice $\theta = a$, is not worth the time to compute the optimal value.]

Exercise 4.2.2 Tuning an Acceptance Algorithm The values of the constants t_j in Example 4.2.2 are good but not necessarily optimal. The best values are affected by the software and hardware used to implement the algorithm. Using software and hardware available to you, code the algorithm and see if detectable improvements are possible.

Exercise 4.2.3 Efficiency of Importance Sampling Suppose that $x \sim N(0, 1)$ and consider the two, alternative, source distributions for importance sampling:

$$x \sim N\left(0, \tfrac{1}{4}\right) \tag{4.11}$$

$$x \sim N(0, 2). \tag{4.12}$$

Suppose that you were to use importance sampling to approximate $E(x)$ for the distribution of interest, $x \sim N(0, 1)$.

(a) For one of the importance sampling distributions, (4.11) or (4.12), a central limit theorem can be used to assess the accuracy of the numerical approximation. Indicate which one, and calculate the relevant variance of the approximation.

(b) For the importance sampling distribution you identified in (a), find the relative numerical efficiency of the approximation.

4.3 MARKOV CHAIN MONTE CARLO

This section discusses a generalization of direct sampling known as Markov chain Monte Carlo (MCMC). The idea is to construct a Markov chain $\{\boldsymbol{\theta}^{(m)}\}$ with state space $\widetilde{\Theta} \supseteq \Theta$ and unique invariant probability density $p(\theta \mid I)$. Following an initial transient or *burn-in* phase, the distribution of $\boldsymbol{\theta}^{(m)}$ is approximately that of the density $p(\boldsymbol{\theta} \mid I)$. The exact sense in which this approximation holds is important, and is taken up in Section 4.5. We continue to assume that $\boldsymbol{\omega}$ can be simulated directly from $p(\boldsymbol{\omega} \mid \boldsymbol{\theta}, I)$, so that given $\{\boldsymbol{\theta}^{(m)}\}$ the corresponding $\boldsymbol{\omega}^{(m)} \sim p(\boldsymbol{\omega} \mid \boldsymbol{\theta}^{(m)}, I)$ can be drawn. We return to the use of $\{\boldsymbol{\omega}^{(m)}\}$ to approximate $E[h(\boldsymbol{\omega}) \mid I]$ in Section 4.5 as well. This section provides an introduction and heuristic motivation of MCMC.

Markov chain methods have a history in mathematical physics dating back to the algorithm of Metropolis et al. (1953). This method, which is described in Hammersly and Handscomb (1964), Section 9.3, and Ripley (1987), Section 4.7, was generalized by Hastings (1970), who focused on statistical problems, and was further explored by Peskun (1973). A version particularly suited to image reconstruction and problems in spatial statistics was introduced by Geman and Geman (1984). This was subsequently shown to have great potential for Bayesian computation by Gelfand and Smith (1990). Their work, combined with data augmentation methods (Tanner and Wong 1987), has proved very successful in the treatment of latent variables in econometrics. Since 1990 application of Markov chain Monte Carlo methods has grown rapidly; new refinements, extensions, and applications appear almost continuously.

This section concentrates on a heuristic development of two widely used MCMC algorithms: the Gibbs sampler and the Metropolis–Hastings algorithm. The general theory of convergence is discussed in Section 4.5. Section 4.6 details some specific variants and combinations of these methods used extensively in the balance of this volume. Section 4.7 turns to the assessment of numerical accuracy. While our main interest is in applying these methods to the posterior density $p(\boldsymbol{\theta}_A \mid \mathbf{y}^o, A)$ they can in principle be used with any density, and so we continue with the generic case of $p(\boldsymbol{\theta} \mid I)$.

4.3.1 The Gibbs Sampler

The Gibbs sampler begins with a partition, or *blocking*, of $\boldsymbol{\theta}$, $\boldsymbol{\theta}' = (\boldsymbol{\theta}'_{(1)}, \ldots, \boldsymbol{\theta}'_{(B)})$. Corresponding to any subvector $\boldsymbol{\theta}_{(b)}$, let $\boldsymbol{\theta}'_{<(b)} = (\boldsymbol{\theta}'_{(1)}, \ldots, \boldsymbol{\theta}'_{(b-1)})$ $(b = 2, \ldots, B)$ and $\boldsymbol{\theta}_{<(1)} = \{\varnothing\}$. Similarly $\boldsymbol{\theta}'_{>(b)} = (\boldsymbol{\theta}'_{(b+1)}, \ldots, \boldsymbol{\theta}'_{(B)})$ $(b = 1, \ldots, B-1)$ and $\boldsymbol{\theta}_{>(B)} = \{\varnothing\}$. Let $\boldsymbol{\theta}'_{-(b)} = (\boldsymbol{\theta}'_{<(b)}, \boldsymbol{\theta}'_{>(b)})$. In application, we generally try to choose the blocking so that it is possible to draw directly from each of the conditional densities $p(\boldsymbol{\theta}_{(b)} \mid \boldsymbol{\theta}_{-(b)}, I)$. In this section we shall assume that it is possible. This assumption will be weakened subsequently in Section 4.6.

Suppose that there existed a single drawing $\boldsymbol{\theta}^{(0)}$, $\boldsymbol{\theta}^{(0)\prime} = (\boldsymbol{\theta}^{(0)\prime}_{(1)}, \ldots, \boldsymbol{\theta}^{(0)\prime}_{(B)})$, from the distribution with pdf $p(\boldsymbol{\theta} \mid I)$. Successively make the drawings

$$\boldsymbol{\theta}^{(1)}_{(b)} \sim p\left(\boldsymbol{\theta}_{(b)} \mid \boldsymbol{\theta}^{(1)}_{<(b)}, \boldsymbol{\theta}^{(0)}_{>(b)}, I\right) \quad (b = 1, \ldots, B). \tag{4.13}$$

This defines a transition process from $\boldsymbol{\theta}^{(0)}$ to $\boldsymbol{\theta}^{(1)}$, $\boldsymbol{\theta}^{(1)\prime} = \left(\boldsymbol{\theta}^{(1)\prime}_{(1)}, \ldots, \boldsymbol{\theta}^{(1)\prime}_{(B)}\right)$. The Gibbs sampler is defined by the choice of blocking, and by the forms of the conditional densities induced by $p(\boldsymbol{\theta} \mid I)$ and the blocking. Since $\boldsymbol{\theta}^{(0)} \sim p(\boldsymbol{\theta} \mid I)$, $\left(\boldsymbol{\theta}^{(1)}_{<(b)}, \boldsymbol{\theta}^{(1)}_{(b)}, \boldsymbol{\theta}^{(0)}_{>(b)}\right) \sim p\left(\boldsymbol{\theta} \mid I\right)$ at the bth step in (4.13). In particular, $\boldsymbol{\theta}^{(1)} \sim p(\boldsymbol{\theta} \mid I)$.

In general, block b of iterate m of the Gibbs sampler is drawn as $\boldsymbol{\theta}^{(m)}_{(b)} \sim p\big(\boldsymbol{\theta}_{(b)} \mid \boldsymbol{\theta}^{(m)}_{<(b)}, \boldsymbol{\theta}^{(m-1)}_{>(b)}, I\big)$ for $b = 1, \ldots, B$ and $m = 1, 2, \ldots$. This produces a sequence

$\{\boldsymbol{\theta}^{(m)}\}$, which is a realization of a Markov chain. The transition density for this chain is

$$p(\boldsymbol{\theta}^{(m)} \mid \boldsymbol{\theta}^{(m-1)}, G) = \prod_{b=1}^{B} p\big(\boldsymbol{\theta}_{(b)}^{(m)} \mid \boldsymbol{\theta}_{<(b)}^{(m)}, \boldsymbol{\theta}_{>(b)}^{(m-1)}, I\big). \tag{4.14}$$

Any single iterate $\boldsymbol{\theta}^{(m)}$ retains the property that it is drawn from the density $p(\boldsymbol{\theta} \mid I)$.

In practice, the Gibbs sampler should use as many blocks as required in order to make the drawings in (4.13) efficiently. On the other hand, it should use no more blocks than necessary because additional blocks usually reduce efficiency; see Exercise 4.5.1 for a simple motivating example. For many problems in econometrics and statistics the blocking is natural and the conditional distributions are familiar.

Example 4.3.1 A Gibbs Sampler in the Normal Linear Regression Model In Example 2.1.2 the independent prior distributions $\boldsymbol{\beta} \mid A \sim N(\underline{\boldsymbol{\beta}}, \underline{\mathbf{H}}^{-1})$ and $\underline{s}^2 h \sim \chi^2(\underline{\nu})$ led to the conditional posterior distributions

$$\boldsymbol{\beta} \mid (h, \mathbf{y}^o, \mathbf{X}, A) \sim N(\overline{\boldsymbol{\beta}}, \overline{\mathbf{H}}^{-1}) \quad \text{and} \quad \overline{s}^2 h \mid (\boldsymbol{\beta}, \mathbf{y}^o, \mathbf{X}, A) \sim \chi^2(\overline{\nu}),$$

with $\overline{\mathbf{H}} = \underline{\mathbf{H}} + h\mathbf{X}'\mathbf{X}$, $\overline{\boldsymbol{\beta}} = \overline{\mathbf{H}}^{-1}(\underline{\mathbf{H}}\underline{\boldsymbol{\beta}} + h\mathbf{X}'\mathbf{y}^o)$, $\overline{s}^2 = \underline{s}^2 + (\mathbf{y}^o - \mathbf{X}\boldsymbol{\beta})'(\mathbf{y}^o - \mathbf{X}\boldsymbol{\beta})$, and $\overline{\nu} = \underline{\nu} + T$. Hence the blocking $\boldsymbol{\theta}_{(1)} = \boldsymbol{\beta}$, $\boldsymbol{\theta}_{(2)} = h$ is natural and convenient.

Of course, if it were possible to make an initial draw from the density $p(\boldsymbol{\theta} \mid I)$, then independent draws directly from $p(\boldsymbol{\theta} \mid I)$ would also be possible. The purpose of that assumption here is to marshal an informal argument that the density $p(\boldsymbol{\theta} \mid I)$ is an invariant density of the Markov chain $p(\boldsymbol{\theta}^{(m)} \mid \boldsymbol{\theta}^{(m-1)}, G)$: that is, if $\boldsymbol{\theta}^{(m)} \sim p(\boldsymbol{\theta} \mid I)$, then $\boldsymbol{\theta}^{(m+s)} \sim p(\boldsymbol{\theta} \mid I)$ for all $s > 0$. An important remaining task is to elucidate conditions for the distribution of $\boldsymbol{\theta}^{(m)}$ to converge in distribution to that of the pdf $p(\boldsymbol{\theta} \mid I)$ given any $\boldsymbol{\theta}^{(0)} \in \Theta$.

A more subtle complication is that even if $\boldsymbol{\theta}^{(0)}$ were drawn from $p(\boldsymbol{\theta} \mid I)$, the argument just given establishes only that any single $\boldsymbol{\theta}^{(m)}$ is also drawn from $p(\boldsymbol{\theta} \mid I)$. It does not establish that a single sequence $\{\boldsymbol{\theta}^{(m)}\}$ is representative of $p(\boldsymbol{\theta} \mid I)$. Consider the example shown in Figure 4.2a, in which $\Theta = \Theta_1 \bigcup \Theta_2$, and the Gibbs sampling algorithm has blocks $\boldsymbol{\theta}_{(1)} = \theta_1$ and $\boldsymbol{\theta}_{(2)} = \theta_2$. If $\boldsymbol{\theta}^{(0)} \in \Theta_1$, then $\boldsymbol{\theta}^{(m)} \in \Theta_1$ for $m = 1, 2, \ldots$. Any single $\boldsymbol{\theta}^{(m)}$ is just as representative of $p(\boldsymbol{\theta} \mid I)$ as is the single drawing $\boldsymbol{\theta}^{(0)}$, but $\{\boldsymbol{\theta}^{(m)}\}$ would not be representative of the distribution of interest. Indeed, it would be misleading. In the example shown in Figure 4.2b, if $\boldsymbol{\theta}^{(0)}$ is the indicated point at the lower left vertex of the triangle closed support of $p(\boldsymbol{\theta} \mid I)$, then $\boldsymbol{\theta}^{(m)} = \boldsymbol{\theta}^{(0)}$ for $m = 1, 2, \ldots$. Clearly neither situation arises in Example 4.3.1, but evidently a careful development of conditions under which $\{\boldsymbol{\theta}^{(m)}\}$ converges in distribution to $p(\boldsymbol{\theta} \mid I)$ is needed. We return to that development in Section 4.5.

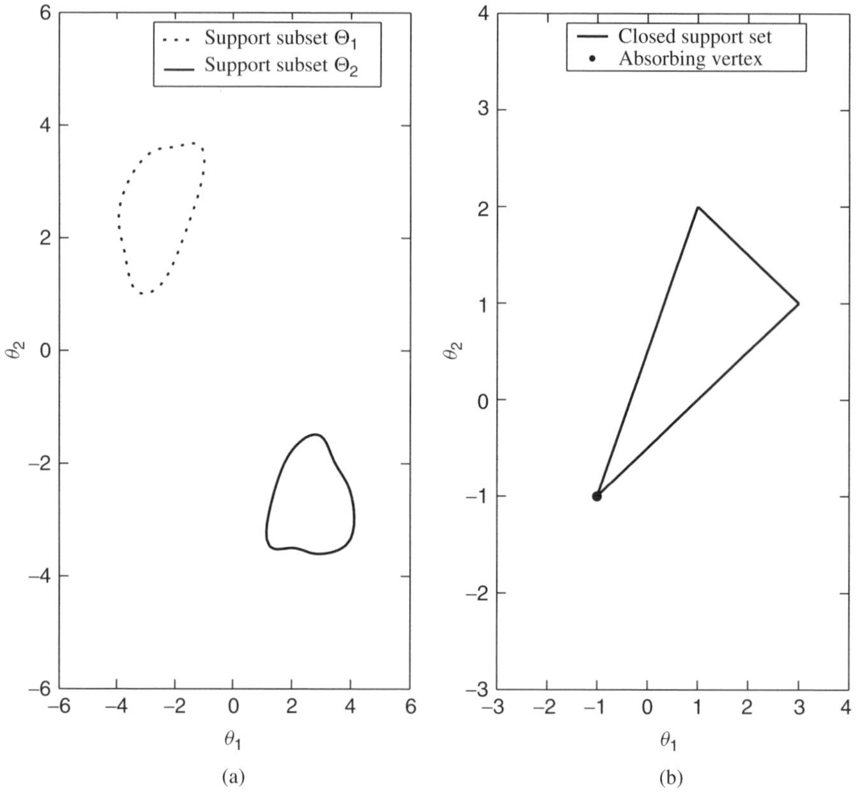

Figure 4.2. Two examples in which a Gibbs sampling Markov chain will be reducible: (a) disjoint support; (b) vertex of closed set support.

4.3.2 The Metropolis–Hastings Algorithm

The Metropolis–Hastings algorithm is defined by an arbitrary transition probability density function $q(\boldsymbol{\theta}^* \mid \boldsymbol{\theta}, H)$ indexed by $\boldsymbol{\theta} \in \Theta$ and with density argument $\boldsymbol{\theta}^*$, and by an arbitrary starting value $\boldsymbol{\theta}^{(0)} \in \Theta$. The random vector $\boldsymbol{\theta}^*$ generated from $q(\boldsymbol{\theta}^* \mid \boldsymbol{\theta}^{(m-1)}, H)$ is a candidate value for $\boldsymbol{\theta}^{(m)}$. The algorithm sets $\boldsymbol{\theta}^{(m)} = \boldsymbol{\theta}^*$ with probability

$$\alpha(\boldsymbol{\theta}^* \mid \boldsymbol{\theta}^{(m-1)}, H) = \min\left\{\frac{p(\boldsymbol{\theta}^* \mid I)/q(\boldsymbol{\theta}^* \mid \boldsymbol{\theta}^{(m-1)}, H)}{p(\boldsymbol{\theta}^{(m-1)} \mid I)/q(\boldsymbol{\theta}^{(m-1)} \mid \boldsymbol{\theta}^*, H)}, 1\right\}. \tag{4.15}$$

Otherwise, the algorithm sets $\boldsymbol{\theta}^{(m)} = \boldsymbol{\theta}^{(m-1)}$. It is common to say that the candidate $\boldsymbol{\theta}^*$ is accepted in the first instance and rejected in the second.

Let

$$u(\boldsymbol{\theta}^* \mid \boldsymbol{\theta}, H) = q(\boldsymbol{\theta}^* \mid \boldsymbol{\theta}, H)\alpha(\boldsymbol{\theta}^* \mid \boldsymbol{\theta}, H), \tag{4.16}$$

a density kernel for all accepted candidates. Denote the unconditional probability of rejection of the candidate drawn when the current state of the Metropolis–Hastings chain is $\boldsymbol{\theta}$ by

$$r(\boldsymbol{\theta} \mid H) = 1 - \int_{\Theta} u(\boldsymbol{\theta}^* \mid \boldsymbol{\theta}, H)\, d\nu(\boldsymbol{\theta}^*). \tag{4.17}$$

This is the probability of rejecting $\boldsymbol{\theta}^*$, given $\boldsymbol{\theta}$, but before $\boldsymbol{\theta}^*$ is actually drawn. The Metropolis–Hastings Markov chain is thus driven by an indexed transition probability measure defined on ν-measurable sets $A \subseteq \Theta$:

$$P(A \mid \boldsymbol{\theta}, H) = \int_A u(\boldsymbol{\theta}^* \mid \boldsymbol{\theta}, H)\, d\nu(\boldsymbol{\theta}^*) + r(\boldsymbol{\theta} \mid H) I_A(\boldsymbol{\theta}).$$

To write the corresponding transition probability density, let $\delta_{\boldsymbol{\theta}}(\boldsymbol{\theta}^*)$ denote the Dirac delta function, a linear operator with the property

$$\int_A \delta_{\boldsymbol{\theta}}(\boldsymbol{\theta}^*) f(\boldsymbol{\theta}^*)\, d\nu(\boldsymbol{\theta}^*) = f(\boldsymbol{\theta}) I_A(\boldsymbol{\theta}). \tag{4.18}$$

Then

$$p(\boldsymbol{\theta}^{(m)} \mid \boldsymbol{\theta}^{(m-1)}, H) = u(\boldsymbol{\theta}^{(m)} \mid \boldsymbol{\theta}^{(m-1)}, H) + r(\boldsymbol{\theta}^{(m-1)} \mid H)\delta_{\boldsymbol{\theta}^{(m-1)}}(\boldsymbol{\theta}^{(m)}). \tag{4.19}$$

This defines a Markov chain indexed by $\boldsymbol{\theta}^{(m-1)}$ that places probability on Θ. The intuition behind this procedure is evident on the right side of (4.15), and is in many respects similar to that in acceptance and importance sampling. If the transition density makes a move from $\boldsymbol{\theta}^{(m-1)}$ to $\boldsymbol{\theta}^*$ quite likely, relative to $p(\boldsymbol{\theta}^* \mid I)$, and a move back from $\boldsymbol{\theta}^*$ to $\boldsymbol{\theta}^{(m-1)}$ quite unlikely, relative to $p(\boldsymbol{\theta}^{(m-1)} \mid I)$, then the algorithm will place a low probability on actually making the transition and a high probability on staying at $\boldsymbol{\theta}^{(m-1)}$. In the same situation, a prospective move from $\boldsymbol{\theta}^*$ to $\boldsymbol{\theta}^{(m-1)}$ will always be made because draws of $\boldsymbol{\theta}^{(m-1)}$ are made infrequently relative to the density of interest $p(\boldsymbol{\theta} \mid I)$.

This is the most general form of the Metropolis–Hastings algorithm, which is due to Hastings (1970). The Metropolis et al. (1953) form takes $q(\boldsymbol{\theta}^* \mid \boldsymbol{\theta}, H) = q(\boldsymbol{\theta} \mid \boldsymbol{\theta}^*, H)$, which leads to the simplification

$$\alpha(\boldsymbol{\theta}^* \mid \boldsymbol{\theta}^{(m-1)}, H) = \min[p(\boldsymbol{\theta}^* \mid I)/p(\boldsymbol{\theta}^{(m-1)} \mid I), 1].$$

A leading instance of that algorithm is the *random-walk Metropolis chain* in which $q(\boldsymbol{\theta}^* \mid \boldsymbol{\theta}, H) = q(\boldsymbol{\theta}^* - \boldsymbol{\theta} \mid H)$, the latter density being symmetric about $\mathbf{0}$; see Example 4.3.2.

Another special case is the *Metropolis independence chain* (Tierney 1994), in which $q(\boldsymbol{\theta}^* \mid \boldsymbol{\theta}, H) = q(\boldsymbol{\theta}^* \mid H)$. This leads to

$$\alpha(\boldsymbol{\theta}^* \mid \boldsymbol{\theta}^{(m-1)}, H) = \min[w(\boldsymbol{\theta}^*)/w(\boldsymbol{\theta}^{(m-1)}), 1],$$

where $w(\boldsymbol{\theta}) = p(\boldsymbol{\theta} \mid I)/q(\boldsymbol{\theta} \mid H)$. The independence chain is closely related to acceptance sampling and importance sampling. But rather than place a low probability of acceptance or a low weight on a draw that is unlikely relative to the distribution of interest, the independence chain assigns a low probability of accepting that candidate $\boldsymbol{\theta}^*$ as the next draw $\boldsymbol{\theta}^{(m)}$.

There is a simple two-step argument that motivates the convergence of the sequence $\{\boldsymbol{\theta}^{(m)}\}$, generated by the Metropolis–Hastings algorithm, to the distribution of interest. [This approach is due to Chib and Greenberg (1995).] First, note that if a transition probability density function $p(\boldsymbol{\theta}^{(m)} \mid \boldsymbol{\theta}^{(m-1)}, T)$ satisfies the *reversibility condition*

$$p(\boldsymbol{\theta}^{(m-1)} \mid I)p(\boldsymbol{\theta}^{(m)} \mid \boldsymbol{\theta}^{(m-1)}, T) = p(\boldsymbol{\theta}^{(m)} \mid I)p(\boldsymbol{\theta}^{(m-1)} \mid \boldsymbol{\theta}^{(m)}, T)$$

with respect to $p(\boldsymbol{\theta} \mid I)$, then

$$\begin{aligned}&\int_\Theta p(\boldsymbol{\theta}^{(m-1)} \mid I)p(\boldsymbol{\theta}^{(m)} \mid \boldsymbol{\theta}^{(m-1)}, T)\,d\nu(\boldsymbol{\theta}^{(m-1)})\\ &\quad= \int_\Theta p(\boldsymbol{\theta}^{(m)} \mid I)p(\boldsymbol{\theta}^{(m-1)} \mid \boldsymbol{\theta}^{(m)}, T)\,d\nu(\boldsymbol{\theta}^{(m-1)})\\ &\quad= p(\boldsymbol{\theta}^{(m)} \mid I)\int_\Theta p(\boldsymbol{\theta}^{(m-1)} \mid \boldsymbol{\theta}^{(m)}, T)\,d\nu(\boldsymbol{\theta}^{(m-1)}) = p(\boldsymbol{\theta}^{(m)} \mid I).\end{aligned} \tag{4.20}$$

Expression (4.20) indicates that if $\boldsymbol{\theta}^{(m-1)} \sim p(\boldsymbol{\theta} \mid I)$, then the same is true of $\boldsymbol{\theta}^{(m)}$. The density $p(\boldsymbol{\theta} \mid I)$ is an invariant density of the Markov chain with transition density $p(\boldsymbol{\theta}^{(m)} \mid \boldsymbol{\theta}^{(m-1)}, T)$. This concept is developed more formally in Section 4.5.

The second step in this argument is to consider the implications of the requirement that the Metropolis–Hastings transition density $p(\boldsymbol{\theta}^{(m)} \mid \boldsymbol{\theta}^{(m-1)}, H)$ be reversible with respect to $p(\boldsymbol{\theta} \mid I)$:

$$p(\boldsymbol{\theta}^{(m-1)} \mid I)p(\boldsymbol{\theta}^{(m)} \mid \boldsymbol{\theta}^{(m-1)}, H) = p(\boldsymbol{\theta}^{(m)} \mid I)p(\boldsymbol{\theta}^{(m-1)} \mid \boldsymbol{\theta}^{(m)}, H).$$

For $\boldsymbol{\theta}^{(m-1)} = \boldsymbol{\theta}^{(m)}$ the requirement holds trivially. For $\boldsymbol{\theta}^{(m-1)} \neq \boldsymbol{\theta}^{(m)}$ it implies that

$$\begin{aligned}&p(\boldsymbol{\theta}^{(m-1)} \mid I)q(\boldsymbol{\theta}^* \mid \boldsymbol{\theta}^{(m-1)}, H)\alpha(\boldsymbol{\theta}^* \mid \boldsymbol{\theta}^{(m-1)}, H)\\ &\quad= p(\boldsymbol{\theta}^* \mid I)q(\boldsymbol{\theta}^{(m-1)} \mid \boldsymbol{\theta}^*, H)\alpha(\boldsymbol{\theta}^{(m-1)} \mid \boldsymbol{\theta}^*, H).\end{aligned} \tag{4.21}$$

Suppose without loss of generality that

$$p(\boldsymbol{\theta}^{(m-1)} \mid I)q(\boldsymbol{\theta}^* \mid \boldsymbol{\theta}^{(m-1)}, H) > p(\boldsymbol{\theta}^* \mid I)q(\boldsymbol{\theta}^{(m-1)} \mid \boldsymbol{\theta}^*, H).$$

If $\alpha(\boldsymbol{\theta}^{(m-1)} \mid \boldsymbol{\theta}^*, H) = 1$ and

$$\alpha(\boldsymbol{\theta}^* \mid \boldsymbol{\theta}^{(m-1)}, H) = \frac{p(\boldsymbol{\theta}^* \mid I)q(\boldsymbol{\theta}^{(m-1)} \mid \boldsymbol{\theta}^*, H)}{p(\boldsymbol{\theta}^{(m-1)} \mid I)q(\boldsymbol{\theta}^* \mid \boldsymbol{\theta}^{(m-1)}, H)},$$

then (4.21) is satisfied.

Example 4.3.2 Restricted Normal Linear Regression Model—Another Approach Consider once again the situation of Example 4.2.1, a normal linear regression model with fixed precision h, and the prior $\boldsymbol{\beta} \sim N(\underline{\boldsymbol{\beta}}, \underline{\mathbf{H}}^{-1})$ truncated to a set $S \subseteq \mathbb{R}^k$. In the acceptance sampling algorithm developed in that example, $\boldsymbol{\beta}$ is drawn from the source distribution $N(\overline{\boldsymbol{\beta}}, \overline{\mathbf{H}}^{-1})$, accepted if $\boldsymbol{\beta} \in S$, and rejected if $\boldsymbol{\beta} \notin S$. The probability of acceptance is

$$(2\pi)^{-k/2} \left|\overline{\mathbf{H}}\right|^{1/2} \int_S \exp[-(\boldsymbol{\beta} - \overline{\boldsymbol{\beta}})'\overline{\mathbf{H}}(\boldsymbol{\beta} - \overline{\boldsymbol{\beta}})/2]\, d\boldsymbol{\beta},$$

and the algorithm will be impractical if this is quite small.

As an alternative, consider a Metropolis–Hastings algorithm in which the transition density $q(\boldsymbol{\beta}^* \mid \boldsymbol{\beta}^{(m)}, H)$ corresponds to the distribution $N(\boldsymbol{\beta}^{(m)}, \mathbf{V})$:

$$q(\boldsymbol{\beta}^* \mid \boldsymbol{\beta}^{(m)}, H) = (2\pi)^{-k/2} |\mathbf{V}|^{-1/2} \exp[-(\boldsymbol{\beta}^* - \boldsymbol{\beta}^{(m)})' \mathbf{V}^{-1}(\boldsymbol{\beta}^* - \boldsymbol{\beta}^{(m)})/2].$$

This is an example of a random-walk Metropolis chain. It shares the property $q(\boldsymbol{\beta}^* \mid \boldsymbol{\beta}^{(m)}, H) = q(\boldsymbol{\beta}^{(m)} \mid \boldsymbol{\beta}^*, H)$ of the algorithm developed in Metropolis et al. (1953). The acceptance probability is

$$\alpha(\boldsymbol{\beta}^* \mid \boldsymbol{\beta}^{(m)}, H) = \frac{\exp[-(\boldsymbol{\beta}^* - \overline{\boldsymbol{\beta}})'\overline{\mathbf{H}}(\boldsymbol{\beta}^* - \overline{\boldsymbol{\beta}})/2]}{\exp[-(\boldsymbol{\beta}^{(m)} - \overline{\boldsymbol{\beta}})'\overline{\mathbf{H}}(\boldsymbol{\beta}^{(m)} - \overline{\boldsymbol{\beta}})/2]}$$

if $\boldsymbol{\beta}^* \in S$ and 0 if $\boldsymbol{\beta}^* \notin S$. This algorithm can succeed where the acceptance algorithm is impractical if $\mathbf{V}$ is chosen carefully. If $\mathbf{V}$ is too large, most draws will not be in S and will therefore be rejected. As a consequence, in order to generate M distinct draws, many times more candidates will need to be drawn. If $\mathbf{V}$ is too small, most draws will be accepted but the distance moved will be quite small and a very large number of iterations will be required to cover S adequately. For the algorithm to succeed, $\mathbf{V}$ must also be scaled appropriately in all dimensions.

Exercise 4.3.1 The Behrens–Fisher Problem Here are several Bayesian variants of this problem. In each variant

$$\mathbf{y}_1 = (y_{11}, \ldots, y_{T_1,1})', \quad \mathbf{y}_2 = (y_{12}, \ldots, y_{T_2,2})$$

$$\begin{pmatrix} \mathbf{y}_1 \\ \mathbf{y}_2 \end{pmatrix} \sim N\left\{ \begin{pmatrix} \boldsymbol{\iota}_{T_1}\mu_1 \\ \boldsymbol{\iota}_{T_2}\mu_2 \end{pmatrix}, \begin{bmatrix} \sigma_1^2\mathbf{I}_{T_1} & \mathbf{0} \\ \mathbf{0} & \sigma_2^2\mathbf{I}_{T_2} \end{bmatrix} \right\}$$

where $\boldsymbol{\iota}_n$ denotes an $n \times 1$ vector $(1, \ldots, 1)'$.

(a) Suppose that the prior distribution is

$$\underline{s}_1^2/\sigma_1^2 \sim \chi^2(\underline{\nu}_1), \quad \underline{s}_2^2/\sigma_2^2 \sim \chi^2(\underline{\nu}_2),$$
$$\mu_1 = \mu_2 = \mu \sim N(\underline{\mu}, \underline{h}^{-1}).$$

(The random variables σ_1^2, σ_2^2, and μ are mutually independent.) Construct a Gibbs sampling algorithm whose invariant distribution is the posterior distribution of $(\mu, \sigma_1^2, \sigma_2^2)$. Be completely explicit about all conditional distributions.

(b) Suppose that the prior distribution is

$$\underline{s}_1^2/\sigma_1^2 \sim \chi^2(\underline{\nu}_1),\ \ \underline{s}_2^2/\sigma_2^2 \sim \chi^2(\underline{\nu}_2),$$
$$\boldsymbol{\mu} = (\mu_1, \mu_2)' \sim N(\underline{\boldsymbol{\mu}}, \underline{\mathbf{H}}^{-1}).$$

(The random variables σ_1^2, σ_2^2, and $\boldsymbol{\mu}$ are mutually independent.) Construct a Gibbs sampling algorithm whose invariant distribution is the posterior distribution of $(\boldsymbol{\mu}, \sigma_1^2, \sigma_2^2)$. Be completely explicit about all conditional distributions.

(c) Suppose that the prior distribution is

$$\underline{s}_1^2/\sigma_1^2 \sim \chi^2(\underline{\nu}_1),\ \ \underline{s}_2^2/\sigma_2^2 \sim \chi^2(\underline{\nu}_2),$$
$$\widetilde{\mu} = (\mu_1 + \mu_2)/2 \sim N(\underline{\widetilde{\mu}}, \underline{\widetilde{h}}^{-1}),$$
$$\mu_1 - \mu_2 \sim N(0, \underline{h}^{-1}) \text{ truncated to } \mu_1 - \mu_2 \geq 0.$$

(The random variables σ_1^2, σ_2^2, $\widetilde{\mu}$, and $\mu_1 - \mu_2$ are mutually independent.) Construct a Gibbs sampling algorithm whose invariant distribution is the posterior distribution of $(\mu_1, \mu_2, \sigma_1^2, \sigma_2^2)$. Be completely explicit about all conditional distributions.

Exercise 4.3.2 Gibbs Sampling in a Nonlinear Regression Model Consider the second-order autoregressive model

$$y_t - \mu = \beta_1(y_{t-1} - \mu) + \beta_2(y_{t-2} - \mu) + \varepsilon_t,\ \ \varepsilon_t \overset{\text{i.i.d.}}{\sim} N(0, h^{-1})\ \ (t = 1, \ldots, T).$$

The values y_0 and y_{-1} are fixed. (Equivalently, they are ancillary statistics for μ, $\boldsymbol{\beta} = (\beta_1, \beta_2)'$, and h.) The prior distribution is

$$\mu \sim N(\underline{\mu}, \underline{h}^{-1}),\ \ \boldsymbol{\beta} \sim N(\underline{\boldsymbol{\beta}}, \underline{\mathbf{H}}_{\beta}^{-1}),\ \ \underline{s}^2 h \sim \chi^2(\underline{\nu}).$$

(In the prior distribution, μ, $\boldsymbol{\beta}$, and h are mutually independent.)

(a) Write the likelihood function for μ, $\boldsymbol{\beta}$, and h and the respective prior density kernels for μ, $\boldsymbol{\beta}$, and h.

(b) Design a Gibbs sampling algorithm to draw from the posterior distribution of μ, $\boldsymbol{\beta}$, and h.

(c) Now suppose that, in addition, we impose the constraint that the roots of the polynomial

$$1 - \beta_1 z - \beta_2 z^2 = 0$$

satisfy $|z| > 1$. (This is a technical condition guaranteeing that $\{y_t\}$ does not "blow up" as $t \to \infty$; more precisely, it ensures that $\{y_t\}$ is asymptotically stationary.) Thus $\boldsymbol{\beta} \in S \subseteq \mathbb{R}^2$, and the prior density kernel for $\boldsymbol{\beta}$ found in (a) is now multiplied by $I_S(\boldsymbol{\beta})$. Modify the algorithm that you designed in (b) to simulate from the posterior distribution in this model.

(For continuation, see Exercise 4.4.2.)

Exercise 4.3.3 Shape Constraints in Regression Hildreth (1954) had observables as follows:

Fertilizer per Acre	Number of Observations	Output per Acre
20	T_1	$(y_{11}, \ldots, y_{T1})$
40	T_2	$(y_{12}, \ldots, y_{T2})$
60	T_3	$(y_{13}, \ldots, y_{T3})$
80	T_4	$(y_{14}, \ldots, y_{T4})$
100	T_5	$(y_{15}, \ldots, y_{T5})$
120	T_6	$(y_{16}, \ldots, y_{T6})$

His model is $y_{tj} \sim N(\mu_j, \sigma^2)$ with all y_{tj} mutually independent.

(a) Given the prior distribution

$$\boldsymbol{\mu} = (\mu_1, \ldots, \mu_6)' \sim N(\underline{\boldsymbol{\mu}}, \underline{\mathbf{H}}^{-1}),\ \underline{s}^2/\sigma^2 \sim \chi^2(\underline{\nu}),$$

where $\boldsymbol{\mu}$ and σ^2 are independent, how would you construct a Gibbs sampling algorithm whose invariant distribution is the posterior distribution of $\boldsymbol{\mu}$ and σ^2?

(b) Suppose that the prior distribution is the same as in (a), except that in addition you believe $\mu_1 < \cdots < \mu_6$. The restrictions could be handled by appending an acceptance step to the algorithm in (a), but this could be quite inefficient. Show how to construct a Gibbs sampler with $B = 7$ blocks whose invariant distribution is the posterior distribution of $\boldsymbol{\mu}$ and σ^2.

(c) Suppose that the prior distribution is the same as in (b), except that in addition you believe that the expected output per acre (given fertilizer per acre) is a strictly concave function of fertilizer per acre. How would you construct a 7-block Gibbs sampling algorithm whose invariant distribution is the posterior distribution of $\boldsymbol{\mu}$ and σ^2?

4.4 VARIANCE REDUCTION

All the Monte Carlo methods for evaluating $E[h(\boldsymbol{\omega}) \mid I]$ considered to this point generate an artificial sample $(\boldsymbol{\theta}_A^{(m)}, \boldsymbol{\omega}^{(m)})$ $(m = 1, 2, \ldots)$ and a sequence of weights

$w(\boldsymbol{\theta}_A^{(m)})$ with the property that

$$\overline{h}^{(M)} = \sum_{m=1}^{M} w(\boldsymbol{\theta}_A^{(m)}) h(\omega^{(m)}) \Big/ \sum_{m=1}^{M} w(\boldsymbol{\theta}_A^{(m)}) \stackrel{\text{a.s.}}{\rightarrow} E[h(\boldsymbol{\omega}) \mid I].$$

Often it is possible to find a function $h^*(\boldsymbol{\theta}_A, \boldsymbol{\omega})$ with the properties

$$E[h^*(\boldsymbol{\theta}_A, \boldsymbol{\omega}) \mid I] = E[h(\boldsymbol{\omega}) \mid I], \tag{4.22}$$

$$\text{var}[h^*(\boldsymbol{\theta}_A, \boldsymbol{\omega}) \mid I] < \text{var}[h(\boldsymbol{\omega}) \mid I], \tag{4.23}$$

$$\overline{h}^{*(M)} = \frac{\sum_{m=1}^{M} w(\boldsymbol{\theta}_A^{(m)}) h^*(\boldsymbol{\theta}_A^{(m)}, \boldsymbol{\omega}^{(m)})}{\sum_{m=1}^{M} w(\boldsymbol{\theta}_A^{(m)})} \stackrel{\text{a.s.}}{\rightarrow} E[h(\boldsymbol{\omega}) \mid I]. \tag{4.24}$$

Typically (4.24) can be derived from (4.22). Then (4.23) suggests that it is likely that the numerical standard error associated with $\overline{h}^{*(M)}$ will be less than that associated with $\overline{h}^{(M)}$. The extent of the reduction in numerical standard error, or whether any reduction at all will necessarily occur, can be difficult to establish as an analytical proposition in typical applications. What is more important is that the relative accuracies of the two approximations can be evaluated as a practical matter using the central limit theorem appropriate to the method: Theorem 4.1.1 for direct sampling, Theorem 4.2.2 for importance sampling, or Theorems 4.7.1 and 4.7.3 for MCMC sampling. Even more important is the fact that there are systematic methods of constructing h^* that will supply the potentially superior alternative in the first place. This section describes two such methods.

4.4.1 Concentrated Expectations

The essentials of the principle of concentrated expectations can be appreciated in the generic problem of evaluating the integral $\iint f(x, y) p(x, y)\, dx\, dy$, in which $p(x, y)$ is a density function. Direct sampling would entail $(x^{(m)}, y^{(m)}) \stackrel{\text{i.i.d.}}{\sim} p(x, y)$ and the approximation $M^{-1} \sum_{m=1}^{M} f(x^{(m)}, y^{(m)})$. Suppose that we can evaluate $E[f(x, y) \mid x] = \int f(x, y) p(x, y)\, dy / \int p(x, y)\, dy$ analytically. By the law of iterated expectations [see Casella and Berger (2002), Theorem 4.4.3]

$$E\{E[f(x, y) \mid x]\} = E[f(x, y)],$$

and by the Rao–Blackwell theorem (Casella and Berger 2002, Theorem 7.3.17)

$$\text{var}\{E[f(x, y) \mid x]\} \leq \text{var}[f(x, y)]$$

as long as the latter variance exists and is finite. Hence the numerical standard error associated with $M^{-1} \sum_{m=1}^{M} E[f(x^{(m)}, y) \mid x^{(m)}]$ is no greater, and is generally smaller, than that associated with $M^{-1} \sum_{m=1}^{M} f(x^{(m)}, y^{(m)})$. The gain comes from performing part of the integration analytically rather than entirely numerically. The following result places this idea in context.

Theorem 4.4.1 Concentrated Expectations in Posterior Simulation Suppose that the sequence $\{\boldsymbol{\theta}^{(m)}, \omega_1^{(m)}\}$ is independently and identically distributed with $\boldsymbol{\theta}^{(m)} \sim p(\boldsymbol{\theta}^{(m)} \mid I)$ and $\omega_1^{(m)} \sim p(\omega \mid \boldsymbol{\theta}^{(m)}, I)$. Let $\boldsymbol{\theta}' = (\boldsymbol{\theta}_1', \boldsymbol{\theta}_2')$, $\omega_2^{(m)} = E(\omega \mid \boldsymbol{\theta}^{(m)}, I)$, $\omega_3^{(m)} = E(\omega \mid \boldsymbol{\theta}_1^{(m)}, I)$, $\overline{\omega} = E(\omega \mid I)$, and $\overline{\omega}_j^{(M)} = M^{-1} \sum_{m=1}^{M} \omega_j^{(m)}$ $(j = 1, 2, 3)$. Then

$$M^{1/2}(\overline{\omega}_j^{(M)} - \overline{\omega}) \xrightarrow{d} N(0, \tau_j^2) \ (j = 1, 2, 3) \tag{4.25}$$

and

$$\tau_3^2 \leq \tau_2^2 \leq \tau_1^2. \tag{4.26}$$

Proof: Because $\{\boldsymbol{\theta}^{(m)}, \omega_1^{(m)}\}$ is i.i.d., so are $\{\omega_j^{(m)}\}$ $(j = 1, 2, 3)$. By the law of iterated expectations $E(\omega_j^{(m)} \mid I) = E(\omega \mid I)$ $(j = 1, 2, 3; m = 1, \ldots, M)$. By the Rao–Blackwell theorem

$$\text{var}(\omega_3^{(m)} \mid I) \leq \text{var}(\omega_2^{(m)} \mid I) \leq \text{var}(\omega_1^{(m)} \mid I). \qquad \blacksquare$$

The conditions in Theorem 4.4.1 are those of direct sampling, and hence the immediate applicability of this result is rather limited. The treatment of convergence of MCMC algorithms in Sections 4.5 and 4.7 applies directly to (4.25), and consequently the methods there can be used to approximate τ_j^2 and thereby evaluate the accuracy of the alternative approximations $\overline{\omega}_j^{(M)} (j = 1, 2, 3)$. Similarly, the method of concentrated expectations can be used in combination with importance sampling, and the methods of Section 4.2.2 can be used to assess accuracy. All of these methods provide an estimate $\widehat{\tau}_j^2$, and this is generally lower when the method of concentrated expectations is applied than when it is not. Whether or not (4.26) applies for these algorithms in general is not known. Liu et al. (1994) and McKeague and Wefelmeyer (2000) have shown that it does for certain MCMC algorithms, given some regularity conditions beyond those discussed in Section 4.5.

Example 4.4.1 Concentrated Expectations and Prediction in the Normal Linear Regression Model In the context of the normal linear model and Gibbs sampling algorithm of Example 4.3.1, suppose $\omega = y_{T+1} = \boldsymbol{\beta}'\mathbf{x}_{T+1} + \varepsilon_{T+1}$, where the covariate vector $\mathbf{x}_{T+1}$ is known. Apply Theorem 4.4.1 using $\boldsymbol{\theta}_{A1} = h$ to obtain $\omega_1^{(m)} \sim N(\boldsymbol{\beta}^{(m)\prime}\mathbf{x}_{T+1}, h^{(m)-1})$, $\omega_2^{(m)} = \boldsymbol{\beta}^{(m)\prime}\mathbf{x}_{T+1}$ and $\omega_3^{(m)} = \overline{\boldsymbol{\beta}}^{(m)\prime}\mathbf{x}_{T+1}$, where

$$\overline{\boldsymbol{\beta}}^{(m)} = (\underline{\mathbf{H}} + h^{(m)}\mathbf{X}'\mathbf{X})^{-1}(\underline{\mathbf{H}\boldsymbol{\beta}} + h^{(m)}\mathbf{X}'\mathbf{y}^o).$$

The numerical standard error of $\overline{\omega}_3^M$ will be much smaller than that of $\overline{\omega}_2^M$ in typical applications; see Exercise 4.4.2 for an example. Moreover, $\overline{\omega}_3^M$ requires less computing than does either $\overline{\omega}_2^M$ or $\overline{\omega}_1^M$.

This example illustrates the fact that virtually every Gibbs sampling algorithm provides potential applications of the principle of concentrated expectations. Since $p(\boldsymbol{\theta}_{(b)} \mid \boldsymbol{\theta}_{-(b)}, I)$ is the density of a tractable distribution, in general $E(\boldsymbol{\theta}_{(b)} \mid \boldsymbol{\theta}_{-(b)}, I)$ will be known analytically. Using $E(\boldsymbol{\theta}_{(b)} \mid \boldsymbol{\theta}_{-(b)}^{(m)}, I)$ rather than $\boldsymbol{\theta}_{(b)}^{(m)}$ to approximate $E(\boldsymbol{\theta}_{(b)} \mid I)$ will typically pay a dividend in the form of reduced numerical standard error. As Example 4.4.1 illustrates, this property of Gibbs sampling algorithms can be used to improve the simulation approximation of other posterior moments, as well.

4.4.2 Antithetic Sampling

The principle of antithetic variates in Monte Carlo integration dates at least to Hammersly and Morton (1956). In the original formulation, an i.i.d. sequence of random vectors $(\omega^{(1,m)}, \omega^{(2,m)})'$ is generated from a sampling scheme R. The marginal distribution of $\{\omega^{j,m}\} \mid R$ $(j = 1, 2)$ is the same as that of an i.i.d. sequence drawn from the distribution I, but $\text{cov}(\omega^{(1,m)}, \omega^{(2,m)} \mid R) < 0$. Then

$$\text{var}\left[\sum_{m=1}^{M}(\omega^{(1,m)} + \omega^{(2,m)})/2M \mid R\right] < \tfrac{1}{2}\,\text{var}\left(\sum_{m=1}^{M}\omega^{(1,m)}/M \mid I\right).$$

This is the relevant comparison if the computation time for generating $\{\omega^{(1,m)}\}$ and $\{\omega^{(2,m)}\}$ is double that for $\{\omega^{(1,m)}\}$ alone. In the limiting case $\text{cov}(\omega^{(1,m)}, \omega^{(2,m)} \mid R) = -\text{var}(\omega^{(1,m)} \mid I)$ the approximation becomes exact, but then also $E(\omega \mid I)$ is likely to be known. To approximate $E[h(\omega) \mid I]$, we use

$$\sum_{m=1}^{M}[h(\omega^{(1,m)}) + h(\omega^{(2,m)})]/2M \tag{4.27}$$

in place of $\sum_{m=1}^{M} h(\omega^{(1,m)})/M$. It need not be the case that the alternative provides a more efficient approximation. Loosely speaking, if h is roughly linear over most of the support of ω and $\text{cov}(\omega^{(1,m)}, \omega^{(2,m)} \mid R) < 0$, there will be an improvement. For example, if $\omega^{(1,m)} \sim N(\mu, \sigma^2)$, with μ and σ^2 known, $E[h(\omega^{(1,m)}) \mid I]$ cannot be derived analytically, and h is roughly linear, then taking $\omega^{(2,m)} = \mu - (\omega^{(1,m)} - \mu) = 2\mu - \omega^{(1,m)}$ may provide a substantially more efficient approximation. The idea extends immediately to random vectors $\boldsymbol{\omega}$.

This principle applies in the more sophisticated sampling methods taken up in this chapter, as well. The only challenge is in finding a sequence $\boldsymbol{\omega}^{(2,m)}$ corresponding to $\boldsymbol{\omega}^{(1,m)}$ that may yield improved approximations. Once this is done, the application of the idea amounts to using (4.27) in direct sampling and MCMC algorithms; some straightforward changes in weighting, developed in Exercise 4.4.3, may be required for importance sampling. In the case of Gibbs sampling algorithms, the individual blocks often provide the basis for the alternative sequence $\boldsymbol{\omega}^{(2,m)}$, as illustrated in the following example.

Example 4.4.2 Antithetic Gibbs Sampling in a Nonlinear Regression Model In the context of the nonlinear regression model and Gibbs sampling algorithm of Exercise 4.3.2, parts (a) and (b), suppose that the function of interest is the amplitude of the smallest root of $1 - \beta_1 z - \beta_2 z^2$; denote this root by $\omega = f(\boldsymbol{\beta})$, where $\boldsymbol{\beta} = (\beta_1, \beta_2)'$. In the posterior simulator of Exercise 4.3.2(b), $\boldsymbol{\beta}^{(1,m)}$ is drawn from a normal distribution with mean $\overline{\boldsymbol{\beta}}^{(m)}$. Let $\omega^{(1,m)} = f(\boldsymbol{\beta}^{(1,m)})$, $\boldsymbol{\beta}^{(2,m)} = 2\overline{\boldsymbol{\beta}}^{(m)} - \boldsymbol{\beta}^{(1,m)}$, and $\omega^{(2,m)} = f(\boldsymbol{\beta}^{(2,m)})$. In the approximation of $E(\omega \mid I)$, use $(\omega^{(1,m)} + \omega^{(2,m)})/2$ in place of $\omega^{(1,m)}$. The methods developed in Section 4.7 for the evaluation of numerical accuracy apply directly, so the efficiencies of the two alternative approximations of $E(\omega \mid \mathbf{y}^o, A)$ will be apparent.

As sample size T increases, the posterior distribution of $T^{1/2}(\boldsymbol{\theta}_A - \boldsymbol{\theta}_A^*)$ typically becomes symmetric about zero for some $\boldsymbol{\theta}_A^* \in \mathbb{R}^k$. For example, this will happen if Theorem 3.4.3 applies. If the function of interest is smooth, then in the limit the approximation problem becomes that of evaluating the mean of a linear function of symmetrically distributed random variables. Therefore gains to antithetic sampling should increase with sample size, given suitable regularity conditions, a phenomenon known as *antithetic acceleration.*

These conditions were developed formally in Geweke (1988) for posterior moments of the form $\overline{g} = E[g(\boldsymbol{\theta}_A) \mid \mathbf{Y}_T, A]$. For each T, by direct sampling

$$\boldsymbol{\theta}_{AT}^{(1,m)} \overset{\text{i.i.d.}}{\sim} p(\boldsymbol{\theta}_A \mid \mathbf{Y}_T^o, A)(m = 1, 2, \ldots),$$

and the antithetic sample is

$$\boldsymbol{\theta}_{AT}^{(2,m)} = 2E[\boldsymbol{\theta}_A \mid \mathbf{Y}_T^o, A] - \boldsymbol{\theta}_{AT}^{(1,m)}.$$

The direct approximation of $E[g(\boldsymbol{\theta}_A) \mid \mathbf{Y}_T^o, A]$ is $\overline{g}_T^{(M)} = \sum_{m=1}^M g(\boldsymbol{\theta}_{AT}^{(1,m)})/M$ and the antithetic approximation is

$$\overline{g}_T^{*(M)} = \sum_{m=1}^{M} \left[g(\boldsymbol{\theta}_{AT}^{(1,m)}) + g(\boldsymbol{\theta}_{AT}^{(2,m)})\right]/2M.$$

The regularity conditions in Geweke (1988) include continuous twice differentiability of g in a neighborhood of $\boldsymbol{\theta}_A^*$, with $\boldsymbol{\alpha} = g'(\boldsymbol{\theta}_A^*)$ and $\mathbf{B} = (1/2)g''(\boldsymbol{\theta}_A^*)$, as well as

$$\lim_{T\to\infty} T \operatorname{var}(\boldsymbol{\theta}_A \mid \mathbf{Y}_T, A) = \boldsymbol{\Sigma} \text{ and } \lim_{T\to\infty} T^2 \operatorname{var}(\boldsymbol{\theta}_A' \mathbf{B} \boldsymbol{\theta}_A \mid \mathbf{Y}_T, A) = \delta > 0,$$

conditions similar to those in Theorem 3.4.3. Then $M^{1/2}(\overline{g}_T^{(M)} - \overline{g}) \xrightarrow{d} N(0, \tau_T^2)$, $M^{1/2}(\overline{g}_T^{*(M)} - \overline{g}) \xrightarrow{d} N(0, \tau_T^{*2})$, and

$$\lim_{T\to\infty} T\tau_T^{*2}/\tau_T^2 = \delta/\boldsymbol{\alpha}'\boldsymbol{\Sigma}\boldsymbol{\alpha}.$$

There are no parallel results for importance or MCMC sampling, but this is no impediment to the use of antithetic sampling. The additional demands are usually modest, as suggested by Example 4.4.2, and the accuracy of the approximation can be evaluated using the method described in Section 4.7.

Exercise 4.4.1 Improving the Approximation of a Posterior Probability Reconsider the problem of approximating (4.1) in Example 4.1.1.

- **(a)** Express $P(\omega \leq a \mid \mathbf{x}^*, \boldsymbol{\beta}, h, A)$ in terms of $\Phi(\cdot)$, the cdf of the standard normal distribution. (Most mathematical applications software can evaluate Φ efficiently.)
- **(b)** Use the result in (a) to find a numerical approximation to (4.1) with variance less than that of $p^{(M)}$, the approximation suggested in Example 4.1.1.

Exercise 4.4.2 Variance Reduction Methods and Forecasting Consider the nonlinear regression model of Exercise 4.3.2 and the Gibbs sampling algorithm developed there. This model can be used to forecast $\boldsymbol{\omega} = (y_{T+1}, \ldots, y_{T+F})'$. If the loss function is quadratic in $\boldsymbol{\omega}$, then the appropriate forecast is $\widehat{\mathbf{y}} = E(\boldsymbol{\omega} \mid Y_T^o, A)$. This could be accomplished by drawing

$$\omega_1^{(m)} \sim N[\mu^{(m)} + \beta_1^{(m)}(y_T^o - \mu^{(m)}) + \beta_2^{(m)}(y_{T-1}^o - \mu^{(m)}),\ h^{(m)-1}],$$
$$\omega_2^{(m)} \sim N[\mu^{(m)} + \beta_1^{(m)}(\omega_1^{(m)} - \mu^{(m)}) + \beta_2^{(m)}(y_T^o - \mu^{(m)}),\ h^{(m)-1}],$$
$$\omega_s^{(m)} \sim N[\mu^{(m)} + \beta_1^{(m)}(\omega_{s-1}^{(m)} - \mu^{(m)}) + \beta_2^{(m)}(\omega_{s-2}^{(m)} - \mu^{(m)}),\ h^{(m)-1}],$$

$s = 3, \ldots, F$. Then the simulation approximation of $\widehat{\mathbf{y}}$ is $M^{-1}\sum_{m=1}^{M}\boldsymbol{\omega}^{(m)}$. This exercise utilizes variance reduction methods to improve on this procedure.

- **(a)** Use the law of iterated expectations to show that conditional on $(\mathbf{Y}_T^o, \boldsymbol{\beta}, \mu, A)$

$$E(y_{T+1} - \mu) = \beta_1(y_T^o - \mu) + \beta_2(y_{T-1}^o - \mu),$$
$$E(y_{T+2} - \mu) = \beta_1 E(y_{T+1} - \mu) + \beta_2(y_T^o - \mu),$$
$$E(y_{T+s} - \mu) = \beta_1 E(y_{T+s-1} - \mu) + \beta_2(y_{T+s-2} - \mu)$$

for $s = 3, \ldots, F$. Use this result, the principle of concentrated expectations, and the original simulation sample $\{\mu^{(m)}, \boldsymbol{\beta}^{(m)}, h^{(m)}\}$ to improve the approximation of $\widehat{\mathbf{y}}$.

(b) Apply the principle of concentrated expectations to $E[\boldsymbol{\beta} \mid Y_T^o, \mu, h, A]$ to further improve the approximation of $\widehat{y}_{T+1}$. Does this idea extend to $\widehat{y}_{T+2}$?

(c) Show how antithetic samples of $\boldsymbol{\beta} \mid (\mathbf{Y}_T^o, \mu, h, A)$ could be used to improve the approximation of $\widehat{y}_{T+s}$ $(s = 2, \ldots, F)$. Would you expect the improvement to be greater for $\widehat{y}_{T+2}$ or for $\widehat{y}_{T+F}$?

Exercise 4.4.3 Antithetic Importance Sampling In many importance sampling algorithms the source density $p(\boldsymbol{\theta} \mid S)$ is symmetric about $\boldsymbol{\mu} = E(\boldsymbol{\theta} \mid S)$, and consequently the construction of the antithetic sequence $\boldsymbol{\theta}^{(2,m)} = 2\boldsymbol{\mu} - \boldsymbol{\theta}^{(m)}$ is trivial. Corresponding to $\boldsymbol{\theta}^{(m)}$, $\boldsymbol{\omega}^{(m)} \sim p(\boldsymbol{\omega} \mid \boldsymbol{\theta}^{(m)}, I)$, and corresponding to $\boldsymbol{\theta}^{(2,m)}$, $\boldsymbol{\omega}^{(2,m)} \sim p(\boldsymbol{\omega} \mid \boldsymbol{\theta}^{(2,m)}, I)$ $(m = 1, 2, \ldots)$. Define $\overline{h}$, $w(\boldsymbol{\theta})$, $\overline{h}^{(M)}$, and τ^2 as in Theorem 4.2.2.

(a) Show that

$$\overline{h}^{*(M)} = \frac{\sum_{m=1}^{M} [w(\boldsymbol{\theta}^{(m)})h(\boldsymbol{\omega}^{(m)}) + w(\boldsymbol{\theta}^{(2,m)})h(\boldsymbol{\omega}^{(2,m)})]}{\sum_{m=1}^{M} [w(\boldsymbol{\theta}^{(m)}) + w(\boldsymbol{\theta}^{(2,m)})]} \xrightarrow{\text{a.s.}} \overline{h}.$$

(b) State conditions under which $M^{1/2}(\overline{h}^{*(M)} - \overline{h}^{*}) \xrightarrow{d} N(0, \tau^{*2})$.

(c) Show that $\tau^{*2}/\tau^2 < \frac{1}{2}$ if and only if

$$\text{cov}\{[h(\boldsymbol{\omega}^{(m)}) - \overline{h}]w(\boldsymbol{\theta}^{(m)}),\ [h(\boldsymbol{\omega}^{(2,m)}) - \overline{h}]w(\boldsymbol{\theta}^{(2,m)}) \mid S\} < 0.$$

Under what conditions is this inequality likely to hold?

4.5 SOME CONTINUOUS STATE SPACE MARKOV CHAIN THEORY

The informal treatment of the Gibbs sampler and the Hastings–Metropolis algorithm in Section 4.3 leaves unresolved important questions about the conditions under which the simulation sample $\{\boldsymbol{\theta}^{(1)}, \ldots, \boldsymbol{\theta}^{(M)}\}$ will become representative of $p(\boldsymbol{\theta} \mid I)$ as $M \to \infty$. This section turns to that question. Much of the treatment here draws heavily on the work of Tierney (1991, 1994), who first used the theory of continuous state space Markov chains to demonstrate convergence, and Roberts and Smith (1994), who elucidated sufficient conditions for convergence that turn out to be applicable in a wide variety of problems in econometrics and statistics.

Let C denote a generic Markov chain $\{\boldsymbol{\theta}^{(m)}\}$ defined on $\Theta \times \Theta$ by a transition kernel $u(\boldsymbol{\theta}^* \mid \boldsymbol{\theta}, C)$ with the property that

$$r(\boldsymbol{\theta} \mid C) = 1 - \int_{\Theta} u(\boldsymbol{\theta}^* \mid \boldsymbol{\theta}, C)\, d\nu(\boldsymbol{\theta}^*) \in [0, 1)\ \forall\ \boldsymbol{\theta} \in \Theta.$$

For any ν-measurable set $A \subseteq \Theta$,

$$P(\boldsymbol{\theta}^{(m)} \in A \mid \boldsymbol{\theta}^{(m-1)}, C) = \int_A u(\boldsymbol{\theta} \mid \boldsymbol{\theta}^{(m-1)}, C)\, d\nu(\boldsymbol{\theta}) + r(\boldsymbol{\theta}^{(m-1)} \mid C) I_A(\boldsymbol{\theta}^{(m-1)}).$$

In the case of the Gibbs sampler discussed in the previous section the transition density function $p(\boldsymbol{\theta}^* \mid \boldsymbol{\theta}, G)$ is defined in (4.14), and $r(\boldsymbol{\theta} \mid C) = 0 \ \forall\ \boldsymbol{\theta} \in \Theta$. In the case of the Hastings–Metropolis algorithm, $u(\boldsymbol{\theta}^* \mid \boldsymbol{\theta}, C) = u(\boldsymbol{\theta}^* \mid \boldsymbol{\theta}, H)$, defined in (4.16), and $r(\boldsymbol{\theta} \mid C) = r(\boldsymbol{\theta} \mid H)$, defined in (4.17). The transition kernel u is *substochastic*; it is proportional to the probability density of the accepted candidates only. The corresponding substochastic kernel over m steps is then defined iteratively:

$$u^{(m)}(\boldsymbol{\theta}^{(m)} \mid \boldsymbol{\theta}^{(0)}, C) = \int_\Theta u^{(m-1)}(\boldsymbol{\theta} \mid \boldsymbol{\theta}^{(0)}, C) u(\boldsymbol{\theta}^{(m)} \mid \boldsymbol{\theta}, C)\, d\nu(\boldsymbol{\theta}) + u^{(m-1)}(\boldsymbol{\theta}^{(m)} \mid \boldsymbol{\theta}^{(0)}, C) r(\boldsymbol{\theta}^{(m)} \mid C) + [r(\boldsymbol{\theta}^{(0)} \mid C)]^{m-1} u(\boldsymbol{\theta}^{(m)} \mid \boldsymbol{\theta}^{(0)}, C).$$

This describes all m-step transitions that involve at least one accepted move.

Definition 4.5.1 An *invariant kernel* for a Markov chain with transition kernel $u(\boldsymbol{\theta}^* \mid \boldsymbol{\theta}, C)$ is a nonnegative function $k(\boldsymbol{\theta} \mid C)$ with support Θ that satisfies

$$\int_\Theta k(\boldsymbol{\theta} \mid C) \left[\int_A u(\boldsymbol{\theta}^* \mid \boldsymbol{\theta}, C)\, d\nu(\boldsymbol{\theta}^*) + r(\boldsymbol{\theta} \mid C) I_A(\boldsymbol{\theta}) \right] d\nu(\boldsymbol{\theta}) = \int_A k(\boldsymbol{\theta} \mid C)\, d\nu(\boldsymbol{\theta}) = K(A \mid C)$$

for all ν-measurable A.

Definition 4.5.2 The transition kernel $u(\boldsymbol{\theta}^* \mid \boldsymbol{\theta}, C)$ is *p-irreducible* if for all $\boldsymbol{\theta}^{(0)} \in \Theta$, $K(A \mid C) > 0$ implies that $P(\boldsymbol{\theta}^{(m)} \in A \mid \boldsymbol{\theta}^{(0)}, C) > 0$ for some $m \geq 1$.

Situations like the one shown in Figure 4.2a, where the support is disconnected and the Markov chain is the Gibbs sampler, cannot arise if u is p-irreducible. Referring to Figure 4.2a, note that if $\boldsymbol{\theta}^{(0)} \in \Theta_1$, it is impossible that $\boldsymbol{\theta}^{(m)} \in \Theta_2$ for any $m > 0$. At best there are two invariant distributions, one for Θ_1 reached if $\boldsymbol{\theta}^{(0)} \in \Theta_1$, and one for Θ_2 reached if $\boldsymbol{\theta}^{(0)} \in \Theta_2$.

Definition 4.5.3 The transition kernel $u(\boldsymbol{\theta}^* \mid \boldsymbol{\theta}, C)$ is *aperiodic* if there exists no ν-measurable partition $\Theta = \bigcup_{s=0}^{r-1} \Theta_s$ $(r \geq 2)$ such that for some $\boldsymbol{\theta}^{(0)} \in \Theta$

$$P(\boldsymbol{\theta}^{(m)} \in \Theta_{m \bmod (r)} \mid \boldsymbol{\theta}^{(0)}, C) = 1 \ \forall\ m.$$

Definition 4.5.4 The transition kernel $u(\boldsymbol{\theta}^* \mid \boldsymbol{\theta}, C)$ is *Harris recurrent* if for all ν-measurable A with $\int_A k(\boldsymbol{\theta} \mid C)\, d\nu(\boldsymbol{\theta}) > 0$, all $\boldsymbol{\theta}^{(0)} \in \Theta$, and all $L < \infty$

$$\lim_{M\to\infty} P\left[\sum_{m=1}^{M} I_A(\boldsymbol{\theta}^{(m)}) \le L \mid C\right] = 0. \tag{4.28}$$

It follows at once that if a transition kernel is Harris recurrent, then it is p-irreducible.

An invariant kernel $k(\boldsymbol{\theta} \mid C)$ is defined only up to an arbitrary scaling constant. Recall from Section 3.2 that if a kernel is finitely integrable, then it is proper. In this case there is a unique probability density $p(\boldsymbol{\theta} \mid C)$ corresponding to $k(\boldsymbol{\theta} \mid C)$.

Definition 4.5.5 If an aperiodic and Harris recurrent transition kernel $u(\boldsymbol{\theta}^* \mid \boldsymbol{\theta}, C)$ has a proper invariant kernel, then u is *ergodic*.

Theorem 4.5.1 Convergence of Continuous State Markov Chains Suppose that $k(\boldsymbol{\theta} \mid C)$ is an invariant kernel of the transition kernel $u(\boldsymbol{\theta}^* \mid \boldsymbol{\theta}, C)$.

(a) If u is p-irreducible, then the invariant kernel is unique up to a scaling factor.

(b) If u is p-irreducible and aperiodic and the invariant kernel is proper, then there exists a set $\widetilde{\Theta} \subseteq \Theta$ with $\int_{\widetilde{\Theta}} p(\boldsymbol{\theta} \mid C)\, d\nu(\boldsymbol{\theta}) = 1$ such that if $\boldsymbol{\theta}^{(0)} \in \widetilde{\Theta}$, then

$$\lim_{m\to\infty} \int_{\widetilde{\Theta}} \left| u^{(m)}(\boldsymbol{\theta} \mid \boldsymbol{\theta}^{(0)}, C) - p(\boldsymbol{\theta} \mid C) \right| d\nu(\boldsymbol{\theta}) = 0. \tag{4.29}$$

If u is ergodic (i.e., if it is also Harris recurrent), then $\widetilde{\Theta} = \Theta$.

(c) If u is ergodic, then for all $\boldsymbol{\theta}^{(0)} \in \Theta$ and all functions $g(\boldsymbol{\theta})$ such that $\int_\Theta |g(\boldsymbol{\theta})|\, p(\boldsymbol{\theta} \mid C)\, d\nu(\boldsymbol{\theta}) < \infty$, we obtain

$$M^{-1} \sum_{m=1}^{M} g(\boldsymbol{\theta}^{(m)}) \overset{\text{a.s.}}{\rightarrow} \int_\Theta g(\boldsymbol{\theta}) p(\boldsymbol{\theta} \mid C)\, d\nu(\boldsymbol{\theta}). \tag{4.30}$$

Proof: Conclusions (a) and (b) follow immediately from Theorem 1 and conclusion (c) from Theorem 3, both in Tierney (1994). ■

Observe that if the transition kernel u is ergodic, then the invariant distribution is unique and (4.29) and (4.30) both obtain.

In using simulation methods for Bayesian inference we are concerned with vectors of interest $\boldsymbol{\omega}$, and functions $h(\boldsymbol{\omega})$, that are not deterministic functions of $\boldsymbol{\theta}$. Theorem 4.5.1 can be extended immediately to include these cases.

Theorem 4.5.2 Convergence of a Vector of Interest with Continuous State Markov Chains Suppose that $\{\boldsymbol{\theta}^{(m)}\}$ is ergodic with invariant density $p(\boldsymbol{\theta} \mid I)$

and $\boldsymbol{\omega}^{(m)} \sim p(\boldsymbol{\omega} \mid \theta^{(m)}, I)$. Then $\{\boldsymbol{\theta}^{(m)}, \boldsymbol{\omega}^{(m)}\}$ is ergodic with invariant density $p(\boldsymbol{\theta} \mid I)p(\boldsymbol{\omega} \mid \theta, I)$.

Proof: Since $\{\boldsymbol{\theta}^{(m)}\}$ is aperiodic, $\{\boldsymbol{\theta}^{(m)}, \boldsymbol{\omega}^{(m)}\}$ must be also. Let G be any subset of $\Theta \times \Omega$ such that $\int_G p(\boldsymbol{\theta} \mid I)p(\boldsymbol{\omega} \mid \boldsymbol{\theta}, I)\, d\nu(\boldsymbol{\omega})\, d\nu(\boldsymbol{\theta}) > 0$, and let N be any positive integer. Let $D = A \times B \subseteq G$, such that $p(\boldsymbol{\omega} \in B \mid \boldsymbol{\theta}, I) \geq \varepsilon > 0\, \forall\, \boldsymbol{\theta} \in A$. Then

$$\sum_{m=1}^{M} I_G(\boldsymbol{\theta}^{(m)}, \boldsymbol{\omega}^{(m)}) \geq \sum_{m=1}^{M} I_D(\boldsymbol{\theta}^{(m)}, \boldsymbol{\omega}^{(m)})$$

but

$$\lim_{M \to \infty} P\left[\frac{\sum_{m=1}^{M} I_D(\boldsymbol{\theta}^{(m)}, \boldsymbol{\omega}^{(m)})}{\sum_{m=1}^{M} I_A(\boldsymbol{\theta}^{(m)})} > \frac{\varepsilon}{2} \mid C\right] = 1.$$

Since (4.28) is true for $L = [2N/\varepsilon] + 1$, $\{\boldsymbol{\theta}^{(m)}, \boldsymbol{\omega}^{(m)}\}$ is Harris recurrent. ■

An ergodic Markov chain can be used to compute simulation-consistent approximations of Bayes actions, in exactly the same way as was the case for direct sampling (Theorem 4.1.2).

Theorem 4.5.3 Approximation of Bayes Actions by MCMC Sampling Suppose that in the Markov chain C the sequence $\{\boldsymbol{\theta}^{(m)}, \boldsymbol{\omega}^{(m)}\}$ is ergodic with invariant density $p(\boldsymbol{\theta} \mid I)p(\boldsymbol{\omega} \mid \theta, I)$. Let $L(\mathbf{a}, \boldsymbol{\omega}) \geq 0$ be a loss function defined on $A \times \Omega$ and suppose that the risk function

$$R(\mathbf{a}) = \int_{\Omega} \int_{\Theta} L(\mathbf{a}, \boldsymbol{\omega}) p(\boldsymbol{\theta}) p(\boldsymbol{\omega} \mid \boldsymbol{\theta})\, d\boldsymbol{\theta}\, d\boldsymbol{\omega}$$

has a strict global minimum at $\widehat{\mathbf{a}} \in A \subseteq \mathbb{R}^m$. Suppose further that for a suitably defined open neighborhood of $\widehat{\mathbf{a}}$, $N(\widehat{\mathbf{a}})$:

1. $M^{-1} \sum_{m=1}^{M} L(\mathbf{a}, \boldsymbol{\omega}^{(m)}) \xrightarrow{p} R(\mathbf{a})$ uniformly on $N(\widehat{\mathbf{a}})$.
2. $\partial L(\mathbf{a}, \boldsymbol{\omega})/\partial \mathbf{a}$ exists and is a continuous function of $\mathbf{a}$, for all $\boldsymbol{\omega} \in \Omega$ and all $\mathbf{a} \in N(\widehat{\mathbf{a}})$.

Let A_M be the set of all roots of $M^{-1} \sum_{m=1}^{M} \partial L(\mathbf{a}, \boldsymbol{\omega}^{(m)})/\partial \mathbf{a} = \mathbf{0}$. Then for any $\varepsilon > 0$, $\lim_{M \to \infty} P[\inf_{\mathbf{a} \in A_M} (\mathbf{a} - \widehat{\mathbf{a}})'(\mathbf{a} - \widehat{\mathbf{a}}) > \varepsilon \mid C] = 0$.

Proof: The result follows from Amemiya (1985), Theorem 4.1.2. ■

Condition 1 can make this result somewhat more awkward to apply than Theorems 4.1.2 or 4.2.3. Although it must be verified for the problem at hand, it can be expected to apply widely. For example, if in the invariant density $p(\boldsymbol{\theta} \mid I)p(\boldsymbol{\omega} \mid \boldsymbol{\theta}, I)$ has moments at least of order n and $L(\mathbf{a}, \boldsymbol{\omega})$ is a polynomial of order n in $\boldsymbol{\omega}$, then condition 1 is satisfied.

Section 4.3 demonstrated how to construct an MCMC algorithm for which specified $p(\boldsymbol{\theta} \mid I)$ is an invariant density. In order to use the realization from such an algorithm to provide a simulation-consistent approximation of a moment under $p(\boldsymbol{\theta} \mid I)$, it is necessary to show that the transition density of the algorithm is ergodic. Direct application of Theorem 4.5.1 can be somewhat tedious. For the Gibbs sampler and the Hastings–Metropolis algorithm there are respective sufficient conditions for ergodicity that are easier to verify. These conditions are stronger than those in Theorem 4.5.1, but they are often satisfied in practice.

4.5.1 Convergence of the Gibbs Sampler

Suppose that a Gibbs sampler is constructed from a specified probability density $p(\boldsymbol{\theta} \mid I)$ as described in Section 4.3.1, producing the transition density $p(\boldsymbol{\theta}^{(m)} \mid \boldsymbol{\theta}^{(m-1)}, G)$ defined in (4.14). If $\boldsymbol{\theta}^{(0)} \in \Theta$, then $p(\boldsymbol{\theta} \mid I)$ is an invariant density of $p(\boldsymbol{\theta}^{(m)} \mid \boldsymbol{\theta}^{(m-1)}, G)$ as shown in Section 4.3.1. It remains to establish that the density of interest $p(\boldsymbol{\theta} \mid I)$ is the *unique* invariant density of the Markov chain, and the sense in which the chain converges. The following result is immediate and is often easy to apply.

Corollary 4.5.1 A First Sufficient Condition for Convergence of the Gibbs Sampler Suppose that for every point $\boldsymbol{\theta} \in \Theta$ and every ν-measurable $A \subseteq \Theta$

$$\int_A p(\boldsymbol{\theta} \mid I)\, d\nu(\boldsymbol{\theta}) > 0 \Rightarrow \int_A p(\boldsymbol{\theta}^* \mid \boldsymbol{\theta}, G)\, d\nu(\boldsymbol{\theta}^*) > 0.$$

Then the transition kernel of the Gibbs sampler G is ergodic.

Example 4.5.1 Convergence of the Gibbs Sampler in the Normal Linear Regression Model Corollary 4.5.1 establishes the ergodicity of the Gibbs sampler in Example 4.3.1. To turn to a common but more difficult variant, consider the special case of the restricted normal linear regression model of Examples 4.2.1 and 4.3.2, in which $S = \{\boldsymbol{\beta} : a_i \leq \beta_i \leq w_i (i = 1, \ldots, k)\}$, where $-\infty \leq a_i < w_i \leq \infty (i = 1, \ldots, k)$. The Gibbs sampler can be used to draw from the posterior distribution, with a full blocking on each β_j as well as h. Corollary 4.5.1 establishes convergence (but Theorem 4.5.4, below, does not). Section 5.3 treats this model, and some variants on it, in detail.

An alternative to Corollary 4.5.1 is provided by Roberts and Smith (1994).

Theorem 4.5.4 A Second Sufficient Condition for Convergence of the Gibbs Sampler Suppose that $\boldsymbol{\theta} \mid I$ is absolutely continuous and the following three conditions are satisfied:

1. The invariant density $p(\boldsymbol{\theta} \mid I)$ is lower semicontinuous; that is, for all $\boldsymbol{\theta}$ with $p(\boldsymbol{\theta} \mid I) > 0$, there exists an open neighborhood $N_{\boldsymbol{\theta}}$ of $\boldsymbol{\theta}$ and $\varepsilon > 0$ such that for all $\boldsymbol{\theta}^* \in N_{\boldsymbol{\theta}}$, $p(\boldsymbol{\theta}^* \mid I) \geq 0$.
2. For every point $\boldsymbol{\theta}^* \in \Theta$ and each block b of the Gibbs sampler, there exists an open neighborhood $N(\boldsymbol{\theta}^*_{(-b)})$ of $\boldsymbol{\theta}^*_{(-b)}$ and a bounded function $c(\boldsymbol{\theta}_{(-b)})$ such that for all $\boldsymbol{\theta}_{(-b)} \in N(\boldsymbol{\theta}^*_{(-b)})$

$$\int_{\Theta_{(b)}} p(\boldsymbol{\theta}^*_{(<b)}, \boldsymbol{\theta}_{(b)}, \boldsymbol{\theta}^*_{(>b)} \mid I)\, d\theta_{(b)} \leq c(\boldsymbol{\theta}^*_{(-b)}).$$

3. Θ is connected.

Then the transition kernel of the Gibbs sampler is ergodic.

Proof: See Theorem 2 of Roberts and Smith (1994). ■

Theorem 4.5.4 rules out situations like the one shown in Figure 4.2b, where the support of the posterior density is a closed set. For any point $\boldsymbol{\theta}$ on the boundary there is no open neighborhood $N_{\boldsymbol{\theta}}$ such that for all $\boldsymbol{\theta}^* \in N_{\boldsymbol{\theta}}$, $p(\boldsymbol{\theta}^* \mid I)$ is bounded away from 0.

Example 4.5.2 Convergence of the Gibbs Sampler in a Normal Linear Regression Model with Weak Inequality Constraints Consider the normal linear regression model

$$y_t \sim N(\beta_1 + \beta_2 x_{t2} + \beta_3 x_{t3},\ h^{-1}) \quad (t = 1, \ldots, T),$$
$$\boldsymbol{\beta}' = (\beta_1, \beta_2, \beta_3)' \sim N(\underline{\boldsymbol{\beta}}, \underline{\mathbf{H}}^{-1}),\ \ \underline{s}^2 h \sim \chi^2(\underline{\nu}),$$

subject to the constraint $0 < \beta_2 + \beta_3 < 1$. If a Gibbs sampler with the four blocks $\beta_1, \beta_2, \beta_3, h$ is used, then Theorem 4.5.4 establishes ergodicity but Corollary 4.5.1 does not. On the other hand, suppose that the model is recast as

$$y_t \sim N[\gamma_1 + \gamma_2 x_{t2} + \gamma_3 (x_{t3} - x_{t2}),\ h^{-1}] \quad (t = 1, \ldots, T),$$
$$\boldsymbol{\gamma}' = (\gamma_1, \gamma_2, \gamma_3)' \sim N(\mathbf{A}\underline{\boldsymbol{\beta}}, \mathbf{A}\underline{\mathbf{H}}^{-1}\mathbf{A}'),\ \ \underline{s}^2 h \sim \chi^2(\underline{\nu}),$$

subject to the constraint $0 < \gamma_2 < 1$, where

$$\mathbf{A} = \begin{bmatrix} 1 & 0 & 0 \\ 0 & 1 & 1 \\ 0 & 0 & 1 \end{bmatrix}.$$

Then a Gibbs sampler with the three blocks (γ_1, γ_3), γ_2, and h can be used, and either Corollary 4.5.1 or Theorem 4.5.4 establishes ergodicity. This is a particular case of linear inequality constraints in the normal linear regression model taken up in Section 5.3.

Tierney (1994) discusses weaker conditions for convergence of the Gibbs sampler. However, the conditions stated here are satisfied for a wide range of problems in econometrics and statistics and are substantially easier to verify.

4.5.2 Convergence of the Metropolis–Hastings Algorithm

Tierney (1994) and Roberts and Smith (1994) show that the convergence properties of the Metropolis–Hastings algorithm are inherited from those of $q(\boldsymbol{\theta}^* \mid \boldsymbol{\theta}, H)$; if q is aperiodic and p-irreducible, then so is the Metropolis–Hastings algorithm. This feature leads to a sufficient condition for convergence analogous to Corollary 4.5.1.

Theorem 4.5.5 A First Sufficient Condition for Convergence of the Metropolis–Hastings Algorithm Suppose that for every point $\boldsymbol{\theta} \in \Theta$ and every $A \subseteq \Theta$ with the property $\int_A p(\boldsymbol{\theta} \mid I)\, d\nu(\boldsymbol{\theta}) > 0$, it is the case that

$$\int_A q(\boldsymbol{\theta}^* \mid \boldsymbol{\theta}, H)\, d\nu(\boldsymbol{\theta}^*) > 0.$$

Then the transition kernel of the Metropolis–Hastings algorithm is ergodic.

Proof: See Tierney (1994), Corollary 2. ■

The condition in Theorem 4.5.5 may be restated as requiring that if $\boldsymbol{\theta}$ is in the support of $p(\boldsymbol{\theta} \mid I)$, then the support of $q(\boldsymbol{\theta}^* \mid \boldsymbol{\theta}, H)$ includes the support of $p(\boldsymbol{\theta} \mid I)$. This condition is satisfied for many Metropolis–Hastings algorithms, which are therefore ergodic.

Example 4.5.3 Some Generically Ergodic Metropolis–Hastings Algorithms Example 4.3.2 introduced the specific case of a random-walk Metropolis–Hastings chain in which $q(\boldsymbol{\theta}^* \mid \boldsymbol{\theta}, H)$ is the $N(\boldsymbol{\theta}, \mathbf{V})$ density. Since the support of this density is $\mathbb{R}^k$, this algorithm satisfies the condition in Theorem 4.5.5. Any Metropolis independence chain $q(\boldsymbol{\theta}^* \mid \boldsymbol{\theta}, H) = q(\boldsymbol{\theta}^* \mid H)$ in which the support of q includes the support of $p(\boldsymbol{\theta} \mid I)$ is also ergodic. As should be clear from previous discussion—Example 4.3.2 in the former case and Section 4.2 in the latter—this condition provides no assurance that the algorithm is sufficiently efficient to be practical. Further analytical work, trial computations, or both are needed to provide a form of the algorithm that is practical in each case.

A complementary sufficient condition for convergence of Metropolis–Hastings chains is provided by the following result, which is analogous to Theorem 4.5.4 for the Gibbs sampler.

Theorem 4.5.6 A Second Sufficient Condition for Convergence of the Metropolis–Hastings Algorithm Suppose that for all pairs $(\boldsymbol{\theta}, \boldsymbol{\theta}^*) \in \Theta \times \Theta$, $p(\boldsymbol{\theta} \mid I)$ and $q(\boldsymbol{\theta}^* \mid \boldsymbol{\theta}, H)$ are positive and continuous. Then the Metropolis–Hastings transition kernel is ergodic.

Proof: See Mengersen and Tweedie (1996), Lemmas 1.1 and 1.2. ■

Exercise 4.5.1 Simulation from the Bivariate Normal Distribution (This is a continuation of Exercise 4.1.2.) Suppose $(x, y)' \overset{\text{i.i.d.}}{\sim} N(\boldsymbol{\mu}, \boldsymbol{\Sigma})$ with

$$\boldsymbol{\mu} = \begin{pmatrix} \mu_1 \\ \mu_2 \end{pmatrix}, \quad \Sigma = \begin{bmatrix} \sigma_{11} & \sigma_{12} \\ \sigma_{12} & \sigma_{22} \end{bmatrix}.$$

Let $\varepsilon_j^{(m)} (j = 1, 2; m = 1, 2, \ldots)$ denote mutually independent, standard normal random variables. The function of interest is

$$f(x, y) = \begin{cases} (x^2 + y^2)^{1/2} \text{ if } x^2 + y^2 < 1 \\ 1 \text{ if } x^2 + y^2 \geq 1a \end{cases}.$$

(a) Consider the simulation

$$x^{(m)} = \mu_1 + (\sigma_{12}/\sigma_{22})(y^{(m-1)} - \mu_2) + (\sigma_{11} - \sigma_{12}^2/\sigma_{22})^{1/2} \varepsilon_1^{(m)},$$

$$y^{(m)} = \mu_2 + (\sigma_{12}/\sigma_{11})(x^{(m)} - \mu_1) + (\sigma_{22} - \sigma_{12}^2/\sigma_{11})^{1/2} \varepsilon_2^{(m)},$$

with $y^{(0)} \sim N(\mu_2, \sigma_{22})$. Show that

$$M^{-1} \sum_{m=1}^{M} f(x^{(m)}, y^{(m)}) \overset{\text{a.s.}}{\to} E[f(x, y)].$$

(b) Derive the correlation coefficient between $x^{(m)}$ and $x^{(m+1)}$ in (a). [Recall that in the algorithm in Exercise 4.1.2, $x^{(m)}$ and $x^{(m+1)}$ are uncorrelated.]

Exercise 4.5.2 Convergence of MCMC with an Inequality-Constrained Support Return to the algorithms you constructed in Exercise 4.3.3.

(a) Show that the Markov chain is ergodic, where the posterior is the unique invariant distribution in each case.

(b) Suppose that the inequality restrictions in parts (b) and (c) of Exercise 4.3.3 were weak rather than strong. Modify the Markov chain appropriately, and show that it is ergodic.

Exercise 4.5.3 Identification, Proper Posteriors, and MCMC This problem asks you to carry out some exercises with a simple normal model. The main lesson to be drawn is in part (e): MCMC algorithms can be constructed in models where the posterior distribution does not even exist, if one is careless. Part (b) illustrates how a model that is unidentified in the conventional sense that $p(\mathbf{y} \mid \boldsymbol{\theta}_{A1}, A) = p(\mathbf{y} \mid \boldsymbol{\theta}_{A2}, A)$ for $\boldsymbol{\theta}_{A1} \neq \boldsymbol{\theta}_{A2}$ can still have a posterior distribution for all parameters, given a proper prior. [For more on identification in a Bayesian context, see Poirier (1998).] If you have had exposure to the ideas of integration and cointegration in time series, then the sense in which an invariant distribution does not exist in part (e) should be quite clear.

(a) Suppose $y_t \mid (\mu, A) \overset{\text{i.i.d.}}{\sim} N(\mu, 1)$ $(t = 1, \ldots, T)$ and $\mu \mid A \sim N(0, \lambda^{-1})$. Derive $\mu \mid (\mathbf{y}^o, A)$.

(b) Now consider the model

$$\boldsymbol{\mu} \mid B \sim N(\mathbf{0}, \lambda^{-1}\mathbf{I}_2), \quad y_t \mid (\boldsymbol{\mu}, B) \overset{\text{i.i.d.}}{\sim} N(\mu_1 + \mu_2, 1).$$

Derive $\boldsymbol{\mu} \mid (\mathbf{y}^o, B)$.

(c) Consider the model $y_t \mid (\boldsymbol{\mu}, C) \overset{\text{i.i.d.}}{\sim} N(\mu_1 + \mu_2, 1)$ with an improper prior distribution $p(\boldsymbol{\mu} \mid A) \propto 1 \ \forall \ \boldsymbol{\mu} \in \mathbb{R}^2$. Show that the posterior distribution is improper.

(d) Derive the distributions $\mu_1 \mid (\mathbf{y}^o, \mu_2, C)$ and $\mu_2 \mid (\mathbf{y}^o, \mu_1, C)$. Note that each is proper.

(e) Construct a Gibbs sampling algorithm based on your work in (d):

$$\mu_1^{(m)} \sim p(\mu_1 \mid \mathbf{y}^o, \mu_2^{(m-1)}, C),$$
$$\mu_2^{(m)} \sim p(\mu_2 \mid \mathbf{y}^o, \mu_1^{(m)}, C).$$

Show that

(i) $\mu_2^{(m)} = \mu_2^{(m-1)} + \zeta_m$ where $\zeta_m \overset{\text{i.i.d.}}{\sim} N(0, 2T^{-1})$.

(ii) $\mu_1^{(m)} + \mu_2^{(m)} \overset{\text{i.i.d.}}{\sim} N(\overline{y}^o, T^{-1})$.

Exercise 4.5.4 Economic Decisionmaking A stochastic production relation is modeled as

$$y_t = \beta_1 + \beta_2 x_{t1} + \beta_3 x_{t2} + \beta_4 x_{t1}^2 + \beta_5 x_{t2}^2 + \beta_6 x_{t1} x_{t2} + \varepsilon_t, \qquad (4.31)$$
$$\varepsilon_t \overset{\text{i.i.d.}}{\sim} N(0, h^{-1}),$$

where y_t is output and x_{t1} and x_{t2} are the two inputs treated as ancillary statistics in the model. The prior distribution has two independent components:

$$\boldsymbol{\beta} = (\beta_1, \beta_2, \beta_3, \beta_4, \beta_5, \beta_6)' \sim N(\underline{\boldsymbol{\beta}}, \underline{\mathbf{H}}^{-1}) \qquad (4.32)$$
$$\underline{s}^2 h \sim \chi^2(\underline{\nu}).$$

The distribution in (4.32) is subject to the further restrictions that

$$E(y \mid \mathbf{x}) = \beta_1 + \beta_2 x_1 + \beta_3 x_2 + \beta_4 x_1^2 + \beta_5 x_2^2 + \beta_6 x_1 x_2$$

is a strictly concave function for all $\mathbf{x} = (x_1, x_2)' > \mathbf{0}$, and that there exists $\mathbf{x}^* > \mathbf{0}$ such that $E(y \mid \mathbf{x}^*) \geq E(y \mid \mathbf{x})$ for all $\mathbf{x} > \mathbf{0}$.

(a) Carefully express the posterior density kernel for this model.

(b) We could use any one of several posterior simulators in this model. Describe two different posterior simulators. What considerations will be important in determining which is more efficient? (A good answer will have nontrivial differences in the two simulators, plus a substantial discussion of efficiency.)

(c) Suppose that you have completed (a) and (b) and have posterior simulator output $\{\boldsymbol{\beta}^{(m)}, h^{(m)}\}$ $(m = 1, \ldots, M)$ using all the data for periods 1 through T. The manager of the firm using the production relation (4.31) knows output price p_{T+1} and input prices $r_{T+1,1}$ and $r_{T+1,2}$ for $x_{T+1,1}$ and $x_{T+1,2}$, respectively, in period $T+1$. He must choose $x_{T+1,1}$ and $x_{T+1,2}$ before observing ε_{T+1}. (Inputs cannot be negative.) The manager's objective is to choose $x_{T+1,1}$ and $x_{T+1,2}$ so as to maximize

$$E\{U(\pi_{T+1}) \mid [(y_t, x_{t1}, x_{t2})\ (t = 1, \ldots, T), p_{T+1}, r_{T+1,1}, r_{T+1,2}]\},$$

where $\pi_{T+1} = p_{T+1}y_{T+1} - r_{T+1,1}x_{T+1,1} - r_{T+1,2}x_{T+1,2}$ and $U(\cdot)$ is a monotone increasing, strictly concave function with first and second derivatives that are easy to compute. Indicate how you would solve the manager's problem using the output of the posterior simulator described in (b).

4.6 HYBRID MARKOV CHAIN MONTE CARLO METHODS

The utility of Monte Carlo methods in Bayesian inference stems in great part from combinations of algorithms. Example 4.2.4 showed that a hybrid importance and acceptance sampling algorithm could be more efficient than either algorithm alone. The addition of MCMC algorithms widens the scope for combination. These hybrid algorithms not only increase efficiency. More importantly, they can provide elegant yet practical solutions of difficult problems in the construction of posterior simulators. This section examines two such hybrid algorithms.

4.6.1 Transition Mixtures

In the context of the Metropolis–Hastings algorithm, suppose that there are J different transition probability densities $q(\boldsymbol{\theta}^* \mid \boldsymbol{\theta}, H_j)$ that might be used. A transition mixture chooses randomly between the J densities $q(\boldsymbol{\theta}^* \mid \boldsymbol{\theta}, H_j)$ with respective choice probabilities π_j assigned to the densities. The probabilities π_j are constant and do not depend on $\boldsymbol{\theta}$:

$$q(\boldsymbol{\theta}^* \mid \boldsymbol{\theta}, H) = \sum_{j=1}^{J} \pi_j q(\boldsymbol{\theta}^* \mid \boldsymbol{\theta}, H_j).$$

Once density j is selected, a candidate $\boldsymbol{\theta}^*$ is drawn from $q(\boldsymbol{\theta}^* \mid \boldsymbol{\theta}, H_j)$, and is accepted with probability

$$\alpha(\boldsymbol{\theta}^* \mid \boldsymbol{\theta}, H_j) = \min\left[\frac{p(\boldsymbol{\theta}^* \mid I)/q(\boldsymbol{\theta}^* \mid \boldsymbol{\theta}, H_j)}{p(\boldsymbol{\theta} \mid I)/q(\boldsymbol{\theta} \mid \boldsymbol{\theta}^*, H_j)}, 1\right]. \tag{4.33}$$

Note that only the chosen transition density enters (4.33): the other densities and the choice probabilities π_j are irrelevant to acceptance or rejection of the candidate once the transition density has been selected.

To see that $p(\boldsymbol{\theta} \mid I)$ is an invariant density of the transition mixture, note that the reversibility condition is

$$p(\boldsymbol{\theta} \mid I) \sum_{j=1}^{J} \pi_j q(\boldsymbol{\theta}^* \mid \boldsymbol{\theta}, H_j) \alpha(\boldsymbol{\theta}^* \mid \boldsymbol{\theta}, H_j)$$

$$= p(\boldsymbol{\theta}^* \mid I) \sum_{j=1}^{J} \pi_j q(\boldsymbol{\theta} \mid \boldsymbol{\theta}^*, H_j) \alpha(\boldsymbol{\theta} \mid \boldsymbol{\theta}^*, H_j).$$

This condition holds if

$$p(\boldsymbol{\theta} \mid I) q(\boldsymbol{\theta}^* \mid \boldsymbol{\theta}, H_j) \alpha(\boldsymbol{\theta}^* \mid \boldsymbol{\theta}, H_j)$$
$$= p(\boldsymbol{\theta}^* \mid I) q(\boldsymbol{\theta} \mid \boldsymbol{\theta}^*, H_j) \alpha(\boldsymbol{\theta} \mid \boldsymbol{\theta}^*, H_j) \quad (j = 1, \ldots, J). \tag{4.34}$$

Condition (4.34) leads to (4.33), just as (4.21) led to (4.15).

Transition mixtures can be powerful tools in building posterior simulators that are ergodic and robust to ill-behaved posterior distributions. To see how ergodicity arises, note that if the support of at least one $q(\boldsymbol{\theta}^* \mid \boldsymbol{\theta}, H_j)$ includes the support of $p(\boldsymbol{\theta} \mid I)$, then the same is true of the transition mixture. Theorem 4.5.5 then implies that the transition mixture kernel is ergodic. [Tierney (1994) shows that it is sufficient that just one of the transition kernels be ergodic.]

4.6.2 Metropolis within Gibbs

Suppose that in attempting to implement a Gibbs sampling algorithm, a conditional density $p(\boldsymbol{\theta}_{(b)} \mid \boldsymbol{\theta}_{-(b)}, I)$ is intractable. The density is not of any known form, and efficient acceptance sampling algorithms are not at hand. This problem can be addressed by applying the Metropolis–Hastings algorithm in block b of the Gibbs sampler while treating the other blocks in the usual way. Specifically, let $q(\boldsymbol{\theta}^*_{(b)} \mid \boldsymbol{\theta}, H_b)$ be the density (indexed by $\boldsymbol{\theta}$) from which candidate $\boldsymbol{\theta}^*_{(b)}$ is drawn. At iteration m, block b, of the Gibbs sampler draw $\boldsymbol{\theta}^*_{(b)} \sim q(\boldsymbol{\theta}^*_{(b)} \mid \boldsymbol{\theta}^{(m)}_{<(b)}, \boldsymbol{\theta}^{(m-1)}_{>(b-1)}, H_b)$, and set $\boldsymbol{\theta}^{(m)}_{(b)} = \boldsymbol{\theta}^*_{(b)}$ with probability

$$\alpha\big(\boldsymbol{\theta}^*_{(b)} \mid \boldsymbol{\theta}^{(m)}_{<(b)}, \boldsymbol{\theta}^{(m-1)}_{>(b-1)}, H_b\big)$$
$$= \min\left\{ \frac{p\big(\boldsymbol{\theta}^{(m)}_{<(b)}, \boldsymbol{\theta}^*_{(b)}, \boldsymbol{\theta}^{(m-1)}_{>(b)} \mid I\big) / q\big(\boldsymbol{\theta}^*_{(b)} \mid \boldsymbol{\theta}^{(m)}_{<(b)}, \boldsymbol{\theta}^{(m-1)}_{>(b-1)}, H_b\big)}{p\big(\boldsymbol{\theta}^{(m)}_{<(b)}, \boldsymbol{\theta}^{(m-1)}_{>(b-1)} \mid I\big) / q\big(\boldsymbol{\theta}^{(m-1)}_{(b)} \mid \boldsymbol{\theta}^{(m)}_{<(b)}, \boldsymbol{\theta}^*_{(b)}, \boldsymbol{\theta}^{(m-1)}_{>(b)}, H_b\big)}, \; 1 \right\}$$

If $\boldsymbol{\theta}^{(m)}_{(b)}$ is not set to $\boldsymbol{\theta}^*_{(b)}$, then $\boldsymbol{\theta}^{(m)}_{(b)} = \boldsymbol{\theta}^{(m-1)}_{(b)}$. The procedure for $\boldsymbol{\theta}_{(b)}$ is exactly the same as for a standard Metropolis step, except that $\boldsymbol{\theta}_{(-b)}$ also enters the density p and transition density q. It is usually called a *Metropolis within Gibbs step.*

To see that $p(\boldsymbol{\theta} \mid I)$ is an invariant density of this Markov chain, consider the simple case of two blocks with a Metropolis within Gibbs step in the second block. Adapting the notation of (4.19), describe the Metropolis step for the second block by

$$p(\boldsymbol{\theta}^*_{(2)} \mid \boldsymbol{\theta}_{(1)}, \boldsymbol{\theta}_{(2)}, H_2) = u(\boldsymbol{\theta}^*_{(2)} \mid \boldsymbol{\theta}_{(1)}, \boldsymbol{\theta}_{(2)}, H_2) + r(\boldsymbol{\theta}_{(2)} \mid \boldsymbol{\theta}_{(1)}, H_2)\delta_{\boldsymbol{\theta}_{(2)}}(\boldsymbol{\theta}^*_{(2)})$$

where

$$u(\boldsymbol{\theta}^*_{(2)} \mid \boldsymbol{\theta}_{(1)}, \boldsymbol{\theta}_{(2)}, H_2) = \alpha(\boldsymbol{\theta}^*_{(2)} \mid \boldsymbol{\theta}_{(1)}, \boldsymbol{\theta}_{(2)}, H_2)q(\boldsymbol{\theta}^*_{(2)} \mid \boldsymbol{\theta}_{(1)}, \boldsymbol{\theta}_{(2)}, H_2)$$

and

$$r(\boldsymbol{\theta}_{(2)} \mid \boldsymbol{\theta}_{(1)}, H_2) = 1 - \int_{\Theta_2} u(\boldsymbol{\theta}^*_{(2)} \mid \boldsymbol{\theta}_{(1)}, \boldsymbol{\theta}_{(2)}, H_2)\, d\nu(\boldsymbol{\theta}^*_{(2)}). \tag{4.35}$$

The one-step transition density for the entire chain is

$$p(\boldsymbol{\theta}^* \mid \boldsymbol{\theta}, G) = p(\boldsymbol{\theta}^*_{(1)} \mid \boldsymbol{\theta}_{(2)}, I)p(\boldsymbol{\theta}^*_{(2)} \mid \boldsymbol{\theta}_{(1)}, \boldsymbol{\theta}_{(2)}, H_2)$$

Then $p(\boldsymbol{\theta} \mid I)$ is an invariant density of $p(\boldsymbol{\theta}^* \mid \boldsymbol{\theta}, G)$ if

$$\int_{\Theta} p(\boldsymbol{\theta} \mid I)p(\boldsymbol{\theta}^* \mid \boldsymbol{\theta}, G)\, d\nu(\boldsymbol{\theta}) = p(\boldsymbol{\theta}^* \mid I). \tag{4.36}$$

To establish (4.36), begin by expanding the left side:

$$\int_{\Theta} p(\boldsymbol{\theta} \mid I)p(\boldsymbol{\theta}^* \mid \boldsymbol{\theta}, G)\, d\nu(\boldsymbol{\theta}) = \int_{\Theta_2}\int_{\Theta_1} p(\boldsymbol{\theta}_{(1)}, \boldsymbol{\theta}_{(2)} \mid I)\, d\nu(\boldsymbol{\theta}_{(1)})p(\boldsymbol{\theta}^*_{(1)} \mid \boldsymbol{\theta}_{(2)}, I)$$
$$\cdot [u(\boldsymbol{\theta}^*_{(2)} \mid \boldsymbol{\theta}^*_{(1)}, \boldsymbol{\theta}_{(2)}, H_2) + r(\boldsymbol{\theta}_{(2)} \mid \boldsymbol{\theta}^*_{(1)}, H_2)\delta_{\boldsymbol{\theta}_{(2)}}(\boldsymbol{\theta}^*_{(2)})]\, d\nu(\boldsymbol{\theta}_{(2)})$$
$$= \int_{\Theta_2} p(\boldsymbol{\theta}_{(2)} \mid I)p(\boldsymbol{\theta}^*_{(1)} \mid \boldsymbol{\theta}_{(2)} \mid I)u(\boldsymbol{\theta}^*_{(2)} \mid \boldsymbol{\theta}^*_{(1)}, \boldsymbol{\theta}_{(2)}, H_2)\, d\nu(\boldsymbol{\theta}_{(2)}) \tag{4.37}$$
$$+ \int_{\Theta_2} p(\boldsymbol{\theta}_{(2)} \mid I)p(\boldsymbol{\theta}^*_{(1)} \mid \boldsymbol{\theta}_{(2)} \mid I)r(\boldsymbol{\theta}_{(2)} \mid \boldsymbol{\theta}^*_{(1)}, H_2)\delta_{\boldsymbol{\theta}_{(2)}}(\boldsymbol{\theta}^*_{(2)})\, d\nu(\boldsymbol{\theta}_{(2)}). \tag{4.38}$$

In (4.37) and (4.38) we have utilized the fact that

$$p(\boldsymbol{\theta}_{(2)} \mid I) = \int_{\Theta_1} p(\boldsymbol{\theta}_{(1)}, \boldsymbol{\theta}_{(2)} \mid I)\, d\nu(\boldsymbol{\theta}_{(1)}).$$

Using Bayes rule (4.37) is the same as

$$p(\boldsymbol{\theta}^*_{(1)} \mid I) \int_{\Theta_2} p(\boldsymbol{\theta}_{(2)} \mid \boldsymbol{\theta}^*_{(1)}, I)u(\boldsymbol{\theta}^*_{(2)} \mid \boldsymbol{\theta}^*_{(1)}, \boldsymbol{\theta}_{(2)}, H_2)\, d\nu(\boldsymbol{\theta}_{(2)}). \tag{4.39}$$

Carrying out the integration in (4.38) yields

$$p(\boldsymbol{\theta}^*_{(2)} \mid I)p(\boldsymbol{\theta}^*_{(1)} \mid \boldsymbol{\theta}^*_{(2)} \mid I)r(\boldsymbol{\theta}^*_{(2)} \mid \boldsymbol{\theta}^*_{(1)}, H_2). \tag{4.40}$$

Recalling the reversibility of the Metropolis step, we obtain

$$p(\boldsymbol{\theta}_{(2)} \mid \boldsymbol{\theta}^*_{(1)}, I)u(\boldsymbol{\theta}^*_{(2)} \mid \boldsymbol{\theta}^*_{(1)}, \boldsymbol{\theta}_{(2)}, H_2) = p(\boldsymbol{\theta}^*_{(2)} \mid \boldsymbol{\theta}^*_{(1)}, I)u(\boldsymbol{\theta}_{(2)} \mid \boldsymbol{\theta}^*_{(1)}, \boldsymbol{\theta}^*_{(2)}, H_2),$$

and so (4.39) becomes

$$p(\boldsymbol{\theta}^*_{(1)} \mid I)p(\boldsymbol{\theta}^*_{(2)} \mid \boldsymbol{\theta}^*_{(1)}, I) \int_{\Theta_2} u(\boldsymbol{\theta}_{(2)} \mid \boldsymbol{\theta}^*_{(1)}, \boldsymbol{\theta}^*_{(2)}, H_2)\, d\nu(\boldsymbol{\theta}_{(2)}). \tag{4.41}$$

We can express (4.40) as

$$p(\boldsymbol{\theta}^*_{(1)}, \boldsymbol{\theta}^*_{(2)} \mid I)r(\boldsymbol{\theta}^*_{(2)} \mid \boldsymbol{\theta}^*_{(1)}, H_2). \tag{4.42}$$

Finally, recalling (4.35), the sum of (4.41) and (4.42) is $p(\boldsymbol{\theta}^*_{(1)}, \boldsymbol{\theta}^*_{(2)} \mid I)$, thus establishing (4.36).

This demonstration of invariance applies to the Gibbs sampler with b blocks, with a Metropolis within Gibbs step for one block, simply through the convention that Metropolis within Gibbs is used in the last block of each iteration. Metropolis within Gibbs steps can be used for several blocks, as well. The argument for invariance proceeds by mathematical induction, and the details are the same. Ergodicity can generally be established in the same way as for Gibbs samplers generally; Corollary 4.5.1 often applies. Section 7.1 provides a specific example of the Metropolis within Gibbs algorithm.

4.7 NUMERICAL ACCURACY AND CONVERGENCE IN MARKOV CHAIN MONTE CARLO

In any practical application we are concerned with the discrepancy between a posterior moment $\overline{h}$ and its numerical approximation from a posterior simulator with M iterations, $\overline{h}^{(M)}$. If the sequence $\{h(\boldsymbol{\omega}^{(m)})\}$ were i.i.d., this discrepancy could be evaluated by means of a conventional central limit theorem and the resulting numerical standard error (NSE, defined on p. 107) and the efficiency of the algorithm could be assessed using the estimated relative numerical efficiency (RNE, defined on p. 117). Serial correlation in $\{h(\boldsymbol{\omega}^{(m)})\}$ is inherent in Markov chain Monte Carlo, however, and so the need to evaluate numerical accuracy of the approximation $\overline{h}^{(M)}$ must be evaluated afresh. The serial dependence in MCMC algorithms also raises the prospect that if the initial value $\boldsymbol{\theta}^{(0)}$ is remote from the posterior distribution, then, although $\overline{h}^{(M)} \to \overline{h}$, early values of $h(\boldsymbol{\omega}^{(m)})$ may be atypical and approximations would be improved by discarding these early values. The issue here is what constitutes "early" and eliminating the possibility that M is still "early" in the sequence. This issue is often referred to as "convergence." There is a substantial literature on the practical aspects of these issues. Key works include Gelman and Rubin (1992), Geweke (1992), Geyer (1992), Cowles and Carlin (1996), Gelman (1996), Brooks and Gelman (1998), and Brooks and Roberts (1998).

To illustrate the issues, consider the Gibbs sampling algorithm for a normal distribution of the random variables (x, y), one block for x and one for y. (See Exercise 4.5.1.) Figure 4.3 illustrates the first 400 iterations from two distributions, each with mean zero and unit variances for x and y. In the first distribution [panels (a) and (b)] the correlation between x and y is $\rho = 0.90$ and in the second distribution [panels (c) and (d)] it is $\rho = 0.99$. Highest-density regions of size 0.2, 0.4, 0.6, 0.8, and 0.95 are indicated in panels (a) and (c). Panels (a) and (c) show (x, y) values from every other iteration, with the first 100 iterations indicated by a cross and the last 300 by a solid point. Panels (b) and (d) show the values of x in each iteration. In each case the starting value is $x = y = 8$. Values this far from the mean are improbable for either distribution, and this will often be the case in research applications. The values chosen here represent that situation, and

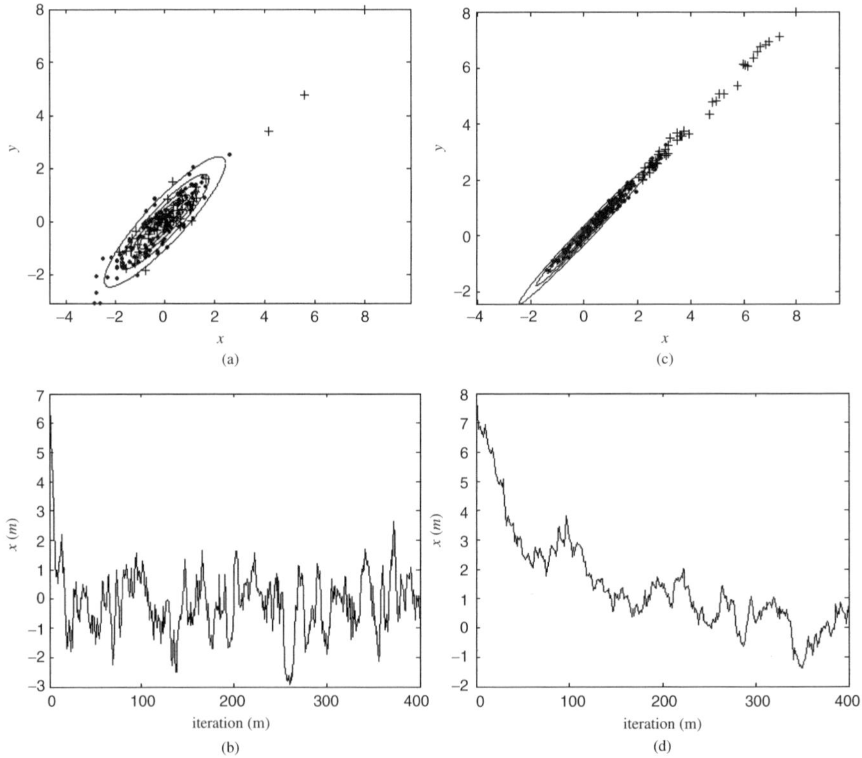

Figure 4.3. Output of a Gibbs sampler for a bivariate normal (x, y), blocked in x and y: $\rho = 0.9$, every second of 400 iterations (a) and x values of all iterations (b); $\rho = 0.99$, every second of 400 iterations (c) and x values of all iterations (d).

illustrate the implications for convergence. In the first case, $\rho = 0.90$, values of x and y have become representative of the bivariate distribution well before the 100th iteration. In the second case, $\rho = 0.99$, all the first 100 iterates are in the positive orthant, and in the next 300 draws there are more iterates in the positive than the negative orthant. This comparison illustrates the point that convergence questions arise from the serial dependence in the Markov chain: the greater the serial dependence, the longer it will take for the chain to become representative of the invariant distribution, other things the same.

We will proceed by first developing some tools for computing NSE and RNE, under the assumption that the sequence $\{\boldsymbol{\theta}^{(m)}, \boldsymbol{\omega}^{(m)}\}$ is stationary. Since the initial simulation $\boldsymbol{\theta}^{(0)}$ is not drawn from the posterior distribution and the sequence is serially dependent, this is not literally true. However, the analysis of the stationary case leads to some analytical tools that are useful in addressing the convergence question. The foundation for evaluating numerical accuracy is a central limit theorem for continuous state space Markov chains.

Definition 4.7.1 Suppose that a Markov chain C has n-step transition probability $P^n(A \mid \boldsymbol{\theta}, C) = P(\boldsymbol{\theta}^{(n)} \in A \mid \boldsymbol{\theta}^{(0)} = \boldsymbol{\theta}, C)$, defined on all ν-measurable sets A, and unique invariant probability $P(A \mid C)$. The Markov chain is *uniformly ergodic* if

$$\sup_{\boldsymbol{\theta} \in \Theta} \left\{ \sup_{A} \left| P^n(A \mid \boldsymbol{\theta}, C) - P(A \mid C) \right| \right\} \le L r^n \tag{4.43}$$

for some $L > 0$ and some positive $r < 1$.

Tierney (1994, p. 1714) derives two results that are useful in establishing uniform ergodicity. First, an independence Metropolis chain with bounded weight function $w(\boldsymbol{\theta}) = p(\boldsymbol{\theta} \mid I)/p(\boldsymbol{\theta} \mid H)$ is uniformly ergodic. (Recalling the similarity between the independence Metropolis kernel and importance sampling, and the discussion about bounded weight functions following Theorem 4.2.2, this result is not surprising.) Second, if one kernel in a transition mixture (Section 4.6.1) is uniformly ergodic, then the mixture kernel itself is uniformly ergodic. Thus for any Markov chain $\{\boldsymbol{\theta}^{(m)}\}$ we could in principle guarantee (4.43) by mixing the chain with an independence Metropolis kernel with a bounded weight function, as long as the posterior mean and variance were known to exist. If the likelihood function is bounded, then the prior distribution itself will provide such an independence transition kernel. The practical difficulty with this approach is that in most problems draws from the prior will rarely be accepted, and it is often difficult to find an independence kernel that overcomes this difficulty. Thus attention typically focuses directly on the unmixed MCMC algorithm.

Theorem 4.7.1 A Central Limit Theorem for MCMC Approximation of Moments Suppose that $\{\boldsymbol{\theta}^{(m)}\}$ is uniformly ergodic with unique invariant density $p(\boldsymbol{\theta} \mid I)$. Let $\boldsymbol{\omega}^{(m)} \sim p(\boldsymbol{\omega} \mid \boldsymbol{\theta}^{(m)}, I)$, suppose that $E[h(\boldsymbol{\omega}) \mid I] = \overline{h}$ and $\text{var}[h(\boldsymbol{\omega}) \mid I]$

exist and are finite, and let $\overline{h}^{(M)} = M^{-1}\sum_{m=1}^{M} h(\boldsymbol{\omega}^{(m)})$. Then there exists finite τ^2 such that

$$M^{1/2}(\overline{h}^{(M)} - \overline{h}) \xrightarrow{d} N(0, \tau^2). \tag{4.44}$$

Proof: The same proof used for Theorem 4.5.2 shows that uniform ergodicity of $\{\boldsymbol{\theta}^{(m)}\}$ implies uniform ergodicity of $\{\boldsymbol{\theta}^{(m)}, \boldsymbol{\omega}^{(m)}\}$. The result is then Theorem 5 of Tierney (1994). ■

Theorem 4.7.2 A Central Limit Theorem for MCMC Approximation of Bayes Actions In addition to the assumptions of Theorem 4.5.3, suppose further that $\boldsymbol{\theta}^{(m)}$ is uniformly ergodic, and that for a suitably defined open neighborhood of $\widehat{\mathbf{a}}$, $N(\widehat{\mathbf{a}})$:

1. $\partial^2 L(\mathbf{a}, \boldsymbol{\omega})/\partial \mathbf{a}\, \partial \mathbf{a}'$ exists and is a continuous function of $\mathbf{a}$, for all $\boldsymbol{\omega} \in \Omega$ and all $\mathbf{a} \in N(\widehat{\mathbf{a}})$.
2. $M^{-1/2}\sum_{m=1}^{M} \partial L(\mathbf{a}, \boldsymbol{\omega}^{(m)})/\partial \mathbf{a}|_{\mathbf{a}=\widehat{\mathbf{a}}} \xrightarrow{d} N(\mathbf{0}, \mathbf{B})$, where $\mathbf{B}$ is a nonnegative definite matrix.
3. $\mathbf{H} = E[\partial^2 L(\mathbf{a}, \boldsymbol{\omega})/\partial \mathbf{a}\, \partial \mathbf{a}'|_{\mathbf{a}=\widehat{\mathbf{a}}} \mid I]$ exists and is finite and nonsingular.
4. For any $\varepsilon > 0$, there exists M_ε such that

$$P\left[\sup_{\mathbf{a}\in N(\widehat{\mathbf{a}})} \left|\partial^3 L(\mathbf{a}, \boldsymbol{\omega})/\partial a_i\, \partial a_j\, \partial a_k\right| < M_\varepsilon \mid I\right] \geq 1 - \varepsilon$$

for all $i, j, k = 1, \ldots, m$.

Then if $\widehat{\mathbf{a}}_M$ is any element of A_M such that $\widehat{\mathbf{a}}_M \xrightarrow{p} \widehat{\mathbf{a}}$,

$$M^{1/2}(\widehat{\mathbf{a}}_M - \widehat{\mathbf{a}}) \xrightarrow{d} N(\mathbf{0}, \mathbf{H}^{-1}\mathbf{B}\mathbf{H}^{-1}). \tag{4.45}$$

Proof: The result is an application of Theorem 4.1.3 of Amemiya (1985). ■

To apply (4.44) or (4.45) in assessing numerical accuracy it is necessary to find a statistic $\widehat{\tau}^{2(M)} \xrightarrow{\text{a.s.}} \tau^2$ as was done for independence and importance sampling. An analogous approximation for the matrix $\mathbf{B}$ in condition 2 of Theorem 4.7.2 is also necessary. There are several approaches to this task. If we are willing to replicate the MCMC computations beginning with a randomly chosen $\boldsymbol{\theta}_A^{(0)}$ each time, comparison of the results provides a basis for approximation of τ^2; see Chan and Geyer (1994) and Chauveau and Diebolt (2000). The same ends can be accomplished by initiating several chains in the midst of the original chain, a process known as "splitting and regeneration"; see Mykland et al. (1995) and Robert (1995). The approach we take here uses a single chain, building on the following result, which is a staple of time series econometrics.

Theorem 4.7.3 Numerical Standard Errors for MCMC Suppose that $\{h(\boldsymbol{\omega}^{(m)})\}$ in Theorem 4.7.1 is a stationary process with autocovariance function

$$c_j = \text{cov}[h(\boldsymbol{\omega}^{(m)}), h(\boldsymbol{\omega}^{(m-j)})] \quad (j = 0, \pm 1, \pm 2, \ldots)$$

and spectral density function $S(\lambda) = \sum_{j=-\infty}^{\infty} c_j \cos(\lambda j)$. If $S(\lambda)$ is bounded uniformly both above and away from zero on $[0, \pi]$, then in (4.44), $\tau^2 = S(0) = \sum_{j=-\infty}^{\infty} c_j$. If

$$\widehat{c}_j^{(M)} = M^{-1} \sum_{m=j+1}^{M} [h(\boldsymbol{\omega}^{(m)}) - \overline{h}^{(M)}][h(\boldsymbol{\omega}^{(m-j)}) - \overline{h}^{(M)}]$$

and $L(M)$ is an integer-valued function for which $\lim_{M\to\infty} L(M) = \infty$ while $\lim_{M\to\infty} L(M)^2/M = 0$, then

$$\widehat{\tau}^{2(M)} = \widehat{S}(0) = \widehat{c}_0^{(M)} + 2\sum_{s=1}^{L-1}[(L-s)/L]\widehat{c}_s^{(M)} \overset{\text{a.s.}}{\to} S(0) = \tau^2. \tag{4.46}$$

Proof: See Newey and West (1987). ■

The condition on the spectral density function in Theorem 4.7.3 guarantees $c_j = \text{cov}[h(\boldsymbol{\omega}^{(m)}), h(\boldsymbol{\omega}^{(m-j)})]$ decays rapidly enough with increasing j that it is possible to obtain a consistent approximation of $\tau^2 = S(0) = \sum_{j=-\infty}^{\infty} c_j$. Given a modest strengthening of this condition, it is possible to investigate the question of whether the mean of $h(\boldsymbol{\omega}^{(m)})$ is the same over different segments of the entire simulation $\{h(\boldsymbol{\omega}^{(m)})\}$.

Theorem 4.7.4 Separated Partial Means Test for MCMC In addition to the assumptions of Theorems 4.7.1 and 4.7.3, suppose also that

$$\left|c_j\right| < c_0\rho^j \quad (j = 1, 2, \ldots) \tag{4.47}$$

for some $\rho \in [0, 1)$. Let p be a fixed positive integer. For each M such that $M_p = M/2p$ is an integer, define the p separated partial means:

$$\overline{h}_{j,p}^{(M)} = M_p^{-1} \sum_{m=1}^{M_p} h(\boldsymbol{\omega}^{(m+M(2j-1)/2p)}) \quad (j = 1, \ldots, p).$$

Let $\widehat{\tau}_{j,p}^{2(M)}$ be the estimate of τ^2 computed for $\overline{h}_{j,p}^{(M)}$ described in Theorem 4.7.3 $(j = 1, \ldots, p)$. Define the $(p-1) \times 1$ vector $\overline{\mathbf{h}}_p^{(M)}$ with jth element $\overline{h}_{j+1,p}^{(M)} - \overline{h}_{j,p}^{(M)}$, and the $(p-1) \times (p-1)$ tridiagonal matrix $\widehat{\mathbf{V}}_p^{(M)}$ in which $\widehat{v}_{jj}^{(M)} = M_p^{-1}(\widehat{\tau}_{j,p}^{2(M)} + \widehat{\tau}_{j+1,p}^{2(M)})$ and $v_{j,j-1} = v_{j-1,j} = -M_p^{-1}\widehat{\tau}_{j,p}^{2(M)}$. Then

$$\overline{\mathbf{h}}_p^{(M)\prime}[\widehat{\mathbf{V}}_p^{(M)}]^{-1}\overline{\mathbf{h}}_p^{(M)} \xrightarrow{d} \chi^2(p-1). \tag{4.48}$$

Proof: Theorem 4.7.1 implies that any linear combination of $\overline{h}_{j,p}^{(M)}$ has a limiting normal distribution, and consequently [see Rao (1965), Theorem 2c.5(iv)] $\overline{\mathbf{h}}_p^{(M)}$ has a limiting multivariate normal distribution. It remains only to show that $\lim_{M_p \to \infty} M_p \mathrm{cov}[\overline{h}_{j,p}^{(M)}, \overline{h}_{k,p}^{(M)}] = 0 \ \forall \ j \neq k$. By virtue of (4.47)

$$M_p \left|\mathrm{cov}[\overline{h}_{j,p}^{(M)}, \overline{h}_{k,p}^{(M)}]\right| < \frac{c_0}{M_p} \sum_{m=1}^{M_p} \sum_{n=1}^{M_p} \rho^{2|j-k|M_p+n-m} \leq \frac{c_0}{M_p} \sum_{m=1}^{M_p} \sum_{n=1}^{M_p} \rho^{2M_p+n-m}$$

$$< c_0 \sum_{m=1}^{M_p} \rho^{2M_p-m} = c_0\rho^{M_p} \sum_{m=1}^{M_p} \rho^{m-1} < \frac{c_0\rho^{M_p}}{(1-\rho)} \to 0. \qquad \blacksquare$$

Application of the partial means test involves choosing p as well as M. The theorem requires that p be fixed. Thus, for example, if we choose $p = 4$, then if $M = 1{,}000$ the partial means are based on iterations 126–250, 376–500, 626–750, and 876–1000, whereas if $M = 40{,}000$, the partial means are based on 5001–10,000, 15,001–20,000, 25,001–30,000, and 35,001–40,000. The test will have power in two situations of particular concern in the application of MCMC. In the first $h(\boldsymbol{\omega}^{(m)})$ exhibits nonstationary or near-nonstationary behavior, such as that in a random walk or random walk with drift. In this situation RNE computed from the entire sequence $\{h(\boldsymbol{\omega}^{(m)})\}$ will also be quite low. The problem may be that serial correlation is still strong with a separation of M iterations, or, perhaps, that a limiting distribution does not exist (see Exercise 4.5.3).

The second situation in which the separated partial means test has power helps to address the convergence question and identify a number of initial iterations B to discard before computing a final approximation $\overline{h}_{M-B}$. In this case the separated partial means test may fail because the first partial mean $\overline{h}_{1,p}^{(M)}$ reflects sensitivity to initial conditions and is therefore atypical of the rest of the sequence. This may be confirmed by examining a plot of the sequence $\{h(\boldsymbol{\omega}^{(m)})\}$ or of a sequence of separated partial means. We may also conduct an obvious variant of the separated partial means test that compares a smaller number M_1 of early iterations with a larger number M_2 of later iterations, taking care that the two groups are separated by omitted iterations, typically at least M_1. The Bayesian analysis, computation, and communication (BACC) software, introduced in Section 5.1, handles the choice of L in Theorem 4.7.3 and makes these kinds of comparisons easy.

The separated partial means test with $p = 2$, applied to the 400 iterations of the Gibbs sampler illustrated in panels (a) and (b) of Figure 4.3, yields a value of 1.79; the corresponding p value from the $\chi^2(1)$ distribution is .181. Since the test compares the means for iterations 101–200 and 301–400, this is consistent with omitting the first 100 iterations and proceeding with the remainder as being negligibly influenced by the starting value and representative of the invariant distribution. When the same test is applied in the situation illustrated in panels (c) and (d) of Figure 4.3, the outcome is 29.29; the corresponding p value is 6.24×10^{-8}. This strong rejection is consistent with visual inspection of panels (c) and (d) and our earlier conclusions about convergence problems when $\rho = 0.99$. In this same situation, using $M = 10{,}000$ iterations and a separated partial means test with $p = 4$, the separated partial means test statistic is 1.66, near the median of the $\chi^2(3)$ distribution. This result would support a decision to discard the first 1250 iterations, and proceed with the remaining 8750 for further analysis.

These procedures are all based on a single sequence, or run, of MCMC draws. Assessment of accuracy and convergence from single runs is inherently limited. Figure 4.2a illustrates an extreme case inwhich a single run would never detect the fact that the chain is reducible. Practical problems can arise from near-reducibility of the Markov chain. Consider the Gibbs sampler with blocks $\boldsymbol{\theta}_{(1)} = \theta_1$ and $\boldsymbol{\theta}_{(2)} = \theta_2$ in the case of a multimodal bivariate posterior density like the one portrayed in Figure 4.4. In that case there is substantial serial correlation and sensitivity to the initial condition, since the probability that $\boldsymbol{\theta}^{(m)}$ will be near one of the two major modes conditional on $\boldsymbol{\theta}^{(m-j)}$ being near the other is quite small, even if j is quite large. If it is possible to conduct multiple runs of MCMC draws with random initial draws $\boldsymbol{\theta}^{(0)}$, then such problems can be detected, but only if the draws $\boldsymbol{\theta}^{(0)}$ are sufficiently dispersed that they have significant probability of being near each

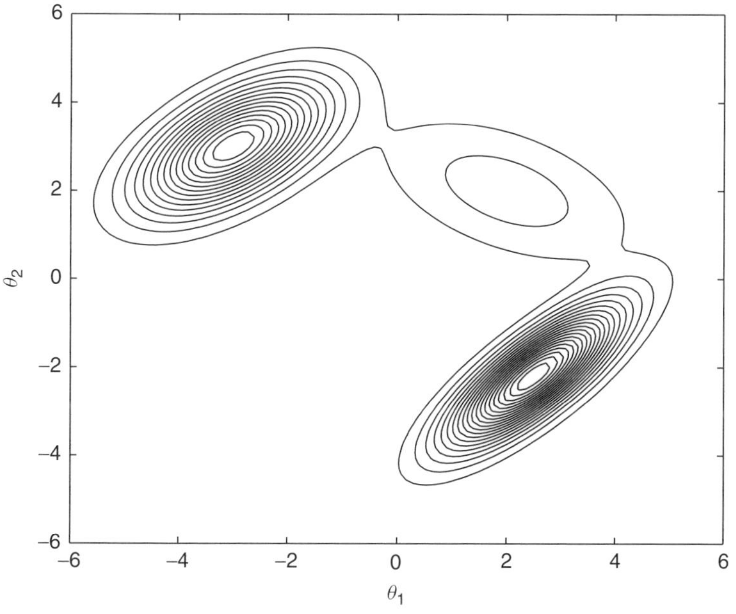

Figure 4.4. A multimodal probability density function ill-suited to Gibbs sampling.

of the two major modes in situations like the one illustrated in Figure 4.4. Theorem 4.7.4 may be applied to independent runs from initial conditions, using the means of the different runs (perhaps after discarding some initial draws) in place of the separated partial means. In this application the theorem exploits the fact that $\overline{h}_M$ together with its NSE provides a prediction of the approximation based on an independent realization of the same Markov chain.

CHAPTER 5

Linear Models

Chapter 2 developed the idea of a complete model, or a complete sequence of models, as an abstract and flexible framework for Bayesian inference. The specification of a complete sequence of models $A = \{A_1, \ldots, A_J\}$ is

$$p(A_j \mid A), \tag{5.1}$$

$$p(\boldsymbol{\theta}_{A_j} \mid A_j), \tag{5.2}$$

$$p(\mathbf{y} \mid \boldsymbol{\theta}_{Aj}, A_j), \tag{5.3}$$

$$p(\boldsymbol{\omega} \mid \mathbf{y}, \boldsymbol{\theta}_{A_j}, A_j), \tag{5.4}$$

where $j = 1, \ldots, J$. In (5.1)–(5.4) A_j denotes the model j, $\boldsymbol{\theta}_{A_j}$ the $k_{A_j} \times 1$ vector of unobservables in model j, $\mathbf{y}$ the vector of observables common to all J models, and $\boldsymbol{\omega}$ the common vector of interest. In many applications the investigator's final or intermediate objective is to determine $p(\boldsymbol{\omega} \mid \mathbf{y}^o, A)$, where $\mathbf{y}^o$ is the observed value of $\mathbf{y}$ (the data). This and the next two chapters present some specifications of functional forms (5.3) for the conditional distribution of observables.

This chapter concentrates on practical issues surrounding the use of the linear model and some important extensions of that model. It begins in Section 5.1 by introducing mathematical applications software incorporating the posterior simulators described in the previous chapter, and illustrates its use in the context of the normal linear model first introduced in Example 2.1.2. This illustration also shows how the output from a posterior simulator can be used in a decisionmaking context. The chapter continues with the seemingly unrelated regressions model, which has played a central role in econometrics and can be regarded as a multivariate generalization of the normal linear model. Section 5.3 takes up the common and related problems of choosing a subset of covariates from a large set of potential covariates, and enforcing inequality constraints on the coefficients of a linear model. Finally, Section 5.4 develops two ways in which the normal linear model can be extended

Contemporary Bayesian Econometrics and Statistics, by John Geweke

to specify regressions that are nonlinear in the covariates, while remaining a special case of the model first introduced in Example 2.1.2.

5.1 BACC AND THE NORMAL LINEAR REGRESSION MODEL

The Bayesian analysis, computation, and communication (BACC) software provides convenient tools for using the models described in this chapter and many of the posterior simulation methods described in Chapter 4. An important feature of the BACC software is that it implements its tools as extensions of the mathematical applications Matlab, Gauss, Splus, and R, running under several variants of Windows, Unix, and Linux operating systems. It provides a seamless integration of special-purpose BACC commands with built-in general-purpose commands for computation, graphics, and program flow control. The user has available a number of models, which describe the joint distribution of observables and unobservables (5.2)–(5.3). The user creates model instances by selecting one of the models and supplying values for its known quantities—the data, and the fixed parameters of the prior distribution. BACC incorporates most of the models discussed in this and the next two chapters. It provides prior and posterior simulation using the methods described in Chapter 4, as well as simulation facilities for model comparison, specification analysis, communication, and robustness analysis described in Chapter 8.

The BACC software is available free of charge through the companion Website described in the preface. This site also provides complete instructions for installation, documentation, and tutorials. The online help features of Matlab, Gauss, Splus, and R include the BACC extensions once they are installed. The online appendix provides code and output for all the examples in this book that use BACC software. The code is heavily annotated to introduce the reader to the use of mathematical applications software for Bayesian analysis generally, and to BACC in particular. After installing BACC and running the test programs described in the companion Website, it is instructive to execute the code for the examples in this section. By editing the code for the examples, the reader can rapidly gain familiarity and confidence with the software. Editing the code for the examples is also the easiest way to approach many of the exercises that require computing.

Example 5.1.1 The Impact of Class Size on Test Scores (The online appendix contains data, annotated code, and output for this example.) An important decision made by school boards and school district superintendents is the ratio of students to teachers. A lower ratio is thought to improve education, including test scores on standardized examinations that are increasingly used to evaluate school districts. A lower ratio certainly requires spending more money on teachers' salaries and supporting infrastructure such as classrooms. Some states have systematically collected district data on test scores, student : teacher ratios, and other factors that may affect educational outcomes including test scores.

The Massachusetts Comprehensive Assessment System (MCAS) test is administered to all fourth-graders in Massachusetts public schools each spring. The test score data used in this example is the average total score from the 1998 examination in each of the 220 Massachusetts elementary school districts. These data were obtained from the Massachusetts Department of Education, as were data on the student : teacher ratio (`str`), the percentage of students still learning English (`el`), and the percentage of students receiving a subsidized lunch in each district (`lunch`). In addition data on the average district per capita income were obtained from the 1990 U.S. census and are coded as the logarithm of income in thousands of dollars (`income`).

This example illustrates the use of mathematical applications software and BACC in creating a complete model to support decisionmaking regarding student : teacher ratios. To get the most out of the example, the reader should follow the code and execute the commands while reading. We begin by finding some summary statistics to gain familiarity with the data, as follows:

	Mean	Median	SD[a]	Minimum	Maximum
`str`	17.34	17.1	2.278	11.4	27.0
`el`	1.13	0	2.90	0	24.49
`lunch`	15.32	10.55	15.06	0.40	76.20
`income`	2.89	2.84	0.27	2.27	3.85
`score`	709.83	711	15.13	658	740

[a]Standard deviation.

The sample distributions of the percentage of students still learning English and the percentage of students receiving a subsidized lunch are strongly positively skewed. The student : teacher ratio is rounded to the nearest one-tenth in the data, ranges from 11.4 to 27, and has a somewhat positively skewed distribution. After transformation to logarithms average district income has a nearly symmetric distribution. Average district income is almost 5 times higher in some districts than in others. The distribution of test scores is close to normal with a mean of about 710 and standard deviation of about 15.

This example uses a normal linear regression model with `str`, `el`, `lunch`, and `income` as covariates, and `score` as the outcome. (Examples 5.4.1–5.4.4 examine and elaborate on this specification.) The prior distribution about to be described reflects the assumption that the model approximates a relationship between average test score and the covariates that applies at the observed values of the covariates, and at those points permits a substantial range of behavior. For each observed combination of covariates $\mathbf{x}_t$, the scalar $E(y_t \mid \mathbf{x}_t, A) = \boldsymbol{\beta}'\mathbf{x}_t$ is a random variable a priori. The prior takes $E(\boldsymbol{\beta}'\mathbf{x}_t \mid A) = \underline{\mu} = E(y_t \mid A) = 710$, which implies that the prior mean of the intercept β_1 is $\underline{\mu}$, while the prior mean of each covariate coefficient is 0. The prior distribution of the linear combinations $\boldsymbol{\beta}'\mathbf{x}_t$ is multivariate normal, with $\text{cov}(\boldsymbol{\beta}'\mathbf{x}_t, \boldsymbol{\beta}'\mathbf{x}_s) = 0$ for all combinations $t \neq s$. If sample size T were the same as the number of covariates k, then the variance associated with each $\boldsymbol{\beta}'\mathbf{x}_t$ would be $\underline{\sigma}^2 = \text{var}(y_t \mid A)$. In order to retain the same notional sample size regardless of T, $\text{var}(\boldsymbol{\beta}'\mathbf{x}_t) = T\underline{\sigma}^2/k$. Thus in general the prior distribution of $\boldsymbol{\beta}$ is

$$\boldsymbol{\beta} \mid (\mathbf{X}, A) \sim N(\underline{\boldsymbol{\beta}}, \underline{\mathbf{H}}^{-1}),$$

with

$$\underline{\mathbf{H}} = (k/T\underline{\sigma}^2)\mathbf{X}'\mathbf{X}, \ \underline{\boldsymbol{\beta}} = (\mathbf{X}'\mathbf{X})^{-1}\mathbf{X}'\iota_T\underline{\mu} = (\underline{\mu}, 0, \ldots, 0)'.$$

(The idea of using the covariate matrix to construct a prior distribution for regression coefficients was introduced in econometrics by Zellner (1986b).) The prior distribution of the precision h derives from choosing the hyperparameter $\underline{\sigma}^2 = \text{var}(y_t \mid A) = 15^2$, and considering the population multiple correlation coefficient $R^2 = 1 - (\underline{\sigma}^2 h)^{-1}$. If $(3/\underline{\sigma}^2)h \mid A \sim \chi^2(1)$, then $1 - [\underline{\sigma}^2 E(h \mid A)]^{-1}$ is about two-thirds, and the prior probability that R^2 exceeds .90 is approximately .25.

Recall from Example 4.3.1 that the Gibbs sampling algorithm for the normal linear model, which is used by BACC, has excellent convergence properties and almost no serial correlation. Hence we use only 100 burn-in iterations, followed by 100,000 draws from the posterior distribution. (This should all take only a few seconds on a desktop or laptop computer.) BACC can provide detailed information about the approximation of any function of interest, illustrated here for the coefficient of `str`, which is central in subsequent analysis (mean, −0.6843; standard deviation, 0.2636):

	Accuracy of Approximation	
Method	Numerical Standard Error	Relative Numerical Efficiency
Assuming no serial correlation	8.3364×10^{-4}	1.0000
Autocovariance function tapered to 1.00%	7.6772×10^{-4}	1.1791
Autocovariance function tapered to 2.00%	7.1833×10^{-4}	1.3468
Autocovariance function tapered to 3.00%	7.1066×10^{-4}	1.3760

There are four alternative approximations of the numerical standard error. The first assumes that the function of interest (here, the coefficient of `str`) is serially uncorrelated. This is generally not the case in MCMC algorithms. The other three approximations use the methods described in Section 4.7, specifically in Theorem 4.7.3, with $L = 0.01M$, $L = 0.02M$, and $L = 0.03M$. All of these methods indicate that relative numerical efficiency is close to 1. This reflects the fact that the Gibbs sampling algorithm of Example 4.3.1 for the normal linear model displays very little serial correlation. This is confirmed by the separated partial means test developed in Theorem 4.7.4, applied here to the `str` coefficient.

Method	Test Statistic p Value
Assuming no serial correlation	.896
Autocovariance function tapered to 1.00%	.896
Autocovariance function tapered to 2.00%	.894
Autocovariance function tapered to 3.00%	.895

We can then compare the prior and posterior moments of the unobservables, as follows:

	Prior		Posterior	
Unobservable	Mean	SD	Mean	SD
`intercept` coefficient	710	120.23	682.61	10.551
`str` coefficient	0	3.0163	−0.6843	0.2636
`el` coefficient	0	3.1937	−0.4086	0.2788
`lunch` coefficient	0	0.7746	−0.5173	0.0678
`income` coefficient	0	33.818	16.418	2.964
$h^{-1/2}$	—	—	8.716	0.422

Note that the ratio of prior to posterior standard deviation is the same for each coefficient. This is due to fact that $\underline{\mathbf{H}} \propto \mathbf{X}'\mathbf{X}$, and the ratio is approximately $[1 + (\bar{h}T\underline{\sigma}^2/k)]^{1/2}$.

The marginal likelihood of the normal linear model can be calculated in two ways. A generic simulation method described in Section 8.2.4 provides the approximation −805.8185 of the log marginal likelihood, with a numerical standard error of .0030. An essentially exact calculation, using (2.80) and one-dimensional quadrature, is −805.8189. This value will be important in comparing this model with some variants introduced in Section 5.4.

The sensitivity of the posterior distribution to the prior distribution can be studied in a number of ways, as discussed in Section 3.3. One of the simplest is to vary the hyperparameters of the prior distribution. In a "weak prior" variant of the model, $\underline{\mathbf{H}}$ is reduced by a factor of 5, and in a "strong prior" variant it is increased by a factor of 5. The direction of the impact on the posterior means of the coefficients is predictable because $\underline{\mathbf{H}} \propto \mathbf{X}'\mathbf{X}$.

Coefficient	Least Squares	Posterior Means with Prior "Weak"	Original	"Strong"	Prior Mean
`intercept`	682.4316	682.4691	682.6081	683.4652	710
`str`	−0.6892	−0.6885	−0.6843	−0.6633	0
`el`	−0.4107	−0.4104	−0.4086	−0.3962	0
`lunch`	−0.5215	−0.5205	−0.5173	−0.5022	0
`income`	16.5294	16.5076	16.4178	15.9102	0

The effect on the marginal likelihood cannot be anticipated. It turns out that the log marginal likelihood with the "weak" prior is -808.4703 and with the "strong" prior, is -808.4148. The Bayes factor in favor of the original specification, relative to either alternative, is about 13.9. Recall the discussion at the end of Section 3.2 that as $\underline{\mathbf{H}} \to 0$, marginal likelihood is driven to zero. At the other extreme a dogmatic prior $\boldsymbol{\beta} = \underline{\boldsymbol{\beta}}$ would produce a very poor fit and consequently also a very small marginal likelihood.

Example 5.1.1 incorporates two of the three elements of a complete model—the prior distribution and the observables distribution. These elements provide the posterior distribution, and Example 5.1.1 shows how posterior simulation methods and mathematical applications software can provide a useful representation of the posterior. This representation is exactly what we need to address the decision-making problems that motivate Bayesian analysis in the first place, as discussed in Chapter 1. The next example carries the previous example forward to two variants of a specific decision making problem.

Example 5.1.2 Deciding on Class Size (The online appendix contains data, annotated code, and output for this example.) The determination of class size in public schools is a political and fiscal decision whose details vary from state to state and district to district. Regardless of the details, the decision ultimately made balances the fact that, given the number of students in the district, a lower student : teacher ratio is more costly, against the perception that a lower student : teacher ratio also increases the quality of education. The results in Example 5.1.1 support this perception, to the extent that we are willing to identify higher test scores with increased quality of education. The effect is not large: $E(\beta_2 \mid \mathbf{y}^o, \mathbf{X}, A) = -0.6827$ implies that the difference between the highest student : teacher ratio of 27 and the lowest student : teacher ratio of 11.4 in the 220 Massachusetts elementary school districts accounts for a difference in test scores of about 10.6, considerably less than the sample standard deviation in test scores. Moreover, the distribution $\beta_2 \mid (\mathbf{y}^o, \mathbf{X}, A)$ implies substantial uncertainty about this effect. In this example we consider two loss functions that, together with the posterior distribution developed in Example 5.1.1, enable us to derive the corresponding optimal student : teacher ratio—the Bayes action.

The first loss function assigns a dollar value to the test score of each student. (This could represent either a social consensus manifest in local school boards, or it could be an explicit subsidy provided from the state of Massachusetts to local school districts.) Let T be the number of teachers in the school district and S the number of students. Suppose that the cost of each teacher is c, including not only the teacher's salary and benefits but also the annual cost of the additional facilities and support required for each teacher. Let d be the value placed by the school district on each test point for each student in each year. Then the loss function is $cT - dS\,\omega$, where ω is the average test score in the district. (The notation reflects the fact that average test score is the vector of interest.) Dividing by S, we obtain the equivalent loss function

$$L(a, \omega) = \frac{c}{a} - d\omega \tag{5.5}$$

where a is the student : teacher ratio (the Bayes action, in this decision problem). We shall refer to (5.5) as the "average score" loss function. In the model of Example 5.1.1, $\omega = \gamma + \beta a + \varepsilon$, where γ denotes the effect of all covariates other than the student : teacher ratio, and is specific to each school district. The expected loss is

$$E[L(a, \omega) \mid \text{data}] = \frac{c}{a} - d(\overline{\gamma} + \overline{\beta} a),$$

where the overbars denote posterior means. Simple calculus shows that the Bayes action is $a = (-c/d\overline{\beta})^{1/2}$. For example, if $c = \$100{,}000$, $d = \$250$, and $\overline{\beta} = -1$, then $a = 20$. The Bayes action depends only on the relative values of c and d.

This computation is simple, and it is easy to use alternative assumptions about c/d and alternative prior distributions. Continuing where we left off in Example 5.1.1, we easily find the following Bayes actions.

	Bayes Action a for $c/d =$		
Prior	150	200	250
"Weak"	14.7604	17.0438	19.0556
Original	14.8056	17.0961	19.1140
"Strong"	15.0379	17.3642	19.4138

Changing the prior precision by a factor of 5 has a small effect on the Bayes action. On the other hand, changes in c and d are important. If the annual cost of a teacher and supporting staff and facilities is \$100,000, then the difference between valuing each test score point per student at \$500 as opposed to \$400 accounts for a decrease in the optimal student : teacher ratio from 19 to 17. (Recall that 17 is the mean student : teacher ratio for all districts.) Similarly, if $c = \$100{,}000$ and $d = \$500$, a 10% increase in teacher costs would result in almost one more student per class, on average.

An alternative approach to examining sensitivity to the prior distribution is to vary the prior over the density ratio class, as discussed in Section 3.3. (Section 8.5 presents the corresponding computations.) For the stated prior density $p(\boldsymbol{\theta}_A \mid A)$, BACC computes upper and lower bounds on posterior moments over all prior densities with a kernel $k(\boldsymbol{\theta}_A \mid A)$ satisfying $r^{-1} \cdot p(\boldsymbol{\theta}_A \mid A) \leq k(\boldsymbol{\theta}_A \mid A) \leq r \cdot k(\boldsymbol{\theta}_A \mid A)$, for specified $r > 1$. In the case of the average score loss function (5.5) the Bayes action a varies inversely with $\overline{\beta}_2$. Therefore bounds on a may be derived from those on $\overline{\beta}_2$. For $r = 2, 4, 8$, we find

Density Ratio Factor r	2	4	8
Lower bound on a	14.0773	15.5338	16.6720
Upper bound on a	15.6607	19.2508	23.0857

In the alternative loss function the school district receives a cash transfer of d dollars per student if the average test score in the district exceeds a target t. Then

$$L(a, \omega) = \frac{c}{a} + dI_{[0,t]}(\omega). \tag{5.6}$$

We shall refer to (5.6) as the "threshold average score" loss function. Conditional on the unobservables β and h, we obtain

$$\begin{aligned} E[L(a, \omega) \mid \gamma, \beta, h] &= \frac{c}{a} + dP(\gamma + \beta a + \varepsilon \leq t) \\ &= \frac{c}{a} + dP(\varepsilon \leq t - \gamma - \beta a) = \frac{c}{a} + d\Phi[h^{1/2}(t - \gamma - \beta a)], \end{aligned}$$

where $\Phi(\cdot)$ is the cdf of the standard normal distribution. Then the applicable risk function is

$$E[L(a, \omega) \mid \mathbf{y}^o, \mathbf{X}, A] = \frac{c}{a} + d \int_{\Theta_A} \Phi[h^{1/2}(t - \gamma - \beta a)] p(\gamma, \beta, h \mid \mathbf{y}^o, \mathbf{X}, A) d\boldsymbol{\theta}_A,$$

where $\boldsymbol{\theta}_A = (\gamma, \beta, h)'$. The first-order condition is

$$\begin{aligned} &\frac{-c}{a^2} - d \int_{\Theta_A} \phi[h^{1/2}(t - \gamma - \beta a)] h^{1/2} \beta p(\gamma, \beta, h \mid \mathbf{y}^o, \mathbf{X}, A) d\boldsymbol{\theta}_A \\ &= \frac{-c}{a^2} - dE\{\phi[h^{1/2}(t - \gamma - \beta a)] h^{1/2} \beta \mid \mathbf{y}^o, \mathbf{X}, A\} = 0. \end{aligned} \tag{5.7}$$

Note that covariates other than the student : teacher ratio affect the Bayes action, given the threshold average score loss function, whereas they did not, given the average score loss function.

Given the output of the posterior simulator, an approximation of (5.7) is

$$\frac{-c}{a^2} - dM^{-1} \sum_{m=1}^{M} \phi[h^{(m)1/2}(t - \gamma^{(m)} - \beta^{(m)}a)]h^{(m)1/2}\beta^{(m)} = 0. \tag{5.8}$$

The root of (5.8) can be found by evaluating the left side of the expression over a suitable range of potential actions and then interpolating. (Theorems 4.5.3 and 4.7.2 provide the formal justification for this procedure.) From (5.8) the result depends on the parameters of the loss function only through the ratio c/d. Suppose the target test score is $t = 710$. Then Bayes actions for a hypothetical school district with sample median values of `el`, `lunch`, and `income`, which enter through γ, are as follows:

c/d	30	40	50
Bayes action a	16.3734	17.9941	19.6036

Exercises 5.1.2 and 5.1.4 explore the sensitivity and interpretation of these results.

Exercise 5.1.1 Prior Sensitivity in Example 5.1.1 This exercise explores in greater detail some of the findings about prior sensitivity in Example 5.1.1.

(a) By experimenting with appropriate variations of the coefficient vector prior precision matrix $\underline{\mathbf{H}}$, demonstrate that marginal likelihood is monotone decreasing for values of $\underline{\mathbf{H}}$ larger than that used in the "strong prior" as well as for values of $\underline{\mathbf{H}}$ smaller than that used in the "weak prior." (It is easiest to answer these questions by modifying the code for Example 5.1.1.)

(b) Why does the Bayes action a for the average score loss function (5.5) increase as the prior precision of $\boldsymbol{\beta}$ increases? (Remember that $\underline{\boldsymbol{\beta}} = (\underline{\mu}, 0, \ldots, 0)'$ in all of these prior distributions.)

(c) In the examples given for the average score loss function (5.5), why did the variation of the prior distribution within the density ratio class have a much more substantial impact on the Bayes action than did the variation in $\underline{\mathbf{H}}$?

Exercise 5.1.2 Prior Sensitivity in Example 5.1.2 Explore the sensitivity to the prior distribution of the Bayes action a for the loss function (5.6) to variations in the prior distribution, in the same way as was done for the loss function (5.5) in Example 5.1.2.

Exercise 5.1.3 Predictive Distributions Consider two hypothetical school districts, one with `str` $= 15$ and the other with `str` $= 20$. The values of all other covariates are the sample medians. Let ω denote the difference between the average test score in the school district with `str` $= 20$ and the district with `str` $= 15$.

(a) Find the predictive mean and standard deviation of ω.

(b) Fix the values of the coefficients $\boldsymbol{\beta}$ at the least-squares estimates $\mathbf{b}$ and fix the value of $\sigma^2 = h^{-1}$ at $\sigma^2 = (\mathbf{y}^o - \mathbf{Xb})'(\mathbf{y}^o - \mathbf{Xb})/(T - k)$. Find the corresponding mean and standard deviation of ω. This computation ignores the uncertainty about the unobservables in the model. Compare the results with those in part (a).

(c) Find $P(\omega > 0 \mid \mathbf{y}^o, \mathbf{X}, A)$.

Exercise 5.1.4 Interpreting Observed Student : Teacher Ratios in Example 5.1.2 There is substantial variation in the average student : teacher ratio across school districts. This exercise investigates the extent to which these ratios can be interpreted as Bayes actions in the context of Example 5.1.2.

(a) In the case of the average score loss function (5.5), the Bayes action does not depend on the values of the covariates (`el`, `lunch`, `income`). If `str` is a Bayes action, then observed variation in `str` can be due only to the variation in c/d. Find the value of c/d corresponding to the student : teacher ratio in each district and display the result as a histogram.

(b) In the case of the threshold average score loss function (5.6) the Bayes action depends on `el`,`lunch`, and `income`. Determine the Bayes action corresponding to $c/d = 40$ in each school district.

(c) Are the Bayes actions in part (b) positively correlated with the observed student : teacher ratios across school districts?

(d) Is there a value of c/d in each school district that fully accounts for the observed values of `str` in the case of the threshold average score loss function (5.6)? If so, is there more, or less, variation in this ratio than was the case in part (a) of this exercise?

Exercise 5.1.5 A Second Decision Problem in Example 5.1.2 The data spreadsheet provided for Examples 5.1.1 and 5.1.2 also includes, in column 16, the average teacher's salary in each school district. In this example, assume that the cost of each teacher is twice the teacher's salary.

(a) Assuming that `str` is the Bayes action, find the value of d in each school district, for the average score loss function (5.5).

(b) Are the values of d found in (a) systematically related to the covariates `el`, `lunch`, and `income`?

5.2 SEEMINGLY UNRELATED REGRESSIONS MODELS

The seemingly unrelated regressions (SUR) model developed in Zellner (1962) is perhaps the most widely used econometric model after linear regression. The reason is that it provides a simple and useful representation of systems of demand equations that arise in neoclassical static theories of producer and consumer behavior.

Two widely applied models in the theory of production provide examples of the SUR model. If the $m \times 1$ vector $\mathbf{w}$ denotes factor prices facing a producer of output y with cost function $c(\mathbf{w}, y)$, then from Shephard's lemma (Varian 1992, Section 5.4) the corresponding $m \times 1$ vector of factor demands is $\mathbf{x}(\mathbf{w}, y) = \partial c(\mathbf{w}, y)/\partial \mathbf{w}$. Given a functional form for $c(\mathbf{w}, y)$, factor demands can be derived explicitly.

The generalized Leontieff (or Diewert) cost function (Varian 1992, Section 12.10) is

$$c(\mathbf{w}, y) = y \sum_{i=1}^{m} \sum_{j=1}^{m} b_{ij} w_i^{1/2} w_j^{1/2} + \sum_{i=1}^{m} w_i \varepsilon_i$$

where $b_{ij} = b_{ji}$, and, defining $\boldsymbol{\varepsilon} = (\varepsilon_1, \ldots, \varepsilon_m)'$, $\boldsymbol{\varepsilon} \mid (\mathbf{w}, y) \sim N(\mathbf{0}, \boldsymbol{\Sigma})$. Then

$$x_i(\mathbf{w}, y) = y \sum_{j=1}^{m} b_{ij} (w_j/w_i)^{1/2} + \varepsilon_i \quad (i = 1, \ldots, m).$$

Note that there are m equations, each of which individually satisfies the observables specification in the normal linear regression model, but that most parameters appear in two equations and the disturbance terms in the different equations are allowed to be correlated.

The translog cost function (Varian 1992, Section 8.4) is

$$\log c(\mathbf{w}, y) = a_0 + \sum_{i=1}^{m} a_i \log w_i + \frac{1}{2} \sum_{i=1}^{m} \sum_{j=1}^{m} b_{ij} \log w_i \log w_j + \log y + \sum_{i=1}^{m} \log(w_i) \varepsilon_i$$

in which $\sum_{i=1}^{m} a_i = 1$, $\boldsymbol{\varepsilon} \mid (\mathbf{w}, y) \sim N(\mathbf{0}, \boldsymbol{\Sigma})$, and for all $i = 1, \ldots, m$:

$$b_{ij} = b_{ji} \ (j = 1, \ldots, m), \quad \sum_{j=1}^{m} b_{ij} = 0.$$

Since $\partial \log c(\mathbf{w}, y)/\partial \log w_i = [\partial c(\mathbf{w}, y)/\partial w_i] \cdot (w_i/c) \ (i = 1, \ldots, m)$, the cost share of the ith factor is

$$\frac{w_i x_i(\mathbf{w}, y)}{c(\mathbf{w}, y)} = a_i + \sum_{j=1}^{m} b_{ij} \log w_j + \varepsilon_i \ (i = 1, \ldots, m).$$

Once again there are m equations, each of which individually satisfies the observables specification of the normal linear regression model. Most parameters appear in two equations, and the disturbance terms in the different equations are allowed to be correlated.

In the seemingly unrelated regressions model, A, the relations of interest are

$$\underset{T\times 1}{\mathbf{y}_j} = \underset{T\times k}{\mathbf{Z}_j}\boldsymbol{\beta} + \boldsymbol{\varepsilon}_j \quad (j = 1, \ldots, m).$$

Let

$$\underset{Tm\times 1}{\mathbf{y}} = \begin{pmatrix} \mathbf{y}_1 \\ \vdots \\ \mathbf{y}_m \end{pmatrix}, \quad \underset{Tm\times k}{\mathbf{Z}} = \begin{bmatrix} \mathbf{Z}_1 \\ \vdots \\ \mathbf{Z}_m \end{bmatrix}, \quad \underset{Tm\times 1}{\boldsymbol{\varepsilon}} = \begin{pmatrix} \boldsymbol{\varepsilon}_1 \\ \vdots \\ \boldsymbol{\varepsilon}_m \end{pmatrix}.$$

The disturbance vector $\boldsymbol{\varepsilon}$ is normally distributed. Components of $\boldsymbol{\varepsilon}$ are uncorrelated across the observations $t = 1, \ldots, T$, but may be correlated across the m equations; thus, $\text{cov}(\boldsymbol{\varepsilon}_i, \boldsymbol{\varepsilon}_j) = \sigma_{ij}\mathbf{I}_T$ and the $m \times m$ matrix $\boldsymbol{\Sigma} = [\sigma_{ij}]$ is positive definite. Then $\boldsymbol{\varepsilon} \mid (\boldsymbol{\beta}, \boldsymbol{\Sigma}, \mathbf{Z}, A) \sim N(\mathbf{0}, \boldsymbol{\Sigma} \otimes \mathbf{I}_T)$.

A special case of the SUR model, sometimes the focus of textbook discussions [see, e.g., Greene (2003), Section 14.2] is

$$\underset{T\times 1}{\mathbf{y}_j} = \underset{T\times k_j}{\mathbf{X}_j}\boldsymbol{\beta}_j + \boldsymbol{\varepsilon}_j \quad (j = 1, \ldots, m). \tag{5.9}$$

In this case

$$\mathbf{Z}_j = \left[\underset{T\times\sum_{i=1}^{j-1}k_i}{\mathbf{0}} \quad \mathbf{X}_j \quad \underset{T\times\sum_{i=j+1}^{m}k_i}{\mathbf{0}} \right] \quad \text{and} \quad \underset{Tm\times\sum_{i=1}^{m}k_i}{\mathbf{Z}} = \begin{bmatrix} \mathbf{X}_1 & \cdots & \mathbf{0} \\ \vdots & \ddots & \vdots \\ \mathbf{0} & \cdots & \mathbf{X}_m \end{bmatrix}.$$

To take another special case, for the Leontieff cost function with two inputs

$$\underset{2T\times 3}{\mathbf{Z}} = \begin{bmatrix} y_1 & y_1(w_{21}/w_{11})^{1/2} & 0 \\ \vdots & \vdots & \vdots \\ y_T & y_T(w_{2T}/w_{1T})^{1/2} & 0 \\ 0 & y_1(w_{11}/w_{21})^{1/2} & y_1 \\ \vdots & \vdots & \vdots \\ 0 & y_T(w_{1T}/w_{2T})^{1/2} & y_T \end{bmatrix} \quad \text{and} \quad \boldsymbol{\beta} = \begin{pmatrix} b_{11} \\ b_{12} \\ b_{22} \end{pmatrix}.$$

In general the SUR model may be written

$$\mathbf{y} = \mathbf{Z}\boldsymbol{\beta} + \boldsymbol{\varepsilon}, \tag{5.10}$$

$$\boldsymbol{\varepsilon} \mid (\boldsymbol{\beta}, \mathbf{H}, \mathbf{Z}, A) \sim N(\mathbf{0}, \mathbf{H}^{-1} \otimes \mathbf{I}_T). \tag{5.11}$$

The $m \times m$ matrix $\mathbf{H} = \boldsymbol{\Sigma}^{-1}$ is the precision matrix of the disturbance vector $(\varepsilon_{1t}, \ldots, \varepsilon_{mt})'$. The formulation (5.10) permits linear cross-equation constraints on the coefficients, whereas the more specific case (5.9) does not.

From (5.10)–(5.11) the pdf of the observables vector $\mathbf{y}$ is

$$p(\mathbf{y} \mid \boldsymbol{\beta}, \mathbf{H}, \mathbf{Z}, A) = (2\pi)^{-Tm/2} |\mathbf{H}|^{T/2} \exp[-(\mathbf{y} - \mathbf{Z}\boldsymbol{\beta})'(\mathbf{H} \otimes \mathbf{I}_T)(\mathbf{y} - \mathbf{Z}\boldsymbol{\beta})/2]. \tag{5.12}$$

Define the residual cross-product terms $s_{ij} = (\mathbf{y}_i - \mathbf{Z}_i\boldsymbol{\beta})'(\mathbf{y}_j - \mathbf{Z}_j\boldsymbol{\beta})$ and $m \times m$ matrix $\mathbf{S} = [s_{ij}]$, and observe

$$(\mathbf{y} - \mathbf{Z}\boldsymbol{\beta})'(\mathbf{H} \otimes \mathbf{I}_T)(\mathbf{y} - \mathbf{Z}\boldsymbol{\beta}) = \sum_{i=1}^{m}\sum_{j=1}^{m}(\mathbf{y}_i - \mathbf{Z}_i\boldsymbol{\beta})'(\mathbf{y}_j - \mathbf{Z}_j\boldsymbol{\beta})h_{ij} = \text{tr}\ \mathbf{SH}. \tag{5.13}$$

An alternative expression for (5.12) is therefore

$$p(\mathbf{y} \mid \boldsymbol{\beta}, \mathbf{H}, \mathbf{Z}, A) = (2\pi)^{-Tm/2} |\mathbf{H}|^{T/2} \exp(-\text{tr}\ \mathbf{SH}/2). \tag{5.14}$$

Define $\widehat{\boldsymbol{\beta}} = [\mathbf{Z}'(\mathbf{H} \otimes \mathbf{I}_T)\mathbf{Z}]^{-1}\mathbf{Z}'(\mathbf{H} \otimes \mathbf{I}_T)\mathbf{y}$ and observe $\mathbf{Z}'(\mathbf{H} \otimes \mathbf{I}_T)(\mathbf{y} - \mathbf{Z}\widehat{\boldsymbol{\beta}}) = \mathbf{0}$. Hence (5.12) may also be expressed

$$p(\mathbf{y} \mid \boldsymbol{\beta}, \mathbf{H}, \mathbf{Z}, A) = (2\pi)^{-Tm/2} |\mathbf{H}|^{T/2} \exp[-(\mathbf{y} - \mathbf{Z}\widehat{\boldsymbol{\beta}})'(\mathbf{H} \otimes \mathbf{I}_T)(\mathbf{y} - \mathbf{Z}\widehat{\boldsymbol{\beta}})/2]$$
$$\cdot \exp[-(\boldsymbol{\beta}-\widehat{\boldsymbol{\beta}})'\mathbf{Z}'(\mathbf{H} \otimes \mathbf{I}_T)\mathbf{Z}(\boldsymbol{\beta}-\widehat{\boldsymbol{\beta}})/2]. \tag{5.15}$$

If $\mathbf{y}^o$ replaces $\mathbf{y}$ in (5.14) or (5.15), then these expressions, interpreted as functions of $\boldsymbol{\beta}$ and $\mathbf{H}$, provide alternative representations of the likelihood function. From (5.15) it then follows that the conditionally conjugate prior distribution for $\boldsymbol{\beta}$ is $\boldsymbol{\beta} \sim \mathbf{N}(\underline{\boldsymbol{\beta}}, \underline{\mathbf{H}}_{\boldsymbol{\beta}}^{-1})$:

$$p(\boldsymbol{\beta} \mid A) = (2\pi)^{-k/2} \left|\underline{\mathbf{H}}_{\boldsymbol{\beta}}\right|^{1/2} \exp[-(\boldsymbol{\beta} - \underline{\boldsymbol{\beta}})'\underline{\mathbf{H}}_{\boldsymbol{\beta}}(\boldsymbol{\beta} - \underline{\boldsymbol{\beta}})/2]. \tag{5.16}$$

The conditionally conjugate prior density function for $\mathbf{H}$ has the functional form (5.14). The kernel is that of the *Wishart distribution* [see Johnson and Kotz (1972), Chapter 38 or Anderson (1984), Section 7.2] for $m \times m$ random positive definite matrices $\mathbf{A}$:

$$p(\mathbf{A} \mid \boldsymbol{\Sigma}) = 2^{-\nu m/2}\pi^{-m(m-1)/4} |\boldsymbol{\Sigma}|^{-\nu/2} \left\{\prod_{i=1}^{m} \Gamma[(\nu + 1 - i)/2]\right\}^{-1}$$
$$\cdot |\mathbf{A}|^{(\nu-1-m)/2} \exp(-\text{tr}\boldsymbol{\Sigma}^{-1}\mathbf{A}/2). \tag{5.17}$$

The corresponding specific distribution is the Wishart distribution with positive definite matrix parameter $\boldsymbol{\Sigma}$ and degrees of freedom parameter $\nu \geq m$, usually denoted $\mathbf{A} \sim W(\boldsymbol{\Sigma}, \nu)$. It is the distribution of the random matrix $\mathbf{A} = \sum_{t=1}^{\nu} \mathbf{x}_t\mathbf{x}_t'$ if ν is an integer and $\mathbf{x}_t \overset{\text{i.i.d.}}{\sim} N(\mathbf{0}, \boldsymbol{\Sigma})$, and thus the marginal distribution of a_{ii} is $\sigma_{ii}^{-1}a_{ii} \sim \chi^2(\nu)$. This genesis of the Wishart distribution provides a simulation method if ν is an integer. The following algorithm, due to Anderson (1984), Section 7.2, does not require ν to be an integer and is more efficient unless ν is a small integer:

1. Compute the lower triangular Choleski factorization $\mathbf{P}$ of $\boldsymbol{\Sigma}$, $\boldsymbol{\Sigma} = \mathbf{PP}'$.
2. Simulate a lower triangular $m \times m$ matrix $\mathbf{B}$ with mutually independent elements: $b_{ii}^2 \sim \chi^2(\nu - i + 1)$ $(i = 1, \ldots, m)$, and $b_{ij} \sim N(0, 1)(i > j)$.
3. Let $\mathbf{C} = \mathbf{BB}'$; then $\mathbf{C} \sim W(\mathbf{I}_m, \nu)$.
4. Set $\mathbf{A} = \mathbf{PCP}'$.

The conditionally conjugate prior distribution can be expressed $\mathbf{H}\,|A \sim W(\underline{\mathbf{S}}^{-1}, \underline{\nu})$, and then

$$p(\mathbf{H}\,|A) = 2^{-\underline{\nu} m/2}\pi^{-m(m-1)/4}\,|\underline{\mathbf{S}}|^{\underline{\nu}/2}\left\{\prod_{i=1}^{m}\Gamma[(\underline{\nu}+1-i)/2]\right\}^{-1} \quad (5.18)$$

$$\cdot\,|\mathbf{H}|^{(\underline{\nu}-1-m)/2}\exp(-\text{tr }\underline{\mathbf{S}}\mathbf{H}/2). \quad (5.19)$$

The "notional data" interpretation of this prior distribution is the information about precision from $\underline{\nu}$ i.i.d. m-variate normal observations with sums of squares and cross-products matrix $\underline{\mathbf{S}}$. The prior mean of $\mathbf{H}$ is $\underline{\nu}\underline{\mathbf{S}}^{-1}$, and $(\underline{s}^{jj})^{-1}h_{jj} \mid A \sim \chi^2(\underline{\nu})$ $(j = 1, \ldots, m)$. This prior distribution is therefore a generalization of the prior distribution of h in the univariate normal linear model (2.12).

From (5.15) and (5.16) the conditional posterior density kernel for $\boldsymbol{\beta}$ is

$$p(\boldsymbol{\beta} \mid \mathbf{H}, \mathbf{y}^o, \mathbf{Z}, A) \propto \exp\{-[(\boldsymbol{\beta}-\widehat{\boldsymbol{\beta}})'\mathbf{Z}'(\mathbf{H}\otimes\mathbf{I}_T)\mathbf{Z}(\boldsymbol{\beta}-\widehat{\boldsymbol{\beta}}) + (\boldsymbol{\beta}-\underline{\boldsymbol{\beta}})'\underline{\mathbf{H}}_{\beta}(\boldsymbol{\beta}-\underline{\boldsymbol{\beta}})]/2\}.$$

Hence the conditional posterior distribution is $\boldsymbol{\beta}\,|(\mathbf{H}, \mathbf{y}^o, \mathbf{Z}, A) \sim N(\overline{\boldsymbol{\beta}}, \overline{\mathbf{H}}_{\beta}^{-1})$, with

$$\overline{\mathbf{H}}_{\beta} = \underline{\mathbf{H}}_{\beta} + \mathbf{Z}'(\mathbf{H}\otimes\mathbf{I}_T)\mathbf{Z}$$

and

$$\overline{\boldsymbol{\beta}} = \overline{\mathbf{H}}_{\beta}^{-1}[\underline{\mathbf{H}}_{\beta}\underline{\boldsymbol{\beta}} + \mathbf{Z}'(\mathbf{H}\otimes\mathbf{I}_T)\mathbf{y}^o] = \overline{\mathbf{H}}_{\beta}^{-1}[\underline{\mathbf{H}}_{\beta}\underline{\boldsymbol{\beta}} + \mathbf{Z}'(\mathbf{H}\otimes\mathbf{I}_T)\mathbf{Z}\widehat{\boldsymbol{\beta}}].$$

From (5.14) and (5.19) the conditional posterior density kernel for $\mathbf{H}$ is

$$p(\mathbf{H} \mid \boldsymbol{\beta}, \mathbf{y}^o, \mathbf{Z}, A) \propto |\mathbf{H}|^{(\underline{\nu}+T-1-m)/2}\exp\{-\text{tr }[(\underline{\mathbf{S}}+\mathbf{S})\mathbf{H}]/2\},$$

whence

$$\mathbf{H}\,|(\boldsymbol{\beta}, \mathbf{y}^o, \mathbf{Z}, A) \sim W[(\underline{\mathbf{S}}+\mathbf{S})^{-1}, \underline{\nu}+T].$$

These conditional posterior distributions provide the Gibbs sampling algorithm, first proposed by Percy (1992). BACC incorporates the seemingly unrelated regressions model using the conjugate prior distribution and posterior simulation algorithm described in this section.

Exercise 5.2.1 Missing Data Suppose that

$$\mathbf{y}_t = (y_{t1}, y_{t2})' \overset{\text{i.i.d.}}{\sim} N(\boldsymbol{\mu}, \boldsymbol{\Sigma}) \; (t = 1, \ldots, T).$$

Some of the observations y_{tj} are missing at random. (Recall the definition in Example 2.2.3.)

(a) Find $p(y_{t1} \mid y_{t2}, \boldsymbol{\mu}, \boldsymbol{\Sigma}, A)$ and $p(y_{t2} \mid y_{t1}, \boldsymbol{\mu}, \boldsymbol{\Sigma}, A)$. (*Hint*: You may find Theorem 5.3.1 useful.)

(b) Develop a Gibbs sampling posterior simulator for this model, assuming conditionally conjugate prior distributions.

(c) In the context of the model in (b), show that if both y_{t1} and y_{t2} are missing for given t, then nothing changes if this observation is simply excluded from the sample. Is the algorithm in (b) more efficient if observation t is included, or if it is excluded?

Exercise 5.2.2 Completing the Argument Derive (5.14) and (5.15) from (5.12).

Exercise 5.2.3 The Wishart and Gamma Distributions Show that the Wishart distribution for 1×1 matrices is the gamma distribution. Specifically, if $h \sim W(1/s^2, \nu)$, then $s^2 h \sim \chi^2(\nu)$.

Exercise 5.2.4 An Auction Application At a second price auction, bids are written and sealed. The object is sold to the highest bidder. The sale price is that bid by the second-highest bidder. Suppose you have data on T auctions, each of which has the same n bidders. The number n is small (say, 4 or 5) while T is large (say, 100–200). For each auction, you know the identity of the winning bidder and the price he offered. You also know the identity of the second-highest bidder, and the price she offered (which in turn is that paid by the winner). You do not know the bids of the other $n-2$ bidders.

Suppose that your model is

$$\widetilde{y}_{it} = \boldsymbol{\beta}_i' \mathbf{x}_{it} + \varepsilon_{it} \; (i = 1, \ldots, n; t = 1, \ldots, T);$$

$$\boldsymbol{\varepsilon}_t = (\varepsilon_{1t}, \ldots, \varepsilon_{nt})'; \;\; \boldsymbol{\varepsilon}_t \mid (\mathbf{H}, A) \sim N(\mathbf{0}, \mathbf{H}^{-1}) \; (t = 1, \ldots, T).$$

The $k \times 1$ vector $\mathbf{x}_{it}$ quantifies characteristics of bidder i and the object offered at auction t. The $n \times 1$ vector $\widetilde{\mathbf{y}}_t = (\widetilde{y}_{1t}, \ldots, \widetilde{y}_{nt})'$ contains the valuations of the n bidders for the object offered at auction t. A standard result in elementary auction theory [see, e.g., Fudenberg and Tirole (1995), Example 1.2] is that each bidder will bid his or her valuation of the object being sold.

You observe all of the $\mathbf{x}_{it}$ $(i = 1, \ldots, n; t = 1, \ldots, T)$. But you observe only the two largest elements of $(\widetilde{y}_{1t}, \ldots, \widetilde{y}_{nt})$ as described above. The prior distribution has two independent components: $\boldsymbol{\beta} \mid A \sim N(\underline{\boldsymbol{\beta}}, \underline{\mathbf{H}}^{-1})$ and $\mathbf{H} \mid A \sim W(\underline{\mathbf{S}}^{-1}, \underline{\nu})$. Construct a posterior simulator that provides random $\boldsymbol{\beta}^{(m)}$ and $\mathbf{H}^{(m)}$ whose invariant distribution is the posterior distribution. [For other auction applications using similar tools, see Sareen (1999, 2003), Bajari and Lee (2003), and Albano and Jouneau-Sion (2004).]

Exercise 5.2.5 Panel Data and Random Coefficients A random coefficients model for panel data applies to each of n households ($i = 1, \ldots, n$) observed in each of T time periods ($t = 1, \ldots, T$). For household i in time period t, we have

$$y_{it} = \boldsymbol{\beta}_i' \mathbf{x}_{it} + \varepsilon_{it}$$

in which $\mathbf{x}_{it}$ is a vector of covariates that may be regarded as fixed or ancillary; $\varepsilon_{it} \mid (h, A) \overset{\text{i.i.d.}}{\sim} N(0, h^{-1})$; and $\boldsymbol{\beta}_i \mid (\boldsymbol{\beta}, \mathbf{H}, A) \overset{\text{i.i.d.}}{\sim} N(\boldsymbol{\beta}, \mathbf{H}^{-1})$. The nT random unobservable disturbances and the n unobservable random vectors $\boldsymbol{\beta}_i$ are mutually independent. The observables are $(\mathbf{x}_{it}', y_{it})$ $(i = 1, \ldots, n;\ t = 1, \ldots, T)$.

(a) Assume that $\boldsymbol{\beta}$, $\mathbf{H}$, and h are independent in the prior distribution. Write down a normal prior density for $\boldsymbol{\beta}$, a Wishart prior density for $\mathbf{H}$, and a gamma prior density for h, including the kernels of these densities.

(b) Express the probability density function for $\boldsymbol{\beta}_1, \ldots, \boldsymbol{\beta}_n$ conditional on $\boldsymbol{\beta}$ and $\mathbf{H}$.

(c) Express the probability density function for y_{it} $(i = 1, \ldots, n; t = 1, \ldots T)$ conditional on $\boldsymbol{\beta}_i$ $(i = 1, \ldots, n)$, $\mathbf{x}_{it}$ $(i = 1, \ldots, n; t = 1, \ldots T)$, and h.

(d) Using the expressions from (a), (b) and (c), write down the posterior density kernel for $\boldsymbol{\beta}$, $\mathbf{H}$, h, and $\boldsymbol{\beta}_1, \ldots, \boldsymbol{\beta}_n$.

(e) Formulate a Gibbs sampling algorithm for the posterior distribution. Clearly indicate each step.

(f) Will the Gibbs sampling algorithm in (e) converge to the posterior distribution? (Briefly justify your answer.)

(g) Suppose that you have available $\mathbf{x}_{1,T+1}$, the covariate vector for the first household in the data set in the next period, but not $y_{1,T+1}$. Assuming that the same model continues to apply to household 1 in period $T + 1$, show how you could use the output of the Gibbs sampler from part (e) to obtain draws from the predictive distribution for $y_{1,T+1}$.

(h) Suppose that you have available $\mathbf{x}_{n+1,T+1}$, the covariate vector in the next period, from a household not in the data set, but not $y_{n+1,T+1}$. Assuming that the same model applies to household $n + 1$ in period $T + 1$, as to the households and time periods in the data set, show how you could use the output of the Gibbs sampler from part (e) to obtain draws from the predictive distribution for $y_{n+1,T+1}$.

Exercise 5.2.6 Hierarchical Priors and Unbalanced Data in the SUR Model Suppose that you are interested in a set of relationships

$$y_{it} = \boldsymbol{\beta}_i' \mathbf{x}_{it} + \varepsilon_{it}.$$

The subscript i denotes countries, the subscript t denotes time, and each covariate vector $\mathbf{x}_{it}$ is $k \times 1$. There are n countries and T time periods, and n is small relative to T. Here are some alternative specifications of the prior distribution, the observables distribution, and the observables.

(a) The prior distribution:

- **(i)** The coefficient vectors $\boldsymbol{\beta}_i$ are all the same.
- **(ii)** The coefficient vectors are similar but not identical.

(b) The observables distribution:

- **(i)** The disturbances ε_{it} are i.i.d. (across both i and t):

$$\varepsilon_{it} \mid (h, A) \sim N(0, h^{-1}).$$

- **(ii)** The disturbance vectors $\boldsymbol{\varepsilon}_t = (\varepsilon_{1t}, \ldots, \varepsilon_{nt})'$ are i.i.d.:

$$\boldsymbol{\varepsilon}_t \mid (\mathbf{H}, A) \sim N(\mathbf{0}, \mathbf{H}^{-1})\ (t = 1, \ldots, T).$$

(c) The observables:

- **(i)** $\{\mathbf{x}_{it}, y_{it}\}$ are observed for all $i = 1, \ldots, n$ and $t = 1, \ldots, T$.
- **(ii)** $\{\mathbf{x}_{it}\}$ are observed for all $i = 1, \ldots, n$ and $t = 1, \ldots, T$. However, for each country i, $\{y_{it}\}$ is observed for the time periods $t_i^1, \ldots, t_i^2$, where $1 \leq t_i^1 < t_i^2 \leq T$.

The objective in this exercise is to carefully complete the model consisting of (a.ii) and (b.ii) with a proper prior distribution, derive the posterior distribution corresponding to this prior distribution and the observed data indicated in (c.ii), and construct a simulator for this posterior distribution that provides simulation-consistent approximations of posterior moments.

If you can do this in one step, then do so. But you may find it easier to consider simpler variants of the model first [say, (a.i) plus (b.ii) plus (c.i)], and then take advantage of the "building block" character of many MCMC algorithms. (The combination of conditions (a.i), (b.ii), and (c.i) is fully covered in this section.)

5.3 LINEAR CONSTRAINTS IN THE LINEAR MODEL

In the normal linear regression model introduced in Example 2.1.2 it is common to impose, or at least entertain, restrictions on the coefficient vector $\boldsymbol{\beta}$ that go well beyond the prior distributions for $\boldsymbol{\beta}$ considered to this point. This is evident in discussions of whether coefficient estimates "have the right sign" and in procedures such as stepwise addition and deletion of covariates. If these restrictions are not incorporated formally in the specification of the model, then it is impossible to provide appropriate statements of uncertainty, taking either a Bayesian or non-Bayesian approach. A case of particular concern is that in which the investigator begins with a long list of potential covariates, and then following some data manipulation (e.g., stepwise deletion of variables) presents least-squares coefficient estimates and accompanying standard errors. The latter do not have a classical, sampling-theoretic interpretation, since there is no accounting for outcomes in which other covariates would have been selected. Nor do they have a Bayesian interpretation even in the context of Example 3.2.1, because they correspond to a prior distribution that excludes the deleted covariates with certainty.

Example 4.2.1 introduced the general problem, and discussed why acceptance sampling algorithms like those first proposed to handle the problem in Geweke (1986) are often impractical. Example 4.5.2 and Exercise 4.3.3 took up more specific algorithms in particular cases. This section generalizes the latter approaches. While the generalization here includes many of the cases that arise, it is by no means exhaustive, and approaches similar to the one described here can often be applied in other instances. The analysis focuses on formulating the prior distribution of $\boldsymbol{\beta}$ to incorporate the restrictions, and on the posterior distribution of $\boldsymbol{\beta}$ conditional on all other parameters in the model as well as the data. Consequently, application of the analysis here extends well beyond the linear model, to models that are equivalent to the linear model given suitable conditioning. In particular, the methods in this section can be used in conjunction with the Gibbs sampling algorithms developed in Sections 5.2, 6.2, 6.1, 6.4, and 7.1.

5.3.1 Linear Inequality Constraints

As a motivating example, suppose that the observables are the $k-1$ inputs $x_{t2}^*, \ldots, x_{tk}^*$ and the output y_t^* from each of T firms. The model incorporates a Cobb–Douglas production technology

$$\log y_t^* = \log \beta_{t1} + \sum_{j=2}^{k} \beta_j \log x_{tj}^*.$$

The unobservable β_{t1} varies across firms due to different fixed factors for each firm. For all T firms denote $x_{t1} = 1$, $x_{tj} = \log x_{tj}^*$ $(j = 2, \ldots k)$, and $y_t = \log(y_t^*)$. Take $\mathbf{X} = [x_{tj}]$ and let $\mathbf{y} = (y_1, \ldots, y_T)'$. If model A specifies $\log \beta_{t1} \mid (\beta_1, h, A) \overset{\text{i.i.d.}}{\sim} N(\beta_1, h^{-1})$ and the independent prior distributions $\boldsymbol{\beta} \mid A \sim N(\underline{\boldsymbol{\beta}}, \underline{\mathbf{H}}^{-1})$ and $\underline{s}^2 h \mid A \sim \chi^2(\underline{\nu})$, it is then a special case of the normal linear model of Example 2.1.2. However, the investigator also believes that $\beta_j > 0$ $(j = 2, \ldots, k)$.

This example is an instance of a more general set of restrictions

$$\mathbf{a} < \boldsymbol{\beta} < \mathbf{w} \tag{5.20}$$

in the linear model, where it is understood that elements of $\mathbf{a}$ may include $-\infty$, and those of $\mathbf{w}$ may include $+\infty$, as well as real numbers. In the motivating example $a_1 = -\infty, a_j = 0$ $(j = 2, \ldots, k)$ and $w_j = \infty$ $(j = 1, \ldots, k)$. Thus (5.20) includes the possibility of up to k nonredundant linear inequality restrictions, but no more.

Inequalities are taken as strict in (5.20) without loss of generality, since both prior and posterior distributions of $\boldsymbol{\beta}$ are absolutely continuous. Inequality restrictions of the form

$$\mathbf{a} < \mathbf{D}\boldsymbol{\beta} < \mathbf{w}, \tag{5.21}$$

with $\mathbf{D}$ nonsingular, may be accommodated by appropriate reparametrization of $\boldsymbol{\beta}$ and a corresponding transformation of the columns of $\mathbf{X}$ (see Exercise 5.3.1).

The posterior density kernel is that in Example 2.1.2, truncated to the set $\{\boldsymbol{\beta} : \mathbf{a} < \boldsymbol{\beta} < \mathbf{w}\}$. Consider a Gibbs sampling posterior simulation algorithm that is fully blocked in $h, \beta_1, \ldots, \beta_k$. The conditional posterior distribution of h remains the same, given by (2.25)–(2.26). Conditional on h, $\mathbf{X}$, and $\mathbf{y}$, $\boldsymbol{\beta} \sim N(\overline{\boldsymbol{\beta}}, \overline{\mathbf{H}}^{-1})$, with $\overline{\mathbf{H}}$ and $\overline{\boldsymbol{\beta}}$ given by (2.19) and (2.20), subject to the restrictions (5.20). In constructing the posterior distribution of β_j conditional on $\beta_i (i \neq j)$, h, $\mathbf{X}$, and $\mathbf{y}^o$, the following result for the multivariate normal distribution is useful.

Theorem 5.3.1 Conditional Multivariate Normal Distribution Let

$$\mathbf{z} = \begin{pmatrix} \mathbf{x} \\ \mathbf{y} \end{pmatrix} \sim N(\boldsymbol{\mu}, \boldsymbol{\Sigma}) \quad \text{with} \quad \boldsymbol{\mu} = \begin{pmatrix} \boldsymbol{\mu}_{\mathbf{x}} \\ \boldsymbol{\mu}_{\mathbf{y}} \end{pmatrix} \quad \text{and} \quad \boldsymbol{\Sigma} = \begin{bmatrix} \boldsymbol{\Sigma}_{\mathbf{xx}} & \boldsymbol{\Sigma}_{\mathbf{xy}} \\ \boldsymbol{\Sigma}_{\mathbf{yx}} & \boldsymbol{\Sigma}_{\mathbf{yy}} \end{bmatrix}.$$

Denote the corresponding precision matrix

$$\mathbf{H} = \boldsymbol{\Sigma}^{-1} = \begin{bmatrix} \mathbf{H}_{\mathbf{xx}} & \mathbf{H}_{\mathbf{xy}} \\ \mathbf{H}_{\mathbf{yx}} & \mathbf{H}_{\mathbf{yy}} \end{bmatrix}.$$

Then the distribution of $\mathbf{y}$ conditional on $\mathbf{x}$ is normal with variance

$$\boldsymbol{\Sigma}_{\mathbf{y}\cdot\mathbf{x}} = \boldsymbol{\Sigma}_{\mathbf{yy}} - \boldsymbol{\Sigma}_{\mathbf{yx}}\boldsymbol{\Sigma}_{\mathbf{xx}}^{-1}\boldsymbol{\Sigma}_{\mathbf{xy}} = \mathbf{H}_{\mathbf{yy}}^{-1} \tag{5.22}$$

and mean

$$\boldsymbol{\mu}_{\mathbf{y}\cdot\mathbf{x}} = \boldsymbol{\mu}_{\mathbf{y}} + \boldsymbol{\Sigma}_{\mathbf{yx}}\boldsymbol{\Sigma}_{\mathbf{xx}}^{-1}(\mathbf{x} - \boldsymbol{\mu}_{\mathbf{x}}) = \boldsymbol{\mu}_{\mathbf{y}} - \mathbf{H}_{\mathbf{yy}}^{-1}\mathbf{H}_{\mathbf{yx}}(\mathbf{x} - \boldsymbol{\mu}_{\mathbf{x}}). \tag{5.23}$$

Proof: The first equalities in (5.22) and (5.23) are derived in many texts on multivariate statistics; see Anderson (1984), Section 2.5.1 and Johnson and Kotz (1972), Section 35.3, for examples. The second equalities are then a consequence of a standard expression for the inverse of the partitioned matrix $\boldsymbol{\Sigma}$ (Anderson 1984, Section A.3.2; Greene 2003, Section A.5.3):

$$\begin{bmatrix} (\boldsymbol{\Sigma}_{\mathbf{xx}} - \boldsymbol{\Sigma}_{\mathbf{xy}}\boldsymbol{\Sigma}_{\mathbf{yy}}^{-1}\boldsymbol{\Sigma}_{\mathbf{yx}})^{-1} & -\boldsymbol{\Sigma}_{\mathbf{xx}}^{-1}\boldsymbol{\Sigma}_{\mathbf{xy}}(\boldsymbol{\Sigma}_{\mathbf{yy}} - \boldsymbol{\Sigma}_{\mathbf{yx}}\boldsymbol{\Sigma}_{\mathbf{xx}}^{-1}\boldsymbol{\Sigma}_{\mathbf{xy}})^{-1} \\ -(\boldsymbol{\Sigma}_{\mathbf{yy}} - \boldsymbol{\Sigma}_{\mathbf{yx}}\boldsymbol{\Sigma}_{\mathbf{xx}}^{-1}\boldsymbol{\Sigma}_{\mathbf{xy}})^{-1}\boldsymbol{\Sigma}_{\mathbf{yx}}\boldsymbol{\Sigma}_{\mathbf{xx}}^{-1} & (\boldsymbol{\Sigma}_{\mathbf{yy}} - \boldsymbol{\Sigma}_{\mathbf{yx}}\boldsymbol{\Sigma}_{\mathbf{xx}}^{-1}\boldsymbol{\Sigma}_{\mathbf{xy}})^{-1} \end{bmatrix}. \quad \blacksquare$$

Applying Theorem 5.3.1 to the conditional posterior distribution of β_j, we obtain

$$\beta_j \mid [\beta_i (i \neq j), h, \mathbf{y}^o, \mathbf{X}, A] \sim N\left[\overline{\beta}_j - \overline{h}_{jj}^{-1}\sum_{i \neq j}\overline{h}_{ji}(\beta_i - \overline{\beta}_i),\ \overline{h}_{jj}^{-1}\right]$$

subject to $a_j < \beta_j < w_j$. Draws from these truncated univariate normal distributions may be made efficiently using the algorithm described in Example 4.2.2. This provides the basis for a Gibbs sampling algorithm, first proposed in Geweke (1996b).

5.3.2 Conjectured Linear Restrictions, Linear Inequality Constraints, and Covariate Selection

In the motivating example of Section 5.3.1, suppose that the investigator thinks that β_{t1} may be related to some other observable characteristics of the firm. If the firm is a farm, for instance, these might include a measure of land quality, the educational attainment of the farm manager, and an indicator of whether the manager is also the owner. If the investigator assumes a linear relationship between $\log \beta_{t1}$ and observable characteristics $x_{t,k+1}, \ldots, x_{t,k+p}$, then $\log y_t^* = \sum_{j=1}^{k+p} \beta_j x_{tj} + \varepsilon_t$, where ε_t reflects the unobservable determinants of $\log \beta_{t1}$. The investigator is uncertain whether each observable characteristic of the firm really influences $\log \beta_{t1}$, given all the other characteristics and the inputs, but is willing to assume that the influence is nonnegative. A prior distribution in which $\beta_{k+1}, \ldots, \beta_{k+p}$ are mutually independent and independent of $\beta_1, \ldots, \beta_k$ might then take the form

$$P(\beta_j = 0 \mid A) = \underline{p}_j,$$

$$\beta_j \mid (\beta_j \neq 0, A) \sim N(\underline{\beta}_j, \underline{h}_j^{-1})$$

subject to $\beta_j > 0$ for $j = k+1, \ldots, p$.

To provide a general analytical treatment, assume the specification of the normal linear model of Example 2.1.2, except that in the prior distribution the coefficients β_j are mutually independent: $P(\beta_j = 0 \mid A) = \underline{p}_j$, and if $\beta_j \neq 0$, then $\beta_j \sim N(\underline{\beta}_j, \underline{h}_j^{-1})$ truncated to $a_j < \beta_j < w_j$. As in Section 5.3.1, $a_j = -\infty$, $w_j = \infty$, or both, are possible. The properly normalized prior density of β_j is then

$$p(\beta_j \mid A) = \underline{p}_j \delta_0(\beta_j) + (1 - \underline{p}_j)\lambda_j \exp[-\underline{h}_j(\beta_j - \underline{\beta}_j)^2/2] I_{(a_j, w_j)}(\beta_j),$$

where δ denotes the Dirac delta function defined in (4.18), and

$$\lambda_j = \{\Phi[\underline{h}_j^{1/2}(w_j - \underline{\beta}_j)] - \Phi[\underline{h}_j^{1/2}(a_j - \underline{\beta}_j)]\}^{-1}(2\pi)^{-1/2}\underline{h}_j^{1/2},$$

where Φ is the cdf of the standard normal distribution.

Because it specifies that the elements of β_j are independent a priori, this model does not include the motivating example, nor does the model in Section 5.3.1 correspond to the special case $\underline{p}_j = 0 (j = 1, \ldots, k)$. The modifications of the treatment below are fairly straightforward in each case, but the notation becomes more cumbersome. For treatments of more general cases, see George and McCulloch (1993, 1997), Geweke (1996a), Chipman et al. (1998), and Brown et al. (1999). Concentrating the prior point mass of β_j on $\beta_j = 0$ is innocuous. As formulated, this specification includes the conventional selection of regressors problem, in which the investigator is uncertain of which in a list of covariates really belongs in a model, and may also wish to impose sign restrictions. This could be cast as a model combination problem, but since there are 2^k possible models, this approach becomes unwieldy for more than (say) a half-dozen candidate covariates.

The posterior distribution can be sampled using a Gibbs sampler fully blocked in $\beta_1, \ldots, \beta_k, h$. The conditional posterior distribution of h is the same as that in Example 2.1.2. For each β_j, we obtain

$$p[\beta_j \mid \beta_i(i \neq j), h, \mathbf{y}^o, \mathbf{X}, A] \propto \{\underline{p}_j \delta_0(\beta_j)$$

$$+ (1 - \underline{p}_j)\lambda_j \exp[-\underline{h}_j(\beta_j - \underline{\beta}_j)^2/2] I_{(a_j, w_j)}(\beta_j)\} \exp\left[-h \sum_{t=1}^{T} (z_t - \beta_j x_{tj})^2/2\right] \tag{5.24}$$

where $z_t = y_t - \sum_{i \neq j} \beta_i x_{ti}$. From (5.24), if $\beta_j \neq 0$, then $\beta_j \sim N(\overline{\beta}_j, \overline{h}_j^{-1})$ subject to $\beta_j \in (a_j, w_j)$, where

$$\overline{h}_j = \underline{h}_j + h \sum_{t=1}^{T} x_{tj}^2 \quad \text{and} \quad \overline{\beta}_j = \overline{h}_j^{-1} \left(\underline{h}_j \underline{\beta}_j + h \sum_{T=1}^{T} x_{tj} z_t \right).$$

Furthermore

$$P[\beta_j = 0 \mid \beta_i(i \neq j), h, \mathbf{y}^o, \mathbf{X}, A] \propto \underline{p}_j \exp\left(-h \sum_{t=1}^{T} z_t^2/2\right)$$

and

$$P[\beta_j \neq 0 \mid \beta_i(i \neq j), h, \mathbf{y}^o, \mathbf{X}, A] \propto (1 - \underline{p}_j)\lambda_j$$

$$\cdot \int_{a_j}^{w_j} \exp[-\underline{h}_j(\beta_j - \underline{\beta}_j)^2/2] \exp\left[-h \sum_{t=1}^{T} (z_t - \beta_j x_{tj})^2/2\right] d\beta_j. \tag{5.25}$$

Completing the square in the last line yields the closed-form expression

$$\kappa_j = (2\pi)^{1/2} \overline{h}_j^{-1/2} \{\Phi[\overline{h}_j^{1/2}(w_j - \overline{\beta}_j)] - \Phi[\overline{h}_j^{1/2}(a_j - \overline{\beta}_j)]\}$$

$$\cdot \exp\left[-\left(\underline{h}_j^2 \underline{\beta}_j^2 + h \sum_{t=1}^{T} z_t^2 - \overline{h}_j \overline{\beta}_j^2\right)/2\right] \tag{5.26}$$

for the integral (5.25). Thus

$$\rho_j = P[\beta_j = 0 \mid \beta_i(i \neq j), h, \mathbf{y}^o, \mathbf{X}, A] \tag{5.27}$$

$$= \frac{\underline{p}_j \exp\left(-h \sum_{t=1}^{T} z_t^2/2\right)}{\underline{p}_j \exp\left(-h \sum_{t=1}^{T} z_t^2/2\right) + (1 - \underline{p}_j)\lambda_j \kappa_j}. \tag{5.28}$$

In each iteration of the Gibbs sampling algorithm some of the coefficients in β_j are zero and some are not. A simulation-consistent approximation of $P[\beta_j = 0 \mid \mathbf{y}^o, \mathbf{X}, A]$ is $M^{-1}\sum_{m=1}^{M}\delta(\beta_j^{(m)}, 0)$. [The expression $\delta(\beta_j^{(m)}, 0)$ is an instance of the *Kronecker delta function*, defined $\delta(a, b) = 1$ if $a = b$ and $\delta(a, b) = 0$ if $a \neq b$.] An alternative approximation can be based on the evaluation $\rho_j^{(m)}$ of (5.28) made in each iteration. Since

$$\rho_j^{(m)} = E[\delta_{\beta_j^{(m)},0} \mid \beta_i^{(m)}(i < j), \beta_i^{(m-1)}\ (i > j), h^{(m)}, \mathbf{y}^o, \mathbf{X}, A],$$

$M^{-1}\sum_{m=1}^{M}\rho_j^{(m)}$ is also a simulation-consistent approximation. The principle of concentrated expectations (Theorem 4.4.1) suggests that the variance of this approximation will be smaller. This generally turns out to be the case. It can be especially advantageous when $M \cdot P(\beta_j = 0 \mid \mathbf{y}^o, \mathbf{X}, A) << 1$, in which case $M^{-1}\sum_{m=1}^{M}\delta_{\beta_j^{(m)},0} = 0$ is a likely outcome and assessment of numerical standard error of this approximation is impossible. The same is true when

$$M[1 - P(\beta_j = 0 \mid \mathbf{y}^o, \mathbf{X}, A)] \ll 1.$$

Exercise 5.3.1 Transformation for Inequality Constraints In the motivating example for linear inequality constraints in Section 5.3.1, suppose that the additional inequalities $\beta_j < 1 (j = 2, \ldots, k)$ and $0 < \sum_{j=2}^{m}\beta_j < 1$ (the latter corresponding to diminishing returns to scale) were imposed.

(a) Show that these inequalities cannot be expressed in the form (5.21).

(b) Adapt the methods of Section 5.3.2 to sample from the posterior distribution.

Exercise 5.3.2 Improving Efficiency Suppose that in the inequality-constrained linear model $\boldsymbol{\beta}' = (\boldsymbol{\beta}_1', \boldsymbol{\beta}_2')$, $\mathbf{a}_1 < \boldsymbol{\beta}_1 < \mathbf{w}_1$, where $\boldsymbol{\beta}_1$ is $k_1 \times 1$, but $\boldsymbol{\beta}_2$ is unconstrained. There is a Gibbs sampling algorithms for the posterior distribution that uses $k_1 + 2$ blocks and is generally more efficient than the one described in Section 5.3.1. Describe the algorithm, indicating explicitly how the elements of $\boldsymbol{\beta}_2$ are drawn.

Exercise 5.3.3 Order-Restricted Inference Suppose that

$$\underset{n\times 1}{\mathbf{x}_t} \overset{\text{i.i.d.}}{\sim} N(\boldsymbol{\mu}, \mathbf{H}^{-1}).$$

The prior belief about the precision matrix $\mathbf{H}$ is $\mathbf{H}|A \sim W(\underline{\nu}, \underline{\mathbf{S}}^{-1})$.

(a) Suppose that the prior belief about $\boldsymbol{\mu}$ is that $\mu_1 > \mu_2 > \cdots > \mu_n$ and that, subject to this restriction, $\mu_j - \mu_{j+1} \overset{\text{i.i.d}}{\sim} N(0, \underline{h}^{-1})$. Adapt the Gibbs sampling algorithm of Section 5.3.1 to this model.

(b) Suppose instead that the prior belief about $\boldsymbol{\mu}$ is that $\mu_j - \mu_{j+1}$ are independent and identically distributed. $P(\mu_j - \mu_{j+1} = 0) = \frac{1}{2}$; if $\mu_j - \mu_{j+1} \neq 0$, then $\mu_j - \mu_{j+1} \sim N(0, \underline{h}^{-1})$ subject to $\mu_j - \mu_{j+1} > 0$. Adapt the Gibbs sampling algorithm of Section 5.3.2 to this model.

(c) Consider an alternative model in which $\mu_1 = \cdots = \mu_n$. Express a conditionally conjugate prior distribution for this model. How would you find the Bayes factor in favor of this model, relative to the one in part (b)?

(d) Now suppose the additional complication that some of the components x_{it} are missing at random (recall Example 2.2.3). How would you modify the posterior simulation algorithms in parts (a) and (b) to accommodate this complication?

Exercise 5.3.4 Inequality Constraints and Model Probability In the normal linear regression model $\mathbf{y} \sim N(\mathbf{X}\boldsymbol{\beta}, h^{-1}\mathbf{I}_T)$ the precision parameter h is known to be $h = 1$, and $\mathbf{X}'\mathbf{X} = \mathbf{D}$, a $k \times k$ diagonal matrix $\mathbf{D} = \text{diag}(d_1, \ldots, d_k)$. Moreover, $\mathbf{b} = (\mathbf{X}'\mathbf{X})^{-1}\mathbf{X}'\mathbf{y}^o = \mathbf{0}$. Consider three variants of this model, distinguished by the prior distribution of $\boldsymbol{\beta}$.

- In model 1, $\boldsymbol{\beta} \mid A_1 \sim N(\mathbf{0}, \underline{\mathbf{H}})$. $\underline{\mathbf{H}}$ is a diagonal matrix:

$$\underline{\mathbf{H}} = \text{diag}(\underline{h}_1, \ldots, \underline{h}_k).$$

- In model 2, $\boldsymbol{\beta} \mid A_2 = \mathbf{0}$. There is no uncertainty about $\boldsymbol{\beta}$ in this model.
- In model 3, the coefficients β_i are independently distributed. With probability $\underline{p}_i$, $\beta_i = 0$. With probability $1 - \underline{p}_i$, β_i has a half-normal distribution with precision $\underline{h}_i$:

$$p(\beta_i \mid \beta_i \neq 0, A_3) = (2\underline{h}_i/\pi)^{1/2} \exp(-\underline{h}_i \beta_i^2/2) I_{(0,\infty)}(\beta_i).$$

(a) Find the marginal likelihood in models 1 and 2. [You may find (2.80) useful.]

(b) Find the marginal likelihood in model 3.

(c) Rank the models according to their marginal likelihood, and explain the ordering.

(d) Find $P(\boldsymbol{\beta} = \mathbf{0} \mid \mathbf{y}^o, \mathbf{X}, A_3)$. Is the result surprising, given the ordering in part (c)?

Exercise 5.3.5 Completing the Argument Derive (5.26) from (5.25).

5.4 NONLINEAR REGRESSION

The normal linear model provides a convenient but restrictive representation of the distribution of $\mathbf{y}$ conditional on $\mathbf{X}$. This section weakens the assumption that the regression function is linear in $\mathbf{x}_t$ in favor of the specification that it is a smooth function of $\mathbf{x}_t$, but maintains the normality stipulation. Section 6.4 will weaken the normality assumption. As with all subjective conditions, "smoothness" can be characterized in different ways. This section takes up two different approaches. Somewhat surprisingly, each leads to a posterior kernel identical to that of the normal linear model of Example 2.1.2—but with a different matrix of covariates $\mathbf{X}$, and with a different interpretation of the coefficient vector $\boldsymbol{\beta}$. That nonlinear regression is thus isomorphic to linear regression has two desirable consequences. On the practical level, many Bayesian methods for the normal linear model can be applied in normal nonlinear regression, including the BACC as described in Section 5.1. On the conceptual level, many of the rich elaborations of the normal linear model that have been applied in Bayesian analysis can be applied directly in nonlinear regression. These include the extension to nonnormality in Section 6.4, and the weakening of the assumption that $y_t - E(y_t \mid \mathbf{x}_t)$ is independently distributed is discussed in Sections 7.1 and 7.3.

5.4.1 Nonlinear Regression with Smoothness Priors

The essentials of nonlinear regression with smoothness priors are captured in the simple model

$$y_t = f(x_t) + \varepsilon_t, \varepsilon_t \overset{\text{i.i.d.}}{\sim} N(0, h^{-1}). \tag{5.29}$$

The function $f(\tau)$ is defined on a closed interval $\tau \in [\tau_1, \tau_2]$. In general, the vector of interest $\boldsymbol{\omega}$ will include elements $f(\tau_1^*), \dots, f(\tau_q^*)$, where $\{\tau_1^*, \dots, \tau_q^*\}$ is a collection of distinct points in $[\tau_1, \tau_2]$. The complete model A must therefore specify $p[f(\tau), \tau \in [\tau_1, \tau_2] \mid A]$. The model must also incorporate the idea that f is a smooth function, in the sense that it is differentiable and $df(\tau)/d\tau$ changes slowly with τ.

A convenient and powerful analytic tool for expressing these beliefs is the *Wiener process* $W(\tau)$, defined on $\tau \in [0, \infty)$ with $W(0) = 0$. A standard representation is $W(\tau) = \int_0^\tau dW(u)$, where it is understood that the orthogonal increments $dW(u)$ are normally distributed. [For further discussion, see Doob (1953), Section 2.9 or Hamilton (1994), Section 17.2.] One important property of a Wiener process is $W(\tau + s) - W(\tau) \sim N(0, s)$, for all $\tau \geq 0$ and all $s > 0$; this limits the rapidity with which W can move as a function of τ, and this feature can in turn be controlled by appropriate scaling of W. Another important property is that any pair of increments $W(\tau + s) - W(\tau)$ and $W(\tau' + s') - W(\tau')$ has a bivariate normal distribution. Each increment has mean zero. If $[\tau, \tau + s]$ and $[\tau', \tau' + s']$ do not overlap, then the increments are uncorrelated; if the intervals do overlap, then their covariance is the length of the overlap. In general

$$\text{cov}[W(\tau + s) - W(\tau),\ W(\tau' + s') - W(\tau')] = \int_0^\infty I_{[\tau, \tau+s]}(u) I_{[\tau', \tau'+s']}(u)\, du.$$

More heuristically, movements of $W(\tau)$ over disjoint regions are uncorrelated.

A Wiener process has the properties ascribed to the function $f'(\tau) = df(\tau)/d\tau$, and thus we pursue the idea that $f(\tau) \mid A$ is the integral of such a process. The approach is that of Shiller (1984). Thus, $f(\tau) \mid A = \underline{h}^{-1/2} \int_0^\tau W(u)\, du$, where the hyperparameter $\underline{h}$ controls smoothness. For any two points τ and s

$$f(\tau) \mid A = \underline{h}^{-1/2} \int_0^\tau W(u)\,du = \int_0^\tau \int_0^u dW(r)\,du, \tag{5.30}$$

$$f(s) \mid A = \underline{h}^{-1/2} \int_0^s W(v)\,dv = \int_0^s \int_0^v dW(p)\,dv \tag{5.31}$$

have a joint normal distribution, with $E[f(\tau) \mid A] = E[f(s) \mid A] = 0$. If $s \geq \tau$, then, from (5.30) and (5.31), we obtain

$$\begin{aligned} E[f(\tau)f(s) \mid A] &= \underline{h}^{-1/2} \int_0^\tau \int_0^s \min(u, v)\,dv\,du \\ &= \underline{h}^{-1/2} \int_0^\tau \left[\int_0^u v\,dv + \int_u^s u\,dv \right] du \\ &= \underline{h}^{-1/2} \int_0^\tau \left[\frac{u^2}{2} + u(s-u) \right] du = \frac{\underline{h}^{-1/2}\tau^2}{6}(3s - \tau). \end{aligned} \tag{5.32}$$

Since $\text{var}[f(\tau)] = \underline{h}^{-1/2}\tau^3/3$, the prior variance ascribed to $f(\tau)$ at a point $\tau = s_1$ will depend strongly on the idea that $\tau = 0$ is a special point at which it is known a priori that $f'(0) = 0$. This is an artificial assumption. It arises not from prior ideas about smoothness (the reason for introducing the Wiener process as a model for the prior) but rather from the analytical necessity of an initial condition for $f'(\tau)$. There is a similar problem with the slope of the function $f(\tau)$ between two points s_1 and s_2 $(s_2 > s_1)$, $[f(s_2) - f(s_1)]/(s_2 - s_1)$. From (5.32), we have

$$\begin{aligned} &\text{var}\{[f(s_2) - f(s_1)]/(s_2 - s_1)\} = \\ &\frac{1}{6\underline{h}} \begin{bmatrix} -(s_2 - s_1)^{-1} \\ (s_2 - s_1)^{-1} \end{bmatrix}' \begin{bmatrix} 2s_1^3 & s_1^2(3s_2 - s_1) \\ s_1^2(3s_2 - s_1) & 2s_2^3 \end{bmatrix} \begin{bmatrix} -(s_2 - s_1)^{-1} \\ (s_2 - s_1)^{-1} \end{bmatrix} \\ &= [3s_1 + (s_2 - s_1)]/3\underline{h}, \end{aligned} \tag{5.33}$$

which depends not only on the length of the interval $s_2 - s_1$ but also on the size of s_1. However, for any three points $s_1 < s_2 < s_3$, we find that for the change in the slope of $f(\tau)$

$$\text{var}\left[\frac{f(s_3) - f(s_2)}{s_3 - s_2} - \frac{f(s_2) - f(s_1)}{s_2 - s_1} \right] = \frac{(s_3 - s_1)}{3\underline{h}}, \tag{5.34}$$

which can be derived from (5.32) in the same way that (5.33) was derived. Thus the distribution of any change in slopes does not depend on the artifice of an initial condition for $f'(\tau)$. This fact is not surprising, given that $f'(\tau) \mid A$ is a Wiener process, and changes in the level of a Wiener process over an interval depend only on the length of the interval and not on the distance of the interval from the origin $\tau = 0$. Given $s_4 > s_3$, we have

$$\text{cov}\left[\frac{f(s_3) - f(s_2)}{s_3 - s_2} - \frac{f(s_2) - f(s_1)}{s_2 - s_1}, \right.$$
$$\left. \frac{f(s_4) - f(s_3)}{s_4 - s_3} - \frac{f(s_3) - f(s_2)}{s_3 - s_2}\right] = \frac{(s_3 - s_2)}{6\underline{h}}. \tag{5.35}$$

If $s_6 > s_5 > s_4 \geq s_3 > s_2 > s_1$, then

$$\text{cov}\left[\frac{f(s_3) - f(s_2)}{s_3 - s_2} - \frac{f(s_2) - f(s_1)}{s_2 - s_1}, \frac{f(s_6) - f(s_5)}{s_6 - s_5} - \frac{f(s_5) - f(s_4)}{s_5 - s_4}\right] = 0. \tag{5.36}$$

Without loss of generality, suppose that there are m distinct values of $x_1, \ldots, x_T$. Denote the ordered distinct values by $s_i (i = 1, \ldots, m)$, and define $\mathbf{s} = (s_1, \ldots, s_m)'$ and $\boldsymbol{\beta} = [f(s_1), \ldots, f(s_m)]'$. Then the vector of unobservables in (5.29) is $\boldsymbol{\theta}'_A = (\boldsymbol{\beta}', h)$, and (5.29) may be written $\mathbf{y} = \mathbf{X}\boldsymbol{\beta} + \boldsymbol{\varepsilon}$ with $x_{ti} = 1$ if $x_t = s_i$ and $x_{ti} = 0$ otherwise. The information in the smoothness prior of the form (5.34)–(5.36) may be expressed

$$\mathbf{R}\boldsymbol{\beta} \sim N(\mathbf{0}, \mathbf{G}). \tag{5.37}$$

The matrix $\mathbf{R}$ is $(m - 2) \times m$, with

$$r_{ii} = (s_{i+1} - s_i)^{-1}, r_{i,i+1} = -[(s_{i+1} - s_i)^{-1} + (s_{i+2} - s_{i+1})^{-1}],$$
$$r_{i,i+2} = (s_{i+2} - s_{i+1})^{-1} \quad (i = 1, \ldots, m - 2)$$

and all other elements 0. The matrix $\mathbf{G}$ is $(m - 2) \times (m - 2)$ with

$$g_{ii} = (s_{i+2} - s_i)/3\underline{h}, \quad g_{i,i+1} = g_{i+1,i} = (s_{i+2} - s_{i+1})/6\underline{h} \quad (i = 1, \ldots, m - 2)$$

and all other elements 0.

The derivation of this smoothness prior from a continuous process has substantial practical advantages in enforcing consistency when the prior is updated with additional prior information or with data, and in expressing posterior distributions for $f(\tau)$ at points τ that do not correspond to any s_i. Suppose that we were to add a point s^* between s_i and s_{i+1} in the list $s_1, \ldots, s_m$, perhaps because $f(s^*)$ is an element of the vector of interest $\boldsymbol{\omega}$ but $s^* \neq s_i$ for any $i = 1, \ldots, m$. Incorporating this point directly in the smoothness prior, $f(s_i)$ and $f(s_{i+1})$ are removed from $\boldsymbol{\beta}$ and replaced with $f(s_i)$, $f(s^*)$ and $f(s_{i+1})$. Then the $(i - 2)$th and $(i - 1)$th linear combinations in $\mathbf{R}\boldsymbol{\beta}$ are removed from (5.37) and replaced with

$$(s_i - s_{i-1})^{-1} f(s_{i-1}) - [(s_i - s_{i-1})^{-1} + (s^* - s_i)^{-1}] f(s_i)$$
$$+ (s^* - s_i)^{-1} f(s^*) = \zeta_i^*, \tag{5.38}$$

$$(s^* - s_i)^{-1} f(s_i) - [(s^* - s_i)^{-1} + (s_{i+1} - s^*)^{-1}]^{-1} f(s^*)$$
$$+ (s_{i+1} - s^*)^{-1} f(s_{i+1}) = \zeta_*^*, \tag{5.39}$$

and

$$(s_{i+1} - s^*)^{-1} f(s^*) - [(s_{i+1} - s^*)^{-1} + (s_{i+2} - s_{i+1})^{-1}] f(s_{i+1}) \\ + (s_{i+2} - s_{i+1})^{-1} f(s_{i+2}) = \zeta_{i+1}^*, \tag{5.40}$$

where $(\zeta_i^*, \zeta_*^*, \zeta_{i+1}^*)'$ is normal with mean $\mathbf{0}$ and

$$\text{var}\begin{pmatrix} \zeta_i^* \\ \zeta_*^* \\ \zeta_{i+1}^* \end{pmatrix} = \frac{1}{6\underline{h}} \begin{bmatrix} 2(s^* - s_{i-1}) & s^* - s_i & 0 \\ s^* - s_i & 2(s_{i+1} - s_i) & s_{i+1} - s^* \\ 0 & s_{i+1} - s^* & 2(s_{i+2} - s^*) \end{bmatrix}. \tag{5.41}$$

In the new prior, the marginal distribution of $\boldsymbol{\beta}$ is the same as in the original prior. This can be seen by multiplying (5.39) by

$$[(s^* - s_i)^{-1} + (s_{i+1} - s^*)^{-1}](s^* - s_i)$$

and adding it to (5.38), and then multiplying (5.39) by

$$[(s^* - s_i)^{-1} + (s_{i+1} - s^*)^{-1}](s_{i+1} - s^*)$$

and adding it to (5.40). The resulting two equations, after some algebra, are

$$(s_i - s_{i-1})^{-1} f(s_{i-1}) - [(s_i - s_{i-1})^{-1} + (s_{i+1} - s_i)^{-1}] f(s_i) \\ + (s_{i+1} - s_i)^{-1} f(s_{i+1}) = \zeta_i$$

$$(s_{i+1} - s_i)^{-1} f(s_i) - [(s_{i+1} - s_i)^{-1} + (s_{i+2} - s_{i+1})^{-1}] f(s_{i+1}) \\ + (s_{i+2} - s_{i+1})^{-1} f(s_{i+2}) = \zeta_{i+1}$$

where

$$\zeta_i = \zeta_i^* + [(s^* - s_i)^{-1} + (s_{i+1} - s^*)^{-1}]\zeta_*^* / (s^* - s_i)^{-1},$$
$$\zeta_{i+1} = \zeta_{i+1}^* + [(s^* - s_i)^{-1} + (s_{i+1} - s^*)^{-1}]\zeta_*^* / (s_{i+1} - s^*)^{-1},$$

and from (5.41)

$$\text{var}\begin{pmatrix} \zeta_i \\ \zeta_{i+1} \end{pmatrix} = \frac{1}{6\underline{h}} \begin{bmatrix} 2(s_{i+1} - s_{i-1}) & s_{i+1} - s_i \\ s_{i+1} - s_i & 2(s_{i+2} - s_i) \end{bmatrix}.$$

These are precisely the $(i-2)$th and $(i-1)$th linear combinations of $\mathbf{R}\boldsymbol{\beta}$ that were removed in the first place.

This smoothness prior, taken alone, does not constitute a proper prior distribution for $\boldsymbol{\beta}$. Its contribution to the prior precision of $\boldsymbol{\beta}$ is $\mathbf{R}'\mathbf{G}^{-1}\mathbf{R}$, an $m \times m$ matrix of rank $m-2$. The balance of a proper normal prior distribution for $\boldsymbol{\beta}$ must provide prior information about the level and slope of the function—the information from the Wiener process that we discarded at the outset because it was artificial. If the additional information is normal and independent of the information in the smoothness prior, it can be represented

$$\mathbf{P}\boldsymbol{\beta} \sim N(\mathbf{p}, \mathbf{F}). \tag{5.42}$$

[If the information involves $f(s^*)$, and s^* is not an element of $\mathbf{s}$, then $\mathbf{s}$ must be redefined so as to incorporate the new point s^*, and the smoothness prior expressed in the form (5.37). As demonstrated in (5.38)–(5.41), this has no impact on the smoothness prior itself; the effect is simply to incorporate the fact that the smoothness prior applies to the new point s^*.] Taken together, this information provides a proper prior distribution if rank $\begin{bmatrix}\mathbf{R}' & \mathbf{P}'\end{bmatrix} = m^*$, where m^* is the number of points involved in the prior distribution. Then the full prior distribution is $\boldsymbol{\beta} \sim N(\underline{\boldsymbol{\beta}}, \underline{\mathbf{H}}^{-1})$, with $\underline{\mathbf{H}} = \mathbf{R}'\mathbf{G}^{-1}\mathbf{R} + \mathbf{P}'\mathbf{F}^{-1}\mathbf{P}$ and $\underline{\boldsymbol{\beta}} = \underline{\mathbf{H}}^{-1}\mathbf{P}'\mathbf{F}^{-1}\mathbf{p}$. For a variant on this idea, see Example 5.4.1.

The vector of interest $\boldsymbol{\omega}$ typically includes the function $f(\tau)$ evaluated at points τ not included in $\mathbf{s}$. For example, if information about the posterior distribution of $f(\tau)$ on the interval $[\tau_1, \tau_2]$ is presented graphically, then good resolution requires evaluation of $f(\tau)$ at 100 or more equally spaced points. Thus the posterior simulator must provide

$$p(\boldsymbol{\beta}, h, \boldsymbol{\omega} \mid \mathbf{y}^o, \mathbf{X}, A) \propto p(\mathbf{y}^o \mid \boldsymbol{\beta}, h, \boldsymbol{\omega}, \mathbf{X}, A)p(\boldsymbol{\beta}, h \mid A)p(\boldsymbol{\omega} \mid \boldsymbol{\beta}, h, A).$$

Since the distribution of $\mathbf{y}$ depends on $\boldsymbol{\beta}$ and h but not $\boldsymbol{\omega}$, $p(\mathbf{y}^o \mid \boldsymbol{\beta}, h, \boldsymbol{\omega}, \mathbf{X}, A) = p(\mathbf{y}^o \mid \boldsymbol{\beta}, h, \mathbf{X}, A)$. Because of the consistency of the smoothness prior with respect to addition and deletion of points of evaluation, $p(\boldsymbol{\beta}, h \mid A) = \int_\Omega p(\boldsymbol{\beta}, h, \boldsymbol{\omega} \mid A)\, d\boldsymbol{\omega}$ is the same regardless of the composition of $\boldsymbol{\omega}$. Hence

$$p(\boldsymbol{\beta}, h \mid \mathbf{y}^o, \mathbf{X}, A) \propto p(\mathbf{y}^o \mid \boldsymbol{\beta}, h, \mathbf{X}, A)p(\boldsymbol{\beta}, h \mid A),$$

so the posterior simulator excludes $\boldsymbol{\omega}$ and takes no account of its subsequent specification. Then

$$p(\boldsymbol{\omega} \mid \mathbf{y}^o, \mathbf{X}, A) = \int_{\mathbb{R}^+} \int_{\mathbb{R}^m} p(\boldsymbol{\omega} \mid \boldsymbol{\beta}, h, \mathbf{X}, A)p(\boldsymbol{\beta}, h \mid \mathbf{y}^o, \mathbf{X}, A)\, d\boldsymbol{\beta}\, dh.$$

The smoothness prior distribution expressed as $\mathbf{R}_1\boldsymbol{\beta} + \mathbf{R}_2\boldsymbol{\omega} \sim N(\mathbf{0}, \mathbf{G})$ provides the form of $p(\boldsymbol{\omega} \mid \boldsymbol{\beta}, h, \mathbf{X}, A) = p(\boldsymbol{\omega} \mid \boldsymbol{\beta}, A)$. Since rank $(\mathbf{R}_2) = \dim(\boldsymbol{\omega})$

$$\boldsymbol{\omega} \mid (\boldsymbol{\beta}, A) \sim N[-(\mathbf{R}_2'\mathbf{G}^{-1}\mathbf{R}_2)^{-1}\mathbf{R}_2'\mathbf{G}^{-1}\mathbf{R}_1\boldsymbol{\beta}, (\mathbf{R}_2'\mathbf{G}^{-1}\mathbf{R}_2')^{-1}]. \tag{5.43}$$

Replacing $\boldsymbol{\beta}$ with $\boldsymbol{\beta}^{(m)} \sim p(\boldsymbol{\beta} \mid \mathbf{y}^o, \mathbf{X}, A)$ in this expression provides the simulation algorithm for $\boldsymbol{\omega}$ given the posterior simulator output.

Example 5.4.1 Nonlinearity in Simple Regression (The online appendix contains data, annotated code, and output for this example.) Return to Example 5.1.1, which examined the impact of class size on test scores using the Massachusetts

Comprehensive Assessment System (MCAS) test data and a normal linear regression model. In this example we remove the assumption that the regression of class size on test scores is linear, and replace it with the assumption that the regression is a smooth function of class size, using (5.29) and the methods described in this section. The objectives are to learn about the posterior distribution of the regression function, and collect some evidence about the degree of nonlinearity in the regression function as indicated by the prior hyperparameter $\underline{h}$. Achieving these objectives illustrates the construction of a proper prior distribution in a nonlinear model, and techniques for inferring the posterior distribution of the regression function evaluated at points that do not correspond to values in the data.

To construct a prior distribution, amend (5.29) slightly, by writing

$$y_t = \alpha_1 + \alpha_2 x_t + f(x_t) + \varepsilon_t, \quad \varepsilon_t \overset{\text{i.i.d.}}{\sim} N(0, h^{-1}) \quad (t = 1, \ldots, T). \tag{5.44}$$

Let $s_1 < \cdots < s_m$ denote the m ordered distinct data points corresponding to the $m \times 1$ vector $\boldsymbol{\beta} = [f(s_1,), \ldots, f(s_m)]$. Then (5.44) has the form $\mathbf{y} = \mathbf{X}_1\boldsymbol{\alpha} + \mathbf{X}_2\boldsymbol{\beta} + \boldsymbol{\varepsilon}$, for the suitably arranged $T \times 2$ matrix $\mathbf{X}_1$ and $T \times m$ matrix $\mathbf{X}_2$. In the data set from Example 5.1.1, the student : teacher ratio (`str`) is rounded to the nearest one-tenth and ranges from 11.4 to 27.0. While $T = 220$, $m = 83$. The two restrictions

$$\sum_{i=1}^{m} f(s_i) = 0, \quad f(s_1) = f(s_m) \tag{5.45}$$

identify α_1, α_2, and f in (5.44) without imposing any additional restrictions. [Recall the discussion of identification in Exercise 4.5.3; see also Poirier (1998).] The restrictions (5.45) are equivalent to $\mathbf{P}\boldsymbol{\beta} = \mathbf{0}$, with an obvious definition of $\mathbf{P}$, and are a limiting case of additional information of the form (5.42), in which $\mathbf{F}$ becomes arbitrarily small. These exact restrictions can be imposed by writing $\boldsymbol{\beta} = \mathbf{Q}\boldsymbol{\beta}^*$, with $\mathbf{Q}' = \left[\, \boldsymbol{\iota}_{m-2}(-\frac{1}{2}) \;\; \mathbf{I}_{m-2} \;\; \boldsymbol{\iota}_{m-2}(-\frac{1}{2}) \,\right]$. The restrictions plus the prior information $\mathbf{R}\boldsymbol{\beta} \sim N(\mathbf{0}, \mathbf{G})$ in $\mathbf{y} = \mathbf{X}_1\boldsymbol{\alpha} + \mathbf{X}_2\boldsymbol{\beta} + \boldsymbol{\varepsilon}$ are equivalent to $\boldsymbol{\beta}^* \sim N(\mathbf{0}, \underline{\mathbf{H}}_2^{-1})$, in $\mathbf{y} = \mathbf{X}_1\boldsymbol{\alpha} + \mathbf{X}_2^*\boldsymbol{\beta}^* + \boldsymbol{\varepsilon}$, where $\mathbf{X}_2^* = \mathbf{X}_2\mathbf{Q}$ and $\underline{\mathbf{H}}_2 = \mathbf{Q}'\mathbf{R}'\mathbf{G}^{-1}\mathbf{R}\mathbf{Q}$. If $\boldsymbol{\alpha} \mid A \sim N(\underline{\boldsymbol{\alpha}}, \underline{\mathbf{H}}_1^{-1})$, independent of $\boldsymbol{\beta} \mid A$, then as $\underline{h} \to \infty$ in $\mathbf{G}$, the marginal likelihood of the model must approach that of the linear model $y_t = \alpha_1 + \alpha_2 x_t + \varepsilon_t$ with the same prior distribution for $\boldsymbol{\alpha}$ and h. Proceeding as in Example 5.1.1, we take $\underline{\boldsymbol{\alpha}} = (\underline{\mu}, 0)'$ and $\underline{\mathbf{H}}_1 = (5/T\underline{\sigma}^2)\mathbf{X}'\mathbf{X}$ using the values of $\underline{\mu}$ and $\underline{\sigma}^2$ in that example.

The model thus formulated is a special case of the normal linear regression model of Example 2.1.2, and therefore BACC may be applied. As noted in Example 4.3.1, the Gibbs sampling algorithm for the normal linear model produces drawings from the posterior distribution that are very nearly serially uncorrelated. There is little or no need for burn-in iterations, and only a few iterations are needed to draw values from the posterior distribution of the regression functions. Marginal likelihood can be computed exactly as described in Example 5.1.1. It is only for computing the posterior mean of the regression, displayed as a heavy line in Figure 5.1, that a larger number of iterations is needed, and in this example we use 100.

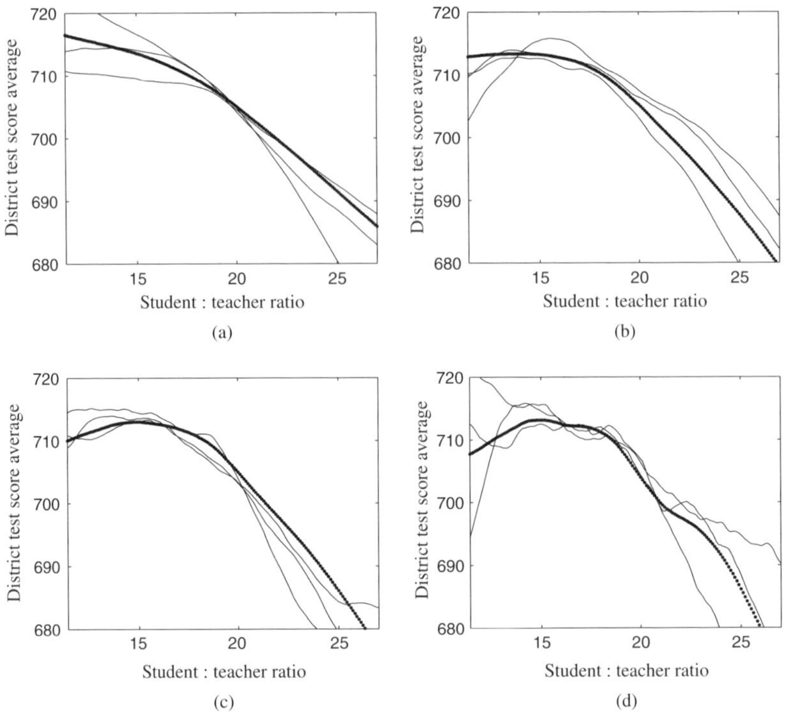

Figure 5.1. Posterior mean of regression of test score on student : teacher ratio (heavy line), and three drawings from the posterior distribution (three light lines) for each of four smoothness hyperparameters $\underline{h}^{-1/2}$: smoothness parameters (a) 0.50; (b) 1.00; (c) 2.00; (d) 5.00.

Recall from (5.34) that

$$\text{var}\{[f(x+1) - f(x)] - [f(x) - f(x-1)]\} = 2/3\underline{h}.$$

Alternative values of the smoothness parameter $\underline{h}^{-1/2}$ produce the following log marginal likelihoods:

$\underline{h}^{-1/2}$	Log Marginal Likelihood
0 (linear model)	−913.6910
0.5	−912.4573
1	−911.5170
2	−911.6266
5	−913.5087

The model with $\underline{h}^{-1/2} = 1$ has the highest marginal likelihood. However, differences in these models are not great; the Bayes factor in favor of this model, as opposed to the linear model, is 8.15.

The regression functions can be plotted using the 157 values in increments of one-tenth over the data range in the student : teacher ratio. Since not all of these values correspond to data points, values for the non–data points are drawn using (5.43). Adding more points to the plot is straightforward and increases computation time only negligibly.

Increasing values of $\underline{h}^{-1/2}$ permit regression functions that are increasingly rough, while as $\underline{h}^{-1/2} \to 0$, the function becomes linear. Note in Figure 5.1 that the posterior mean of the regression function (solid line) appears smoother than the drawings from the posterior distribution (lighter lines) in every case. That is because visual smoothness is inversely related to the absolute value of second differences of a function; Jensen's inequality accounts for the difference. It is important to keep in mind that while the visual appearances of these curves differ, the evidence that discriminates between them is rather weak as indicated by the Bayes factors implicit in the log marginal likelihoods.

In Example 5.1.1 other covariates were found to be important in accounting for the average test score in each district: the percentage of students still learning English, the percentage of students receiving a subsidized lunch, and the log of average district income. Example 5.4.1 omitted these covariates. Reintroducing them in linear fashion poses no essential complications.

Example 5.4.2 Nonlinearity in Multiple Regression (The online appendix contains data, annotated code, and output for this example.) Maintain all the assumptions of Example 5.1.1, except that now the student : teacher ratio enters the model in nonlinear fashion. Then we may express the model as an extension of (5.44):

$$y_t = \alpha_1 + \alpha_2 x_t + f(x_t) + \sum_{j=1}^{n} \gamma_j z_{jt} + \varepsilon_t, \quad \varepsilon_t \overset{\text{i.i.d.}}{\sim} N(0, h^{-1}) \quad (t = 1, \ldots, T). \tag{5.46}$$

In this application $n = 3$ and j indexes the three other covariates. More generally, (5.46) is sometimes called a *semiparametric model*, since it mixes the function $f(x_t)$ ("nonparametric") with the other linear components in the model. The Bayesian approach developed in this section renders this distinction inessential, since we may select from among the continuum of unobservables any finite number of functions of interest and carry out inference in a fashion that is logically consistent over all possible choices of these functions of interest.

We proceed in the same fashion as in the previous example. The prior distribution of the vector $(\alpha_1, \alpha_2, \gamma_1, \gamma_2, \gamma_3)'$ is the same as the prior distribution of $\boldsymbol{\beta}$ in Example 5.1.1, and the alternative values of the smoothing hyperparameter $\underline{h}$ are the same as those in the previous example. The model comparison exercise produces the following findings:

$\underline{h}^{-1/2}$	Log Marginal Likelihood
0 (linear model)	−805.8189
0.5	−806.2772
1	−807.0044
2	−808.2665
5	−811.1137

The original linear model is the most favored of the five specifications, but the evidence in favor of linearity is slightly weaker than the evidence in favor of nonlinearity in the previous example. The Bayes factor in favor of the linear model, versus the nonlinear specification with $\underline{h}^{-1/2} = 1$ is 3.27. On the other hand, the evidence that the three additional covariates should be included is overwhelming, as indicated by comparison with the log marginal likelihoods in the previous example.

Figure 5.2 provides some aspects of the posterior distribution of the regression function, evaluated at the student : teacher ratio values indicated on the horizontal axes, and sample median values of all of the other covariates. Whereas a student : teacher ratio of 25 as opposed to 15 accounted for a 20–30-point difference in test scores in the absence of the other covariates, here it accounts for somewhat less than 10 points, which in turn is roughly consistent with the posterior mean of −0.68 for the student : teacher ratio coefficient in Example 5.1.1. The predominant characteristic of the nonlinearity of the regression function in Example 5.4.1 was a kink at a student : teacher ratio of ~17, a feature that is exhibited weakly in Figure 5.2. Moreover, the uncertainty about departures from nonlinearity clearly overwhelm any such systematic tendency, as indicated by the lighter lines in the figure. This is consistent with the Bayes factor in favor of the linear specification as opposed to any one of the nonlinear specifications.

The impact of the nonlinear specification in student : teacher ratio has a negligible impact on the posterior distribution of the other coefficients of the other covariates. Comparison with the results in Example 5.1.1 shows the following:

Student : Teacher Ratio Functional Form	Linear		Nonlinear	
Coefficient Posterior	Mean	SD	Mean	SD
Learning English	−0.4086	0.2788	−0.3990	0.2811
Subsidized lunch	−0.5173	0.0678	−0.5117	0.0673
Log income	16.418	2.964	17.005	2.902

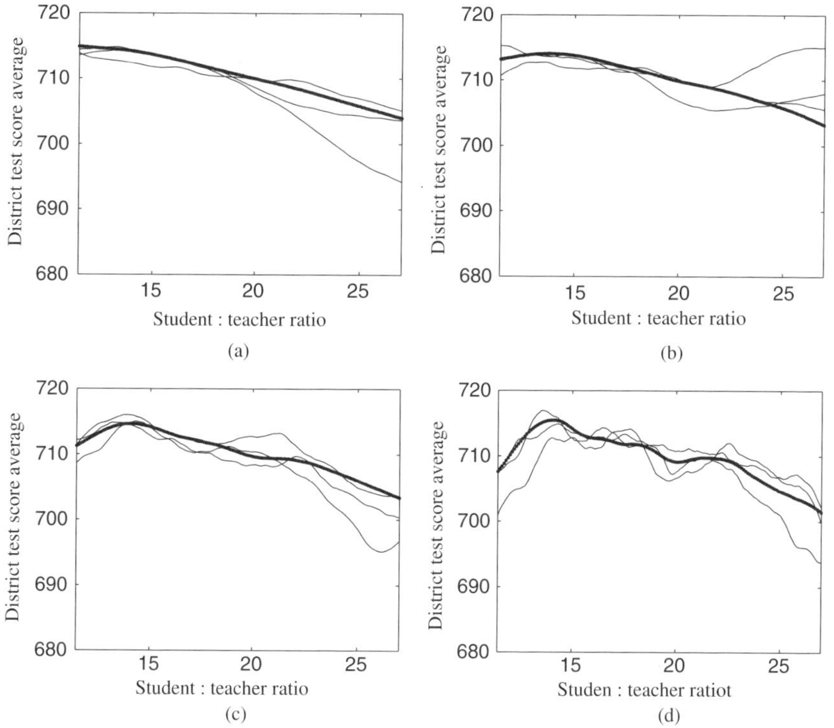

Figure 5.2. Posterior mean of regression function for test score nonlinear in student : teacher ratio (heavy line) and linear in other covariates, and three drawings from the posterior distribution (three light lines) of the nonlinear student : teacher portion for each of four smoothness hyperparameters $\underline{h}^{-1/2}$: smoothness parameters (a) 0.50; (b) 1.00; (c) 2.00; (d) 5.00.

5.4.2 Nonlinear Regression with Basis Functions

A sequence of normal linear models $A_j (j = 1, 2, \ldots)$ captures the essentials of nonlinear regression with basis functions. In model A_j, we obtain

$$y_t = f_j(\mathbf{x}_t) + \varepsilon_t = \sum_{i=1}^{k_j} \beta_{ji}\phi_{ji}(\mathbf{x}_t) + \varepsilon_t = \boldsymbol{\beta}'_j\boldsymbol{\phi}_j(\mathbf{x}_t) + \ \varepsilon_t; \ \ \varepsilon_t \overset{\text{i.i.d.}}{\sim} N(0, h^{-1}). \tag{5.47}$$

As j increases so does k_j, and the function f_j becomes, loosely speaking, more flexible. For example, if $\mathbf{x}_t$ is 2×1 and the basis functions are monomials, then a sequence might be

$$\phi_{j1}(\mathbf{x}) = 1 \text{ for all } j = 1, 2, \ldots$$

$$\phi_{j2}(\mathbf{x}) = x_1, \ \phi_{j3}(\mathbf{x}) = x_2 \text{ for all } j = 2, 3, \ldots$$

$$\phi_{j4}(\mathbf{x}) = x_1^2,\ \phi_{j5}(\mathbf{x}) = x_1x_2,\ \phi_{j6}(\mathbf{x}) = x_2^2, \text{ for all } j = 3, 4, \ldots \quad (5.48)$$

$$\phi_{j7}(\mathbf{x}) = x_1^3,\ \phi_{j8}(\mathbf{x}) = x_1^2x_2,\ \phi_{j9}(\mathbf{x}) = x_1x_2^2,\ \phi_{j10}(\mathbf{x}) = x_2^3, \ldots$$

for all $j = 4, 5, \ldots$ and so on. For this sequence $k_j = j(j+1)/2$. By the Weirstrass theorem, there exists a sequence of such basis functions $\{\boldsymbol{\phi}_j\}$ such that $\lim_{j\to\infty} \sup_{\mathbf{x}\in C} \left|f(\mathbf{x}) - f_j(\mathbf{x})\right| = 0$ for any compact set $C \subseteq \mathbb{R}^2$. Other systems of basis functions that have proved useful in econometric applications include Fourier sequences (Gallant 1981) and Muntz–Szasz expansions (Barnett and Jonas 1983). In all of these systems, the functions $\boldsymbol{\phi}_j$ are chosen so that $f(\mathbf{x})$ is forced to be smooth when j is small, and the smoothness assumption is weakened steadily as j increases. Precisely what is meant by "smooth" and "weakened" depends on the particular system of basis functions.

The approaches to nonlinear regression in (5.29) and (5.47) are complementary. Basis functions can be applied when the domain of f is multidimensional in much the same way as when it is unidimensional, especially for monomial basis functions. On the other hand, the ordering of the covariates x_t exploited in developing the approach in Section 5.4.1 cannot be extended to multidimensional $\mathbf{x}_t$. For any given order of expansion j, basis functions force the function f to be smooth no matter how strong the evidence to the contrary in the data, whereas for a given smoothness prior hyperparameter $\underline{h}$ in Section 5.4.1, a sufficiently strong departure from smoothness in the data can place substantial posterior probability on regression functions that are not smooth. This latter contrast is mitigated, to some extent, by the fact that each entails a portfolio of models, indexed by $\underline{h}$ in the case of smoothness priors and by j in the case of basis functions. In each case the models may be compared or averaged as described in Section 2.6.1. The less smooth is the function f, the smaller will be the hyperparameter $\underline{h}$ favored by Bayes factors in the former case and the larger will be the order of expansion j favored in the latter.

In formulating prior distributions of the coefficient vector $\boldsymbol{\beta}$ in nonlinear regression with basis functions, it is useful to think in terms of the function f. This is especially important in comparing variants with different numbers of basis functions, because it ensures comparable priors. For example, the prior distribution consisting of the components $f(\mathbf{a}_i) \overset{\text{i.i.d.}}{\sim} N(\mu, \tau^2)$ $(i = 1, \ldots, n)$ implies the prior distribution consisting of the components

$$\boldsymbol{\beta}_j'\boldsymbol{\phi}_j(\mathbf{a}_i) \mid A_j \overset{\text{i.i.d.}}{\sim} N(\mu, \tau^2)(i = 1, \ldots, n)$$

when the order of expansion is j. If J is the highest order of expansion considered and $n \geq k_J$ points are chosen appropriately, this approach will provide comparable and proper prior distributions for the coefficients in all orders of expansion. This approach is illustrated in the following example.

Example 5.4.3 Basis Functions for a Single Covariate (The online appendix contains annotated code and output for this example.) In the nonlinear model (5.29), nonlinearity in the function of one variable $f(\cdot)$ may be captured by using a sequence of basis functions with a single argument. This example uses polynomials of increasing order:

$$y_t = \sum_{i=0}^{j} \beta_{ji} x_t^i + \varepsilon_t, \quad \varepsilon_t \overset{\text{iid}}{\sim} N(0, h^{-1}) \quad (t = 1, \ldots, T) \tag{5.49}$$

in conjunction with the sequence of conditionally conjugate prior distributions

$$\underline{\mathbf{H}} = (k_j / T\underline{\sigma}^2)\mathbf{X}'\mathbf{X}, \ \underline{\boldsymbol{\beta}} = (\mathbf{X}'\mathbf{X})^{-1}\mathbf{X}'\iota_T \underline{\mu} = (\underline{\mu}, 0, \ldots, 0)', \tag{5.50}$$

and the same values of $\underline{\sigma}^2$ and $\underline{\mu}$ used in the previous examples. In (5.50) $\mathbf{X}$ is the covariate matrix in (5.49) and has $k_j = j + 1$ columns.

This model may be applied using BACC, much as in Example 5.4.1. The marginal likelihoods are

J	Log Marginal Likelihood
1 (linear model)	−913.6910
2	−911.0031
3	−912.6455
4	−914.1524
5	−915.3689

The results are similar to those in the nonlinear model of Example 5.4.1, in that there is moderate evidence in favor of nonlinearity, with nonlinearity in this case represented by low-order polynomials. Note that the marginal likelihood of the most favored model, a polynomial of order 2, is very nearly the same as the most favored model in Example 5.4.1, which has a smoothing hyperparameter of $\underline{h}^{-1/2} = 1$. The posterior distribution of the regression function reveals both further similarities and important contrasts with the approach taken in that earlier example.

In general the posterior means in Figures 5.1 and 5.3 show functions of similar global orientation and shape, and appear more irregular as more flexibility is allowed. The polynomial basis functions exhibit stronger curvature with less flexible models, and the smoothness prior shows a kink near the student : teacher ratio value of 17 with less flexible models. The most flexible models, in panel (d) of each figure, exhibit posterior means of regression functions that are almost identical. The three drawings from the posterior distribution of regression functions in each panel illustrate the fact that the functions themselves are quite different. In the case of basis functions (Figure 5.3) the second derivative at a point is a deterministic function of second derivatives globally, and therefore a function of information in the data as well as the prior, whereas in the case of smoothness priors (Figure 5.1) it is a function only of the prior distribution. This property of

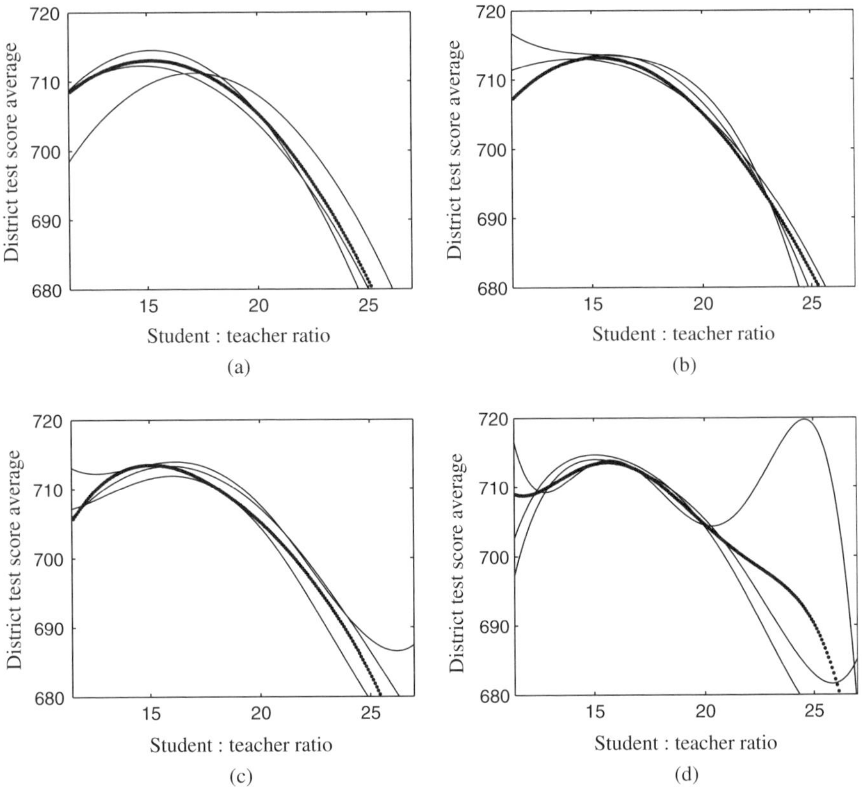

Figure 5.3. Posterior mean of regression of test score on student : teacher ratio (heavy line), and three drawings from the posterior distribution (three light lines) for polynomials of different order: polynomial orders (a) 2; (b) 3; (c) 4; (d) 5.

posterior distributions of regression functions with smoothness priors is obscured in the posterior mean.

Introducing the other covariates, as in Example 5.4.2, produces similar comparisons and contrasts. The marginal likelihoods are

J	Log Marginal Likelihood
1 (linear model)	−805.8189
2	−807.7776
3	−809.8334
4	−810.9572
5	−812.7080

The linear model is, again, favored over any of the alternatives that permit a nonlinear, but separable, effect of the student : teacher ratio on test scores; once

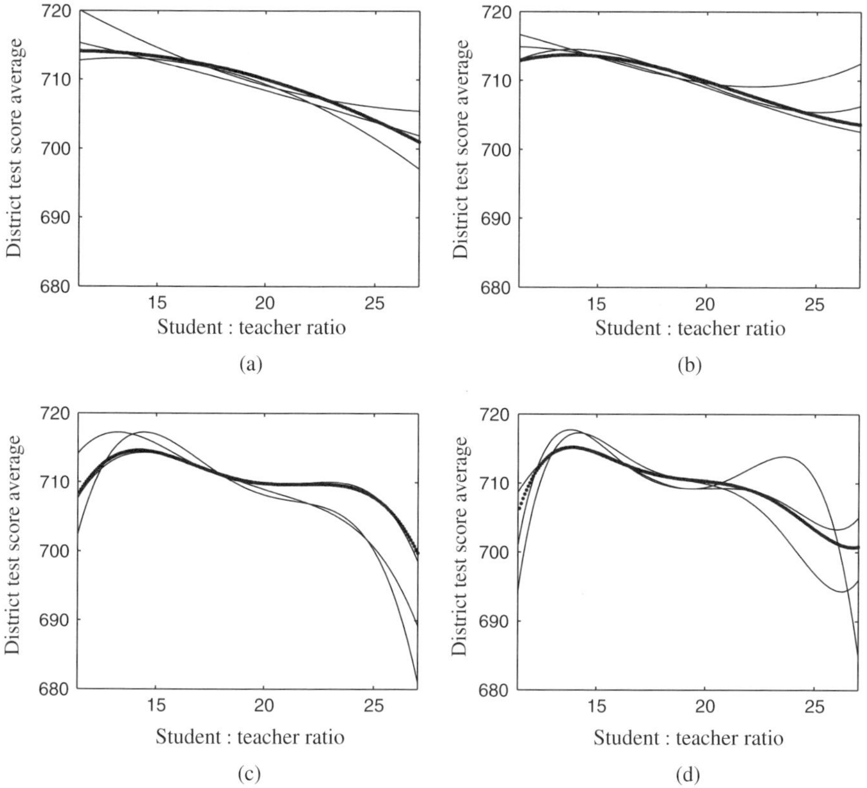

Figure 5.4. Posterior mean of regression function for test score polynomial in student : teacher ratio (heavy line) and linear in other covariates, and three drawings from the posterior distribution (three light lines) of the nonlinear student : teacher portion for different orders of the polynomial: polynomial orders (a) 2; (b) 3; (c) 4; (d) 5.

again, the evidence is not strong. The posterior means of the regression as a function of the student : teacher ratio, using this polynomial expansion, bear a striking resemblance to those using the smoothness priors, as may be seen by comparing Figure 5.4 with Figure 5.2.

An attraction of nonlinear regression with basis functions is that the nonlinear component of the regression can be a vector, as well as a scalar, as indicated in the illustration (5.48) for polynomial basis functions and a vector of dimension 2. For a vector of dimension k and a polynomial expansion of order J, the number of terms in the expansion is of the order k^J. For a given sample size, the larger is k, the lower is the order of expansion that will be reasonable, appraised (appropriately) by marginal likelihoods for the different expansions. In the case of the smoothness priors developed in Section 5.4.1, however, treatment of the vector case is impossible. That is because the approach taken there relies critically on the ordering of the covariates, and vectors of dimension greater than one cannot be ordered.

Example 5.4.4 Basis Functions for a Pair of Covariates (The online appendix contains annotated code and output for this example.) In the context of the test score example with all covariates, consider the polynomial expansion (5.48) using the covariates student : teacher ratio (`str`) and log income (`income`). Continuing to denote the full matrix of covariates by **X**, the prior distribution is (5.50). Consider a few orders of expansion, presented in the following table along with the corresponding marginal likelihood values:

Terms Included:	Log Marginal Likelihood
`str`, `income`	−805.8189
Above plus $\texttt{str} \cdot \texttt{income}$	−805.6166
Above plus $\texttt{str}^2$, $\texttt{income}^2$	−808.9928
Above plus $\texttt{str}^{3-i} \cdot \texttt{income}^i \, (i = 0, 1, 2, 3)$	−814.0507

Bayes factors favor the model with the single interaction term: the values are 1.22 against the linear model, and 29.26 against the model with quadratic terms.

As always, it is important to examine the implications of alternative models, including those with low posterior probabilities, if it is thought that the inclusion of these models in the analysis may affect the decision at hand. Using BACC and mathematical applications software, we can produce the representations of the posterior mean of the regression function for the model with the single interaction term, and the one that adds the two quadratic terms, shown in Figure 5.5. In both cases, the relationship between the student : teacher ratio and the district average test score depends strongly on average income in the district. The lower the average income in a school district, the more rapidly the expected average test score drops as student : teacher ratio increases. As incomes increase, this relationship is attenuated, and for log per capita incomes above about 3.05 [$\exp(3.05) \times \$1000 = \$21,115$, about one-quarter of the sample], it is in fact reversed, so that expected test scores increase as the student : teacher ratio increases. In the context of decision problems like the ones taken up in Example 5.1.2 and Exercise 5.1.5, this is a serious complication. Exercise 5.4.5 pursues this issue.

Figure 5.5 conveys the posterior expectation of the regression function, but no other aspects of its distribution. Interactive displays can convey useful descriptions of uncertainty about a function; code for some displays may be found in the online appendix for this example. In some decision problems, like the one stemming from the loss function (5.5) in Example 5.1.2, uncertainty doesn't matter, but in others, such as the one stemming from the loss function (5.6) in the same example, it does. For the models studied in Figure 5.5 it is easy to find the posterior probability that the regression function is monotone decreasing in student : teacher ratio, for all values of log income and student teacher ratio included in the grid; it is .065 in the case of the model with a single interaction term and .031 with the model that adds the quadratic terms.

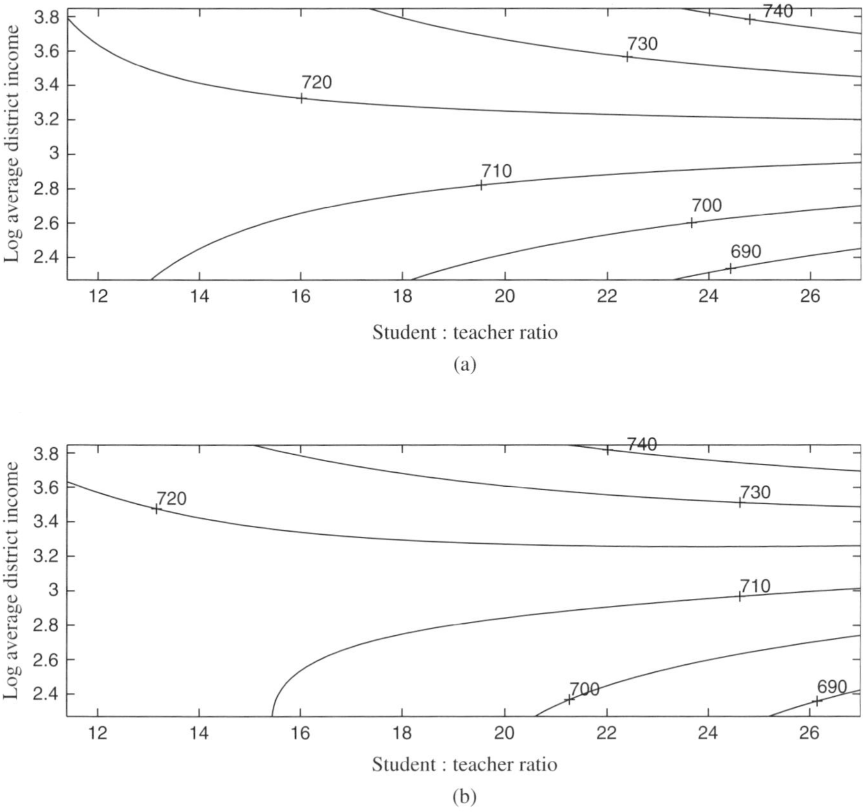

Figure 5.5. Contour plots of the posterior mean of the regression function, evaluated at combinations of two covariates as indicated, and sample median values of all other covariates: (a) linear terms plus str·logincome interaction; (b) linear terms plus polynomials to order 2.

Exercise 5.4.1 Completing the Argument Derive (5.34), (5.35), and (5.36). You may find a symbolic processor, like Maple or Mathematica, helpful.

Exercise 5.4.2 Credible Sets In non-Bayesian nonparametric regression it is common to find an approximate confidence interval for $f(x)$ for each of many closely spaced values of x over a given range. In the Bayesian nonlinear model of Section 5.4.1 we can find an exact credible interval for $f(x)$ for these same values. In each case we may plot the intervals as a function of x, and the resulting picture provides a representation of uncertainty about the function.

(a) Choose either Example 5.4.1 or 5.4.2, and plot 80% credible intervals over the range of student : teacher ratio in the data, 11.4–27.

(b) Find the posterior probability that $f(x)$ is contained in the intervals found in (a), for all x in the range 11.4–27.

(c) Find a constant c such that

$$P\{f(x) \in [\overline{f}(x) - c, \overline{f}(x) + c] \ \forall \ x \in [11.4, 27]\} = 0.8,$$

where

$$\overline{f}(x) = E[f(x) \mid (\mathbf{y}^o, \mathbf{X}, A)].$$

Plot the functions $\overline{f}(x) - c$, $\overline{f}(x)$, and $\overline{f}(x) + c$ on the same graph with the credible intervals found in (a). Compare and discuss.

Exercise 5.4.3 Inference about Shape The posterior means of the regression functions in Figures 5.1 through 5.4 are monotone decreasing and concave over the range shown in many cases.

(a) Show that the posterior probability that the regression function has this shape is zero in all cases, in Examples 5.4.1 and 5.4.2.

(b) Find the posterior probability that the regression function has this shape for polynomials of order 2, 3, 4 and 5 in Examples 5.4.3 and 5.4.4 (For continuation, see Exercise 8.2.5.)

Exercise 5.4.4 Model Combination Figures 5.1 and 5.2, together with the marginal likelihoods that accompany these models, indicate substantial uncertainty over models, and perhaps about shape as well. Select the model corresponding to one of the figures for this exercise.

(a) Assign equal prior probability to the linear model and each of the four nonlinear models. Conditional on this specification, find and plot the posterior expectation of the regression function as a heavy line, and draws from the posterior distribution as lighter lines, in the same way as done in the figures.

(b) An alternative approach to uncertainty about the smoothing hyperparameter $\underline{h}$ is to embed it in a hierarchical prior. Construct such a prior, and then find the posterior distribution of the regression function as well as $\underline{h}^{-1/2}$. Compare the result with those for the examples in Section 5.4.1.

Exercise 5.4.5 Decisionmaking in the Nonlinear Regression Model Example 5.1.2 introduced a decision problem about the student : teacher ratio using the alternative loss functions (5.5) and (5.6). That example provided solutions of those problems assuming a linear regression function.

(a) Using the smoothing approach of Example 5.4.2, or the polynomial approach of Example 5.4.3 with the full set of covariates, solve the same decision problems, for the same combinations of c and d considered in Example 5.4.2. Compare your answers with those in that example.

(b) Now consider the same problem using the polynomial in student : teacher ratio and log income of Example 5.4.4. Recall the complication implied by the results in Figure 5.5 for school districts with per capita incomes in the top quartile of the sample, implying that these districts would choose very high student : teacher ratios. You might address this problem by model averaging, by imposing shape constraints, or in some other way. After doing so, find the optimal student : teacher ratio for the same combinations of c and d considered in Example 5.4.2, and some alternative, representative values of income.

Exercise 5.4.6 Specification of the Earnings Regression Function Example 6.4.1 introduces the regression of log earnings on age and education, emphasizing the nonnormality of the distribution of the regression residuals. That example assumes that the regression function of log earnings on age and education is an interactive polynomial of order 4 in age and 2 in education. In answering the following questions, assume that the distribution of the regression residual is i.i.d. normal.

(a) What is the evidence in the data on the adequacy of this choice? Do marginal likelihoods for different orders of polynomial expansion indicate a clear-cut choice, or do they suggest model averaging using several alternative specifications?

(b) From the models investigated in (a), select the three with the three highest marginal likelihoods. Is there much difference among the corresponding functions $E(y \mid a, e, \mathbf{y}^o, A)$ as indicated, for example, by representations like those in Figure 5.5?

CHAPTER 6

Modeling with Latent Variables

Latent, or unobserved, variables are often important components in econometric and statistical models. They occur in these models for a variety of reasons.

1. The model may pertain to a substantial number of heterogeneous entities, each with its own set of parameters. It is natural to regard these parameters as latent variables having a distribution that, in turn, is characterized by unknown parameters. (See Example 3.1.1.) Section 3.1 showed that this formulation is equivalent to a hierarchical prior distribution for the parameters of the heterogeneous entities. The techniques described in this chapter have been applied with noted success in this context, particularly in marketing; see, for example, Ainslee and Rossi (1998), Allenby and Rossi (1999), and Kim et al. (2003), as well as Rossi and Allenby (2003) for an overview.
2. If data are missing because of complications in data collection, it is natural to regard the missing data as latent. The original model must be supplemented with one for the process by which data are recorded or not. The latter model turns out to be simple if data are missing at random (Example 2.2.3). [For specific examples, see Exercises 5.2.1 and 5.3.3(d).]
3. Outcomes may not be fully observed, because of the way data are recorded, or because observable behavior reveals only certain characteristics of the data. It is natural to regard the outcomes as latent variables. The complete model has two components: the first linking the incompletely observed outcomes to parameters or other unobservables in the usual way, and the second linking data to incompletely observed outcomes. Section 6.1 provides a general approach for this setting, and Section 6.2 applies it to the modeling of discrete outcomes.
4. In a mixture model the distribution of a random variable depends on a latent state. The state can be either continuous or discrete, and in either case a supplementary model describes the distribution of the state. Mixture models

Contemporary Bayesian Econometrics and Statistics, by John Geweke

provide versatile tools for extending simple distributions, such as the normal, to classes that are richer, more flexible, and more realistic. Section 6.4 enriches the normal linear model in this way.

A common characteristic of all these models is a hierarchy, or layering, in which the distribution of latent variables is accounted for by a stated set of assumptions and unobservables (parameters), and the distribution of observables is then driven by the latent variables together with some combination of the same and additional unobservables. (Exercise 3.1.2 developed the essence of this structure.) This hierarchy leads to significant simplification in the posterior distribution that can be exploited by the Gibbs sampler and related posterior simulators, as illustrated in Exercises 5.2.1 and 5.3.3(d). This simplification plays a critical role in the models described in this chapter as well.

6.1 CENSORED NORMAL LINEAR MODELS

An important source of earnings data in the United States is the records kept by the Social Security Administration. These records track individuals' annual earnings subject to social security tax. The tax is applied to all labor earnings, up to a limit that changes from year to year. Thus if t indexes an individual in a particular year, a known upper limit y_t^* applies to social security earnings y_t for observation t. If we denote actual earnings by $\widetilde{y}_t$, then

$$y_t = \begin{cases} \widetilde{y}_t & \text{if } \widetilde{y}_t \le y_t^* \\ y_t^* & \text{if } \widetilde{y}_t > y_t^* \end{cases}.$$

This is an example of censoring of a measurement. In this case values that exceed a known threshold are replaced by the threshold, a process often called "right censoring." It is important that the threshold be known. Censored measurement of the outcome variable in the normal linear model provides an important special case of the censored normal linear model.

As a second motivating example, suppose that T households indexed by t have preferences over a specific good x and all other goods z given by the utility function

$$U_t(x, z) = \log(a_t + x) + b\log(z).$$

The unobservable a_t is specific to household t. Only nonnegative amounts of x and z may be consumed. Clearly, each household t must consume some positive amount z_t of z. If $a_t < 0$, then household t must consume more than $-a_t$ units of x, whereas if $a_t > 0$, household t may choose not to consume x at all. Suppose that in each household t the sum of expenditures on x and z is fixed, and that these goods have prices p_{xt} and p_{zt}, respectively. If household t consumes a positive amount x_t of x, then $(\partial U_t/\partial x_t)/(\partial U_t/\partial p_t) = p_{xt}/p_{zt}$, and in this case

$$z_t/b(a_t + x_t) = p_{xt}/p_{zt} \Longrightarrow x_t = z_t p_{zt}/bp_{xt} - a_t.$$

If $a_t > z_t p_{zt}/bp_{xt}$, then household t does not consume x. If the unobservable a_t is independently and identically distributed across households, $a_t \sim N(\mu, \sigma^2)$, then

$$x_t = \max[-\mu + b^{-1}(p_{z_t} z_t)/p_{x_t} + \varepsilon_t,\ 0],\ \ \varepsilon_t \overset{\text{i.i.d.}}{\sim} N(0, \sigma^2)\ \ (t = 1, \ldots, T).$$

Models of this kind have been important in marketing and the success of methods like the ones described in this section; see, for example, Kim et al. (2003).

Note that the second example produces a left-censored outcome, whereas the first produces a right-censored outcome. In the second example the censoring points are all the same (zero), whereas in the first they vary (y_t^*) but are known. In order to handle both of these examples as special cases of a more general model, this section develops a general version of the censored normal linear model. This treatment is extended to nonnormal censored linear models in Section 6.4.3. BACC supports censored linear models at the level described in this chapter.

A censored normal linear model begins with

$$\underset{T\times 1}{\widetilde{\mathbf{y}}} = \underset{T\times k}{\mathbf{X}}\boldsymbol{\beta} + \boldsymbol{\varepsilon},\quad \boldsymbol{\varepsilon} \sim N(\mathbf{0}, h^{-1}\mathbf{I}_T), \tag{6.1}$$

in which $\widetilde{\mathbf{y}} = (\widetilde{y}_1, \ldots, \widetilde{y}_T)'$ is a vector of latent variables. There is a known, set-valued function

$$C_t = c_t(\widetilde{y}_t) \tag{6.2}$$

mapping each possible outcome $\widetilde{y}_t$ into exactly one set C_t. The observable set C_t contains the latent variable $\widetilde{y}_t$. It may be a (half) open or (half) closed interval. Important special cases are intervals including $-\infty$ or ∞ as endpoints, and singletons. Denote the collection $\mathbf{C} = (C_1, \ldots, C_T)$.

In the first motivating example $\widetilde{y}_t$ is actual earnings. If $\widetilde{y}_t < y_t^*$, then C_t is the singleton $\widetilde{y}_t$; otherwise $C_t = [y_t^*, \infty)$. In the second motivating example

$$\widetilde{y}_t = -\mu + b^{-1}(p_{z_t} z_t)/p_{x_t} + \varepsilon_t, \varepsilon_t \sim N(0, \tau).$$

If $\widetilde{y}_t > 0$, then $C_t = \widetilde{y}_t$; otherwise $C_t = (-\infty, 0]$.

In the conditionally conjugate prior distribution $\boldsymbol{\beta}$ and h are independently distributed:

$$\boldsymbol{\beta} \mid A \sim N(\underline{\boldsymbol{\beta}}, \underline{\mathbf{H}}^{-1}),\ \ \underline{s}^2 h \mid A \sim \chi^2(\underline{\nu}). \tag{6.3}$$

Utilizing (6.1), (6.2), and (6.3), the joint distribution of observables and unobservables in the censored linear model is

$$p(\boldsymbol{\beta}, h, \widetilde{\mathbf{y}}, \mathbf{C} \mid \mathbf{X}, A) = p(\boldsymbol{\beta} \mid A)p(h \mid A)p(\widetilde{\mathbf{y}} \mid \boldsymbol{\beta}, h, \mathbf{X}, A)p(\mathbf{C} \mid \widetilde{\mathbf{y}}, A).$$

From (6.2)

$$p(\mathbf{C} \mid \widetilde{\mathbf{y}}, A) = \prod_{t=1}^{T} p(C_t \mid \widetilde{y}_t, A) = \prod_{t=1}^{T} I_{C_t}(\widetilde{y}_t). \tag{6.4}$$

[Observe that $p(C_t \mid \widetilde{y}_t, A)$ in (6.4) is a probability function and not a probability density function; it places all its mass on a single observable outcome set indexed by $\widetilde{y}_t$.] Hence $p(\boldsymbol{\beta}, h, \widetilde{\mathbf{y}}, \mathbf{C} \mid \mathbf{X}, A) \propto$

$$\exp[-(\boldsymbol{\beta} - \underline{\boldsymbol{\beta}})'\underline{\mathbf{H}}(\boldsymbol{\beta} - \underline{\boldsymbol{\beta}})/2]h^{(\underline{\nu}-2)/2} \exp(-\underline{s}^2 h/2) \tag{6.5a}$$

$$\cdot \exp[-h(\widetilde{\mathbf{y}} - \mathbf{X}\boldsymbol{\beta})'(\widetilde{\mathbf{y}} - \mathbf{X}\boldsymbol{\beta})/2] \tag{6.5b}$$

$$\cdot \prod_{t=1}^{T} I_{C_t}(\widetilde{y}_t). \tag{6.5c}$$

The kernel of (6.5a)–(6.5c) in $(\boldsymbol{\beta}, h, \widetilde{\mathbf{y}})$, with the observed outcome $\mathbf{C}^o$ replacing the observable $\mathbf{C}$, is the kernel of the posterior distribution. The conditional posterior distributions are simple. The kernels of $p(\boldsymbol{\beta} \mid h, \widetilde{\mathbf{y}}, \mathbf{y}^o)$ and $p(h \mid \boldsymbol{\beta}, \widetilde{\mathbf{y}}, \mathbf{y}^o)$ are determined by (6.5a)–(6.5b) alone, from which

$$\boldsymbol{\beta} \mid (h, \widetilde{\mathbf{y}}, \mathbf{y}^o, \mathbf{X}, A) \sim N(\overline{\boldsymbol{\beta}}, \overline{\mathbf{H}}), \tag{6.6}$$

where $\overline{\mathbf{H}} = \underline{\mathbf{H}} + h\mathbf{X}'\mathbf{X}$ and $\overline{\boldsymbol{\beta}} = \overline{\mathbf{H}}^{-1}(\underline{\mathbf{H}}\underline{\boldsymbol{\beta}} + h\mathbf{X}'\widetilde{\mathbf{y}})$, and

$$\overline{s}^2 h \mid (\boldsymbol{\beta}, \widetilde{\mathbf{y}}, \mathbf{y}^o, \mathbf{X}, A) \sim \chi^2(\overline{\nu}), \tag{6.7}$$

where $\overline{s}^2 = \underline{s}^2 + (\widetilde{\mathbf{y}} - \mathbf{X}\boldsymbol{\beta})'(\widetilde{\mathbf{y}} - \mathbf{X}\boldsymbol{\beta})$ and $\overline{\nu} = \underline{\nu} + T$. These results reflect the fact, also apparent in (6.1) and (6.3), that with $\widetilde{\mathbf{y}}$ in the conditioning set, thus treating it as observable, this model is precisely the same as the normal linear model introduced in Example 2.1.2. The distributions in (6.6) and (6.7) are the same as those in that example, but with $\widetilde{\mathbf{y}}$ replacing $\mathbf{y}^o$.

The kernel of $p(\widetilde{\mathbf{y}} \mid \boldsymbol{\beta}, h, \mathbf{C}^o, \mathbf{X}, A)$ is determined by (6.5b)–(6.5c). Utilizing the equivalent expression

$$\prod_{t=1}^{T} \exp[-h(\widetilde{y}_t - \boldsymbol{\beta}'\mathbf{x}_t)^2/2]$$

in lieu of (6.5b), it is apparent that $\widetilde{y}_1, \ldots, \widetilde{y}_T$ are conditionally independent, with

$$p(\widetilde{y}_t \mid \boldsymbol{\beta}, h, C_t^o, \mathbf{x}_t, A) \propto \exp[-h(\widetilde{y}_t - \boldsymbol{\beta}'\mathbf{x}_t)^2/2] I_{C_t^o}(\widetilde{y}_t).$$

Thus

$$\widetilde{y}_t \mid (\boldsymbol{\beta}, h, C_t^o, \mathbf{x}_t, A) \sim N(\boldsymbol{\beta}'\mathbf{x}_t, h^{-1}) \text{ subject to } \widetilde{y}_t \in C_t^o. \tag{6.8}$$

In the first motivating example, conditional on $\boldsymbol{\beta}$ and h, $\widetilde{y}_t = y_t^o$ if $y_t < y_t^*$ and $\widetilde{y}_t \sim N(\boldsymbol{\beta}'\mathbf{x}_t, h^{-1})$ subject to $\widetilde{y}_t \geq y_t^*$ if $y_t = y_t^*$. In the second example, conditional on $\boldsymbol{\beta}$ and h, $\widetilde{y}_t = y_t$ if $y_t > 0$ and $\widetilde{y}_t \sim N(\boldsymbol{\beta}'\mathbf{x}_t, h^{-1})$ subject to $\widetilde{y}_t \leq 0$ if $y_t = 0$.

This algorithm is the same as that developed in Example 4.3.1, but with the additional step (6.8). It was first proposed by Chib (1992). The approach can be applied to many kinds of censoring in more complex models as well. See Section 6.4.3, and Cowles et al. (1996) for examples.

Exercise 6.1.1 Categorical Censoring Household income in marketing surveys is often reported in brackets: for example, under \$15,000, \$15,000–\$25,000, ..., over \$95,000.

(a) Suppose that household income is the outcome variable in a normal linear model. Using clear and consistent notation, show that this model, in conjunction with the reporting of income in brackets, is a special case of the censored linear model.

(b) Suppose that the logarithm of household income is the covariate x_{tk} (the last column of $\mathbf{X}$) in a normal linear model. Suppose further that the normal linear model is a correct specification—in the notation of Section 3.4, $p(\mathbf{y} \mid \mathbf{X}, D) = p(\mathbf{y} \mid \boldsymbol{\beta}, h, \mathbf{X}, A)$ for appropriate values $\boldsymbol{\beta} = \boldsymbol{\beta}^*$ and $h = h^*$. Finally, suppose that household income is incorrectly assumed to be at the midpoint of the observed bracket, and \$140,000 is assumed to be household income if the observed bracket is (\$95,000, ∞). Show that the pseudotrue values of $\boldsymbol{\beta}$ and h in this situation are not $\boldsymbol{\beta}^*$ and h^*, respectively.

(c) Let $\mathbf{X}^*$ denote columns 1 through $k-1$ of $\mathbf{X}$, and $\mathbf{X}^{*\prime} = [\mathbf{x}_1^*, \ldots, \mathbf{x}_T^*]$. Formulate a complete model incorporating the assumption that household incomes x_{tk} are conditionally independent given the $\mathbf{x}_t^*$, with $\log(x_{tk}) \sim N(\boldsymbol{\gamma}'\mathbf{x}_t^*, j^{-1})$. Utilize a prior distribution for $\boldsymbol{\gamma}$ and j that is independent of the prior distribution for $\boldsymbol{\beta}$ and h.

(d) Outline a Gibbs sampling algorithm to simulate the posterior distribution of the model in (c).

(e) Given the assumptions made about $p(\mathbf{y} \mid \mathbf{X}, D)$ in (b) and about x_{tk} in (c), are the pseudotrue values of $\boldsymbol{\beta}$ and h in (c) equal to $\boldsymbol{\beta}^*$ and h^*, respectively?

(f) Suppose that in part (b) the level of household income rather than the logarithm of household income were the covariate x_{tk}. How would this affect the Gibbs sampling algorithm in part (d)?

Exercise 6.1.2 Inference for Censored Outcomes For the normal linear regression model

$$\mathbf{y} \mid (\boldsymbol{\beta}, h, \mathbf{X}, A) \sim N(\mathbf{X}\boldsymbol{\beta}, h^{-1}\mathbf{I}_T), \quad \boldsymbol{\beta} \mid A \sim N(\underline{\boldsymbol{\beta}}, \underline{\mathbf{H}}^{-1}), \quad \underline{s}^2 h \sim \chi^2(\underline{\nu}),$$

three independent samples have been collected.

- Sample 1 has $T = T_1$ observations, with observed covariate matrix $\mathbf{X} = \mathbf{X}_1$ and the observed outcome vector $\mathbf{y}_1^o$.

- Sample 2 has $T = T_2$ observations. The covariate matrix $\mathbf{X} = \mathbf{X}_2$ is observed, but the outcome vector $\mathbf{y}_2$ is not. However, for any two observations t and s, we know whether $y_t > y_s$ or $y_s > y_t$. (In other words, we observe the rank ordering of the outcome variables.)
- Sample 3 has $T = T_3$ observations. The covariate matrix $\mathbf{X} = \mathbf{X}_3$ is observed. The outcome vector $\mathbf{y}_3^o$ was also observed, but through an error in data processing the correspondence to the covariate matrix is no longer known. (In other words, we observe the T_3 outcomes but don't know which observations they correspond to.)

(a) Using samples 1 and 2, construct a simulator that draws from the posterior distribution for $\boldsymbol{\beta}$, h, and $\mathbf{y}_2$. Be as explicit as you possibly can about the distributions used in the simulator.

(b) Could you use sample 2 by itself to construct a simulator that would draw from the posterior distribution for $\boldsymbol{\beta}$, h, and $\mathbf{y}_2$? If so, state the algorithm for the simulator. If not, indicate the difficulty and what additional information would be required.

(c) Using samples 1 and 3, construct a simulator that draws from the posterior distribution for $\boldsymbol{\beta}$, h, and $\mathbf{y}_3$. Be as explicit as you possibly can about the distributions used in the simulator.

(d) Show that the simulator constructed in part (c) is ergodic.

6.2 PROBIT LINEAR MODELS

Suppose that T individuals, indexed by t, each allocate income y_t between two goods, x and z, with respective prices p_x and p_z constant across individuals. Good x is continuously divisible, but z can be only 0 or 1. (For example, z might represent a decision to enlist in the military, or not.) Individual t's utility function is

$$U_t(x, z) = z(a_t - b/x) + (1 - z)(r_t - s/x);\ b > 0, s > 0;$$

and he/she consumes out of income y_t. If $z = 0$, then $x = y_t/p_x$ and $U_t(x, z) = r_t - sp_x/y_t$. If $z = 1$, then $x = (y_t - p_z)/p_x$ and $U_t(x, z) = a_t - bp_x/(y_t - p_z)$. Individual t chooses $z_t = 1$ if $-bp_x/(y_t - p_z) + sp_x/y_t + a_t - r_t > 0$.

The econometrician observes p_x, p_z, and $(y_1, z_1), \ldots, (y_T, z_T)$. She does not observe $(a_1, r_1), \ldots, (a_T, r_T)$, but suppose that she is willing to take

$$(a_t - r_t) \mid (p_x, p_z, y_1, \ldots, y_T) \overset{\text{i.i.d.}}{\sim} N(\mu, \sigma^2)$$

as representative of the distribution of these unobservables across individuals, and to regard income y_t as ancillary. Define the unobservable

$$\widetilde{z}_t = \mu - bp_x/(y_t - p_z) + sp_x/y_t + \varepsilon_t;\ \varepsilon_t \overset{\text{i.i.d.}}{\sim} N(0, \sigma^2) \qquad (6.9)$$

and note that $z_t = I_{(0,\infty)}(\widetilde{z}_t)$. From (6.9), we have

$$P(z_t = 1 \mid b, s, \mu, \sigma^2, p_x, p_y, y_t, A) = \Phi\{\sigma^{-1}[-bp_x/(y_t - p_z) + sp_x/y_t + \mu]\}. \tag{6.10}$$

Because nothing is changed if b, s, μ, and σ are scaled by a common positive factor, the normalization $\sigma^2 = 1$ is convenient.

In a probit model, A, the observables are a $T \times k$ matrix of covariates $\mathbf{X} = [\mathbf{x}_1, \ldots, \mathbf{x}_T]'$ and a corresponding set of T binary outcomes with

$$P(\text{first binary outcome} \mid \mathbf{x}_t, A) = \Phi(\boldsymbol{\beta}'\mathbf{x}_t), \tag{6.11}$$

$$P(\text{second binary outcome} \mid \mathbf{x}_t, A) = 1 - \Phi(\boldsymbol{\beta}'\mathbf{x}_t). \tag{6.12}$$

In the motivating example the three covariates are a constant term, $-p_x/(y_t - p_z)$, and p_x/y_t; the first binary outcome is $z_t = 1$ and the second binary outcome is $z_t = 0$.

The substance of the probit model is (6.11)–(6.12), regardless of how the binary outcome is coded. If we introduce the latent variables $\widetilde{y}_t = \boldsymbol{\beta}'\mathbf{x}_t + \varepsilon_t$, $\varepsilon_t \sim N(0, 1)$, then the first outcome corresponds to $\widetilde{y}_t \leq 0$ and the second to $\widetilde{y}_t > 0$. In this context the natural outcome coding is the set-valued function $C_t = c_t(\widetilde{y}_t)$, with $C_t = (-\infty, 0]$ for the first outcome and $C_t = (0, \infty)$ for the second. The conditionally conjugate prior distribution is $\boldsymbol{\beta} \sim N(\underline{\boldsymbol{\beta}}, \underline{\mathbf{H}}^{-1})$.

The joint distribution of observables and unobservables in the probit model is

$$p(\boldsymbol{\beta}, \widetilde{\mathbf{y}}, C \mid \mathbf{X}, A) = p(\boldsymbol{\beta} \mid A)p(\widetilde{\mathbf{y}}, \boldsymbol{\beta} \mid \mathbf{X}, A)p(\mathbf{C} \mid \widetilde{\mathbf{y}}, A)$$

$$\propto \exp[-(\boldsymbol{\beta} - \underline{\boldsymbol{\beta}})'\underline{\mathbf{H}}(\boldsymbol{\beta} - \underline{\boldsymbol{\beta}})/2] \exp[-(\widetilde{\mathbf{y}} - \mathbf{X}\boldsymbol{\beta})'(\widetilde{\mathbf{y}} - \mathbf{X}\boldsymbol{\beta})/2] \prod_{t=1}^{T} I_{C_t}(\widetilde{y}_t).$$

This expression is identical to (6.5a)–(6.5c) for the censored linear model, except that here $h = 1$. Proceeding as in the analysis of that model, we obtain

$$\boldsymbol{\beta} \mid (\widetilde{\mathbf{y}}, \mathbf{C}, A) \sim N(\overline{\boldsymbol{\beta}}, \overline{\mathbf{H}}^{-1})$$

where $\overline{\mathbf{H}} = \underline{\mathbf{H}} + \mathbf{X}'\mathbf{X}$ and $\overline{\boldsymbol{\beta}} = \overline{\mathbf{H}}^{-1}(\underline{\mathbf{H}}\underline{\boldsymbol{\beta}} + \mathbf{X}'\widetilde{\mathbf{y}})$. In the distribution of $\widetilde{\mathbf{y}}$ conditional on $(\mathbf{C}, \boldsymbol{\beta}, \mathbf{X}, A)$ the elements $\widetilde{y}_t$, known as probits, are independent:

$$p(\widetilde{y}_t \mid \boldsymbol{\beta}, C_t, \mathbf{X}, A) \propto \exp[-(\widetilde{y}_t - \boldsymbol{\beta}'\mathbf{x}_t)^2/2] I_{C_t}(\widetilde{y}_t).$$

These conditional posterior distributions are the basis of a very simple Gibbs sampling algorithm, first proposed in Albert and Chib (1993b). BACC incorporates the probit linear model using the conjugate prior distribution and posterior simulation algorithm described in this section. This approach can be extended to situations involving more complex choices. When one choice is made from several alternatives, the logical extension is the multinomial probit model, for which Bayesian Markov chain Monte Carlo methods have been developed by McCulloch and Rossi

(1994), Geweke et al. (1994, 1997), and McCulloch et al. (2000). When there are several related dichotomous choices, the natural extension is the multivariate probit model; see Chib and Greenberg (1998).

Exercise 6.2.1 Normalization and Representation in the Probit Model Consider the motivating example at the start of this section.

(a) Show that if the same constant is added to a_t and r_t, an individual's consumption will not change. What does this say about the econometrician's decision to assume only a distribution for $a_t - r_t$, rather than assume a nondegenerate bivariate distribution for (a_t, r_t)?

(b) Show that if the parameters b, s, μ, and σ in (6.9) or (6.10) are scaled by a common positive constant, then there is no change in the distribution of the observables $(y_1, z_1), \ldots, (y_T, z_T)$.

(c) Why is it preferable to resolve the indeterminacy in scaling by taking $\sigma^2 = 1$ rather than $b = 1$ or $d = 1$?

(d) Derive (6.10) from (6.9).

Exercise 6.2.2 Ordered Probit Model In this model, $\widetilde{\mathbf{y}} \mid (\boldsymbol{\beta}, h, \mathbf{X}, A) \sim N(\mathbf{X}\boldsymbol{\beta}, h^{-1}\mathbf{I}_T)$. However, the variables $\widetilde{y}_t$ are unobservable. Instead, we observe the outcome $y_t = -1$ if $\widetilde{y}_t < 0$, $y_t = 0$ if $\widetilde{y}_t \in [0, 1]$, and $y_t = 1$ if $\widetilde{y}_t > 1$. (In actual application this could be coding for the outcomes "negative," "indeterminate," and "positive" in a medical test.)

(a) Are the "cutoff" values of 0 and 1 for $\widetilde{y}_t$ arbitrary? For example, would it have been less restrictive to choose $y_t = -1$ if $\widetilde{y}_t < c_1$, $y_t = 0$ if $\widetilde{y}_t \in [c_1, c_2]$, and $y_t = 1$ if $\widetilde{y}_t > c_2$, treating c_1 and c_2 as unobservable parameters?

(b) Derive a Gibbs sampling algorithm whose blocks are $\boldsymbol{\beta}$, h, and the unobservable $\widetilde{y}_t$. Show that it is ergodic with the unique invariant distribution being the posterior distribution of $\boldsymbol{\beta}$, h, and the unobservable $\widetilde{y}_t$.

(c) Could this idea be extended to $n > 3$ ordered outcomes?

6.3 THE INDEPENDENT FINITE STATE MODEL

Often economic agents or entities can be characterized as being in one of a small number of possible states. For example, a sample of individuals from a population at a specific point in time might be classified as registered to vote with one of a few political party preferences, registered without preference, or unregistered; or households might be classified by the number of individuals in the household. If the probability that a particular individual (household) is in a particular state is independent of other observables and of the states of all other individuals (households), then the distribution of classifications is an independent finite state model.

This model is extremely simple. While we are sometimes interested in it directly, it is interesting primarily because it can arise as a constituent of a more complicated model. The independent finite state model will appear again in models with mixtures (Section 6.4) and the first-order Markov finite state model (Section 7.2).

In the independent finite state model there are m possible states of the world that are occupied by n agents over T time periods; $m > 1$, $n > 0$, $T > 0$. For any agent k, let the integer s_{kt} indicate the state occupied at time t; $1 \leq s_{kt} \leq m$. The independent finite state model specifies that the NT random variables s_{kt} are mutually independent, and

$$P[s_{kt} = j \mid \pi_1, \ldots, \pi_m, A] = \pi_j \quad (k = 1, \ldots, n; t = 1, \ldots, T; j = 1, \ldots, m);$$

of course, $\sum_{i=1}^{m} \pi_j = 1$. The observables can be collected in the $n \times T$ matrix $\mathbf{S} = [s_{kt}]$ and the unobservables arranged in the $m \times 1$ vector $\boldsymbol{\pi} = (\pi_1, \ldots, \pi_m)'$.

Turning to inference, define

$$n_j = \sum_{k=1}^{n} \sum_{t=1}^{T} \delta(s_{kt}, j), \tag{6.13}$$

the number of times that $s_{kt} = j$ occurs in the sample. [The expression $\delta(s_{kt}, j)$ in (6.13) is an instance of the *Kronecker delta function*, defined $\delta(a, b) = 1$ if $a = b$ and $\delta(a, b) = 0$ if $a \neq b$.] From (6.3), we obtain

$$P(\mathbf{S} \mid \boldsymbol{\pi}, A) = \prod_{k=1}^{n} \prod_{t=1}^{T} \pi_{s_{kt}} = \prod_{j=1}^{m} \pi_j^{n_j}. \tag{6.14}$$

The number of occurrences n_j $(j = 1, \ldots, m)$ constitutes a vector of sufficient statistics in this model. The likelihood function provides the kernel of the conjugate prior distribution of $\boldsymbol{\pi}$. It is that of the *Dirichlet distribution* (Kotz et al. 2000, Section 49.1) for random variables $x_1, \ldots, x_m$ jointly distributed on the $(m-1)$-*dimensional unit simplex*

$$\left\{ x_i : x_i \geq 0 \ (i = 1, \ldots, m); \ \sum_{i=1}^{m} x_i = 1 \right\},$$

with m positive parameters $a_1, \ldots, a_m$ and density

$$p(x_1, \ldots, x_m \mid a_1, \ldots, a_m) = \left[\Gamma\left(\sum_{i=1}^{m} a_i \right) \Big/ \prod_{i=1}^{m} \Gamma(a_i) \right] \prod_{i=1}^{m} x_i^{(a_i - 1)}. \tag{6.15}$$

[Note that the beta(a_1, a_2) distribution (Casella and Berger 2002, Section 3.3) is the special case $m = 2$.] Thus the conjugate prior distribution of $\boldsymbol{\pi}$ is the Dirichlet distribution with density

$$p(\boldsymbol{\pi} \mid \mathbf{a}) = \left[\Gamma\left(\sum_{j=1}^{m} a_j\right) \bigg/ \prod_{j=1}^{m} \Gamma(a_j)\right] \prod_{j=1}^{m} \pi_j^{(a_j-1)}. \tag{6.16}$$

Like all conjugate prior distributions, this one has a notional data interpretation; the information in the prior is analogous to that in $\sum_{j=1}^{m} a_j - m$ observations in the context of the independent finite state model, with $a_j - 1$ occurrences of state j $(j = 1, \ldots, m)$.

The product of (6.14) and (6.16) is the posterior density kernel in standard form:

$$p(\boldsymbol{\pi} \mid \mathbf{S}) \propto \left[\Gamma\left(\sum_{j=1}^{m} a_j\right) \bigg/ \prod_{j=1}^{m} \Gamma(a_j)\right] \prod_{j=1}^{m} \pi_j^{(a_j+n_j-1)}. \tag{6.17}$$

Comparing (6.17) with (6.15), it follows that the posterior distribution of $\boldsymbol{\pi}$ is Dirichlet with parameters $a_j + n_j$ $(j = 1, \ldots m)$. The posterior density combines prior and sample information in a direct and obvious way. This is a consequence of the fact that the Dirichlet distribution is a member of the exponential family (Definition 2.3.3) and the general result (2.60) for posterior distributions with conjugate priors in that family.

The corresponding marginal likelihood is the integral of the right side of (6.17) with respect to $\boldsymbol{\pi}$. The value of the integral can be read from the normalizing constant in (6.15), and thus

$$p(\mathbf{S} \mid A) = \frac{\Gamma\left(\sum_{j=1}^{m} a_j\right) \prod_{j=1}^{m} \Gamma(a_j + n_j)}{\Gamma\left[\sum_{j=1}^{m} (a_j + n_j)\right] \prod_{j=1}^{m} \Gamma(a_j)}. \tag{6.18}$$

As a consequence of the simplicity of the independent finite state model, the posterior density and marginal likelihood have closed-form analytical expressions. In a direct application of the model, there may be no need for posterior simulation; see Exercise 6.3.1. When the independent finite state model is a constituent of a more complex model, however, it is useful to be able to draw from the Dirichlet distribution whose general form is (6.15). This can be done using a technique given in Devroye (1986), pp. 593–596—construct the independent random variables $d_i \sim \chi^2(2a_i)$ $(i = 1, \ldots, m)$ and then take $x_i = d_i / \sum_{j=1}^{m} d_j$ $(j = 1, \ldots, m)$.

BACC incorporates the independent finite state model using the conjugate prior distribution and posterior simulation algorithm described in this section.

Exercise 6.3.1 Properties of the Posterior Distribution Suppose that $\mathbf{S} = \mathbf{S}^o$ is observed in the context of the independent finite state model A with observables probability distribution (6.14) and prior pdf (6.16).

(a) Derive an analytical expression for the moment $E\left(\prod_{j=1}^{m} \pi_j^{k_j} \mid \mathbf{S}^o, A\right)$.

(b) From the result in (a) express the posterior mean and variance of π_j.

(c) What is the value of the posterior mode?

(d) Show that the marginal distribution of π_i is beta$(a_i, \sum_{j \neq i} a_j)$. [*Hint*: Consider the construction of Devroye (1986) and the reproductive property of the chi square distribution.]

(e) Determine the marginal posterior density of any pair of parameters (π_i, π_k) $(i \neq k)$.

Exercise 6.3.2 Empty Cells in the Independent Finite State Model Suppose that in a sample $\mathbf{S}^o$, $n_1 = 0$.

(a) What is the maximum likelihood estimate of π_1? Can you state the asymptotic standard error associated with this estimate?

(b) What is the marginal posterior distribution of π_1, and what are the mean and standard deviation of this distribution?

6.4 MODELING WITH MIXTURES OF NORMAL DISTRIBUTIONS

The models for continuously distributed observables treated up to this point have repeatedly exploited the specifications

$$\boldsymbol{\beta} \sim N(\underline{\boldsymbol{\beta}}, \underline{\mathbf{H}}_{\beta}^{-1}), \tag{6.19}$$

$$\boldsymbol{\varepsilon} \sim N(\mathbf{0}, h^{-1}\mathbf{I}_T), \tag{6.20}$$

in the relation

$$\mathbf{y} = \mathbf{X}\boldsymbol{\beta} + \boldsymbol{\varepsilon} \tag{6.21}$$

between the observables $\mathbf{X}$ and $\mathbf{y}$. As first noted in Section 1.1.2, the specification (6.20) is quite poor in some applications. It has nonetheless been maintained, to this point, because (1) the combination (6.19)–(6.20) leads to a normal conditional posterior distribution for $\boldsymbol{\beta}$ in all these models, and (2) we can generalize both (6.19) and (6.20), building on the tools developed in the process of treating (6.19)–(6.20) in the context of (6.21). This section turns to the generalization of (6.20), while that of (6.19) is taken up in Section 8.4. The keys to extending (6.20) are using latent variables and successive conditioning to create mixtures of normal distributions. In the process of posterior simulation the successive conditioning in the Gibbs sampler and related MCMC procedures recovers the underlying normal distributions.

6.4.1 The Independent Student-*t* Linear Model

In many applications of the normal linear model there is substantial evidence that the probability of an unusually large or small value of the outcome y_t is substantially greater than indicated by a normal distribution. This is a well-documented phenomenon in the case of financial asset returns; Section 1.1.2 provides an example.

An alternative to (6.20)–(6.21) that better accommodates this phenomenon is

$$y_t = \boldsymbol{\beta}'\mathbf{x}_t + \varepsilon_t, \tag{6.22}$$

$$\varepsilon_t \mid (h, \lambda, A) \overset{\text{i.i.d.}}{\sim} t(0, h^{-1}; \lambda) \; (t = 1, \ldots, T). \tag{6.23}$$

That is, each disturbance ε_t has a Student-*t* distribution with location parameter 0, scale parameter h^{-1}, and λ degrees of freedom. A standard representation for ε_t is

$$\varepsilon_t = \widetilde{h}_t^{-1/2}\eta_t, \tag{6.24}$$

with

$$\eta_t \mid (h, A) \overset{\text{i.i.d.}}{\sim} N(0, h^{-1}) \; (t = 1, \ldots, T), \tag{6.25}$$

$$\lambda \cdot \widetilde{h}_t \mid (\lambda, A) \overset{\text{i.i.d.}}{\sim} \chi^2(\lambda) \; (t = 1, \ldots, T). \tag{6.26}$$

The $2T$ random variables $\boldsymbol{\eta} = (\eta_1, \ldots, \eta_T)'$ and $\widetilde{\mathbf{h}} = (\widetilde{h}_1, \ldots, \widetilde{h}_T)'$ are mutually independent. A convenient prior distribution for λ, which may easily be generalized using the methods of Section 8.4, is the exponential with mean $\underline{\lambda}$:

$$\lambda \sim \exp(\underline{\lambda}), \tag{6.27}$$

$$p(\lambda) = \underline{\lambda}^{-1}\exp(-\lambda/\underline{\lambda}). \tag{6.28}$$

Smaller values for $\underline{\lambda}$ reflect assumptions that the distribution is more leptokurtic.

From (6.22), (6.24), and (6.25)

$$p(\mathbf{y} \mid \mathbf{X}, \boldsymbol{\beta}, \widetilde{\mathbf{h}}, h, A) \propto h^{T/2}\left(\prod_{t=1}^{T}\widetilde{h}_t^{1/2}\right)\exp\left[-h\sum_{t=1}^{T}\widetilde{h}_t(y_t - \boldsymbol{\beta}'\mathbf{x}_t)^2/2\right], \tag{6.29}$$

and from (6.26), we have

$$p(\widetilde{h}_t \mid \lambda, A) = [2^{\lambda/2}\Gamma(\lambda/2)]^{-1}\lambda^{\lambda/2}\widetilde{h}_t^{(\lambda-2)/2}\exp(-\lambda\widetilde{h}_t/2) \; (t = 1, \ldots, T). \tag{6.30}$$

Completing the model with $\boldsymbol{\beta} \mid A \sim N(\underline{\boldsymbol{\beta}}, \underline{\mathbf{H}}_\beta^{-1})$ and $\underline{s}^2 h \mid A \sim \chi^2(\underline{\nu})$, we obtain

$$p(\boldsymbol{\beta} \mid A) \propto \exp[-(\boldsymbol{\beta} - \underline{\boldsymbol{\beta}})'\underline{\mathbf{H}}_\beta(\boldsymbol{\beta} - \underline{\boldsymbol{\beta}})/2], \tag{6.31}$$

$$p(h \mid A) \propto h^{(\underline{\nu}-2)/2}\exp(-\underline{s}^2 h/2). \tag{6.32}$$

In view of (6.29), it proves convenient to define the "reweighted" observables:

$$\mathbf{y}_t^* = \widetilde{h}_t^{1/2} y_t, \ \mathbf{x}_t^* = \mathbf{x}_t h_t^{1/2} \ (t = 1, \ldots, T)$$
$$\mathbf{y}^* = (y_1^*, \ldots, y_T^*)', \ \mathbf{X}^* = [\mathbf{x}_1^*, \ldots, \mathbf{x}_T^*]'.$$

Examining (6.31) and (6.29), it is apparent that the conditional distribution $p(\boldsymbol{\beta} \mid \widetilde{\mathbf{h}}, h, \mathbf{X}, \mathbf{y}^o, A)$ has exactly the same form as in the normal linear model of Example 2.1.2, but with $\mathbf{X}^*$ and $\mathbf{y}^{*o}$ in place of $\mathbf{X}$ and $\mathbf{y}^o$:

$$\boldsymbol{\beta} \mid (\lambda, \widetilde{\mathbf{h}}, h, \mathbf{y}^o, A) \sim N(\overline{\boldsymbol{\beta}}, \overline{\mathbf{H}}_\beta^{-1}),$$
$$\overline{\mathbf{H}}_\beta = \underline{\mathbf{H}}_\beta + h\mathbf{X}^{*\prime}\mathbf{X}^*, \ \overline{\boldsymbol{\beta}} = \overline{\mathbf{H}}_\beta^{-1}(\underline{\mathbf{H}}_\beta \underline{\boldsymbol{\beta}} + \mathbf{X}^{*\prime}\mathbf{y}^{*o}).$$

Similarly, from (6.32) and (6.29), we obtain

$$\overline{s}^2 h \mid (\lambda, \boldsymbol{\beta}, \widetilde{\mathbf{h}}, \mathbf{y}^o, A) \sim \chi^2(\overline{\nu}),$$
$$\overline{s}^2 = \underline{s}^2 + (\mathbf{y}^{*o} - \mathbf{X}^*\boldsymbol{\beta})'(\mathbf{y}^{*o} - \mathbf{X}^*\boldsymbol{\beta}), \ \overline{\nu} = \underline{\nu} + T.$$

Note that the degrees of freedom parameter λ is not involved in either distribution. This is a consequence of the equivalent interpretation of $\widetilde{\mathbf{h}}$ as a vector of parameters, and (6.26)–(6.27) as a hierarchical prior distribution for $\widetilde{\mathbf{h}}$.

The product of (6.29) and (6.30) provides the kernel of $p(\widetilde{\mathbf{h}} \mid \lambda, \boldsymbol{\beta}, h, \mathbf{X}, \mathbf{y}^o, A)$, in which $\widetilde{h}_1, \ldots, \widetilde{h}_T$ are conditionally independent, specifically

$$p(\widetilde{h}_t \mid \lambda, \boldsymbol{\beta}, h, \mathbf{y}^o, \mathbf{X}, A) \propto \widetilde{h}_t^{(\lambda-1)/2} \exp\{-[\lambda + h(y_t - \boldsymbol{\beta}'\mathbf{x}_t)^2]\widetilde{h}_t/2\},$$

implying

$$[\lambda + h(y_t - \boldsymbol{\beta}'\mathbf{x}_t)^2] \ \widetilde{h}_t \mid (\lambda, \boldsymbol{\beta}, h, \mathbf{y}^o, \mathbf{X}, A) \sim \chi^2(\lambda + 1).$$

The conditional posterior density kernel of λ is proportional to the product of (6.28) and (6.30):

$$p(\lambda \mid \boldsymbol{\beta}, h, \widetilde{\mathbf{h}}, \mathbf{y}^o, A) \propto [2^{\lambda/2}\Gamma(\lambda/2)]^{-T}\lambda^{T\lambda/2}\left(\prod_{t=1}^{T} \widetilde{h}_t^{(\lambda-2)/2}\right) \tag{6.33a}$$

$$\cdot \exp\left[-\left(\underline{\lambda} + \frac{1}{2}\sum_{t=1}^{T}\widetilde{h}_t\right)\lambda\right] = k(\lambda) \tag{6.33b}$$

Clearly the kernel $k(\lambda)$ does not correspond to any common distribution. The second component (6.33b) is the kernel of an exponential distribution that could be used as a candidate distribution in a Metropolis within Gibbs step. However the first component (6.33a) is also relatively quite informative for λ, leading to acceptance probabilities that in general are very low. Instead, take the candidate density $q(\lambda)$ to be that of a univariate normal distribution, with mean at the mode $\widehat{\lambda}$ of $k(\lambda)$ and precision equal to $-d^2 \log k(\lambda)/d\lambda^2 \mid_{\lambda=\widehat{\lambda}}$. This method is used in BACC, for models with Student-t distributions. Typically the degrees of freedom parameter, λ, exhibits more serial correlation in the Gibbs sampling algorithm than do other parameters in the model, but the separated partial means test indicates satisfactory performance in simulations that can be computed in a few minutes; for more detail, see Exercise 6.4.4.

6.4.2 Normal Mixture Linear Models

The normal mixture linear model begins with (6.22) and then introduces the latent state vector $\widetilde{\mathbf{s}} = (\widetilde{s}_1, \ldots, \widetilde{s}_T)'$. Conditional on $\mathbf{X}$, $\widetilde{\mathbf{s}}$ obeys the independent finite state model of Section 6.3 with parameter vector $\boldsymbol{\pi} = (\pi_1, \ldots, \pi_m)'$

$$p(\widetilde{\mathbf{s}} \mid \pi, A) = \prod_{t=1}^{T} \pi_{\widetilde{s}_t} = \prod_{j=1}^{m} \pi_j^{T_j} \tag{6.34}$$

where $T_j = \sum_{t=1}^{T} \delta(\widetilde{s}_t, j)$ is the number of observations t for which $\widetilde{s}_t = j$.

Corresponding to each of the m states j, there is a mean parameter α_j and a positive precision parameter h_j; let $\boldsymbol{\alpha} = (\alpha_1, \ldots, \alpha_m)'$ and $\mathbf{h} = (h_1, \ldots, h_m)'$. Conditional on $\widetilde{s}_t = j$, $\varepsilon_t \sim N[\alpha_j, (h \cdot h_j)^{-1}]$. Thus

$$\begin{aligned} p[y_t \mid \boldsymbol{\beta}, h, \boldsymbol{\pi}, \boldsymbol{\alpha}, \mathbf{h}, \widetilde{s}_t = j, A] &= (2\pi)^{-1/2}(h \cdot h_j)^{1/2} \\ &\cdot \exp[-h \cdot h_j(y_t - \alpha_j - \boldsymbol{\beta}'\mathbf{x}_t)^2/2] \ (t = 1, \ldots, T). \end{aligned} \tag{6.35}$$

The disturbances ε_t are i.i.d. and follow a *discrete normal mixture distribution*:

$$p(\varepsilon_t \mid h, \boldsymbol{\pi}, \boldsymbol{\alpha}, \mathbf{h}, A) = (2\pi)^{-1/2} h^{1/2} \sum_{j=1}^{m} \pi_j h_j^{1/2} \exp[-h \cdot h_j(\varepsilon_t - \alpha_j)^2/2].$$

Clearly h and $\mathbf{h}$ are unidentified, in the sense described in Exercise 4.5.3, as is $\boldsymbol{\alpha}$ if $\mathbf{X}$ contains a column of units, and states are not identified with respect to permutation of the state index. Identification issues will be taken up subsequently in the context of prior distributions.

The mixture of normals distribution is very flexible. Figure 6.1 provides several examples. For the special case in which the means α_j are all the same, the normal mixture distribution is known as the "scale mixture of normals distribution." That distribution is symmetric, is unimodal, and must be leptokurtic; that is, the coefficient of kurtosis $K = E[\varepsilon_t - E(\varepsilon_t)]^4/\text{var}(\varepsilon_t)^2 > 3$, its value if ε_t is normally distributed (see Exercise 6.4.1). Panels (a) and (f) of Figure 6.1 provide examples of scale mixture of normals distributions. If the means α_j are not all the same, then the normal mixture distribution can be skewed, as illustrated in panels (c) and (d). It can also be platykurtic ($K < 3$), as is the case in panels (b) and (e). Of course, these distributions can be multimodal [panel (e)]. With a sufficient number

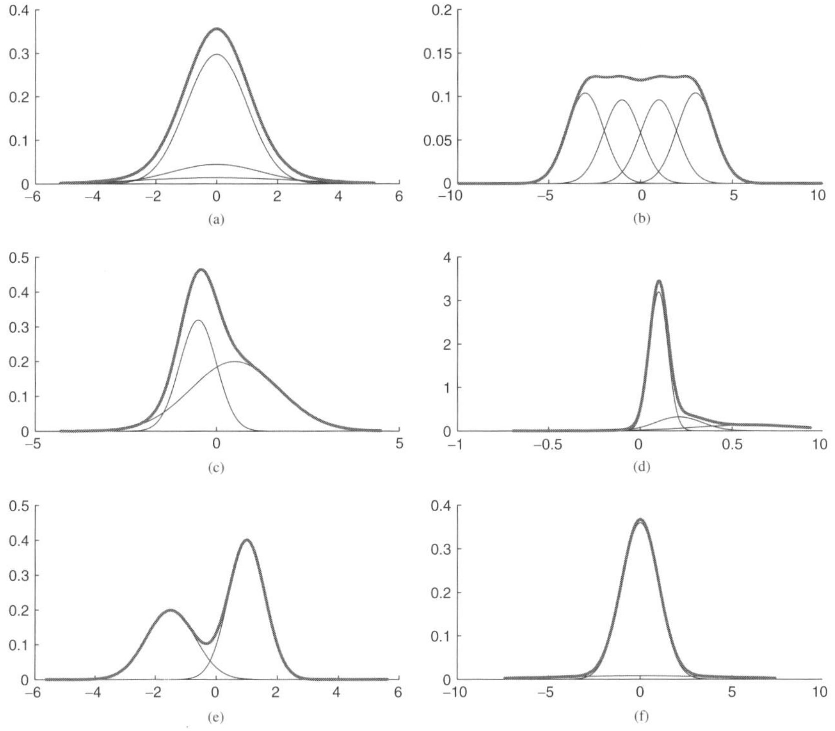

Figure 6.1. Several normal mixture probability density functions: (a) leptokurtic distribution, $t(5)$ moments; (b) platykurtic distribution; (c) mildly right-skewed distribution; (d) severely right-skewed distribution; (e) bimodal distribution; (f) $N(0, 1)$ plus 10% $N(0, 5^2)$ outliers. In each panel the heavy line indicates a normal mixture probability density function. The lighter lines indicate the component normal density functions, each scaled by its probability.

of components, the normal mixture distribution can mimic distributions that are quite different from the normal, like the uniform [panel (b)].

The conditionally conjugate prior densities in the normal mixture linear model are (6.31) for $\boldsymbol{\beta}$ and (6.32) for h. The other parameters in the model all pertain to the normal mixture distribution of ε_t. The choice of the prior distribution of these parameters is driven by three considerations:

1. Priors should be conditionally conjugate and proper. Conditionally conjugate priors simplify simulation from the posterior, as first noted in Section 2.3, and these prior can be revised by the reweighting methods discussed in Section 8.4. As discussed in Section 3.2, improper priors lead to difficulties in model comparison. In the normal mixture model, improper priors can lead to the even more serious complication that the posterior distribution itself is improper; see West and Harrison (1989), Section 12.3.4, Roeder and Wasserman (1997), and Geweke and Keane (2001).

2. Since we have introduced no information to the effect that two states indexed by i and j should not be indexed by j and i instead, the states $\widetilde{s}_t$ are symmetric a priori. Prior distributions must incorporate this symmetry. A consequence is that the posterior distribution will also be symmetric, since interchanging the states does not affect the likelihood function. The posterior distribution will have $m!$ symmetric components. In applications in which the different states have a substantive interpretation, this creates serious conceptual complications (Celeux et al. 2000) and requires some explicit modifications of the posterior simulator like those suggested by Fruhwirth-Schnatter (2001). If, however, the only function of the latent states is to permit flexibility in the functional form of the probability density function, as is the case here, these issues are moot.
3. It is easier to specify a prior distribution with a smaller number of hyperparameters than a larger number. On the other hand, the range of hyperparameters must not be so narrow as to unduly compromise the flexibility of the normal mixture distribution.

These considerations lead to a Dirichlet distribution with parameters $a_1 = \cdots = a_m = r$ for $\boldsymbol{\pi}$

$$p(\boldsymbol{\pi} \mid A) = \Gamma(mr)\Gamma(r)^{-m} \prod_{j=1}^{m} \pi_j^{r-1}, \tag{6.36}$$

independent gamma distributions for the components of $\mathbf{h}$, $\underline{\nu}_{.} h_j \overset{\text{i.i.d.}}{\sim} \chi^2(\underline{\nu}_{.})$ $(j = 1, \ldots, m)$

$$p(\mathbf{h} \mid A) = 2^{-m\underline{\nu}/2}\Gamma(\underline{\nu}_{.}/2)^{-m} \underline{\nu}_{.}^{m\underline{\nu}_{.}/2} \prod_{j=1}^{m} h_j^{(-\underline{\nu}_{.}-2)/2} \exp(-\underline{\nu}_{.} h_j/2), \tag{6.37}$$

and $\boldsymbol{\alpha} \mid (h, A) \sim N[\mathbf{0}, (\underline{h}_{\alpha} \cdot h)^{-1}\mathbf{I}_m]$, so that

$$p(\boldsymbol{\alpha} \mid h, A) = (2\pi)^{-m/2}(\underline{h}_{\alpha} h)^{m/2} \exp(-\underline{h}_{\alpha} h \boldsymbol{\alpha}'\boldsymbol{\alpha}/2). \tag{6.38}$$

The specification $E(\boldsymbol{\alpha}) = \mathbf{0}$ resolves the identification issues with respect to $\boldsymbol{\alpha}$ and $\boldsymbol{\beta}$ given that (as is usually the case) $\mathbf{X}$ has a column of units. The prior variance in $\boldsymbol{\beta}$ conveys uncertainty about the location of the distribution of $\mathbf{y}$ given $\mathbf{X}$. The prior distribution of $\boldsymbol{\alpha}$ is scale dependent on $h^{-1/2}$; that is, it states prior beliefs about the shape of the distribution. Keeping in mind that $E(\mathbf{h}) = \boldsymbol{\iota}_m$, a prior distribution with $\underline{h}_{\alpha}^{-1/2} = 5$ implies a prior probability of multimodality that is near 1, whereas $h_{\alpha}^{-1/2} = \frac{1}{5}$ makes this probability negligibly small. Keeping in mind that $E(\boldsymbol{\alpha}) = \mathbf{0}$, choice of $\underline{\nu}_{.}$ governs the prior probability of tail thickness in the mixture normal density relative to the normal. In the prior distribution the ratio $h_j/h_k \sim F(\underline{\nu}_{.}, \underline{\nu}_{.})$ for all $j \neq k$. If $\underline{\nu}_{.} = 1$, the prior probability of component variance ratios at least as great as those shown in Figure 6.1f is significant, whereas if $\underline{\nu}_{.} = 5$, it is negligible.

Posterior inference in the normal mixture model utilizes five blocks: $\boldsymbol{\gamma}' = (\boldsymbol{\alpha}', \boldsymbol{\beta}')$, h, $\boldsymbol{\pi}$, $\mathbf{h}$, and $\widetilde{\mathbf{s}}$. It is useful to define

$$\underset{T\times m}{\widetilde{\mathbf{Z}}(\widetilde{\mathbf{s}})} = \widetilde{\mathbf{Z}} = [\widetilde{\mathbf{z}}_1, \ldots, \widetilde{\mathbf{z}}_T]' = [\delta(\widetilde{s}_t, j)], \quad \underset{T\times(m+k)}{\widetilde{\mathbf{W}}} = \begin{bmatrix} \widetilde{\mathbf{Z}} & \mathbf{X} \end{bmatrix},$$

$$\underset{(m+k)\times 1}{\underline{\boldsymbol{\gamma}}} = \begin{pmatrix} \mathbf{0} \\ \underline{\boldsymbol{\beta}} \end{pmatrix}, \quad \underset{(m+k)\times(m+k)}{\underline{\mathbf{H}}_{\gamma}(h)} = \underline{\mathbf{H}}_{\gamma} = \begin{bmatrix} \underline{h}_{\alpha} h \mathbf{I}_m & \mathbf{0} \\ \mathbf{0} & \underline{\mathbf{H}}_{\beta} \end{bmatrix},$$

$$\underset{T\times T}{\widetilde{\mathbf{Q}}(\widetilde{\mathbf{s}})} = \widetilde{\mathbf{Q}} = \operatorname{diag}(h_{\widetilde{s}_1}, \ldots, h_{\widetilde{s}_T}).$$

With this notation, (6.35) is equivalent to

$$p(\mathbf{y} \mid \boldsymbol{\gamma}, h, \boldsymbol{\pi}, \mathbf{h}, \widetilde{\mathbf{s}}, \mathbf{X}) = (2\pi)^{-T/2} h^{T/2} \left|\widetilde{\mathbf{Q}}\right|^{1/2} \tag{6.39}$$

$$\cdot \exp[-h(\mathbf{y} - \widetilde{\mathbf{W}}\boldsymbol{\gamma})'\widetilde{\mathbf{Q}}(\mathbf{y} - \widetilde{\mathbf{W}}\boldsymbol{\gamma})/2]. \tag{6.40}$$

The kernel of the conditional posterior density of $\boldsymbol{\gamma}$ is the product of (6.31), (6.38), and (6.40), from which the conditional posterior distribution is

$$\boldsymbol{\gamma} \mid (h, \mathbf{h}, \widetilde{\mathbf{s}}, \mathbf{y}^o, \mathbf{X}, A) \sim N(\overline{\boldsymbol{\gamma}}, \overline{\mathbf{H}}_{\gamma});$$

$$\overline{\mathbf{H}}_{\gamma} = \underline{\mathbf{H}}_{\gamma} + h\widetilde{\mathbf{W}}'\widetilde{\mathbf{Q}}\widetilde{\mathbf{W}}, \ \overline{\boldsymbol{\gamma}} = \overline{\mathbf{H}}_{\gamma}^{-1}[\underline{\mathbf{H}}_{\gamma}\underline{\boldsymbol{\gamma}} + +h\widetilde{\mathbf{W}}'\widetilde{\mathbf{Q}}\mathbf{y}^o]. \tag{6.41}$$

The conditional posterior density of h is the product of (6.32), (6.38), and (6.35). This kernel corresponds to the conditional posterior distribution

$$\left[\underline{s}^2 + \underline{h}_{\alpha}\boldsymbol{\alpha}'\boldsymbol{\alpha} + \sum_{t=1}^{T} h_{\widetilde{s}_t}(y_t - \alpha_{\widetilde{s}_t} - \boldsymbol{\beta}'\mathbf{x}_t)^2\right] h \mid (\boldsymbol{\gamma}, \mathbf{h}, \widetilde{\mathbf{s}}, \mathbf{y}^o, \mathbf{X}, A) \sim \chi^2(\underline{\nu} + m + T). \tag{6.42}$$

The conditional posterior density kernel of $\boldsymbol{\pi}$ is the product of (6.34) and (6.36), $\prod_{j=1}^{m} \pi_j^{r+T_j-1}$, and thus the conditional posterior distribution is Dirichlet with parameters $r + T_j$ $(j = 1, \ldots, m)$.

The conditional posterior density kernel of $\mathbf{h}$ is the product of (6.37) and (6.35), which implies

$$\left[\underline{\nu}_{\cdot}^2 + h\sum_{t=1}^{T} \delta(\widetilde{s}_t, j)(y_t - \alpha_j - \boldsymbol{\beta}'\mathbf{x}_t)^2\right] h_j \mid (\boldsymbol{\gamma}, h, \widetilde{\mathbf{s}}, \mathbf{y}^o, \mathbf{X}, A)$$

$$\sim \chi^2(\underline{\nu}_{\cdot} + T_j) \quad (j = 1, \ldots, m). \tag{6.43}$$

The conditional posterior density kernel for the state assignments $\widetilde{\mathbf{s}}$ is the product of (6.34) and (6.35) taken over $t = 1, \ldots, T$. Thus the states $\widetilde{s}_t$ are conditionally independent, with

$$P(\widetilde{s}_t = j) \mid (\boldsymbol{\gamma}, h, \boldsymbol{\pi}, \mathbf{h}, \mathbf{y}^o, \mathbf{X}, A)$$
$$\propto \pi_j h_j^{1/2} \exp[-h \cdot h_j (y_t - \alpha_j - \boldsymbol{\beta}'\mathbf{x}_t)^2/2] \quad (j = 1, \ldots, m). \tag{6.44}$$

Draws from these independent finite state distributions are straightforward.

BACC incorporates the normal mixture linear model using the conditionally conjugate prior distribution and posterior simulation algorithm described in this section. The output of the posterior simulator will, in general, reflect some switching between labels assigned to states. This poses no complications for problems in which functions of interest depend on the parameters $\boldsymbol{\alpha}$, $\mathbf{h}$, and $\boldsymbol{\pi}$ only through the pdf of $\mathbf{y}_t - \boldsymbol{\beta}'\mathbf{x}_t$, as is always the case when the mixture of normals model is introduced solely to provide a flexible representation of this distribution. That is the case in the following example.

Example 6.4.1 A Normal Mixture Linear Model for Earnings (The online appendix contains data, annotated code, and output for this example.) There is a long and well-established literature that studies the relationship between earnings and the determinants of earnings suggested by lifecycle human capital models. Going back at least to the work of Mincer (1958), the essence of these models is that an individual's productivity, or human capital, is an increasing function of formal education and work experience. By far the most common measure of formal education is years of schooling, and the most common measure of experience is age. The panel study of income dynamics (PSID) is a household-based panel that has collected information on earnings and other aspects of economic activity. This example uses data collected in 1993 on the ages, levels of education, and earnings of 2698 white men between the ages of 25 and 65 who had earnings of at least \$1000. It focuses on the distribution of earnings conditional on age and education.

Economic theory provides little guidance on the functional form of this conditional distribution. Consistent with much of the human capital literature, the expectation of the logarithm of earnings (y_t) is assumed to be a polynomial function of age (a_t) and education (e_t), including all terms up to order 4 in age and 2 in education; thus, $E(y_t \mid a_t, e_t, A) = \sum_{i=0}^{4} \sum_{j=0}^{2} \beta_{ij} a_t^i e_t^j$. If the polynomial terms are organized in a 15×1 vector $\mathbf{x}_t$ and the coefficients are arranged correspondingly in a 15×1 vector $\boldsymbol{\beta}$, then we may write $E(y_t \mid \mathbf{x}_t, \boldsymbol{\beta}, A) = \boldsymbol{\beta}'\mathbf{x}_t$, consistent with the general notation adopted for the linear model in Example 2.1.2. Section 5.4 considers the question of specification of the order of the polynomial in the context of nonlinear regression; see in particular Exercise 5.4.6. Examples 8.3.1 and 8.3.3 return to the issue of the adequacy of this formulation of the regression function. The prior distribution derives from the assumption $\boldsymbol{\beta}'\mathbf{x}_t \mid A \overset{\text{i.i.d.}}{\sim} N(\mu, \tau^2)(t = 1, \ldots, T)$, where $\mu = 10.5$, roughly the sample average of log earnings; $\tau^2 = T\sigma^2$, where T is sample size and $\sigma = 0.7$, roughly the sample standard deviation of log earnings conditional on age and education. The prior distribution of the precision parameter h is $\frac{8}{3}h \mid A \sim \chi^2(10)$. The mode of this prior distribution is $h = 3$, and the standard deviation is ~1.7. This completes the specification of a normal linear model, which is useful as a comparison benchmark with the mixture models.

The mixture models require the specification of the prior hyperparameters m, $\underline{h}_\alpha$, $\underline{\nu}_\cdot$, and r. We consider mixtures of two ($m = 2$) and three ($m = 3$) normal distributions, and in each case take $\underline{h}_\alpha = 0.4$, $\underline{\nu}_\cdot = 3$, and $r = 1$. This allows enough spread in the distributions to ensure a small but nonnegligible probability of multimodality in the density, while placing substantial prior probability on large ratios of component variances h_i/h_j. In this and other complex models, prior distributions can best be understood through their implications for the relevant functions of interest. Example 8.3.1 pursues this method in detail for this application.

The normal model and 2 mixture models have almost identical implications for the posterior regression function $E(y \mid a, e, A)$. Figure 6.2a provides one representation, for the mixture of three normal distributions. At all levels of education, expected log earnings is a concave function of age, with a peak at age 50 for college graduates ($e = 16$) and the early 50s for those who did not graduate from high school ($e < 12$). On the other hand, expected log earnings do not drop as rapidly after their peak for college graduates as they do for high school graduates, but rise more rapidly for young men. Returns to education, measured as the difference between expected log earnings for college and high school graduates, steadily increase with age.

The data strongly favor the mixture models as compared with a normal model. The log marginal likelihood for the latter model is −3056.1, whereas it is −2762.0 for the mixture of two normals and −2757.2 for the mixture of three normals. The remaining panels of Figure 6.2 provide more detail for the conditional distributions and some insight into the marginal likelihood values. Panel (b) provides the posterior expectation of the pdf of the residual term $y - \boldsymbol{\beta}'\mathbf{x}$ in the normal model. The darker line in panel (c) does the same for the mixture of two normals model and in (d), for the mixture of three normals model. The corresponding lighter line in each of these panels provides the posterior expectation of a normal density with the same mean and variance as in the mixture of normals distribution. The striking similarity of the normal densities in panels (b), (c), and (d) can be interpreted as a consequence of Theorem 3.4.2 and Example 3.4.3; the pseudotrue normal model will have the mean and variance of the assumed data generating process D. [For a variant on this approach that uses all three models simultaneously in a single MCMC algorithm, see Richardson and Green (1997).]

The normal mixture distributions in panels (c) and (d) have mean zero. They are strongly negatively skewed, as indicated by the thicker left tail and the positive mode. They are strongly leptokurtic, as indicated by a modal value substantially higher than that of the normal distribution, as well as the thicker tails. The mixture of normals densities in panels (c) and (d) are strikingly similar. The model and data do little to exploit the additional flexibility provided by a third component to modify the two-component density. Nevertheless, the sample of 2698 observations provides decisive evidence in favor of the mixture of three normals model over the mixture of two normals model, with a Bayes factor of 121.5.

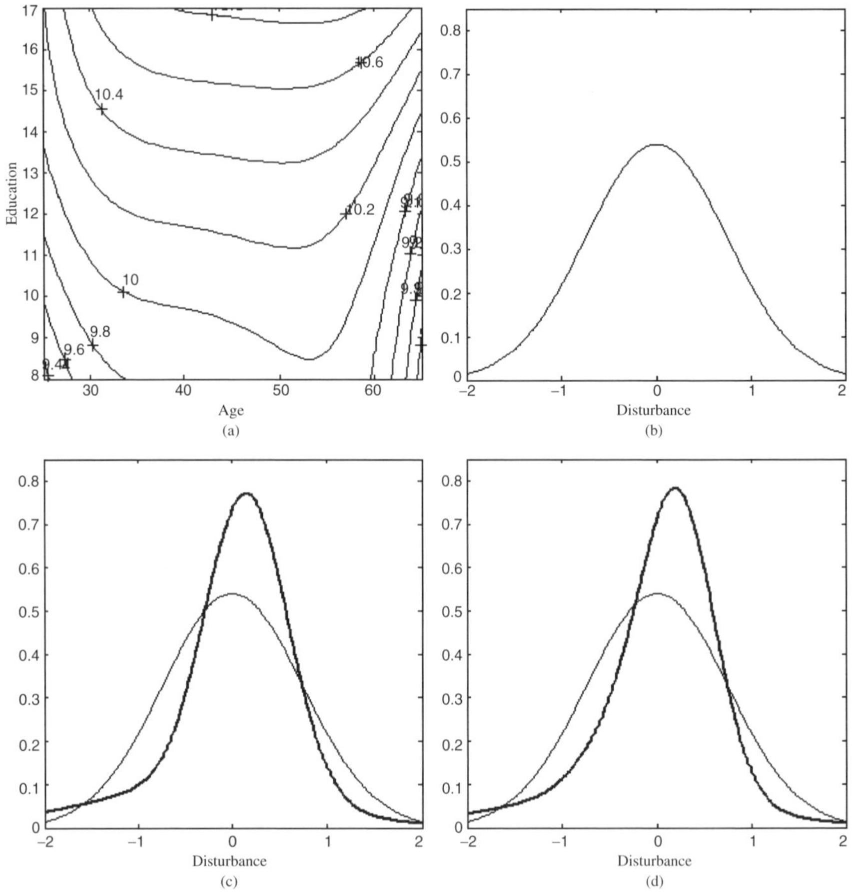

Figure 6.2. Some aspects of the posterior distribution of log earnings conditional on age and education: (a) expected log earnings conditional on age and education; disturbance pdf assuming normal distribution (b), mixture of two normals (c), mixture of three normals (d).

Distributions can be summarized in many ways. The coefficients of skewness and kurtosis are the most common representations of the third and fourth moments. In any given application there may be more substantive summaries, as well. In this example, the conditional distributions indicate the magnitude of inequality in earnings, given age and education. A widely used measure of inequality is the Gini coefficient, which derives from the Lorenz curve. The Lorenz curve $L(p)$, defined on the unit interval, is the fraction of total earnings accruing to individuals in earnings quantile p or lower. If all individuals have the same earnings, then $L(p) = p$ and in general $L(p) \leq p$. The Gini coefficient is $G = 2\int_0^1 [p - L(p)]\, dp$; $G \in [0, 1]$ with $G = 0$ if and only if all individuals have the same earnings and $G = 1$ if and only if all earnings accrue to one individual. We consider two other measures of inequality: P, the fraction of men with earnings less than one-half of median earnings; and R, the fraction of earnings accruing to men in the top decile of the earnings distribution.

The distribution of $y - \boldsymbol{\beta}'\mathbf{x}$ in the mixture of three normals model is more negatively skewed and more leptokurtic than in the mixture of two normals model, but only slightly so. By contrast, the coefficient of skewness is 0 and the coefficient of kurtosis is 3 in a normal model. The measures of inequality are nearly identical in the two mixture models. These results reconfirm the fact that addition of a third component to a mixture of two normals distribution changes essentially nothing in this example. By contrast, all three measures of inequality are substantially higher in the normal model. Panels (c) and (d) of Figure 6.2 indicate the reason why. The assumed normal distribution, in capturing the mean and variance of the mixture of normals distribution, shifts mass away from the mean, rather than toward the mean, except in the extreme tails of the distribution. [See panels (c) and (d) of Figure 6.2.] The former effect dominates the latter in all three measures of inequality, whereas the converse is true in the coefficients of skewness and kurtosis.

Measure	Model	Median	Interquartile Range
Coefficient of skewness	Mixture (2)	−1.063	(−1.527, −0.868)
	Mixture (3)	−1.340	(−1.599, −0.981)
Coefficient of kurtosis	Mixture (2)	6.263	(5.871, 6.722)
	Mixture (3)	6.529	(6.177, 6.896)
Gini coefficient G	Normal	0.398	(0.393, 0.404)
	Mixture (2)	0.345	(0.337, 0.353)
	Mixture (3)	0.345	(0.337, 0.355)
Low earnings P	Normal	0.174	(0.168, 0.179)
	Mixture (2)	0.142	(0.137, 0.148)
	Mixture (3)	0.150	(0.145, 0.156)
High earnings R	Normal	0.294	(0.290, 0.299)
	Mixture (2)	0.264	(0.256, 0.272)
	Mixture (3)	0.266	(0.256, 0.276)

6.4.3 Generalizing the Observable Outcomes

Recall that in the censored linear model (Section 6.1) the continuous outcome variable $\widetilde{y}_t$ is latent (unobservable). The observable outcome is a set-valued function $C_t = c_t(\widetilde{y}_t)$, mapping each possible outcome into exactly one set (6.2) having the property $\widetilde{y}_t \in C_t$. The probit model (Section 6.2), the censored linear model, and more general censoring of outcome variables (see Exercise 6.1.1) are special cases. The fully observed linear model of Example 2.1.2 is the trivial special case $C_t = \widetilde{y}_t$.

This same strategy may be used to extend both the Student-t and normal mixture linear models. The distribution of $\widetilde{y}_t$ conditional on all parameters, latent variables and $\mathbf{X}$, but not $C = \{C_1, \ldots, C_T\}$, is

$$\widetilde{y}_t \mid (\boldsymbol{\beta}, h, \widetilde{\mathbf{h}}, \mathbf{X}, A) \sim N[\boldsymbol{\beta}'\mathbf{x}_t, (h \cdot \widetilde{h}_t)^{-1}] \tag{6.45}$$

in the extension of the Student-t linear model, and

$$\widetilde{y}_t \mid (\boldsymbol{\gamma}, h, \mathbf{h}, \widetilde{\mathbf{s}}, \mathbf{X}, A) \sim N[\boldsymbol{\gamma}'\mathbf{w}_t, (h \cdot h_{\widetilde{s}_t})^{-1}] \tag{6.46}$$

in the extension of the normal mixture linear model. The probability distribution of observables is given by (6.4), repeated here for reference:

$$p(C \mid \widetilde{\mathbf{y}}) = \prod_{t=1}^{T} p(C_t \mid \widetilde{y}_t) = \prod_{t=1}^{T} I_{C_t}(\widetilde{y}_t). \tag{6.47}$$

In posterior simulation the generalization introduced in Section 6.1 requires only one additional step, here, as it did there. In the case of the Student-t model, draw from (6.45) subject to the constraint $\widetilde{y}_t \in C_t^o$, and in the normal mixture model draw from (6.46) subject to the same restriction.

This generalization leads to a wide class of models, many on the frontier of current econometric research; for further discussion, see Section 5 of Geweke and Keane (2001), and for applications see Chib and Greenberg (1995) and Rossi et al. (2001). All of the variants discussed in this section are incorporated in BACC.

Exercise 6.4.1 Properties of the Normal Mixture Linear Model Consider the disturbance term, $\varepsilon_t = y_t - \boldsymbol{\beta}'\mathbf{x}_t$, in this model.

- **(a)** Show that if $\alpha_1 = \cdots = \alpha_m = 0$, then the distribution of the disturbance is symmetric, unimodal, and leptokurtic.
- **(b)** Show that if $h_1 = \cdots = h_m$, then the distribution of the disturbance can be either leptokurtic or platykurtic.
- **(c)** Show that if X is any random variable, there exists a sequence of random variables X_n, each with a normal mixture distribution, such that $X_n \xrightarrow{d} X$.

Exercise 6.4.2 Ergodicity Consider the posterior simulation algorithm for the normal mixture linear model presented in this section.

- **(a)** Show that the Markov chain is ergodic.
- **(b)** Is the Markov chain uniformly ergodic? If so, indicate why. If it is difficult to demonstrate uniform ergodicity, try to develop a modified algorithm that is uniformly ergodic using the methods described in Section 4.6.1. [Diebolt and Robert (1994) investigate ergodicity problems for this and similar algorithms in mixture models.]

Exercise 6.4.3 Interval Data A historian is using Army recruiting records from the American Revolutionary War to learn about the distribution of height (y_{1t}) and weight (y_{2t}) among young men in the late 1700s in colonial America. She has the following information.

(a) For n_1 individuals who were accepted as Army recruits, she knows both height and weight.

(b) For n_2 individuals who were accepted as Army recruits, she knows height but not weight.

(c) For n_3 individuals who were accepted as Army recruits, she has weight but not height.

(d) She knows that there were n_4 individuals accepted as Army recruits, but for these individuals show knows neither height nor weight.

(e) She knows that n_5 individuals were rejected as recruits because they failed to meet height and weight standards.

(f) She knows that the height standard was $c_{11} \leq y_{1t} \leq c_{12}$ and the weight standard was $c_{21} \leq y_{2t} \leq c_{22}$; she knows all four c_{ij} values.

The historian is willing to make the following assumptions:

1. The population from which recruits were either accepted or rejected is a random sample of young men in colonial America.
2. The only reason for rejection was failure to meet height and weight standards.
3. All missing data are missing completely at random (recall Example 2.2.3).
4. For $\mathbf{y}_t = (y_{1t}, y_{2t})'$, $\mathbf{y}_t \overset{\text{i.i.d.}}{\sim} N(\boldsymbol{\mu}, \mathbf{H}^{-1})$.
5. The prior distributions of $\boldsymbol{\mu}$ and $\mathbf{H}$ are independent, the prior for $\boldsymbol{\mu}$ is normal, and the prior for $\mathbf{H}$ is Wishart.

The historian's immediate objective is to construct a posterior simulator for $\boldsymbol{\mu}$ and $\mathbf{H}$. Show how to do this, using a Markov chain Monte Carlo algorithm.

Exercise 6.4.4 Class Size and Test Scores Revisited Example 5.1.1 assumed that the distribution of test scores conditional on covariates is normal. Consider two alternatives: that this distribution is i.i.d. Student-t, and that it is i.i.d. normal mixture with two components.

(a) For each alternative, set up conditionally conjugate prior distributions that are comparable to those in Example 5.1.1. In each case, defend your choices of prior distributions for the additional parameters.

(b) For each alternative distribution, compute the Bayes factor in favor of that alternative, versus the original specification.

(c) Regardless of the Bayes factor in (b), work through the decision problems of Example 5.1.2 under each alternative distribution. Are the results significantly affected? Why or why not?

Exercise 6.4.5 Outliers The problem of "outliers" in regression is a conventional topic in many regression courses. One approach to outliers is to assume that

$$y_t \mid (\boldsymbol{\beta}, h, h^*, A) \sim N(\boldsymbol{\beta}'\mathbf{x}_t, h^{-1}) \tag{6.48}$$

if y_t is not an outlier and

$$y_t \mid (\boldsymbol{\beta}, h, h^*, A) \sim N(\boldsymbol{\beta}'\mathbf{x}_t, h^{*-1}) \tag{6.49}$$

if y_t is an outlier, together with the idea that $h^* \ll h$. Outliers typically constitute only a small fraction of an entire sample.

- **(a)** Suppose that we knew which observations were outliers and which were not. State the conditional posterior distribution for $\boldsymbol{\beta}$. Show that for given h, as $h^* \to 0$, the posterior distribution effectively ignores the outlier observations.
- **(b)** For the rest of this problem, assume that we do not know which observations are outliers and which are not. Carefully state a complete mixture of normals linear model that incorporates all the features of outliers stated at the start of this problem. Be as specific as possible about parameters in the prior distribution that would best incorporate these features.
- **(c)** Briefly describe a Markov chain Monte Carlo algorithm for the posterior distribution in (b). Show how the algorithm yields, as a byproduct, the probability that each observation is an outlier.
- **(d)** Suppose that (6.48)–(6.49) is, indeed, the data-generating process. Suppose without loss of generality that $h = 1$. Sample size T is fixed. Show that as $h^* \to 0$, the algorithm in (c) will correctly classify each observation as an outlier or not.

  [For more on this approach to outliers, see Chaloner and Brant (1988) and Smith and Kohn (1996).]

Exercise 6.4.6 Specification of the Regression Function in Example 6.4.1 Repeat the analysis in Exercise 5.4.6, but assuming a normal mixture linear model.

Exercise 6.4.7 The Earnings Example Extended This exercise extends Example 6.4.1.

- **(a)** The example focused on five properties of the conditional distribution: skewness and kurtosis, and the measures of inequality G, P, and R. The example also showed that earnings vary systematically with age and education. Assuming that the relevant distribution of age and education is given by the sample of 2698 men used in the example, find posterior medians and interquartile ranges for the unconditional distribution of log earnings.
- **(b)** What is the probability that a 40-year-old man with 16 years of education has higher earnings than a 40-year-old man with 12 years of education?

(c) The example reported medians and interquartile ranges for skewness and kurtosis. Try to compute posterior means and variances for skewness and kurtosis, identify the problem that results, and attempt to rectify it. (*Hint*: The prior distribution guarantees the existence of some, but not all, posterior moments.)

CHAPTER 7

Modeling for Time Series

In many decisionmaking problems the vector of interest is future (and therefore unobserved) values of time series. Section 1.1.2 introduced one such problem, assessing value at risk. The common structure of all of these problems is that the distribution of the vector of interest $\boldsymbol{\omega} = (y_{T+1}, \ldots, y_F)'$ is inherent in the model's specification of $p(y_t \mid \mathbf{Y}_{t-1}, \boldsymbol{\theta}_A, A)$ $(t = 1, 2, \ldots)$:

$$p(\boldsymbol{\omega} \mid \mathbf{Y}_T, \boldsymbol{\theta}_A, A) = \prod_{t=T+1}^{F} p(\mathbf{y}_t \mid \mathbf{Y}_{t-1}, \boldsymbol{\theta}_A, A). \tag{7.1}$$

Then, as always

$$p(\boldsymbol{\omega} \mid \mathbf{Y}_T^o, A) = \int_{\Theta_A} p(\boldsymbol{\omega} \mid \mathbf{Y}_T^o, \boldsymbol{\theta}_A, A) p(\boldsymbol{\theta}_A \mid \mathbf{Y}_T^o, A)\, d\nu(\boldsymbol{\theta}_A). \tag{7.2}$$

The central technical problem is construction of the posterior simulator $\boldsymbol{\theta}_A^{(m)} \sim p(\boldsymbol{\theta}_A \mid \mathbf{Y}_T^o, A)$. Given this, draws from $p(\boldsymbol{\omega} \mid \mathbf{Y}_T^o, A)$ require only the forward simulation of the model evident in (7.1). Because of the consistent conditioning in (7.1)–(7.2), uncertainty about parameters or other unobservables $\boldsymbol{\theta}_A$ and uncertainty about the future conditional on $\boldsymbol{\theta}_A$ is integrated in seamless fashion. For further development of this idea and comparison with other methods, see Geweke and Whiteman (in press).

As emphasized in Chapter 1, this conditioning is congruent with the circumstances of the decisionmaker, who must proceed on the basis of information $\mathbf{Y}_T^o$ and A that is available. This fact, combined with the development of posterior simulators that make (7.2) practical, has led to vigorous growth in Bayesian modeling for time series and the application of these models in forecasting, portfolio allocation, and other decisionmaking contexts. Geweke and Whiteman (in press) review this literature. This chapter provides technical detail for three time series models, each illustrating a different significant set of the tools that have proved

Contemporary Bayesian Econometrics and Statistics, by John Geweke

useful in this endeavor. Section 7.1 develops Bayesian methods for autoregressions using the exact likelihood function for the stationary case. Section 7.2 turns to the first-order Markov model, which is the most commonly applied model for discrete time series. Section 7.3 uses this model in combination with latent variables and normal distributions to construct a leading simple yet general model for conditional dependence in time series.

7.1 LINEAR MODELS WITH SERIAL CORRELATION

Suppose that in the linear regression model introduced in Example 2.1.2 and used throughout Chapter 2 the covariates and dependent variable are time series, each measured at a point in time or as averages over successive intervals. If in continuous time these variables move smoothly without jumps, then as the sampling interval becomes shorter and shorter, the assumption that the disturbances $\varepsilon_t = y_t - \boldsymbol{\beta}'\mathbf{x}_t$ are mutually independent becomes untenable.

This section takes up a modification of this model that weakens the assumption of independence, replacing it with the assumption that ε_t obeys a stationary autoregressive process of order p. This modification specifies

$$y_t = \boldsymbol{\beta}'\mathbf{x}_t + \varepsilon_t, \tag{7.3}$$

$$\varepsilon_t = \sum_{s=1}^{p} \phi_s \varepsilon_{t-s} + u_t, \tag{7.4}$$

$$u_t \mid (\varepsilon_{t-1}, \varepsilon_{t-2}, \ldots) \overset{\text{i.i.d.}}{\sim} N(0, h^{-1}), \tag{7.5}$$

for all periods $t = 1, \ldots, T$. Moreover, ε_t is stationary—that is, for any set of $s_1, \ldots, s_m$, the distribution of the vector $(\varepsilon_t, \varepsilon_{t-s_1}, \ldots, \varepsilon_{t-s_m})'$ does not depend on t. The assumption of stationarity is important in addressing the complication introduced by the fact that the observables are $(y_1, \mathbf{x}_1), \ldots, (y_T, \mathbf{x}_T)$, and (7.4) introduces $\varepsilon_{-p+1}, \ldots, \varepsilon_0$ in addition to $\boldsymbol{\beta}$, h, and $\boldsymbol{\phi} = (\phi_1, \ldots, \phi_p)'$. A necessary condition for stationarity is $\boldsymbol{\phi} \in S_p \subseteq \mathbb{R}^p$, where

$$S_p = \left\{ \boldsymbol{\phi} : \left| 1 - \sum_{s=1}^{p} \phi_s z^s \right| \neq 0 \, \forall \, z : |z| \leq 1 \right\}$$

and z is complex.

Motivated by (7.4), define

$$y_t^* = y_t - \sum_{s=1}^{p} \phi_s y_{t-s} \quad \text{and} \quad \mathbf{x}_t^* = \mathbf{x}_t - \sum_{s=1}^{p} \mathbf{x}_{t-s}\phi_s \quad (t = p+1, \ldots, T),$$

and take $\mathbf{y}^* = (y_{p+1}^*, \ldots, y_T^*)'$, $\mathbf{X}^* = [\mathbf{x}_{p+1}^*, \ldots, \mathbf{x}_T^*]'$. Then

$$\mathbf{y}^* \mid (\boldsymbol{\beta}, \boldsymbol{\phi}, h, \mathbf{X}, A) \sim N(\mathbf{X}^*\boldsymbol{\beta}, h^{-1}\mathbf{I}_{T-p}).$$

Let $\mathbf{y}_p$ denote the first p elements of $\mathbf{y}$ and $\mathbf{X}_p$ the first p rows of $\mathbf{X}$. Then

$$\mathbf{y}_p \mid (\boldsymbol{\beta}, \boldsymbol{\phi}, h, \mathbf{X}, A) \sim N[\mathbf{X}_p\boldsymbol{\beta},\ h^{-1}\mathbf{V}_p(\boldsymbol{\phi})]. \tag{7.6}$$

The $p \times p$ matrix $\mathbf{V}_p(\boldsymbol{\phi})$ will be derived shortly. From (7.5), $\mathbf{y}^*$ and $\mathbf{y}_p$ are independent conditional on $(\boldsymbol{\beta}, \boldsymbol{\phi}, h, \mathbf{X})$, and the Jacobian of the one-to-one transformation between $\mathbf{y}'$ and $(\mathbf{y}_p', \mathbf{y}^{*\prime})$ is one. Hence

$$\begin{aligned} p(\mathbf{y} \mid \boldsymbol{\beta}, \boldsymbol{\phi}, h, \mathbf{X}, A) = {} & (2\pi)^{-T/2} h^{T/2} \left|\mathbf{V}_p(\boldsymbol{\phi})\right|^{-1/2} \\ & \cdot \exp\left\{-h\left[(\mathbf{y}^* - \mathbf{X}^*\boldsymbol{\beta})'(\mathbf{y}^* - \mathbf{X}^*\boldsymbol{\beta})\right.\right. \\ & \left.\left. + (\mathbf{y}_p - \mathbf{X}_p\boldsymbol{\beta})'\mathbf{V}_p(\boldsymbol{\phi})^{-1}(\mathbf{y}_p - \mathbf{X}_p\boldsymbol{\beta})\right]/2\right\}. \end{aligned} \tag{7.7}$$

An alternative expression for the observables density emphasizing the role of $\boldsymbol{\phi}$ begins with $\varepsilon_t = y_t - \boldsymbol{\beta}'\mathbf{x}_t$ from (7.3). Define

$$\boldsymbol{\varepsilon}^* = \begin{pmatrix} \varepsilon_{p+1} \\ \varepsilon_{p+2} \\ \vdots \\ \varepsilon_T \end{pmatrix} \quad \text{and} \quad \mathbf{E} = \begin{bmatrix} \varepsilon_p & \cdots & \varepsilon_1 \\ \varepsilon_{p+1} & \cdots & \varepsilon_2 \\ \vdots & & \vdots \\ \varepsilon_{T-1} & \cdots & \varepsilon_{T-p} \end{bmatrix}. \tag{7.8}$$

Then (7.7) becomes

$$p(\mathbf{y} \mid \boldsymbol{\beta}, \boldsymbol{\phi}, h, \mathbf{X}, A) = (2\pi)^{-T/2} h^{T/2} \exp[-h(\boldsymbol{\varepsilon}^* - \mathbf{E}\boldsymbol{\phi})'(\boldsymbol{\varepsilon}^* - \mathbf{E}\boldsymbol{\phi})/2] \tag{7.9a}$$

$$\cdot \left|\mathbf{V}_p(\boldsymbol{\phi})\right|^{-1/2} \exp[-h(\mathbf{y}_p - \mathbf{X}_p\boldsymbol{\beta})'\mathbf{V}_p(\boldsymbol{\phi})^{-1}(\mathbf{y}_p - \mathbf{X}_p\boldsymbol{\beta})/2]. \tag{7.9b}$$

Because ε_t is stationary, $\text{cov}(\varepsilon_i, \varepsilon_j)$ depends only on $|i-j|$, and so the (i, j) entry of $\mathbf{V}_p(\boldsymbol{\phi})$ may be expressed $v_{|i-j|}$. From (7.4)

$$\begin{aligned} h^{-1}v_j &= \text{cov}(\varepsilon_t, \varepsilon_{t-j} \mid \boldsymbol{\phi}, h, A) \\ &= \sum_{s=1}^{p} \phi_s \text{cov}(\varepsilon_{t-s}, \varepsilon_{t-j} \mid \boldsymbol{\phi}, h, A) + \text{cov}(u_t, \varepsilon_{t-j} \mid \boldsymbol{\phi}, h, A) \\ &= h^{-1}\sum_{s=1}^{p} \phi_s v_{j-s} + \delta(0, j)h^{-1} \end{aligned} \tag{7.10}$$

for all $j \geq 0$. Evaluating (7.10) for $j = 1, \ldots, p$ leads to the p *Yule–Walker equations*

$$\begin{bmatrix} v_0 & v_1 & \cdots & v_{p-1} \\ v_1 & v_0 & \cdots & v_{p-2} \\ \cdots & \cdots & \cdots & \cdots \\ v_{p-1} & v_{p-2} & \cdots & v_0 \end{bmatrix} \begin{pmatrix} \phi_1 \\ \phi_2 \\ \vdots \\ \phi_p \end{pmatrix} = \begin{pmatrix} v_1 \\ v_2 \\ \vdots \\ v_p \end{pmatrix} \tag{7.11}$$

If $\boldsymbol{\phi} \in S_p$, then (7.11) determines $v_0, v_1, \ldots, v_{p-1}$ up to a scaling factor, and the $p \times p$ matrix in (7.11) will be positive definite if $v_0 > 0$ and negative definite if $v_0 < 0$. Evaluating (7.10) for $j = 0$ yields

$$v_0 = \sum_{s=1}^{p} \phi_s v_s + 1, \tag{7.12}$$

which determines the scale factor; $\boldsymbol{\phi} \in S_p$ implies $v_0 > 0$.

The posterior simulator in this model is similar to that first proposed by Chib (1993). [For an extension to autoregressive-moving average models, see Chib and Greenberg (1994), and for extensions involving missing data, see Barnett et al. (1996).] The kernel of (7.7) in $\boldsymbol{\beta}$ indicates that the conditionally conjugate prior distribution of $\boldsymbol{\beta}$ is normal, $\boldsymbol{\beta} \sim N(\underline{\boldsymbol{\beta}}, \underline{\mathbf{H}}_{\beta}^{-1})$:

$$p(\boldsymbol{\beta} \mid A) = (2\pi)^{-k/2} \left|\underline{\mathbf{H}}_{\beta}\right|^{1/2} \exp[-(\boldsymbol{\beta} - \underline{\boldsymbol{\beta}})'\underline{\mathbf{H}}_{\beta}(\boldsymbol{\beta} - \underline{\boldsymbol{\beta}})/2]. \tag{7.13}$$

That of h is a gamma distribution, $\underline{s}^2 h \sim \chi^2(\underline{\nu})$:

$$p(h \mid A) = 2^{-\underline{\nu}/2}\Gamma(\underline{\nu}/2)^{-1}(\underline{s}^2)^{\underline{\nu}/2} h^{(\underline{\nu}-2)/2} \exp(-\underline{s}^2 h/2). \tag{7.14}$$

Examining (7.9a)–(7.9b), it is evident from the presence of $\mathbf{V}_p(\boldsymbol{\phi})$ in (7.9b) that a conditionally conjugate prior distribution for $\boldsymbol{\phi}$ would involve the awkward functional forms $\left|\mathbf{V}_p(\boldsymbol{\phi})\right|$ and $\mathbf{V}_p(\boldsymbol{\phi})^{-1}$. On the other hand, the kernel of (7.9a) in $\boldsymbol{\phi}$ is normal. This suggests a prior distribution $\boldsymbol{\phi} \sim N(\underline{\boldsymbol{\phi}}, \underline{\mathbf{H}}_{\phi}^{-1})$ truncated to the set S_p

$$\begin{aligned} p(\boldsymbol{\phi} \mid A) = {} & (2\pi)^{-p/2} D(\underline{\boldsymbol{\phi}}, \underline{\mathbf{H}}_{\phi}) \left|\underline{\mathbf{H}}_{\phi}\right|^{1/2} \\ & \cdot \exp[-(\boldsymbol{\phi} - \underline{\boldsymbol{\phi}})'\underline{\mathbf{H}}_{\phi}(\boldsymbol{\phi} - \underline{\boldsymbol{\phi}})/2] I_{S_p}(\boldsymbol{\phi}), \end{aligned} \tag{7.15}$$

where

$$D(\underline{\boldsymbol{\phi}}, \underline{\mathbf{H}}_{\phi})^{-1} = (2\pi)^{-p/2} \left|\underline{\mathbf{H}}_{\phi}\right|^{1/2} \int_{S_p} \exp[-(\boldsymbol{\phi} - \underline{\boldsymbol{\phi}})'\underline{\mathbf{H}}_{\phi}(\boldsymbol{\phi} - \underline{\boldsymbol{\phi}})/2]\, d\boldsymbol{\phi}.$$

A Gibbs sampling algorithm with a Metropolis step can simulate the unobservables in this complete model. The posterior density kernel is the product of the prior densities (7.13), (7.14), (7.15), and the observables density expressed in either of the forms (7.7) or (7.9a)–(7.9b). The conditional posterior density kernel of $\boldsymbol{\beta}$, from (7.7) and (7.13), is

$$\begin{aligned} p(\boldsymbol{\beta} \mid h, \boldsymbol{\phi}, \mathbf{y}^o, \mathbf{X}, A) \propto \exp\{-[& (\boldsymbol{\beta} - \underline{\boldsymbol{\beta}})'\underline{\mathbf{H}}_{\beta}(\boldsymbol{\beta} - \underline{\boldsymbol{\beta}}) \\ & + h(\mathbf{y}^{*o} - \mathbf{X}^*\boldsymbol{\beta})'(\mathbf{y}^{*o} - \mathbf{X}^*\boldsymbol{\beta}) \\ & + h(\mathbf{y}_p^o - \mathbf{X}_p\boldsymbol{\beta})'\mathbf{V}_p(\boldsymbol{\phi})^{-1}(\mathbf{y}_p^o - \mathbf{X}_p\boldsymbol{\beta})]/2\}. \end{aligned}$$

Hence

$$\boldsymbol{\beta} \mid (h, \boldsymbol{\phi}, \mathbf{y}^o, \mathbf{X}, A) \sim N(\overline{\boldsymbol{\beta}}, \overline{\mathbf{H}}_{\beta}^{-1});$$
$$\overline{\mathbf{H}}_{\beta} = \underline{\mathbf{H}}_{\beta} + h\mathbf{X}^{*\prime}\mathbf{X}^{*} + h\mathbf{X}_p^{\prime}\mathbf{V}_p(\boldsymbol{\phi})^{-1}\mathbf{X}_p,$$
$$\overline{\boldsymbol{\beta}} = \overline{\mathbf{H}}_{\beta}^{-1}[\underline{\mathbf{H}}_{\beta}\underline{\boldsymbol{\beta}} + h\mathbf{X}^{*\prime}\mathbf{y}^{*o} + h\mathbf{X}_p^{\prime}\mathbf{V}_p(\boldsymbol{\phi})^{-1}\mathbf{y}_p^o].$$

Similarly, the posterior density kernel for h, from (7.7) and (7.14), shows

$$\overline{s}^2 h \mid (\boldsymbol{\beta}, \boldsymbol{\phi}, \mathbf{y}^o, \mathbf{X}, A) \sim \chi^2(\overline{\nu}); \quad \overline{\nu} = \underline{\nu} + T,$$
$$\overline{s}^2 = \underline{s}^2 + (\mathbf{y}^{*o} - \mathbf{X}^{*}\boldsymbol{\beta})^{\prime}(\mathbf{y}^{*o} - \mathbf{X}^{*}\boldsymbol{\beta}) + (\mathbf{y}_p^o - \mathbf{X}_p\boldsymbol{\beta})^{\prime}\mathbf{V}_p(\boldsymbol{\phi})^{-1}(\mathbf{y}_p^o - \mathbf{X}_p\boldsymbol{\beta}).$$

From (7.9a)–(7.9b) and (7.15) the conditional posterior density kernel of $\boldsymbol{\phi}$ is

$$p(\boldsymbol{\phi} \mid \boldsymbol{\beta}, h, \mathbf{y}^o, \mathbf{X}, A) \propto \exp\{-[(\boldsymbol{\phi} - \underline{\boldsymbol{\phi}})^{\prime}\underline{\mathbf{H}}_{\phi}(\boldsymbol{\phi} - \underline{\boldsymbol{\phi}})$$
$$+ h(\boldsymbol{\varepsilon}^{*o} - \mathbf{E}^o\boldsymbol{\phi})^{\prime}(\boldsymbol{\varepsilon}^{*o} - \mathbf{E}^o\boldsymbol{\phi})]/2\} \tag{7.16a}$$
$$\cdot\, r(\boldsymbol{\beta}, h, \boldsymbol{\phi}) I_{S_p}(\boldsymbol{\phi}), \tag{7.16b}$$

where $\boldsymbol{\varepsilon}^{*o}$ and $\mathbf{E}^{*o}$ are defined by substituting $\varepsilon_t^o = y_t^o - \boldsymbol{\beta}^{\prime}\mathbf{x}_t$ for ε_t in (7.8) $(t = 1, \ldots, T)$ and $r(\boldsymbol{\beta}, h, \boldsymbol{\phi})$ is expression (7.9b) after substituting $\mathbf{y}^o$ for $\mathbf{y}$. The distribution corresponding to the kernel of (7.16a) in $\boldsymbol{\phi}$ is

$$\boldsymbol{\phi} \mid (\boldsymbol{\beta}, h, \mathbf{y}^o, \mathbf{X}, A) \sim N(\overline{\boldsymbol{\phi}}, \overline{\mathbf{H}}_{\phi}^{-1}), \tag{7.17}$$

where

$$\overline{\mathbf{H}}_{\phi} = \underline{\mathbf{H}}_{\phi} + h\mathbf{E}^{o\prime}\mathbf{E}^{o}, \quad \overline{\boldsymbol{\phi}} = \overline{\mathbf{H}}_{\phi}^{-1}(\underline{\mathbf{H}}_{\phi}\underline{\boldsymbol{\phi}} + h\mathbf{E}^{o\prime}\boldsymbol{\varepsilon}^{*o}).$$

At iteration m a Metropolis within Gibbs step (see Section 4.6.2) for $\boldsymbol{\phi}$ draws a candidate $\boldsymbol{\phi}^{*}$ from the distribution (7.17), using the current values $\boldsymbol{\beta}^{(m)}$ of $\boldsymbol{\beta}$ and $h^{(m)}$ of h. From (7.16b) the acceptance probability for the candidate is

$$\min\left[\frac{r(\boldsymbol{\beta}^{(m)}, h^{(m)}, \boldsymbol{\phi}^{*}) I_{S_p}(\boldsymbol{\phi}^{*})}{r(\boldsymbol{\beta}^{(m)}, h^{(m)}, \boldsymbol{\phi}^{(m-1)})}, \; 1\right].$$

BACC incorporates the linear model with serial correlation using the conjugate prior distribution and posterior simulation algorithm described in this section.

Exercise 7.1.1 A Linear Model with Serial Correlation and Missing Data Suppose $y_t = \boldsymbol{\beta}^{\prime}\mathbf{x}_t + \varepsilon_t$ $(t = 1, \ldots, T)$. The disturbance ε_t is stationary and obeys the first-order autoregression

$$\varepsilon_t = \rho\varepsilon_{t-1} + u_t, \quad u_t \mid (\varepsilon_{t-1}, \varepsilon_{t-2}, \ldots) \overset{\text{i.i.d.}}{\sim} N(0, h^{-1}).$$

Prior beliefs are given by the three independent distributions

$$\boldsymbol{\beta} \sim N(\underline{\boldsymbol{\beta}}, \underline{\mathbf{H}}_{\beta}^{-1}), \quad \underline{s}^2 h \sim \chi^2(\underline{\nu}), \quad \rho \sim \text{uniform}\,(-1, 1).$$

(a) Design a Markov chain Monte Carlo algorithm for Bayesian inference in this model.

(b) Now suppose that some of the observables y_t are missing at random (recall Example 2.2.3). The covariates $\mathbf{x}_t$ are always observed. Modify the Gibbs sampling algorithm designed in part (a) to accommodate this complication.

Exercise 7.1.2 Marginal Likelihood in the Linear Model with Serial Correlation The prior pdf of $\boldsymbol{\phi}$ is (7.15).

(a) Explain how to approximate $D(\underline{\boldsymbol{\phi}}, \underline{\mathbf{H}}_{\phi})$ by means of direct simulation.

(b) Why is the value of $D(\underline{\boldsymbol{\phi}}, \underline{\mathbf{H}}_{\phi})$ important in evaluating the marginal likelihood of this model?

7.2 THE FIRST-ORDER MARKOV FINITE STATE MODEL

Often economic agents or entities can be characterized as moving through states over time, being in exactly one of m states in each time period. For example, an individual might be employed, unemployed, or out of the labor force; or, an individual might be married or not married. If the probability of an entity being in a particular state in a period depends only on the state occupied by that entity in the previous period, the model is a *first-order Markov finite state model*. These models are of interest not only for their own sake but also because they frequently arise as important constituents of more complicated models, for example, the Markov mixture of normals model discussed in Section 7.3.

There are two variants of this model. In the *nonstationary first-order Markov model* the probability distribution of agents or entities over states is different from one period to the next, but, given weak side conditions presented below, converges to a limiting invariant distribution. In the *stationary first-order Markov model* the unconditional probability distributions across states are the same in each period. In both variants of this model individual agents move among states and the dynamics of this movement are nontrivial and usually a focal point of study. The stationary first-order Markov model corresponds more closely to assumptions about behavior in many economic applications, and when the first-order Markov model is used as a constituent of more complicated models, stationarity may be essential.

In either variant, the first-order Markov model may be regarded as a generalization of the independent finite state model. The observables are the same: s_{kt}, the state occupied by entity k at time t, collected in the $n \times T$ matrix $\mathbf{S}$. As in that model, the state transitions between time periods t are mutually independent and identically distributed across agents k.

In both the stationary and nonstationary models, for any agent $k = 1, \ldots, n$, state $j = 1, \ldots, m$ and time period $t = 2, \ldots, T$, we have

$$P[s_{kt} = j \mid s_{k,t-1} = i, s_{ku}(u < t-1), A]$$
$$= P(s_{kt} = j \mid s_{k,t-1} = i, A) = p_{ij}. \tag{7.18}$$

Let

$$\mathbf{p}_j = (p_{j1}, \ldots, p_{jm})' \tag{7.19}$$

and

$$\mathbf{P} = [p_{ij}] = [\mathbf{p}_1, \ldots, \mathbf{p}_m]'. \tag{7.20}$$

In the nonstationary model, the initial period distribution for any agent k ($k = 1, \ldots, n$) is

$$P(s_{k1} = j \mid A) = \pi_{1j}. \tag{7.21}$$

(This section returns to the stationary model in more detail subsequently.)

Expression (7.18) provides the probability distribution across states for an agent, conditional on that agent's history. Not conditioning on this history, denote

$$P(s_{kt} = j \mid A) = \pi_{tj}. \tag{7.22}$$

Then from (7.18), (7.21), and (7.22), $\pi_{tj} = \sum_{i=1}^{m} p_{ij}\pi_{t-1,i}$ ($j = 1, \ldots, m$); equivalently, $\boldsymbol{\pi}'_t = \boldsymbol{\pi}'_{t-1}\mathbf{P}$, where $\boldsymbol{\pi}_t = (\pi_{t1}, \ldots, \pi_{tm})'$. For any $s < t$, $\boldsymbol{\pi}'_t = \boldsymbol{\pi}'_{t-s}\mathbf{P}^s$, and in particular

$$\boldsymbol{\pi}'_t = \boldsymbol{\pi}'_1\mathbf{P}^{t-1}. \tag{7.23}$$

The eigenvalues and eigenvectors of the transition matrix $\mathbf{P}$ are important for the properties of the model. Denote the eigenvalues by $\lambda_1, \ldots, \lambda_m$, ordered so that $|\lambda_1| \geq \cdots \geq |\lambda_m|$. We shall assume that $\mathbf{P}$ is diagonable; that is, it may be represented $\mathbf{P} = \mathbf{C}\boldsymbol{\Lambda}\mathbf{C}^{-1}$, where the columns of $\mathbf{C}$ are right eigenvectors of $\mathbf{P}$ and the rows of $\mathbf{C}^{-1}$ are left eigenvectors of $\mathbf{P}$. [If the prior distribution of $\mathbf{P}$ is absolutely continuous—as is the case for the prior distribution employed subsequently in this section—then $\mathbf{P}$ is diagonable with probability one. Necessary and sufficient conditions for a nonsymmetric matrix to be diagonable can be found in many linear algebra texts, for example, Schott (1997), Section 4.4.] The eigenvalues of $\mathbf{P}$ cannot exceed 1 in modulus, because from (7.23) $0 \leq \text{tr}(\mathbf{P}^j) \leq m$ and $\text{tr}(\mathbf{P}^j) = \sum_{i=1}^{m} \lambda_i^j$. But since $\sum_{j=1}^{m} p_{ij} = 1 \ \forall\, i = 1, \ldots, m$, $\mathbf{P}\boldsymbol{\iota}_m = \boldsymbol{\iota}_m$; $\boldsymbol{\iota}_m = (1, \ldots, 1)'$ is a right eigenvector of $\mathbf{P}$ corresponding to an eigenvalue 1, and it is convenient to take $\lambda_1 = 1$.

A probability distribution over the m states $\boldsymbol{\pi}$ is an *invariant distribution* if $\boldsymbol{\pi}' = \boldsymbol{\pi}'\mathbf{P}$. The vector $\boldsymbol{\pi}$ must be a left eigenvector corresponding to an eigenvalue $\lambda = 1$. If $|\lambda_1| > |\lambda_2|$, then this invariant distribution is unique. Suppose instead that $|\lambda_2| = 1$. If $\lambda_2 = 1$, then the Markov chain is *reducible*, with invariant states depending on the initial distribution. Examples are $\mathbf{P} = \mathbf{I}_2$ and

$$\mathbf{P} = \begin{bmatrix} 1 - p_{12} & p_{12} & 0 \\ p_{21} & 1 - p_{21} & 0 \\ 0 & 0 & 1 \end{bmatrix}.$$

If $\lambda_2 = -1$ or $|\lambda_2| = 1$ and λ_2 is complex, then the chain is *periodic*. Examples include

$$\mathbf{P} = \begin{bmatrix} 0 & 1 \\ 1 & 0 \end{bmatrix} \quad \text{and} \quad \mathbf{P} = \begin{bmatrix} 0 & 1 & 0 \\ 0 & 0 & 1 \\ 1 & 0 & 0 \end{bmatrix}.$$

For the prior distribution employed subsequently in this section $|\lambda_2| < 1$ with probability 1. In this case the Markov chain is irreducible and aperiodic, and hence has a unique invariant distribution. The eigenvalue λ_2 then provides an upper bound on the rate of convergence to the invariant distribution, as indicated in the following result.

Theorem 7.2.1 Convergence in the First-Order Markov Model Suppose that the first-order Markov m-state transition matrix $\mathbf{P}$ is diagonable with eigenvalues λ_j and $|\lambda_1| \geq \cdots \geq |\lambda_m|$. Suppose also that $|\lambda_2| < 1$, and denote the unique invariant distribution by $\boldsymbol{\pi}$. Then for any $r : |\lambda_2| < r < 1$, $\lim_{t\to\infty} r^{-t}(\boldsymbol{\pi}_t - \boldsymbol{\pi}) = \mathbf{0}$.

Proof: Let $\mathbf{P}$ have the diagonalization

$$\mathbf{P} = \mathbf{C}\Lambda\mathbf{C}^{-1}, \quad \Lambda = \text{diag}(\lambda_1, \ldots, \lambda_m).$$

Since $\boldsymbol{\pi}_t' = \boldsymbol{\pi}_{t-1}'\mathbf{P} = \boldsymbol{\pi}_1'\mathbf{P}^{t-1}$ and $\boldsymbol{\pi}' = \boldsymbol{\pi}'\mathbf{P} = \boldsymbol{\pi}'\mathbf{P}^{t-1}$, it follows that

$$\begin{aligned} \boldsymbol{\pi}_t' - \boldsymbol{\pi}' &= (\boldsymbol{\pi}_1 - \boldsymbol{\pi})'\mathbf{P}^{t-1} = (\boldsymbol{\pi}_1 - \boldsymbol{\pi})'\mathbf{C}\Lambda^{t-1}\mathbf{C}^{-1} \\ &= (\boldsymbol{\pi}_1 - \boldsymbol{\pi})'\mathbf{C}\widetilde{\Lambda}^{t-1}\mathbf{C}^{-1} = \boldsymbol{\pi}_1'\mathbf{C}\widetilde{\Lambda}^{t-1}\mathbf{C}^{-1} \end{aligned} \tag{7.24}$$

where $\widetilde{\Lambda} = \text{diag}(0, \lambda_2, \ldots, \lambda_m)$. [The third equality in (7.24) follows because the first column of $\mathbf{C}$ is proportional to $\boldsymbol{\iota}_m$ and the last follows because $\boldsymbol{\pi}'$ is proportional to the first row of $\mathbf{C}^{-1}$.] Then

$$r^{-t}(\boldsymbol{\pi}_t - \boldsymbol{\pi})' = r^{-t}\boldsymbol{\pi}_1'\mathbf{C}\widetilde{\Lambda}^{t-1}\mathbf{C}^{-1} = r^{-1}\boldsymbol{\pi}_1'\mathbf{C}(r^{-1}\widetilde{\Lambda})^{t-1}\mathbf{C}^{-1}.$$

Since $\lim_{t\to\infty}(r^{-1}\widetilde{\Lambda})^t = \mathbf{0}$, $\lim_{t\to\infty} r^{-t}(\boldsymbol{\pi}_t - \boldsymbol{\pi}) = \mathbf{0}$. ■

Note that if all λ_j for which $|\lambda_j| = |\lambda_2|$ are real and positive, then, from (7.24), we obtain

$$\lim_{t\to\infty} |\lambda_2|^{-t} (\boldsymbol{\pi}_t - \boldsymbol{\pi})' = |\lambda_2|^{-1} \boldsymbol{\pi}_1' \mathbf{C}[\lim_{t\to\infty} (|\lambda_2|^{-1} \widetilde{\Lambda})^{t-1}]\mathbf{C}^{-1}$$
$$= |\lambda_2|^{-1} \boldsymbol{\pi}_1' \mathbf{CDC}^{-1} = \mathbf{v}'$$

where $\mathbf{D} = \text{diag}(d_1, \ldots, d_m)$ and $d_j = \delta(|\lambda_j|, |\lambda_2|)$. For any positive integer h

$$\lim_{t\to\infty} |\lambda_2|^{-t} (\boldsymbol{\pi}_{t+h} - \boldsymbol{\pi})' = \lim_{t\to\infty} |\lambda_2|^{h-t} (\boldsymbol{\pi}_t - \boldsymbol{\pi})' = |\lambda_2|^{h} \mathbf{v}'.$$

If $h = -\log 2/\log|\lambda_2|$, then $|\lambda_2|^h = \frac{1}{2}$. This value of h is known as the *half-life* of the first-order Markov model. (The definition still applies for second-largest roots that are negative or complex, but in that case the limit does not exist as it has been taken here, and the result is in terms of amplitudes of oscillations about the invariant distribution.)

While the entries p_{ij} of $\mathbf{P}$ completely characterize the first-order Markov finite state model, they are not as directly related to the implied dynamics as some functions of these parameters. The invariant distribution $\boldsymbol{\pi}$ and the convergence bound $|\lambda_2|$ are examples. There are also many measures of mobility between states, including the expected length of stay in state i, $(1 - p_{ii})^{-1}$, and the overall measure of mobility $[m - \text{tr}(\mathbf{P})]/(m - 1)$. For further discussion and properties of these measures, see Geweke et al. (1986).

7.2.1 Inference in the Nonstationary Model

Recall that the observables are collected in $\mathbf{S} = [s_{kt}]$, where s_{kt} is the state occupied by entity k at time t. From (7.18) and (7.21), we obtain

$$P(\mathbf{S} \mid \boldsymbol{\pi}_1, \mathbf{P}, A) = \prod_{k=1}^{n} \left(\pi_{1,s_{k1}} \prod_{t=2}^{T} p_{s_{k,t-1},s_{kt}} \right). \tag{7.25}$$

Let $n_j = \sum_{k=1}^{n} \delta(s_{k1}, j)$ denote the number of agents in state j at $t = 1$. Let $n_{ij} = \sum_{k=1}^{n} \sum_{t=2}^{T} \delta(s_{k,t-1}, i)\delta(s_{kt}, j)$ denote the number of observable transitions from state i in one period to state j in the next period. Then (7.25) may be expressed

$$p(\mathbf{S} \mid \boldsymbol{\pi}_1, \mathbf{P}, A) = \left(\prod_{j=1}^{m} \pi_{1j}^{n_j} \right) \left(\prod_{i=1}^{m} \prod_{j=1}^{m} p_{ij}^{n_{ij}} \right). \tag{7.26}$$

Observe that (7.26) is the product of $m + 1$ components:

$$p(\mathbf{S} \mid \boldsymbol{\pi}_1, \mathbf{P}, A) = \prod_{j=1}^{m} \pi_{1j}^{n_j} \cdot \prod_{j=1}^{m} p_{1j}^{n_{1j}} \cdot \cdots \cdot \prod_{j=1}^{m} p_{mj}^{n_{mj}}. \tag{7.27}$$

Each of these terms has the same functional form as (6.14). Formally, there are $m + 1$ independent finite state models in (7.25) and (7.27): one for the first period, and one for each of the m states on which the transition probabilities are conditioned. Just as in (6.14), the probabilities are nonnegative and must sum to one in each model.

Since the likelihood function in (7.27) has $m + 1$ factors, the conjugate prior distribution will have $m + 1$ corresponding independent components. Moreover, these conjugate prior distributions will all be Dirichlet, as shown in Section 6.3. Thus the conjugate prior density is

$$p(\boldsymbol{\pi}_1 \mid A) = \left[\Gamma\left(\sum_{j=1}^{m} a_j\right) \Big/ \prod_{j=1}^{m} \Gamma(a_j)\right] \prod_{j=1}^{m} \pi_{1j}^{(a_j-1)}, \tag{7.28}$$

$$p(\mathbf{P} \mid A) = \left[\prod_{i=1}^{m} \Gamma\left(\sum_{j=1}^{m} a_{ij}\right) \Big/ \prod_{i=1}^{m}\prod_{j=1}^{m} \Gamma(a_{ij})\right] \prod_{i=1}^{m}\prod_{j=1}^{m} p_{ij}^{(a_{ij}-1)}, \tag{7.29}$$

where $a_i > 0$ $(i = 1, \ldots, m)$ and $a_{ij} > 0$ $(i, j = 1, \ldots, m)$. The support is the Cartesian product of $m + 1$ $(m - 1)$-dimensional unit simplexes, one each for $\boldsymbol{\pi}_1$ and the rows $\mathbf{p}_1, \ldots, \mathbf{p}_m$ of $\mathbf{P}$ [recall (7.19)–(7.20)].

It follows immediately from (7.28)–(7.29) and (7.27) that in the posterior distribution the $m \times 1$ vectors $\boldsymbol{\pi}_1, \mathbf{p}_1, \ldots$ and $\mathbf{p}_m$ are mutually independent, each with a Dirichlet distribution:

$$p(\boldsymbol{\pi}_1, \mathbf{P} \mid \mathbf{S}, A) \propto \prod_{j=1}^{m} \pi_{1j}^{(a_j+n_j^o-1)} \prod_{i=1}^{m}\prod_{j=1}^{m} p_{ij}^{(a_{ij}+n_{ij}^o-1)}. \tag{7.30}$$

The Dirichlet posterior distribution of $\boldsymbol{\pi}_1$ has parameters $a_j + n_j^o$ $(j = 1, \ldots, m)$ and the posterior distribution of $\mathbf{p}_i$ $(i = 1, \ldots, m)$ has parameters $a_{ij} + n_{ij}^o$ $(j = 1, \ldots, m)$. Derivation of a closed-form expression for the marginal likelihood of the model is straightforward and is left to Exercise 7.2.4.

7.2.2 Inference in the Stationary Model

In the stationary model, $\boldsymbol{\pi}_t = \boldsymbol{\pi}$ for all time periods t, equivalent to the restriction $\boldsymbol{\pi}_1 = \boldsymbol{\pi}$. If the transition matrix $\mathbf{P}$ is irreducible and aperiodic, then $|\lambda_2| < 1$ and there is a $1 \times m$ left eigenvector $\mathbf{c}^1$, unique up to an arbitrary scale factor, with the property $\mathbf{c}^1\mathbf{P} = \mathbf{c}^1$. Computation of $\mathbf{c}^1$ given $\mathbf{P}$ is standard, and $\boldsymbol{\pi}'$ is $\mathbf{c}^1$ normalized so that its elements sum to one. (The elements of $\mathbf{c}^1$ will all be nonnegative; see Exercise 7.2.1. Exercise 7.2.2 provides an alternative method for finding $\boldsymbol{\pi}$ given $\mathbf{P}$.) Denote this mapping $\boldsymbol{\pi}(\mathbf{P}) = [\pi_1(\mathbf{P}), \ldots, \pi_m(\mathbf{P})]'$. As long as the rows of $\mathbf{P}$ have absolutely continuous distributions on the unit simplex—as is the case for Dirichlet distribution—then, with probability one, $|\lambda_2| < 1$ and $\boldsymbol{\pi}$ may be computed in this way.

Given the stationarity restriction, the likelihood function is

$$p(\mathbf{S} \mid \mathbf{P}, A) = \left(\prod_{j=1}^{m} \pi_j(\mathbf{P})^{n_j^o}\right)\left(\prod_{i=1}^{m}\prod_{j=1}^{m} p_{ij}^{n_{ij}^o}\right).$$

Retaining the prior density (7.29) for $\mathbf{P}$, the posterior density kernel is

$$p(\mathbf{P} \mid \mathbf{S}^o, A) \propto \left(\prod_{j=1}^{m} \pi_j(\mathbf{P})^{n_j^o}\right)\left(\prod_{i=1}^{m}\prod_{j=1}^{m} p_{ij}^{a_{ij}+n_{ij}^o-1}\right). \tag{7.31}$$

The kernel (7.31) does not correspond to any standard distribution function, but its second component is the product of Dirichlet probability density functions for $\mathbf{p}_1, \ldots, \mathbf{p}_m$, while its first component is bounded above. This suggests three closely related methods of sampling from the posterior distribution.

1. *Importance Sampling*. Draw $\mathbf{p}_i$ from a Dirichlet distribution with parameters $a_{i1} + n_{i1}^o, \ldots, a_{im} + n_{im}^o$ $(i = 1, \ldots, m)$. The weight associated with the draw is $\prod_{j=1}^{m} \pi_j(\mathbf{P})^{n_j^o}$.
2. *Acceptance Sampling*. The largest possible value of the first component on the right side of (7.31) is $\prod_{j=1}^{m} \widehat{\pi}_{1j}^{n_j}$, where $\widehat{\pi}_{1j} = n_j^o / \sum_{i=1}^{m} n_i^o$. The source density is the same as the importance sampling density, and the acceptance probability is $\prod_{j=1}^{m} (\pi_j(\mathbf{P})/\widehat{\pi}_{1j})^{n_j^o}$.
3. *An Independence Metropolis–Hastings Algorithm*. The probability distribution of the candidate $\mathbf{P}^*$ density is the same set of independent Dirichlet distributions used for the draws in importance and acceptance sampling, and the acceptance probability is

$$\min\left\{\prod_{j=1}^{m} [\pi_j(\mathbf{P}^*)/\pi_j(\mathbf{P})]^{n_j},\ 1\right\},$$

where $\mathbf{P}$ is the value in the previous iteration.

The efficiency of these algorithms will depend on how close the observed distribution of entities across states at $t = 1$ is to the invariant distribution corresponding to transition matrices $\mathbf{P}$ that are probable given the subsequent state-to-state transitions. Loosely speaking, the metric for measuring "close" is the probability ratio of the $t = 1$ outcome under the nonstationary model specification to that under the stationary model specification. For an algorithm that can be more efficient than any of these, see Exercise 7.2.3.

BACC incorporates both the stationary and nonstationary first-order Markov finite state models using the conjugate prior distributions and posterior simulation algorithms described in this section.

Exercise 7.2.1 Properties of the Leading Left Eigenvector of P Suppose that $\mathbf{P} = \mathbf{C}\Lambda\mathbf{C}^{-1}$ is irreducible and aperiodic.

(a) Show that $\lim_{t\to\infty} \mathbf{P}^t = \mathbf{c}_1\mathbf{c}^1$, where $\mathbf{c}_1 \propto \boldsymbol{\iota}_m$ is the first column of $\mathbf{C}$ (first right eigenvector of $\mathbf{P}$) and $\mathbf{c}^1$ is the first row of $\mathbf{C}^{-1}$ (first left eigenvector of $\mathbf{P}$).

(b) Show that the elements of $\mathbf{c}^1$ must be nonnegative.

Exercise 7.2.2 Computation of the Invariant Distribution $\boldsymbol{\pi}$ This exercise develops a method of obtaining $\boldsymbol{\pi}$ from $\mathbf{P}$ that avoids computation of the eigenvectors of $\mathbf{P}$.

(a) Show that when $\mathbf{P}$ is irreducible and aperiodic, $\boldsymbol{\pi}$ is the unique solution of the system of $m+1$ linear equations in m unknowns $\mathbf{Ax} = \mathbf{b}$, where

$$\mathbf{A} = \begin{bmatrix} \mathbf{I}_m - \mathbf{P}' \\ \boldsymbol{\iota}_m' \end{bmatrix}, \quad \mathbf{b} = \begin{pmatrix} \mathbf{0} \\ 1 \end{pmatrix}.$$

(b) From (a), deduce $\boldsymbol{\pi} = (\mathbf{A}'\mathbf{A})^{-1}\mathbf{A}'\mathbf{b}$.

(c) Show that $\boldsymbol{\pi}$ is the sum of the columns of $(\mathbf{A}'\mathbf{A})^{-1}$.

Exercise 7.2.3 An Alternative Posterior Simulator Consider the following MCMC algorithm for the stationary first-order Markov finite state model. At each step s, there are m substeps. Let $\mathbf{p}_j^{(s)}$ denote the jth row of $\mathbf{P}$ at the end of step s. At substep j of step s, define

$$\mathbf{P}^{(s,j)} = [\mathbf{p}_1^{(s)}, \ldots, \mathbf{p}_{j-1}^{(s)}, \mathbf{p}_j^{(s-1)}, \ldots, \mathbf{p}_m^{(s-1)}]'.$$

At substep j of step s, draw a candidate $\mathbf{p}_j^*$ from a Dirichlet distribution with parameters $a_{j1} + n_{j1}^o, \ldots, a_{jm} + n_{jm}^o$, and define

$$\mathbf{P}^{(s,j)*} = [\mathbf{p}_1^{(s)}, \ldots, \mathbf{p}_{j-1}^{(s)}, \mathbf{p}_j^*, \mathbf{p}_{j+1}^{(s-1)}, \ldots, \mathbf{p}_m^{(s-1)}]'.$$

Set $\mathbf{p}_j^{(s)} = \mathbf{p}_j^*$ with probability

$$\min\left\{\prod_{j=1}^{m} [\pi_j(\mathbf{P}^{(s,j)*})/\pi_j(\mathbf{P}^{(s,j)})]^{n_j},\ 1\right\}.$$

and otherwise set $\mathbf{p}_j^{(s)} = \mathbf{p}_j^{(s-1)}$.

(a) Show that the invariant distribution of this algorithm is the posterior distribution in the stationary first-order Markov finite state model.

(b) Indicate why this algorithm might be more efficient than the independence Metropolis–Hastings algorithm described in this section.

Exercise 7.2.4 Provide a closed-form expression for the marginal likelihood in the nonstationary first-order Markov model with prior density (7.28)–(7.29), and likelihood function given by (7.27), with the random n_j and n_{ij} replaced by the corresponding observed n_i^o and n_{ij}^o. [*Hint*: Review the derivation of (6.18) from (6.17).]

7.3 MARKOV NORMAL MIXTURE LINEAR MODEL

Section 6.4.2 introduced normal mixture linear models (6.34)–(6.35) to accommodate a nonnormal disturbance term in the linear model. In that model the latent states $\widetilde{s}_t$ are i.i.d., and conditional on each state the disturbance is normally distributed. That model is attractive because it can approximate i.i.d. disturbances with absolutely continuous distributions very well, by incorporating a sufficiently large number of states, while at the same time the posterior distribution can always be blocked into three components for Gibbs sampling using the elementary normal, gamma, and independent finite state distributions.

Example 6.4.1 applied the normal mixture linear model in a situation in which the assumption of normality was easily overturned. In many time series applications, however, the specification that disturbances are i.i.d. is undesirable. This is particularly so in the case of asset return modeling, introduced in Section 1.1.2. Not only are the sample moments of financial returns strongly inconsistent with a normal sampling distribution; these moments also appear to evolve slowly with time. For example, if y_t is the return on a financial asset then sample correlations of y_t^2 and y_{t-j}^2 are typically positive for small values of j, whereas the corresponding population correlations would be zero if $\{y_t\}$ were i.i.d.

It is straightforward to generalize the normal mixture linear model to permit this behavior, by substituting the stationary first-order Markov finite state model of Section 7.2 with a single cross section ($n = 1$) for the independent finite state model of the latent state vector $\widetilde{\mathbf{s}} = (\widetilde{s}_1, \ldots, \widetilde{s}_T)'$ used in Section 6.4.2. This idea dates at least to Lindgren (1978); early Bayesian treatments include Albert and Chib (1993a), McCulloch and Tsay (1994), and Chib (1996). In lieu of (6.34), from (7.18) we then have

$$P[\widetilde{s}_t = j \mid \widetilde{s}_{t-1} = i, \widetilde{s}_u(u < t-1), A] = P(\widetilde{s}_t = j \mid \widetilde{s}_{t-1} = i, A) = p_{ij}.$$

The type of behavior just described for financial asset returns would be exhibited if, for example, $p_{ii} \gg \sum_{j \neq i} p_{ij}$ for at least some states i, while the precisions $h \cdot h_i$ differ substantially across those same states. From the parameters $[p_{ij}]$ define the Markov transition $\mathbf{P}$ as in (7.20). Conditional on $\widetilde{\mathbf{s}}$, the Markov normal mixture linear model is exactly the same as the normal mixture linear model. In particular, (6.35) provides the conditional pdf of y_t, and the conditionally conjugate prior densities of $\boldsymbol{\beta}$, h, $\mathbf{h}$, and $\boldsymbol{\alpha}$ continue to be (6.31), (6.32), (6.37), and (6.38), respectively. From Section 7.2, the rows of the transition matrix $\mathbf{P}$ have independent Dirichlet distributions in the conditionally conjugate prior:

$$p(p_{i1}, \ldots, p_{ii}, \ldots, p_{im} \mid A) \propto p_{ii}^{r_1} \prod_{j \neq i} p_{ij}^{r_2}. \tag{7.32}$$

The distinction between r_1 and r_2 allows the prior to specify more and less plausible degrees of persistence, while retaining the interchangeability of the m states. These states remain interchangeable in the posterior distribution, as well, and the remarks about this same feature in the normal mixture model in Section 6.4.2 apply here also.

The conditional posterior distributions of $\boldsymbol{\gamma}' = (\boldsymbol{\alpha}', \boldsymbol{\beta}')$, h, and $\mathbf{h}$ continue to be (6.41), (6.42), and (6.43), respectively, exactly as in Section 6.4.2. In particular, because $\widetilde{\mathbf{s}}$ is present in all of these conditional distributions, $\boldsymbol{\pi}$ was absent in Section 6.4.2 and $\mathbf{P}$ is absent here. In Section 6.4.2 the conditional posterior distribution of $\mathbf{p}$ was Dirichlet. Here, the conditional posterior distribution of each row of $\mathbf{P}$ is Dirichlet, (7.31), but with the terms n_j^o and n_{ij}^o referring to the latent states on which the distribution is conditioned. Since there is a single time series, $n_j^o = \delta(\widetilde{s}_1, j)$, and $\sum_{i=1}^m \sum_{j=1}^m n_{ij}^o = T - 1$. The principal new complication for posterior simulation introduced is the conditional posterior distribution of the latent states $\widetilde{\mathbf{s}}$. In the normal mixture linear model the states were conditionally independent, leading to the sequence of T independent finite state distributions with probabilities (6.44). In the Markov normal mixture linear model the conditional posterior kernel is

$$p(\widetilde{\mathbf{s}} \mid \boldsymbol{\gamma}, h, \mathbf{P}, \mathbf{h}, \mathbf{y}^o, \mathbf{X}, A) \propto \pi_{\widetilde{s}_1} h_{\widetilde{s}_1}^{1/2} \exp[-h \cdot h_{\widetilde{s}_1} (y_1 - \alpha_{\widetilde{s}_1} - \boldsymbol{\beta}'\mathbf{x}_1)^2/2]$$
$$\cdot \prod_{t=2}^{T} p_{\widetilde{s}_{t-1}\widetilde{s}_t} \cdot h_{\widetilde{s}_t}^{1/2} \exp[-h \cdot h_{\widetilde{s}_t} (y_t - \alpha_{\widetilde{s}_t} - \boldsymbol{\beta}'\mathbf{x}_t)^2/2] \tag{7.33}$$

and thus

$$p(\widetilde{s}_t \mid \widetilde{s}_j (j \neq t), \boldsymbol{\gamma}, h, \mathbf{P}, \mathbf{h}, \mathbf{y}^o, \mathbf{X}, A)$$
$$\propto p_{\widetilde{s}_{t-1}\widetilde{s}_t} p_{\widetilde{s}_t \widetilde{s}_{t+1}} \cdot h_{\widetilde{s}_t}^{1/2} \exp[-h \cdot h_{\widetilde{s}_t} (y_t - \alpha_{\widetilde{s}_t} - \boldsymbol{\beta}'\mathbf{x}_t)^2/2] \tag{7.34}$$

for $t = 2, \ldots, T - 1$. (Expressions for $t = 1$ and $t = T$ are slightly modified.) Draws for $\widetilde{s}_t$ could be made successively from (7.34), but that algorithm induces substantial serial correlation if, as is typically the case, $p_{ii} \gg \sum_{j \neq i} p_{ij}$ for at least some states i.

A more efficient algorithm due to Chib (1996) draws $\widetilde{\mathbf{s}}$ directly from (7.34), and yields several important functions of interest as byproducts. Consistent with our definition of $\mathbf{Y}_T$ in Section 2.1, let $\widetilde{\mathbf{S}}_t = (\widetilde{s}_1, \ldots, \widetilde{s}_t)'$, further define $\mathbf{Y}^t = (\mathbf{y}_t, \ldots, \mathbf{y}_T)'$ and $\widetilde{\mathbf{S}}^t = (\widetilde{s}_t, \ldots, \widetilde{s}_T)'$, and extend the convention $\mathbf{Y}_0 = \{\varnothing\}$ to $\widetilde{\mathbf{S}}_0 = \mathbf{Y}^{T+1} = \widetilde{\mathbf{S}}^{T+1} = \{\varnothing\}$. To render the notation more compact as well as to emphasize the fact that this algorithm applies to first-order Markov mixtures of distributions generally, let $\boldsymbol{\theta}'_{A1} = (\boldsymbol{\gamma}', h, \mathbf{h}')$ and denote

$$p(y_t \mid \mathbf{Y}_{t-1}, \widetilde{s}_t = j, \boldsymbol{\theta}_{A1}, A) \propto h_j^{-1/2} \exp[-h \cdot h_j (y_t - \alpha_j - \boldsymbol{\beta}'\mathbf{x}_t)^2/2]. \quad (7.35)$$

We may write and decompose (7.34) in the form

$$p(\widetilde{\mathbf{S}}_T \mid \mathbf{Y}_T^o, \boldsymbol{\theta}_{A1}, \mathbf{P}, A) = \prod_{t=1}^{T} p(\widetilde{s}_t \mid \mathbf{Y}_T^o, \widetilde{\mathbf{S}}^{t+1}, \boldsymbol{\theta}_{A1}, \mathbf{P}, A). \quad (7.36)$$

For each of the T terms on the right side of (7.36), we obtain

$$\begin{aligned} p(\widetilde{s}_t \mid \mathbf{Y}_T^o, \widetilde{\mathbf{S}}^{t+1}, \boldsymbol{\theta}_{A1}, \mathbf{P}, A) &\propto p(\widetilde{s}_t, \mathbf{Y}_T^o, \widetilde{\mathbf{S}}^{t+1} \mid \boldsymbol{\theta}_{A1}, \mathbf{P}, A) \\ &= p(\widetilde{s}_t, \mathbf{Y}_t^o, \mathbf{Y}^{o,t+1}, \widetilde{\mathbf{S}}^{t+1} \mid \boldsymbol{\theta}_{A1}, \mathbf{P}, A) \\ &= p(\mathbf{Y}^{o,t+1}, \widetilde{\mathbf{S}}^{t+1} \mid \widetilde{s}_t, \mathbf{Y}_t^o, \boldsymbol{\theta}_{A1}, \mathbf{P}, A) p(\widetilde{s}_t, \mathbf{Y}_t^o \mid \boldsymbol{\theta}_{A1}, \mathbf{P}, A) \\ &\propto p(\mathbf{Y}^{o,t+1}, \widetilde{\mathbf{S}}^{t+1} \mid \widetilde{s}_t, \mathbf{Y}_t^o, \boldsymbol{\theta}_{A1}, \mathbf{P}, A) p(\widetilde{s}_t \mid \mathbf{Y}_t^o, \boldsymbol{\theta}_{A1}, \mathbf{P}, A) \\ &= p(\mathbf{Y}^{o,t+1}, \widetilde{\mathbf{S}}^{t+2} \mid \widetilde{s}_t, \widetilde{s}_{t+1}, \mathbf{Y}_t^o, \boldsymbol{\theta}_{A1}, \mathbf{P}, A) \\ &\quad \cdot p(\widetilde{s}_{t+1} \mid \widetilde{s}_t, \mathbf{Y}_t^o, \boldsymbol{\theta}_{A1}, \mathbf{P}, A) p(\widetilde{s}_t \mid \mathbf{Y}_t^o, \boldsymbol{\theta}_{A1}, \mathbf{P}, A) \\ &= p(\mathbf{Y}^{o,t+1}, \widetilde{\mathbf{S}}^{t+2} \mid \widetilde{s}_{t+1}, \mathbf{Y}_t^o, \boldsymbol{\theta}_{A1}, \mathbf{P}, A) \\ &\quad \cdot p(\widetilde{s}_{t+1} \mid \widetilde{s}_t, \mathbf{P}, A) p(\widetilde{s}_t \mid \mathbf{Y}_t^o, \boldsymbol{\theta}_{A1}, \mathbf{P}, A) \\ &\propto p(\widetilde{s}_{t+1} \mid \widetilde{s}_t, \mathbf{P}, A) p(\widetilde{s}_t \mid \mathbf{Y}_t^o, \boldsymbol{\theta}_{A1}, \mathbf{P}, A). \end{aligned} \quad (7.37)$$

We exploit this decomposition to simulate $\widetilde{\mathbf{s}}$. The first of the two terms in (7.37) is simply $p(\widetilde{s}_{t+1} = j \mid \widetilde{s}_t = i, \mathbf{P}, A) = p_{ij}$. The second term may be evaluated in a forward recursion beginning with

$$p(\widetilde{s}_1 = j \mid \mathbf{Y}_1^o, \boldsymbol{\theta}_{A1}, \mathbf{P}, A) \propto p(\widetilde{s}_1 = j \mid \mathbf{P}, A) p(y_1^o \mid \widetilde{s}_1 = j, \boldsymbol{\theta}_{A1}, A).$$

Recall that $p(\widetilde{s}_1 = j \mid \mathbf{P}, A)$ is the unconditional state j probability $\pi_j(\mathbf{P})$, defined in Section 7.2, and (7.35) provides $p(\mathbf{y}_1^o \mid \widetilde{s}_1 = j, \boldsymbol{\theta}_{A1}, \mathbf{P}, A)$. Note also that

$$p(y_1^o \mid \boldsymbol{\theta}_{A1}, \mathbf{P}, A) = \sum_{j=1}^{m} \pi_j(\mathbf{P}) p(y_1^o \mid \widetilde{s}_1 = j). \quad (7.38)$$

Step t of the recursion has two substeps. In the *prediction step*

$$p(\widetilde{s}_t = j \mid \mathbf{Y}_{t-1}^o, \boldsymbol{\theta}_{A1}, \mathbf{P}, A) = \sum_{i=1}^{m} p_{ij} \cdot p(\widetilde{s}_{t-1} = i \mid \mathbf{Y}_{t-1}^o, \boldsymbol{\theta}_A, A). \quad (7.39)$$

The name derives from the fact that, as a byproduct, we can produce the one-step-ahead predictive conditional density

$$p(y_t \mid \mathbf{Y}^o_{t-1}, \boldsymbol{\theta}_{A1}, \mathbf{P}, A) = \sum_{j=1}^{m} p(\widetilde{s}_t = j \mid \mathbf{Y}^o_{t-1}, \boldsymbol{\theta}_{A1}, \mathbf{P}, A) \cdot p(y_t \mid \mathbf{Y}^o_{t-1}, \widetilde{s}_t = j, \boldsymbol{\theta}_{A1}, A). \tag{7.40}$$

In the *update step*

$$p(\widetilde{s}_t = j \mid \mathbf{Y}^o_t, \boldsymbol{\theta}_{A1}, \mathbf{P}, A) \propto p(\widetilde{s}_t = j \mid \mathbf{Y}^o_{t-1}, \boldsymbol{\theta}_{A1}, \mathbf{P}, A) \cdot p(y^o_t \mid \mathbf{Y}^o_{t-1}, \widetilde{s}_t = j, \boldsymbol{\theta}_{A1}, A). \tag{7.41}$$

The name derives from the fact that this step updates the conditional time t state probabilities produced at time $t-1$ in (7.39), producing the *filtered probabilities* in (7.41), so called because they are a function of past values of the observables. Substituting the observed y^o_t for y_t in (7.40) provides the tth component of the likelihood function

$$p(y^o_t \mid \mathbf{Y}^o_{t-1}, \boldsymbol{\theta}_{A1}, \mathbf{P}, A) = \sum_{j=1}^{m} p(\widetilde{s}_t = j \mid \mathbf{Y}^o_{t-1}, \boldsymbol{\theta}_{A1}, \mathbf{P}, A) \cdot p(y^o_t \mid \mathbf{Y}^o_{t-1}, \widetilde{s}_t = j, \boldsymbol{\theta}_{A1}, A), \tag{7.42}$$

and at the end of the recursion $p(\mathbf{y}^o \mid \boldsymbol{\theta}_{A1}, \mathbf{P}, A)$ is provided by the product of (7.38) and (7.42) evaluated for $t = 2, \ldots, T$.

Drawing from $p(\widetilde{\mathbf{S}}_T \mid \mathbf{Y}^o_T, \boldsymbol{\theta}_{A1}, \mathbf{P}, A)$ is now straightforward. The last update step (7.41) provides $p(\widetilde{s}_T = j \mid \mathbf{Y}^o_T, \boldsymbol{\theta}_{A1}, \mathbf{P}, A)$, an m-state distribution. Then successive evaluation of (7.37) for $t = T-1, \ldots, 1$ provides the finite state distributions for the other time periods. These distributions provide the *smoothed probabilities* for the states, so-called because they take into account observations made after the occurrence of each latent state as well as before.

Example 7.3.1 Filtering and Smoothing in the Markov Normal Mixture Linear Model (The online appendix contains annotated code and output for this example.) To appreciate some of the properties of this model, consider a hypothetical simple case in which there are $m = 3$ components, a single covariate $x_t = 1$, and known parameter values $\beta = 0$, $h = 1$:

$$\mathbf{P} = \begin{bmatrix} 0.95 & 0.03 & 0.02 \\ 0.10 & 0.54 & 0.36 \\ 0.05 & 0.57 & 0.38 \end{bmatrix}, \quad \boldsymbol{\alpha} = \begin{pmatrix} 0 \\ 2 \\ -3 \end{pmatrix}, \quad \mathbf{h} = \begin{pmatrix} 1 \\ 0.25 \\ 0.111 \end{pmatrix}. \tag{7.43}$$

The invariant distribution corresponding to $\mathbf{P}$ is $\pi = (0.6154, 0.2308, 0.1538)'$. For these parameter values $E(y_t \mid \widetilde{s}_t = j, A) = 0$ $(j = 1, 2, 3)$, and consequently

$E(y_t \mid A) = 0$ and $E(y_t \mid \mathbf{Y}_{t-1}, A) = 0$. For all i, $p_{i2}/p_{i3} = 1.5$, and consequently $p(y_t \mid \widetilde{s}_t = 2, A) = p(y_t \mid \widetilde{s}_t = 3, A)$. State 1 characterizes periods of low volatility; states 2 and 3, periods of high volatility. In a period of high volatility, a return to a period of low volatility is twice as likely in state 2 (when y_t is usually positive) as it is in state 3 (when y_t is usually negative).

Although parameter values are known, the states are unobserved. Hence at any time T there is uncertainty about $\widetilde{s}_t$ $(t \leq T)$. Consider the situation portrayed in Figure 7.1, in which $T = 200$. Panel (a) shows y_t^o. The squares of these values, in panel (b), indicate clearly that there are alternating periods of low and high volatility. At time t, the r-step-ahead predictive density is

$$p(y_{t+j} \mid \mathbf{Y}_t^o, A) = \sum_{i=1}^{m} p(\widetilde{s}_t = i \mid \mathbf{Y}_t^o, A) \tag{7.44}$$

$$\cdot p(\widetilde{s}_{t+r} = j \mid \widetilde{s}_t = i, A) p(y_{t+r} \mid \widetilde{s}_{t+r} = j, A), \tag{7.45}$$

where $p(y_{t+r} \mid \widetilde{s}_{t+r} = j, A)$ is the normal density with mean α_j and precision $h \cdot h_j$, $p(\widetilde{s}_{t+r} = j \mid \widetilde{s}_t = i, A)$ is the element in row i and column j of P^r, and $p(\widetilde{s}_t = i \mid \mathbf{Y}_t^o, A)$ is given by (7.41). The latter filtered probabilities are shown in panels (c), (e), and (g) of Figure 7.1. In some periods t the value of $\widetilde{s}_t$ is nearly certain. This is especially so when $\left|y_t^o\right|$ is large, exceeding about 2. In several periods, however, there is substantial uncertainty. This uncertainty is reflected in predictive densities (7.45), especially for $r = 1$. [As $r \to \infty$, $p(\widetilde{s}_{t+r} = j \mid \widetilde{s}_t = i) \to \pi_j$ for all i.]

As time passes, much of the uncertainty about $\widetilde{s}_t$ is resolved by means of conditioning on future y_{t+j}^o as well as past y_{t-j}^o $(j > 0)$. The smoothing filter (7.37) provides the conditional probabilities, displayed in panels (d), (f), and (h) of Figure 7.1. Whereas there are 118, out of 600, filtered probabilities between 0.10 and 0.90 in Figure 7.1, there are only 72 smoothed probabilities in this range. The smoothed probabilities may matter for inference about the past in some applications, but they are irrelevant for prediction.

Panel (a) of Figure 7.2 shows the unconditional density

$$p(y_t \mid A) = \sum_{j=1}^{m} \pi_j p(y_t \mid \widetilde{s}_t = j, A),$$

which clearly reflects the negative skewness coefficient (−0.7166) and the excess kurtosis (1.4533) of the unconditional distribution. The predictive distribution $p(y_{t+1} \mid \mathbf{Y}_t^o, A)$ varies considerably, depending on the filtered state probabilities $p(\widetilde{s}_t = i \mid \mathbf{Y}_t^o, A)$, by means of (7.45). The remaining panels of Figure 7.2 provide some examples. In period $t = 15$, the filtered probability of state 1 is nearly 1; $y_{14}^o = -0.5056$ and $y_{15}^o = 0.0302$. The predictive density is nearly identical with the first normal distribution in (7.43). Because $y_{49}^o = -6.2801, P(\widetilde{s}_{49} = 3 \mid \mathbf{Y}_{49}^o, A) \approx 1$, and because $y_{50}^o = 3.1759$, $P(\widetilde{s}_{50} = 2 \mid \mathbf{Y}_{50}^o, A) \approx 1$. Since $p_{i2}/p_{i3} = 1.5$ for all i, the one-step-ahead predictive densities in panels (c) and (d) of Figure 7.2 are nearly

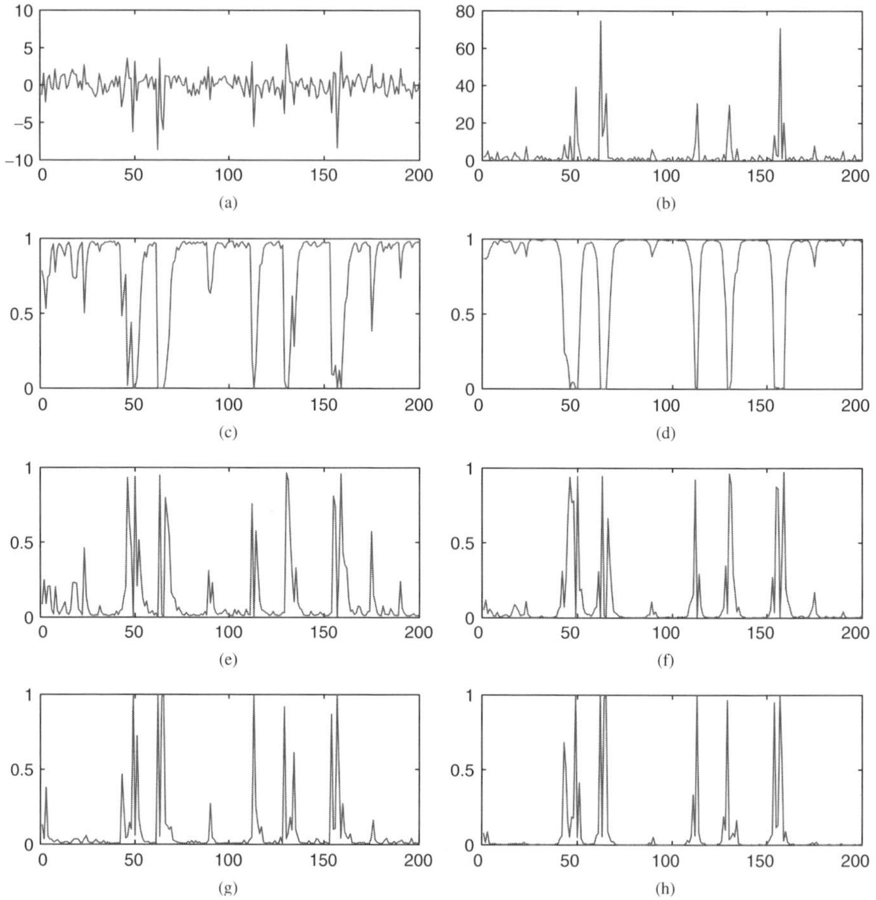

Figure 7.1. A hypothetical time series obeying a known Markov mixture of normals model: (a) observed values; (b) observed values squared; (c) state 1 filtered probabilities; (d) state 1 smoothed probabilities; (e) state 2 filtered probabilities; (f) state 2 smoothed probabilities; (g) state 3 filtered probabilities; (h) state 3 smoothed probabilities.

identical. In periods $t = 68$ and $t = 69$ the filtered probability of state 1 is not close to either zero or one. The consequence is that $p(y_{t+1} \mid \mathbf{Y}_t^o, A)$ is a weighted average of $p(y_{t+1} \mid \widetilde{s}_t = 1, A)$ and $p(y_{t+1} \mid \widetilde{s}_t = 2 \text{ or } 3, A)$.

The conditional mean of y_t is always zero, and the conditional skewness is always negative. Conditional excess kurtosis can be positive (when the filtered probability of state 1 is large) or negative (when it is small); unconditionally, it is positive. These features are consequences of the specific parameter values in (7.43).

Example 7.3.2 The Markov Mixture Model and Value at Risk (The online appendix contains data, annotated code, and output for this example.) Recall the value at risk decision problem introduced in Section 1.1.2. The price of an asset or

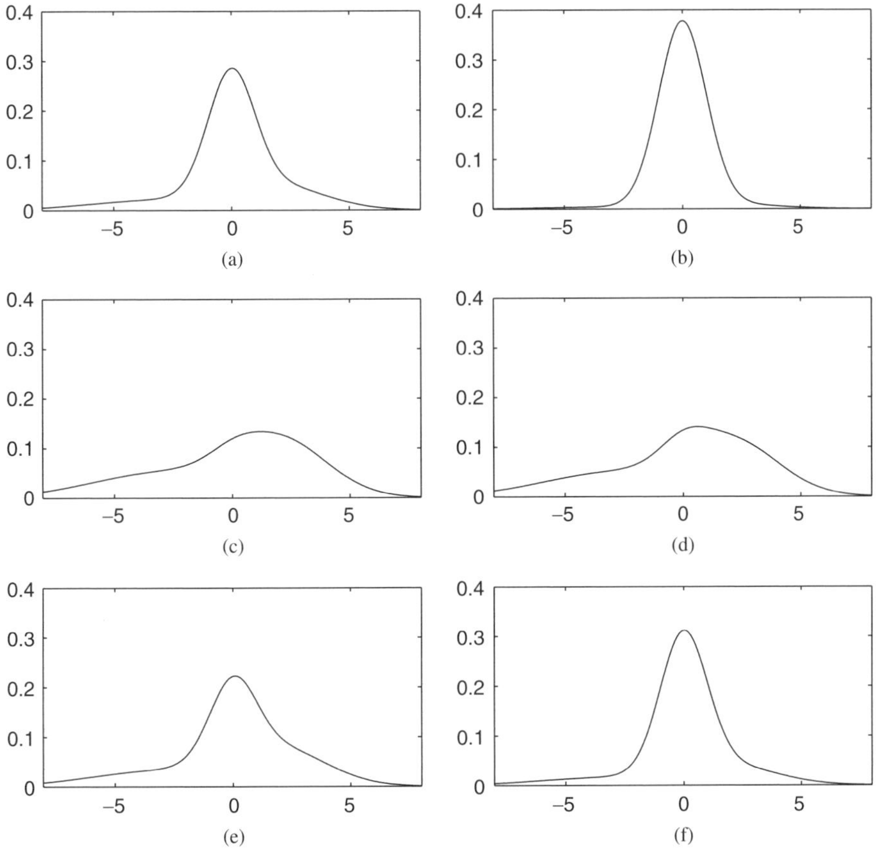

Figure 7.2. Some conditional densities for the hypothetical Markov mixture of normals time series: (a) unconditional predictive density; (b) $t = 15$, $p = (0.975, 0.018, 0.008)$; (c) $t = 49$, $p = (0.000, 0.001, 0.999)$; (d) $t = 50$, $p = (0.001, 0.939, 0.060)$; (e) $t = 68$, $p = (0.357, 0.544, 0.099)$; (f) $t = 69$, $p = (0.716, 0.168, 0.117)$.

portfolio on day t is p_t. For a date or dates $t^* > t$, the decisionmaker must state a value at risk v_{t,t^*} such that $P(p_t - p_{t^*} \geq v_{t,t^*}) = .05$. The probability is, of course, conditional on the information and data available. Letting $r_{t,t^*} = \log(p_{t^*}/p_t)$ denote the return between day t and day t^*, an equivalent problem is to find a return at risk w_{t,t^*} such that $P(r_{t,t^*} \leq -w_{t,t^*}) = .05$.

This example illustrates the process of finding return at risk, conditional on a single series of returns and a Markov normal mixture linear model. The asset is the *Standard and Poors* (S&P) *500* stock price index, for the period March 23, 1978 through December 7, 1984, sample size $T = 1700$. This is the 9th of 10 subsamples of a longer series of the S&P 500 index used by Ryden et al. (1998) in an investigation of the ability of Markov mixture models to account for several features of these data. The Markov mixture model has three states, a constant term as its only covariate, and employs the prior distributions (6.31), (6.32), (6.37), (6.38), and (7.32) with

$$\underline{\beta} = 0,\ \ \underline{h} = 10^6,\ \ \underline{\nu} = 5,\ \underline{s}^2 = 10^3, \tag{7.46}$$

$$\underline{h}_{\alpha} = 1,\ \underline{\nu}_{\cdot} = 3,\ r_1 = 10, \text{ and } r_2 = 1. \tag{7.47}$$

Example 8.3.2 interprets this prior distribution. For comparison purposes only, we also consider a model in which returns are i.i.d. normal, utilizing the prior distribution introduced in Example 2.1.2 with the settings given in (7.46).

The posterior simulator for the Markov mixture model ran 22,000 iterations, the first 2000 of which were discarded. The analysis that follows uses every 20th iteration, for a total of 1000 simulated values from the posterior distribution. The alternative normal model posterior simulator ran 1100 iterations, with analysis based on the last 1000. Panel (a) of Figure 7.3 shows the logarithm of the posterior

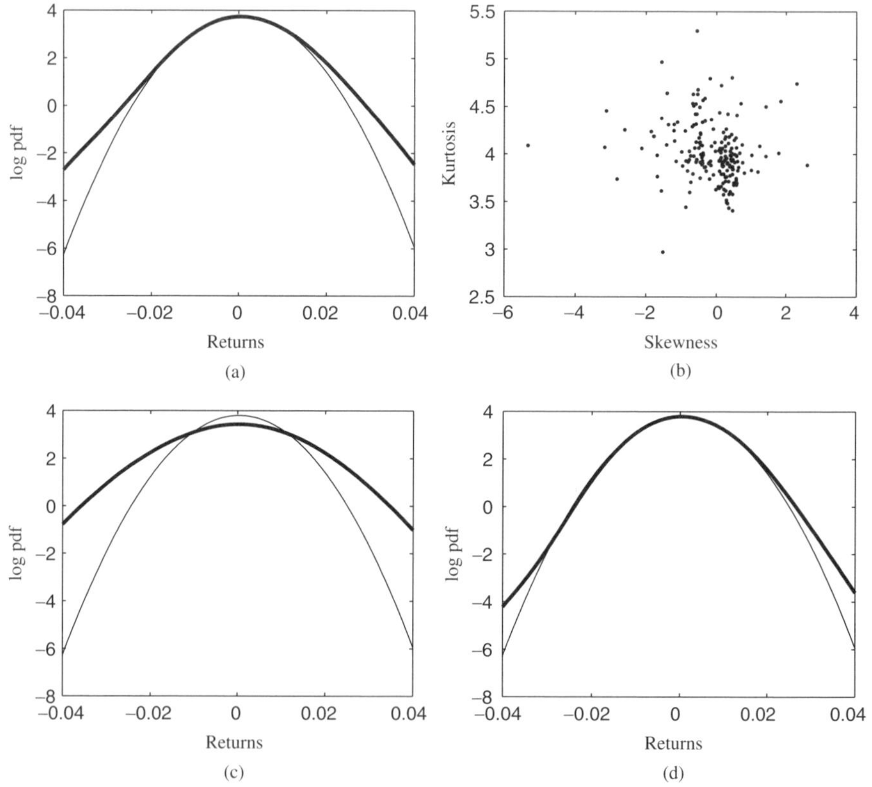

Figure 7.3. Some aspects of the Markov mixture model applied to 1700 daily returns of the S&P 500 index return: (a) unconditional pdf; (b) posterior distribution of unconditional moments; log predictive densities on (c) July 2, 1984 and (d) August 3, 1984.

mean of the unconditional probability density function of returns in the Markov mixture model (heavier curve), together with the log posterior mean of the pdf of returns in the i.i.d. normal model (lighter curve). Panel (b), which displays 200 points from the posterior distribution of skewness and kurtosis, provides another perspective on the shape of the unconditional distribution in the Markov mixture model. It is mildly but decisively leptokurtic, displays little or no skewness, and is centered at a very small positive return.

Panels (c) and (d) in Figure 7.3 show the logarithm of the predictive density for the next day's returns, for two particular dates in the sample period. In the Markov normal mixture linear model predictive densities (heavier curves) vary from one day to the next, as emphasized in Example 7.3.1, whereas in the normal linear model these densities (lighter curves) are always the same as the unconditional pdf. On July 2 the Markov normal mixture model indicates substantially more uncertainty about the next day's return than does the normal model, and careful inspection of the figure in panel (c) shows that the Markov normal mixture predictive distribution is also slightly skewed to the left. On August 3 the distribution of the next day's return is more closely aligned in the two models. The predictive density in the Markov normal mixture model is distinctly leptokurtic, and careful inspection of the figure in panel (d) shows that the distribution is slightly skewed to the right.

From the posterior simulation output, there is a state assignment $\widetilde{s}_T^{(m)}$ for the last date in the sample at iteration m. State assignments can be generated recursively for future dates $T + j$, according to $P(\widetilde{s}_{T+j}^{(m)} = i \mid \widetilde{s}_{T+j-1}^{(m)}, \mathbf{P}^{(m)}, A) = p_{\widetilde{s}_{T+j-1},i}^{(m)}$. A random sample $y_{T+1}^{(m)}, y_{T+2}^{(m)}, \ldots$ can then be generated from these assignments and the other simulated parameters $\boldsymbol{\gamma}^{(m)}$, $h^{(m)}$, and $\mathbf{h}^{(m)}$. The simulated j-day-ahead return is then $\sum_{i=1}^{j} y_{T+i}^{(m)}$. These returns can be simulated several times for each parameter vector drawn from the posterior simulator. Sorting the simulated returns over all iterations and simulations then provides the return at risk. Figure 7.4 was constructed in this way from the 1000 drawings from the posterior distribution and 100 simulations for each drawing.

This figure indicates return at risk using a probability .05 [panels (a) and (b)] as well as .01 [panels (c) and (d)] for total returns up to 10 business days after the dates indicated. The unconditional distribution of returns in the i.i.d. Gaussian model implies that return at risk j days in the future is always the same, depending only on j. Consequently the lighter curves in the left and right panels are identical. In the Markov normal mixture linear model return at risk j days in the future is always changing, but as $j \to \infty$, the predictive density for y_{T+j} approaches the one shown in the panel (a) of Figure 7.3, for all days T on which predictions are made. This is reflected in returns at risk, which must approach the same value at long horizons regardless of the state probabilities at time T. This is evident in Figure 7.3. The one-day-ahead return at risk is higher on July 2 than on August 3, reflecting the predictive densities shown in the panels (c) and (d) of Figure 7.3, but 10-day-ahead is nearly the same for the 2 days. By implication, return at risk rises more rapidly with lengthening horizon starting from August 3 than it does from July 2.

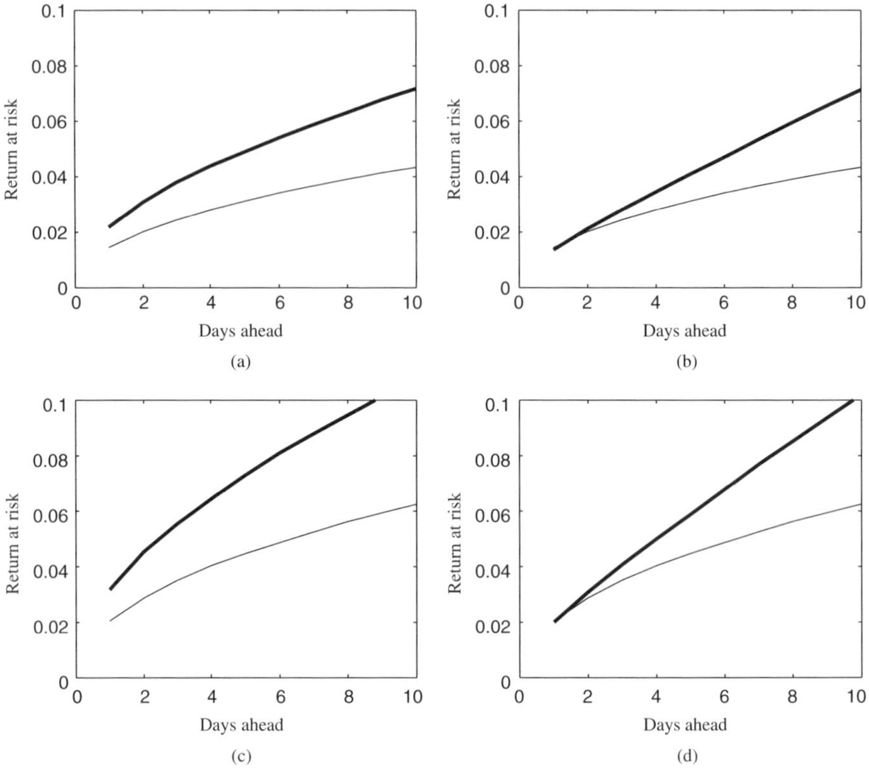

Figure 7.4. Return at risk [(a) $p = 5\%$, 7/2/84; (b) $p = 5\%$, 8/3/84; (c) $p = 1\%$, 7/2/84; (d) $p = 1\%$, 8/3/84] in the Markov normal mixture model (heavier curve) and i.i.d. normal model (lighter curve).

Exercise 7.3.1 There are a number of important technical details that underlie the work in Example 7.3.2 that can be appreciated only by using the code in the online appendix.

(a) From the output of the posterior simulator, look for examples of "label switching." Label switching is indicated by the same, sudden permutation of the parameter vectors $\boldsymbol{\alpha}, \mathbf{h}$ and of the rows and columns of the matrix $\mathbf{P}$ between successive iterations. Note that this has no impact on any function of interest that depends only on future values of the time series to which the model is being applied.

(b) Using the methods of Section 4.7, illustrated in Example 5.1.1, test for convergence of the posterior simulator.

Exercise 7.3.2 Real decisionmakers are likely to want to assess the sensitivity of conclusions to assumptions in the model or models used.

(a) Repeat the analysis of Example 7.3.2 using different numbers of states (m). How sensitive are the findings about return at risk in Figure 7.4 to this specification?

(b) How sensitive are these results to the specification of the hyperparameters of the prior distribution (7.46)–(7.47)? (Before answering this question, you may wish to consult Examples 8.3.2 and 8.3.4.)

Exercise 7.3.3 The online appendix contains a much longer set of returns from the S&P 500 stock price index, and the code indicates how to reconstruct the subsamples used in Ryden et al. (1998).

(a) Repeat the analysis in Example 7.3.2 for some of the other nine subsamples used in Ryden et al. (1998). Are the findings similar for these other periods?

(b) Repeat the analysis in Example 7.3.2 using the entire sequence of S&P 500 returns but with a larger number of states. Interpret the results in the context of the results of part (a). To what extent is there a tendency for states to occur in one part of the sample (e.g., the 1930s or 1980s) and never reappear?

CHAPTER 8

Bayesian Investigation

Multiple stakeholders have competing interests in the decisions that motivate Bayesian inference. Investigators, sometimes working on behalf of stakeholder clients, will carefully examine formal models used to inform these decisions. If a model is taken seriously and is likely to bear on a decision, then its credibility, often as indicated by its implications for observables, will receive close scrutiny. So, too, will the sensitivity of these implications to changes in the specification of the model. It is likely that new models or major variants on existing models will be introduced to cope with specific features of the problem at hand, the decision being addressed, and their interaction.

This chapter presents tools at the disposal of the Bayesian investigator party to this process. The simulation methods set forth in Chapter 4 are well suited to examination of models of the type discussed in Chapters 5 through 7. A seasoned Bayesian investigator can go well beyond the models in these chapters, using the simulation methods of Chapter 4 and the underlying insights of Chapters 2 and 3, to construct the variants required for a specific decision problem. This process often involves combinations and syntheses of simpler models, such as those taken up in Chapters 5 through 7.

The development of a new model, even if it is an apparently straightforward variant on an existing and thoroughly understood model, consumes resources. Good investigators must understand the implications of their complete models for observables, and must be able to reflect the beliefs of their clients in the specification of these models. This chapter addresses this problem, in Section 8.3, by means of forward simulation: that is, drawing unobservables from a candidate prior, followed by observables drawn conditional on unobservables, followed by the vector of interest drawn conditional on both. Section 8.3 also details how the investigator can go some distance in ascertaining whether a model yields sensible implications for observables and vectors of interest, before setting up a posterior simulator.

If a complete model yields sensible prior implications for observables and vectors of interest, the investigator may proceed to write a posterior simulator, a

Contemporary Bayesian Econometrics and Statistics, by John Geweke

task often more time-consuming than writing a forward simulator. This entails developing an algorithm like those taken up in Chapters 5–7, and executing the algorithm with suitable computer code. Section 8.1 develops tests that should be applied at the end of this process, before taking code to the data and problem at hand. These tests have substantial power against logical errors that may arise anywhere following the specification of the model, including expression of densities and conditional densities and their embodiment in computer code.

Given posterior simulators for alternative models, the investigator may turn to formal comparison of these models using Bayes factors and marginal likelihoods as discussed in Section 2.6. Section 8.2 describes several ways in which posterior simulators can be used to approximate marginal likelihoods or Bayes factors.

The process of examining the sensitivity of the results that count—the "bottom line" of uncertainty about the vector of interest—to the necessary but intermediate steps of specifying prior and observables distributions is one that often involves both the investigator and the client. Section 8.4 provides methods that enable a Bayesian investigator, perhaps working with a complex model, a sophisticated posterior simulator, and a very fast computer, to communicate results in a fashion that permits clients with spreadsheet software and laptops to manipulate some of the specifications of the model and examine the implications for vectors of interest. A particular client of interest in this process is the remote client—for example, an anonymous reader of the investigator's published work. Section 8.5 shows how the Bayesian investigator can facilitate this process.

8.1 IMPLEMENTING SIMULATION METHODS

A complete model A provides a prior density of unobservables $p(\boldsymbol{\theta}_A \mid A)$, a conditional density of observables $p(\mathbf{y} \mid \boldsymbol{\theta}_A, A)$, and a conditional density of a vector of interest, $p(\boldsymbol{\omega} \mid \boldsymbol{\theta}_A, \mathbf{y}, A)$. It is usually straightforward to simulate from each of these distributions:

$$\boldsymbol{\theta}_A^{(m)} \sim p(\boldsymbol{\theta}_A \mid A), \tag{8.1}$$

$$\mathbf{y}^{(m)} \sim p(\mathbf{y} \mid \boldsymbol{\theta}_A, A), \tag{8.2}$$

$$\boldsymbol{\omega}^{(m)} \sim p(\boldsymbol{\omega} \mid \mathbf{y}, \boldsymbol{\theta}_A, A). \tag{8.3}$$

If $\boldsymbol{\theta}_A = \boldsymbol{\theta}_A^{(m)}$ in (8.2), then $\mathbf{y}^{(m)} \sim p(\mathbf{y} \mid \mathbf{A})$. If $\mathbf{y} = \mathbf{y}^{(m)}$ and $\boldsymbol{\theta}_A = \boldsymbol{\theta}_A^{(m)}$ in (8.3), then $\boldsymbol{\omega}^{(m)} \sim p(\boldsymbol{\omega} \mid \mathbf{A})$.

As a matter of research strategy, the relative ease of constructing the simulators (8.1), (8.2), and (8.3) suggests that this be done before undertaking the more challenging task of constructing a posterior simulator $\boldsymbol{\theta}_A^{(m)} \sim p(\boldsymbol{\theta}_A \mid \mathbf{y}^o, A)$. The forward simulator can reveal interesting and relevant properties of the model—in particular, its ability to account for salient features, $\boldsymbol{\omega}$, of observables. It can indicate the suitability of the model for the purposes at hand, for example, its ability to replicate important features of the data, as discussed in greater detail in Section 8.3.

8.1.1 Density Ratio Tests

It is also generally straightforward to express the densities that appear on the right sides of (8.1)–(8.3), and to write code that evaluates these densities. This is a requirement, in any event, in order to compute or approximate the marginal likelihood of the model. There is an intimate relationship between the simulators (8.1)–(8.3) and the evaluation of the corresponding densities, which is useful in checking their derivation as well as their expression in computer code.

Theorem 8.1.1 Density Ratio Test Suppose that $\{\mathbf{x}^{(m)}\}$ is an ergodic process with unique invariant density $p(\mathbf{x} \mid I)$ with respect to a measure ν, having support $X \subseteq \mathbb{R}^n$. Let $k(\mathbf{x} \mid I)$ be any kernel of this probability density, and $c_I = \int_X k(\mathbf{x} \mid I)\, d\nu(\mathbf{x})$. Let f be any probability density with respect to ν having support $X^* \subseteq X$. Then

$$M^{-1} \sum_{m=1}^{M} f(\mathbf{x}^{(m)})/k(\mathbf{x}^{(m)} \mid I) \overset{\text{a.s.}}{\to} c_I^{-1}.$$

Proof: For $g(\mathbf{x}) = f(\mathbf{x})/k(\mathbf{x} \mid I)$, we have

$$\begin{aligned} E[g(\mathbf{x}) \mid I] &= \int_X g(\mathbf{x})p(\mathbf{x} \mid I)\, d\nu(\mathbf{x}) = c_I^{-1} \int_X g(\mathbf{x})k(\mathbf{x} \mid I)\, d\nu(\mathbf{x}) \\ &= c_I^{-1} \int_X f(\mathbf{x})\, d\nu(\mathbf{x}) = c_I^{-1}. \end{aligned}$$

Since c_I^{-1} is finite and $\{x^{(m)}\}$ is ergodic, $M^{-1} \sum_{m=1}^{M} g(\mathbf{x}^{(m)}) \overset{\text{a.s.}}{\to} c_I^{-1}$. ∎

In the particular case $k(\mathbf{x} \mid I) = p(\mathbf{x} \mid I)$, we obtain

$$M^{-1} \sum_{m=1}^{M} f(\mathbf{x}^{(m)})/p(\mathbf{x}^{(m)} \mid I) \overset{\text{a.s.}}{\to} 1. \tag{8.4}$$

Theorem 8.1.1 provides a basis for testing simulators and density evaluations. If both have been derived and coded correctly, then (8.4) must hold. It is, of course, not the case that (8.4) must be violated if there are errors. However, in most settings it is difficult to produce errors, even intentional ones, that leave (8.4) in tact. Moreover the derivation and coding of a simulator is typically independent of the derivation and coding of the evaluation of $p(\mathbf{x})$; thus it is unlikely that the same error can enter each in a way that preserves (8.4). Note that "error," here, subsumes everything from misconceptions to mistakes in derivations to "bugs" in computer code.

The less variation in $f(\mathbf{x})/p(\mathbf{x} \mid I)$, the better is the approximation in Theorem 8.1.1. If $\{\mathbf{x}^{(m)}\}$ is uniformly ergodic and $\text{var}[g(\mathbf{x}) \mid I] < \infty$, then Theorem 4.7.1 may be invoked to provide the numerical standard error of the approximation $M^{-1}\sum_{m=1}^{M} g(\mathbf{x}^{(m)})$. The variance requirement amounts to

$$\int_{X^*} [f^2(\mathbf{x})/p(\mathbf{x} \mid I)]\, d\nu(\mathbf{x}) < \infty$$

and will be satisfied if $f(\mathbf{x})/p(\mathbf{x} \mid I)$ is bounded above on X^*. There is a large class of cases in which such a density $f(\mathbf{x})$ may be constructed from the simulations $\mathbf{x}^{(1)}, \ldots, \mathbf{x}^{(M)}$.

Theorem 8.1.2 Constructing Density Ratio Tests Suppose that the $n \times 1$ random vector $\mathbf{x}$ has an absolutely continuous distribution with mean $\boldsymbol{\mu}$, variance $\boldsymbol{\Sigma}$, and probability density p that is bounded above as well as bounded away from zero on all compact sets $A \subseteq \mathbb{R}^n$. Suppose that $\{\mathbf{x}^{(m)}\}$ is ergodic with invariant density $p(\mathbf{x} \mid I)$, and denote the sample mean and variance of $\mathbf{x}^{(m)} (m = 1, \ldots, M)$ by $\boldsymbol{\mu}^{(M)}$ and $\boldsymbol{\Sigma}^{(M)}$. Denote the pdf of a multivariate normal distribution with mean $\boldsymbol{\mu}^{(M)}$ and variance $\boldsymbol{\Sigma}^{(M)}$, truncated to its highest density region $X_\alpha^{(M)}$ of size $100(1-\alpha)\%$, by

$$f_\alpha^{(M)}(\mathbf{x}) = (1-\alpha)^{-1}(2\pi)^{-n/2}\left|\boldsymbol{\Sigma}^{(M)}\right|^{-1/2} \cdot \exp[-(\mathbf{x}-\boldsymbol{\mu}^{(M)})(\boldsymbol{\Sigma}^{(M)})^{-1}(\mathbf{x}-\boldsymbol{\mu}^{(M)})/2] I_{X_\alpha^{(M)}}(\mathbf{x}).$$

Let $k(\mathbf{x} \mid I) = c_I \cdot p(\mathbf{x})$, where $c_I > 0$, be a kernel of $p(\mathbf{x} \mid I)$. Then

$$M^{-1}\sum_{m=1}^{M} f_\alpha^{(M)}(\mathbf{x}^{(m)})/k(\mathbf{x}^{(m)} \mid I) \overset{\text{a.s.}}{\to} c_I^{-1} \tag{8.5}$$

and

$$\overline{\lim}_{M\to\infty}\text{var}\left[M^{-1}\sum_{m=1}^{M} f_\alpha^{(M)}(\mathbf{x}^{(m)})/k(\mathbf{x}^{(m)} \mid I)\right] < \infty.$$

Proof: Let

$$X_\alpha = \{\mathbf{x} : (\mathbf{x}-\boldsymbol{\mu})'\boldsymbol{\Sigma}^{-1}(\mathbf{x}-\boldsymbol{\mu}) \leq \chi_\alpha^2(n)\}$$

and

$$f_\alpha(\mathbf{x}) = (1-\alpha)^{-1}(2\pi)^{-n/2}|\boldsymbol{\Sigma}|^{-1/2}\exp[-(\mathbf{x}-\boldsymbol{\mu})\boldsymbol{\Sigma}^{-1}(\mathbf{x}-\boldsymbol{\mu})/2] I_{X_\alpha}(\mathbf{x}).$$

Given any $\varepsilon > 0$, let

$$\widetilde{X}_\varepsilon^{(M)} = \{\mathbf{x} : \left|f_\alpha^{(M)}(\mathbf{x}) - f_\alpha(\mathbf{x})\right| > \varepsilon\}.$$

Since $\boldsymbol{\mu}^{(M)} \overset{\text{a.s.}}{\to} \boldsymbol{\mu}$ and $\boldsymbol{\Sigma}^{(M)} \overset{\text{a.s.}}{\to} \boldsymbol{\Sigma}$, it follows that $\int_{\widetilde{X}_\varepsilon^{(M)}} d\mathbf{x} \overset{\text{a.s.}}{\to} 0$. The results follow from the conditions that f_α and p are bounded above and bounded away from zero on all compact sets. ■

Since the density $f_\alpha^{(M)}$ is constructed from the simulator output $\{\mathbf{x}^{(m)}\}$, the density ratio test requires no further simulation. The conditions of Theorem 8.1.2 assure $\text{var}[f_\alpha^{(M)}(\mathbf{x})/k_I(\mathbf{x} \mid I) \mid I] < \infty$, and therefore Theorem 4.7.1 may be applied to assess the accuracy of this approximation. When $k(\mathbf{x} \mid I) = p(\mathbf{x} \mid I)$, then $c_I = 1$ and we can formally test of the correctness of the simulator that produces $\{\mathbf{x}^{(m)}\}$ and the code that evaluates $p(\mathbf{x} \mid I)$.

The compact support of f in Theorem 8.1.2, achieved by truncating a multivariate normal density, is important because it bounds the ratio f/p. In many applications f/p will become unbounded as $\alpha \to 0$, and it is always the case that the accuracy of the approximation must deteriorate as $\alpha \to 1$. Thus it may be prudent to conduct a density ratio test with several alternative values of α. Other things the same, the greater the dimension of $\mathbf{x}$, the greater the variation in f/p, and for sufficiently high dimensions the procedure may become impractical. This is generally not a problem in the density ratio test as applied in this section, where the size of $\mathbf{x}$ can be controlled as illustrated in Examples 8.1.1 and 8.1.2. However, this consideration becomes important in Section 8.2.4 in the application of Theorems 8.1.1 and 8.1.2 in approximating the marginal likelihood.

Example 8.1.1 Application of the Density Ratio Test to an Observables Density Consider the mean zero, normal first-order autoregression model, with stationarity imposed: $\rho \in (-1, 1)$, $h > 0$ and

$$y_1 \mid A \sim N[0, h^{-1}(1-\rho^2)^{-1}]; \quad y_t \mid (\mathbf{Y}_{t-1}, A) \sim N(\rho y_{t-1}, h^{-1}) \quad (t = 2, \ldots, T). \tag{8.6}$$

(This is a very special case of the linear model with serial correlation discussed in Section 7.1.) We can simulate observables from this model, given fixed values of ρ and h, without even thinking about the observables density, which is

$$p(y_1, \ldots, y_T \mid h, \rho, A) = (2\pi)^{-T/2} h^{T/2} (1-\rho^2)^{1/2} \cdot \exp\left\{-h\left[y_1^2(1-\rho^2) + \sum_{t=2}^{T}(y_t - \rho y_{t-1})^2\right]\Big/2\right\}. \tag{8.7}$$

To apply the density ratio test, construct $f_\alpha^{(M)}(\cdot)$ as described in Theorem 8.1.2, with $\alpha = 0.5$ and $M = 10{,}000$ simulations.

Errors may be made either in simulating $\mathbf{y}^{(m)}$ or in the evaluation $p(\mathbf{y} \mid h, \rho, A)$. As an alternative to correct simulation, suppose $y_1 \sim N(0, h^{-1})$ in lieu of $N[0, h^{-1}(1-\rho^2)^{-1}]$ in (8.6). As an alternative to the correct evaluation of the probability density, consider omission of the term $h^{T/2}(1-\rho^2)^{1/2}$ (error 1) or $y_1^2(1-\rho^2)$ (error 2) in (8.7).

The outcomes of the density ratio tests are

Log $\left[M^{-1}\sum_{m=1}^{M} f(\mathbf{y}^{(m)})/p(\mathbf{y}^{(m)} \mid h, \rho, A)\right]$ Shown; Standard Errors in Parentheses			
Density evaluation error	None	Error 1	Error 2
Simulation error			
None	−.006 (.010)	.508 (.010)	.259 (.011)
Error	−.342 (.011)	.194 (.011)	−.271 (.011)

The tests easily detect the errors, and could have done so with $M = 100$ iterations rather than $M = 10{,}000$. The tests also appropriately indicate no error when both simulator and data density evaluation are correct. These tests were conducted using the single setting of parameters $\rho = 0.8$, $h = 1$. We might, of course, carry out tests for several alternative settings of parameters. The latter alternative extends density ratio tests to the case in which the normalizing constant of the likelihood function has been omitted, which often occurs in maximum likelihood estimation. Then the limit in (8.4) is not 1.0, but should be the same for different settings of the parameters. Testing this hypothesis is straightforward.

In many instances the support of $p(\mathbf{x} \mid I)$ is a subset of $\mathbb{R}^n$. Theorem 8.1.2 may still be applied, if the random vector $\mathbf{x}$ is transformed so that its support is $\mathbb{R}^n$. The density of the transformed vector involves the Jacobian of transformation, which must also be derived and coded correctly.

Example 8.1.2 Application of the Density Ratio Test to a Prior Density Consider the Wishart distribution of an $m \times m$ positive definite matrix $\mathbf{A}$, with $m \times m$ positive definite matrix parameter $\mathbf{\Sigma}$ and degrees of freedom parameter $\nu \geq m$, denoted $\mathbf{A} \sim W(\mathbf{\Sigma}, \nu)$. Section 5.2 introduced the Wishart as a conditionally conjugate prior distribution in the seemingly unrelated regressions model, and provided its pdf in (5.17) followed by an algorithm for i.i.d. simulations $\mathbf{A}^{(m)} \sim W(\mathbf{\Sigma}, \nu)$.

There are $m(m+1)/2$ distinct elements of $\mathbf{A}$. Since $\mathbf{A}$ is positive definite, the support of $\mathbf{A}$ is not $\mathbb{R}^{m(m+1)/2}$, and the distribution of these elements is seldom well approximated by a normal distribution unless ν is quite large. The density ratio test is more efficient if applied to a transformation of $\mathbf{A}$. Let $\mathbf{A} = \mathbf{C}\mathbf{C}'$ be the Choleski factorization of $\mathbf{A}$, in which $\mathbf{C}$ is lower triangular with positive diagonal elements, and use logarithms of the diagonal elements of $\mathbf{C}$ rather than the values themselves. The Jacobian of transformation for this new set of random variables, which we denote $\boldsymbol{\theta}$, is $2^m \prod_{i=1}^{m} c_{ii}^{(m+2-i)}$.

To illustrate the application of the density ratio test to this distribution, take $m = 4$, $\nu = 10$, and

$$\Sigma = \begin{bmatrix} 4 & -1 & 2 & 0 \\ -1 & 3 & -2 & 1 \\ 2 & -2 & 6 & -1 \\ 0 & 1 & -1 & 3 \end{bmatrix}.$$

In one case the simulation is carried out correctly, whereas in the other $b_{ii}^2 \sim \chi^2(\nu - i)$ in the second step of the Wishart simulation algorithm presented in Section 5.2. Consider two errors in evaluating the density: (1) the exponent $\nu - 1 - m$ appearing in (5.17) is replaced by $\nu - m$ and (2) $\pi^{-m(m-1)/4}$ in (5.17) is omitted. The outcomes of the density ratio tests using $M = 10{,}000$ simulations and $\alpha = 0.5$ in (8.5) are

Log $\left[M^{-1}\sum_{m=1}^{M} f(\boldsymbol{\theta}^{(m)})/p(\boldsymbol{\theta}^{(m)})\right]$ Shown; Standard Errors in Parentheses			
Density evaluation error	None	Error 1	Error 2
Simulation error			
None	−.003 (.011)	6.314 (.014)	3.423 (.010)
Error	−.240 (.014)	5.543 (.023)	3.185 (.013)

As in the previous example, the results are clear.

8.1.2 Joint Distribution Tests

If there are alternative ergodic simulators $\{\mathbf{x}^{(m)}\}$ and $\{\widetilde{\mathbf{x}}^{(m)}\}$ with the same invariant distribution, then

$$M^{-1}\sum_{m=1}^{M} g(\mathbf{x}^{(m)}) - M^{-1}\sum_{m=1}^{M} g(\widetilde{\mathbf{x}}^{(m)}) \overset{\text{a.s.}}{\to} 0$$

as long as $E[g(\mathbf{x}) \mid I]$ is well defined and finite. If $\text{var}[g(\mathbf{x}) \mid I]$ is also well defined and finite, and the two simulators are independent, then a central limit theorem may be applied to test formally the proposition that the invariant distributions are in fact the same.

This test does not require evaluation of the density $p(\mathbf{x} \mid I)$. This is an advantage if we wish only to check the simulator, since it spares the effort of deriving and coding $p(\mathbf{x} \mid I)$, as well as transformation of $\mathbf{x}$ to ensure support in all of $\mathbb{R}^n$ if that is necessary. However, as we shall see in Section 8.2, it is necessary to evaluate the prior density $p(\boldsymbol{\theta}_A \mid A)$ and the data density $p(\mathbf{y}^o \mid \boldsymbol{\theta}_A, A)$ to approximate marginal likelihoods. Eventually, therefore, it is necessary to derive, code, and check the evaluation of the prior and data densities. Density ratio tests, but not joint distribution tests, are applicable at that stage.

An important use of joint distribution tests is in checking the posterior simulator. The joint distribution of observables and unobservables is

$$p(\boldsymbol{\theta}_A, \mathbf{y} \mid A) = p(\boldsymbol{\theta}_A \mid A)p(\mathbf{y} \mid \boldsymbol{\theta}_A, A).$$

We have already noted, below (8.3), that simulation from this joint distribution can be achieved by means of a marginal–conditional simulator

$$\boldsymbol{\theta}_A^{(m)} \sim p(\boldsymbol{\theta}_A^{(m)} \mid A), \mathbf{y}^{(m)} \sim p(\mathbf{y} \mid \boldsymbol{\theta}_A^{(m)}, A) \quad (m = 1, \ldots, M). \tag{8.8}$$

For any function $g(\boldsymbol{\theta}_A, \mathbf{y})$ whose expectation with respect to the joint distribution of $\boldsymbol{\theta}_A$ and $\mathbf{y}$ is well defined and finite, we obtain

$$M^{-1}\sum_{m=1}^{M} g(\boldsymbol{\theta}_A^{(m)}, \mathbf{y}^{(m)}) \overset{\text{a.s.}}{\to} \int_{\Theta_A}\int_{Y} g(\boldsymbol{\theta}_A, \mathbf{y})p(\boldsymbol{\theta}_A, \mathbf{y} \mid A)\, d\nu(\mathbf{y})\, d\nu(\boldsymbol{\theta}_A). \tag{8.9}$$

If $\text{var}[g(\boldsymbol{\theta}_A, \mathbf{y}) \mid A] < \infty$, then the Lindeberg–Lévy central limit theorem can be used to approximate the accuracy of the approximation $M^{-1}\sum_{m=1}^{M} g(\boldsymbol{\theta}_A^{(m)}, \mathbf{y}^{(m)})$ in (8.9), since the sequence $\{\boldsymbol{\theta}_A^{(m)}, \mathbf{y}^{(m)}\}$ is i.i.d. If g is a function of $\boldsymbol{\theta}_A$ but not of $\mathbf{y}$, only the prior simulator is required.

A posterior simulator generates the sequence $\{\widetilde{\boldsymbol{\theta}}_{A,\mathbf{y}^o}^{(m)}\}$ from a Markov chain C:

$$\widetilde{\boldsymbol{\theta}}_{A,\mathbf{y}^o}^{(m)} \sim p(\boldsymbol{\theta}_A \mid \widetilde{\boldsymbol{\theta}}_{A,\mathbf{y}^o}^{(m-1)}, \mathbf{y}^o, C).$$

Consider the successive conditional simulator

$$\begin{aligned}
&\widetilde{\boldsymbol{\theta}}_A^{(0)} \sim p(\boldsymbol{\theta}_A \mid A);\\
&\widetilde{\mathbf{y}}^{(m)} \sim p(\mathbf{y} \mid \widetilde{\boldsymbol{\theta}}_A^{(m-1)}, A), \quad \widetilde{\boldsymbol{\theta}}_A^{(m)} \sim p(\boldsymbol{\theta}_A \mid \widetilde{\boldsymbol{\theta}}_A^{(m-1)}, \mathbf{y}^{(m)}, C)\\
&\qquad (m = 1, \ldots, M).
\end{aligned} \tag{8.10}$$

If we have at hand a demonstration of the (uniform) ergodicity of $\{\widetilde{\boldsymbol{\theta}}_{A,\mathbf{y}^o}^{(m)}\}$ for almost all $\mathbf{y}^o$, then, showing that $\{\widetilde{\boldsymbol{\theta}}_A^{(m)}, \widetilde{\mathbf{y}}^{(m)}\}$ is (uniformly) ergodic with unique invariant density $p(\boldsymbol{\theta}_A \mid A)p(\mathbf{y} \mid \boldsymbol{\theta}_A, A)$ typically involves little, if any, additional work. In this case

$$M^{-1}\sum_{m=1}^{M} g(\widetilde{\boldsymbol{\theta}}_A^{(m)}, \widetilde{\mathbf{y}}^{(m)}) \overset{\text{a.s.}}{\to} \int_{\Theta_A}\int_{Y} g(\boldsymbol{\theta}_A, \mathbf{y})p(\boldsymbol{\theta}_A, \mathbf{y} \mid A)\, d\nu(\mathbf{y})\, d\nu(\boldsymbol{\theta}_A). \tag{8.11}$$

The standard error of approximation in (8.11) can be assessed using Theorems 4.7.1 and 4.7.3.

The thought processes and coding in the simulations of $\{\boldsymbol{\theta}^{(m)}, \mathbf{y}^{(m)}\}$ and $\{\widetilde{\boldsymbol{\theta}}^{(m)}, \widetilde{\mathbf{y}}^{(m)}\}$ are nearly independent. The former involves the prior and observables simulators, and the latter involves the posterior and observables simulators. The observables simulator is common to both, but an error in this simulator will have different consequences for the invariant distributions in the two cases. Consequently a formal comparison of the left sides of (8.9) and (8.11) has power against error in the simulation of observables, as well as error in the simulation of unobservables from the prior or posterior. The simulator $\{\boldsymbol{\theta}_A^{(m)}, \mathbf{y}^{(m)}\}$ is a logical first step in understanding the properties of a new model and in developing prior distributions in any event, as detailed in Section 8.3.1. The marginal effort in producing $\{\widetilde{\boldsymbol{\theta}}_A^{(m)}, \widetilde{\mathbf{y}}^{(m)}\}$ is the addition of a few lines to generate artificial data in posterior simulation code. Thus the cost of making the comparison is relatively low. The gain is that subtle errors producing reasonable but incorrect results will likely be detected.

Example 8.1.3 Joint Distribution Tests of a Posterior Simulator [This example appears in Geweke (2004).] To demonstrate the kinds of errors that these tests can detect in a research (as opposed to illustrative) problem, consider the Student-t mixture model

$$y_t \sim t(\mu_1, h_1^{-1}; \lambda) \quad \text{with probability } p, \tag{8.12}$$

$$y_t \sim t(\mu_2, h_2^{-1}; \lambda) \quad \text{with probability } 1 - p. \tag{8.13}$$

In this problem λ is fixed at $\lambda = 5$, but the model could be extended to make λ an unknown parameter as described in Section 6.4.1. This model, or one like it, can be used to model outliers (see Exercise 6.4.5), and models similar to this one can be used in financial applications similar to the financial decisionmaking problem described in Section 1.1.2, as illustrated in Example 7.3.2.

As discussed in Section 6.4.1, we can exploit the fact that the sequence $\lambda\widetilde{h}_t \sim \chi^2(\lambda)$ followed by $y_t \sim N[\mu_1, (h_1\widetilde{h}_t)^{-1}]$ is equivalent to (8.12). To recapitulate briefly, the model is augmented with $(\widetilde{h}_1, \ldots, \widetilde{h}_T)$ and the latent state vector $(\widetilde{s}_1, \ldots, \widetilde{s}_T)$, with $\widetilde{s}_t = 1$ indicating (8.12) and $\widetilde{s}_t = 2$ indicating (8.13). Then normal priors for μ_1 and μ_2, gamma priors for h_1 and h_2, and a beta prior for p are all conditionally conjugate, and the resulting conditional distributions in a Gibbs sampling algorithm are also of these forms.

This example uses two variants of the Gibbs sampler. In the first (MCMC1), $\widetilde{s}_t$ and $\widetilde{h}_t$ are drawn jointly; in the second (MCMC2), they are drawn separately. In the simulator $\{\boldsymbol{\theta}_A^{(m)}, \mathbf{y}^{(m)}\}$ the five parameters are drawn from the prior, and then $T = 6$ observations are generated. The simulator $\{\widetilde{\boldsymbol{\theta}}_A^{(m)}, \widetilde{\mathbf{y}}^{(m)}\}$ follows (8.10), again using $T = 6$ observations. The joint distribution test is carried out using the 5 first and 15 second moments of the parameter vector $\boldsymbol{\theta}_A' = (\mu_1, \mu_2, h_1, h_2, p)'$. [Since y_t is not involved in the comparison, it is really not necessary to generate $y^{(m)}$ in the marginal–conditional simulator.]

To gather evidence on the power of the posterior simulator joint distribution test, we introduce some errors. The first is an error in simulating from the prior: p is drawn from a beta(1, 1) distribution in the prior, whereas the posterior employs beta(2, 2) prior density. The second is an error in simulating the observables in the successive conditional simulator; the simulator ignores the $\widetilde{h}_t$ from the posterior simulator, and uses instead fresh values to construct y_t. The third error is in the simulation of μ_1 and μ_2 in the Gibbs sampler; they are set equal to their conditional means, with no allowance for their conditional variance. In the fourth error the degrees of freedom in the draw of $\widetilde{h}_t$ is 5, rather than its correct value of 6 in the conditional posterior distribution. The final error is in the generation of $(\widetilde{s}_t, \widetilde{h}_t)$ in MCMC1. The correct algorithm generates $\widetilde{s}_t$ (conditional on all unknowns except

$\widetilde{h}_t$) and then generates $\widetilde{h}_t$ conditional on all unknowns including $\widetilde{s}_t$ just drawn. In the error, $\widetilde{h}_t$ is drawn several steps later in the Gibbs sampling algorithm rather than immediately after $\widetilde{s}_t$.

For each of seven simulators, two correct and five incorrect, $M = 250{,}000$ values are drawn from $p(\boldsymbol{\theta}_A, \mathbf{y} \mid A)$ using both the marginal conditional simulator (8.8) and the successive conditional simulator (8.10). In each case, the 20 moments are computed for each simulator, and a conventional equality of means test is applied, taking care to account for serial correlation in the successive conditional simulator (Theorems 4.7.1 and 4.7.3). Tests carried out at some alternative conventional significance levels produce the following results:

		Rejections (out of 20) at			
Algorithm	Error	$p = .05$	$p = .01$	$p = .005$	$p = .001$
MCMC1	0. None	0	0	0	0
MCMC2	0. None	0	0	0	0
MCMC1	1. Prior simulation of p	4	3	3	2
MCMC1	2. Simulation of $\mathbf{y}$	10	9	9	9
MCMC1	3. $\widetilde{h}_t$ degrees of freedom	5	3	3	3
MCMC1	4. μ variance	11	10	10	9
MCMC1	5. $(\widetilde{s}_t, \widetilde{h}_t)$ draw	7	6	6	6

The correct algorithm clearly passes the joint distribution tests, whereas errors—in the prior, observables, or posterior simulators—are all flagged.

Exercise 8.1.1 More Applications of Joint Distribution Tests In the course of her work, an investigator has created a new prior density $p(\theta_A \mid A)$ for a particular scalar parameter θ_A. She has developed an algorithm for direct sampling from this distribution, and has written the corresponding software.

(a) Suppose that the investigator also has developed and coded an algorithm that evaluates the cdf of this distribution. How might she conduct a joint distribution test of the correctness of both her cdf evaluation and her direct sampling algorithm? (*Hint*: She can also simulate directly from a uniform distribution.)

(b) Suppose instead that the investigator also has developed and coded an algorithm that evaluates the inverse cdf of the distribution in question. How might she conduct a joint distribution test of the correctness of both her inverse cdf evaluation and her direct sampling algorithm?

Exercise 8.1.2 Density Ratio Test for Importance Sampling Theorem 8.1.1 assumes that the simulation algorithm is a Markov chain. Suppose, instead, that the algorithm uses importance sampling. State and prove a variant of Theorem 8.1.1 appropriate to this situation.

8.2 FORMAL MODEL COMPARISON

Model comparison can entail model averaging or hypothesis testing, as discussed in Section 2.6.1. In either case, given a set of models $A_1, \ldots, A_J$, the essential computational task is to approximate the marginal likelihoods $p(\mathbf{y}^o \mid A_j)(j = 1, \ldots, J)$. Analytical evaluations are possible only in very special cases, generally requiring fully conjugate prior distributions; instances include the linear model (Examples 2.3.2 and 2.3.3) and the first-order Markov finite state model (Section 7.2).

In most models, we must compute a good approximation to the marginal likelihood. A key difficulty is that the marginal likelihood cannot be expressed directly as a posterior moment, and consequently the problem cannot be treated directly as a special case of the simulation-consistent approximation of posterior moments developed in Chapter 4. There are specific cases in which the approximation of a Bayes factor is simply a special case of the approximation of a posterior moment. One of these will be important subsequently in Bayesian communication (Section 8.4) and robustness analysis (Section 8.5), and so we develop it here (Section 8.2.1). In the more general case there are methods specifically tailored to the kind of simulator used; here we examine the cases of importance sampling (Section 8.2.2) and Gibbs sampling (Section 8.2.3).

The computation or approximation of Bayes factors is an important current research topic, and this section does not include all approaches to this problem. In particular, Tierney and Kadane (1986, 1989) developed an approximation of the marginal likelihood based on Laplace expansions, and Green (1995) has developed simulation methods that treat several models simultaneously, with the number of drawings from each model proportional to its posterior probability. These approaches have proved very effective in some applications, and less so in others. The surveys of Carlin and Chib (1995) and Han and Carlin (2001) provide accessible introductions to these and other approaches to formal model comparison not discussed in detail here.

8.2.1 Bayes Factors for Modeling with Common Likelihoods

Suppose that the models A_1 and A_2 share the same conditional probability density of observables, $p(\mathbf{y} \mid \boldsymbol{\theta}_A, A)$, but have different prior densities, $p(\boldsymbol{\theta}_A \mid A_1)$ and $p(\boldsymbol{\theta}_A \mid A_2)$. The Bayes factor in favor of A_2 is

$$\frac{\int_{\Theta_{A_2}} p(\boldsymbol{\theta}_A \mid A_2) p(\mathbf{y}^o \mid \boldsymbol{\theta}_A, A)\, d\nu(\boldsymbol{\theta}_A)}{\int_{\Theta_{A_1}} p(\boldsymbol{\theta}_A \mid A_1) p(\mathbf{y}^o \mid \boldsymbol{\theta}_A, A)\, d\nu(\boldsymbol{\theta}_A)}. \tag{8.14}$$

If $\Theta_{A_2} \subseteq \Theta_{A_1}$, then (8.14) may be expressed

$$\frac{\int_{\Theta_{A_1}} [p(\boldsymbol{\theta}_A \mid A_2)/p(\boldsymbol{\theta}_A \mid A_1)]p(\boldsymbol{\theta}_A \mid A_1)p(\mathbf{y}^o \mid \boldsymbol{\theta}_A, A)\,d\nu(\boldsymbol{\theta}_A)}{\int_{\Theta_{A_1}} p(\boldsymbol{\theta}_A \mid A_1)p(\mathbf{y}^o \mid \boldsymbol{\theta}_A, A)\,d\nu(\boldsymbol{\theta}_A)}$$

$$= E[g(\boldsymbol{\theta}_A) \mid \mathbf{y}^o, A_1] \tag{8.15}$$

with $g(\boldsymbol{\theta}_A) = p(\boldsymbol{\theta}_A \mid A_2)/p(\boldsymbol{\theta}_A \mid A_1)$.

The posterior moment in (8.15) is well defined and finite. Given a posterior simulator $\boldsymbol{\theta}_A^{(m)} \sim p(\boldsymbol{\theta}_A \mid \mathbf{y}^o, A_1)$, it may be approximated consistently by

$$M^{-1} \sum_{m=1}^{M} p(\boldsymbol{\theta}_A^{(m)} \mid A_2)/p(\boldsymbol{\theta}_A^{(m)} \mid A_1).$$

If the posterior variance of $g(\boldsymbol{\theta}_A)$ is finite and the simulator is uniformly ergodic, then Theorem 4.7.1 can be the basis for evaluating the numerical accuracy of the approximation. A sufficient condition for finite variance of $g(\boldsymbol{\theta}_A)$ is that the ratio of prior densities $p(\boldsymbol{\theta}_A \mid A_2)/p(\boldsymbol{\theta}_A \mid A_1)$ be bounded on Θ_{A_2}.

Example 8.2.1 Changing the Prior in the Normal Linear Regression Model Consider a normal linear regression model with independent priors $\boldsymbol{\beta} \mid A_1 \sim N(\underline{\boldsymbol{\beta}}, \underline{\mathbf{H}}^{-1})$ and $\underline{s}^2 h \mid A_1 \sim \chi^2(\underline{\nu})$. Denote the corresponding prior pdf by $p(\boldsymbol{\beta}, h \mid A_1)$. Suppose that the Gibbs sampler has been used as a posterior simulator as described in Example 4.3.1 and the simulated values $\{\boldsymbol{\beta}^{(m)}, h^{(m)}\}$ are available for the approximation of posterior moments. An investigator or client entertaining a different prior density $p(\boldsymbol{\beta}, h \mid A_2)$ can approximate the Bayes factor in favor of their model by

$$M^{-1} \sum_{m=1}^{M} p(\boldsymbol{\beta}^{(m)}, h^{(m)} \mid A_2)/p(\boldsymbol{\beta}^{(m)}, h^{(m)} \mid A_1). \tag{8.16}$$

The approximation is simulation consistent. If $p(\boldsymbol{\beta}, h \mid A_2)/p(\boldsymbol{\beta}, h \mid A_1)$ is bounded above then the numerical accuracy of the approximation can be evaluated using a central limit theorem. For a specific example, see Exercise 8.2.2.

8.2.2 Marginal Likelihood Approximation Using Importance Sampling

Suppose that $p(\boldsymbol{\theta}_A \mid S)$, with support Θ_A, is the probability density function (not just a kernel) of an importance sampling distribution for the posterior density $p(\boldsymbol{\theta}_A \mid \mathbf{y}^o, A) \propto p(\boldsymbol{\theta}_A \mid A)p(\mathbf{y}^o \mid \boldsymbol{\theta}_A, A)$. Denote the corresponding weight function $w(\boldsymbol{\theta}_A) = p(\boldsymbol{\theta}_A \mid A)p(\mathbf{y}^o \mid \boldsymbol{\theta}_A, A)/p(\boldsymbol{\theta}_A \mid S)$, as in Section 4.2.2. Then

$$\overline{w}^{(M)} = M^{-1} \sum_{m=1}^{M} w(\boldsymbol{\theta}_A^{(m)}) \xrightarrow{\text{a.s.}} \int_{\theta_A} w(\boldsymbol{\theta}_A) p(\boldsymbol{\theta}_A \mid S)\, d\nu(\boldsymbol{\theta}_A)$$

$$= \int_{\theta_A} p(\boldsymbol{\theta}_A \mid A) p(\mathbf{y}^o \mid \boldsymbol{\theta}_A, A)\, d\nu(\boldsymbol{\theta}_A) = p(\mathbf{y}^o \mid A) = \overline{w},$$

the marginal likelihood for model A. If $w(\boldsymbol{\theta}_A)$ is bounded, then

$$M^{1/2}(\overline{w}^{(M)} - \overline{w}) \xrightarrow{d} N(0, \tau^2), \quad M^{-1} \sum_{m=1}^{M} [w(\boldsymbol{\theta}^{(m)}) - \overline{w}^{(M)}]^2 \xrightarrow{\text{a.s.}} \tau^2.$$

The first published application of this idea appears to be Geweke (1989b); see also Gelfand and Dey (1994) and Raftery (1996).

With slight modification this method may be applied with acceptance sampling or an independence Metropolis chain as well. In the former case, for all candidates $\boldsymbol{\theta}_A^*$, keep a running sum of $w(\boldsymbol{\theta}_A^*) = p(\boldsymbol{\theta}_A^* \mid A) p(\mathbf{y}^o \mid \boldsymbol{\theta}_A^*, A)/p(\boldsymbol{\theta}_A^* \mid S)$, and take $\overline{w}^{(M)}$ to be this sum deflated by the total number of candidates drawn. For an independence Metropolis chain the procedure is identical, except that $q(\boldsymbol{\theta}_A^* \mid H)$ replaces $p(\boldsymbol{\theta}_A^* \mid S)$. At each iteration the running sum is incremented by $w(\boldsymbol{\theta}_A^*)$, $\boldsymbol{\theta}_A^*$ being the candidate.

For all of these algorithms, the only incremental effort beyond what is otherwise required is that $p(\boldsymbol{\theta}_A \mid S)$ must be normalized to a density (not just a kernel), and the posterior density kernel must be expressed in standard form [recall (2.8)].

8.2.3 Marginal Likelihood Approximation Using Gibbs Sampling

In the case of the Gibbs sampler, an entirely different procedure due to Chib (1995) can provide quite accurate evaluations of the marginal likelihood, at the cost of additional simulations. Given the blocking $\boldsymbol{\theta}'_A = (\boldsymbol{\theta}'_{A(1)}, \ldots, \boldsymbol{\theta}'_{A(B)})$, suppose that the conditional probability density functions $p(\boldsymbol{\theta}_{A(b)} \mid \boldsymbol{\theta}_{A-(b)}, \mathbf{y}^o, A)$ can be evaluated in closed form for all blocks b. [This latter requirement is generally satisfied for a pure Gibbs sampler. For an extension of this method to the Metropolis–Hastings algorithm, see Chib and Jeliazkov (2001).]

From the identity $p(\boldsymbol{\theta}_A \mid A) p(\mathbf{y}^o \mid \boldsymbol{\theta}_A, A) = p(\mathbf{y}^o \mid A) p(\boldsymbol{\theta}_A \mid \mathbf{y}^o, A)$, we have

$$p(\mathbf{y}^o \mid A) = p(\boldsymbol{\theta}_A^* \mid A) p(\mathbf{y}^o \mid \boldsymbol{\theta}_A^*, A)/p(\boldsymbol{\theta}_A^* \mid \mathbf{y}^o, A) \tag{8.17}$$

for any *fixed* $\boldsymbol{\theta}_A^* \in \Theta_A$. Typically $p(\boldsymbol{\theta}_A^* \mid A)$ and $p(\mathbf{y}^o \mid \boldsymbol{\theta}_A^*, A)$ can be evaluated in closed form but $p(\boldsymbol{\theta}_A^* \mid \mathbf{y}^o, A)$ cannot. A marginal–conditional decomposition of $p(\boldsymbol{\theta}_A^* \mid \mathbf{y}^o, A)$ is

$$p(\boldsymbol{\theta}_A^* \mid \mathbf{y}^o, A) = p(\boldsymbol{\theta}_{A(1)}^* \mid \mathbf{y}^o, A) p(\boldsymbol{\theta}_{A(2)}^* \mid \boldsymbol{\theta}_{A(1)}^*, \mathbf{y}^o, A) \cdot \cdots \cdot p(\boldsymbol{\theta}_{A(B)}^* \mid \boldsymbol{\theta}_{A<(B)}^*, \mathbf{y}^o, A). \tag{8.18}$$

The first term in the product of B terms on the right side of (8.18) can be approximated from the output of the posterior simulator because

$$M^{-1}\sum_{m=1}^{M} p(\boldsymbol{\theta}^*_{A(1)} \mid \boldsymbol{\theta}^{(m)}_{A>(1)}) \xrightarrow{\text{a.s.}} p(\boldsymbol{\theta}^*_{A(1)} \mid \mathbf{y}^o, A).$$

To approximate $p(\boldsymbol{\theta}^*_{A(b)} \mid \boldsymbol{\theta}^*_{A<(b)}, \mathbf{y}^o, A)$, first execute the Gibbs sampler with the parameters in the first $b-1$ blocks fixed at $\boldsymbol{\theta}^*_{A(1)}, \ldots, \boldsymbol{\theta}^*_{A(b-1)}$. This provides a sequence $\{\boldsymbol{\theta}^{b(m)}_{A>(b-1)}\}$ from the conditional posterior distribution. Then

$$M^{-1}\sum_{m=1}^{M} p(\boldsymbol{\theta}^*_{A(b)} \mid \boldsymbol{\theta}^*_{A<(b)}, \boldsymbol{\theta}^{b(m)}_{>(b)}, \mathbf{y}^o, A) \xrightarrow{\text{a.s.}} p(\boldsymbol{\theta}^*_{A(b)} \mid \boldsymbol{\theta}^*_{A<(b)}, \mathbf{y}^o, A) \qquad (8.19)$$

for blocks $b = 2, \ldots, B-1$. The last term on the right side of (8.18) can be evaluated directly and requires no simulation. These approximations are then used in (8.18) and (8.17) to obtain the approximation to $p(\mathbf{y}^o \mid A)$.

This procedure is generally more efficient the larger is $p(\boldsymbol{\theta}^*_A \mid \mathbf{y}^o, A)$, so it helps to choose $\boldsymbol{\theta}^*_A$ to be near the mode of the posterior density. Of course, we should get the same result, up to numerical standard error, for any choice of $\boldsymbol{\theta}^*_A$. This property is also the basis of a test for accuracy and convergence of Gibbs sampling algorithms proposed by Zellner and Min (1995).

Example 8.2.2 Marginal Likelihood Approximation in the Normal Linear Regression Model In using the Gibbs sampler in the normal linear regression model (Example 4.3.1) there are only two blocks. If we designate the draw for h to be the first block of the Gibbs sampler and the draw for $\boldsymbol{\beta}$ to be the second block, the additional computational burden is negligible. Since there are only two blocks, no auxiliary simulations are needed.

Example 8.2.3 Marginal Likelihood Approximation in the Normal Mixture Linear Model Recall that the posterior density function in the normal mixture linear model (Section 6.4.2) is multimodal, a feature that makes it ill-suited to importance sampling, and can present problems for the density ratio marginal likelihood approximation method described in Section 8.2.4. Let $\boldsymbol{\theta}$ denote the parameters $\boldsymbol{\gamma}$, h, $\boldsymbol{\pi}$ and $\mathbf{h}$. Then

$$p(\mathbf{y}^o \mid A) = \frac{p(\boldsymbol{\theta}^* \mid A)p(\widetilde{\mathbf{s}}^* \mid \boldsymbol{\theta}^*, A)p(\mathbf{y}^o \mid \widetilde{\mathbf{s}}^*, \boldsymbol{\theta}^*, A)}{p(\widetilde{\mathbf{s}}^* \mid \mathbf{y}^o, A)p(\boldsymbol{\theta}^* \mid \widetilde{\mathbf{s}}^*, \mathbf{y}^o)}. \qquad (8.20)$$

Each of the three terms in the numerator of the right side of this expression can be evaluated analytically. The evaluation of terms in the denominator requires the output of the original posterior simulator and two conditional posterior simulators. The first term in the denominator

$$p(\widetilde{\mathbf{s}}^* \mid \mathbf{y}^o, A) = \int_{\Theta} p(\widetilde{\mathbf{s}}^* \mid \boldsymbol{\theta}, \mathbf{y}^o, A)p(\boldsymbol{\theta} \mid \mathbf{y}^o, A)$$

can be approximated by $M^{-1}\sum_{m=1}^{M} p(\widetilde{\mathbf{s}}^* \mid \boldsymbol{\theta}^{(m)}, A)$; note that the terms in this sum are available from the original posterior simulator. The last term in the denominator of (8.20) can be expressed as the product of the following four terms:

$$\begin{aligned} p(\boldsymbol{\pi}^* \mid \widetilde{\mathbf{s}}^*, \mathbf{y}^o, A) &= p(\boldsymbol{\pi}^* \mid \widetilde{\mathbf{s}}^*, A), \\ p(\boldsymbol{\gamma}^* \mid \boldsymbol{\pi}^*, \widetilde{\mathbf{s}}^*, \mathbf{y}^o, A) &= p(\boldsymbol{\gamma}^* \mid \widetilde{\mathbf{s}}^*, \mathbf{y}^o, A), \\ p(\mathbf{h}^* \mid \boldsymbol{\gamma}^*, \boldsymbol{\pi}^*, \widetilde{\mathbf{s}}^*, \mathbf{y}^o, A) &= p(\mathbf{h}^* \mid \boldsymbol{\gamma}^*, \widetilde{\mathbf{s}}^*, \mathbf{y}^o, A), \\ p(h^* \mid \mathbf{h}^*, \boldsymbol{\gamma}^*, \boldsymbol{\pi}^*, \widetilde{\mathbf{s}}^*, \mathbf{y}^o, A) &= p(h^* \mid \mathbf{h}^*, \boldsymbol{\gamma}^*, \widetilde{\mathbf{s}}^*, \mathbf{y}^o, A). \end{aligned}$$

The first and last terms can be evaluated analytically. A second run of the posterior simulator with $\widetilde{\mathbf{s}} = \widetilde{\mathbf{s}}^*$ and the block in $\boldsymbol{\pi}$ omitted produces the simulations $\boldsymbol{\gamma}^{1(m)}$, $h^{1(m)}$, and $\mathbf{h}^{1(m)} (m = 1, \ldots, M)$, and the approximation

$$M^{-1} \sum_{m=1}^{M} p(\boldsymbol{\gamma}^* \mid h^{1(m)}, \mathbf{h}^{1(m)}, \widetilde{\mathbf{s}}^*, \mathbf{y}^o, A)$$

of the second term above. A third run with $\widetilde{\mathbf{s}} = \widetilde{\mathbf{s}}^*$ and $\boldsymbol{\gamma} = \boldsymbol{\gamma}^*$ produces the simulations $h^{2(m)}$ and $\mathbf{h}^{2(m)}$, and the approximation $M^{-1}\sum_{m=1}^{M} p(\mathbf{h}^* \mid h^{2(m)}, \widetilde{\mathbf{s}}^*, \mathbf{y}^o, A)$ of the third term.

8.2.4 Density Ratio Marginal Likelihood Approximation

The marginal likelihood is the integrated posterior density kernel in standard form (2.8)

$$p(\mathbf{y}^o \mid A) = \int_{\Theta_A} p(\boldsymbol{\theta}_A \mid A) p(\mathbf{y}^o \mid \boldsymbol{\theta}_A, A)\, d\nu(\boldsymbol{\theta}_A). \tag{8.21}$$

Given the output of a posterior simulator, $\boldsymbol{\theta}_A^{(m)} \sim p(\boldsymbol{\theta}_A \mid \mathbf{y}^o, A)$, and evaluations of the prior density $p(\boldsymbol{\theta}_A^{(m)} \mid A)$ and data density $p(\mathbf{y}^o \mid \boldsymbol{\theta}_A^{(m)}, A)$, we can use Theorem 8.1.1 to approximate (8.21):

$$M^{-1} \sum_{m=1}^{M} f(\boldsymbol{\theta}_A^{(m)}) / p(\boldsymbol{\theta}_A^{(m)} \mid A) p(\mathbf{y}^o \mid \boldsymbol{\theta}_A^{(m)}, A) \overset{\text{a.s.}}{\to} [p(\mathbf{y}^o \mid A)]^{-1}. \tag{8.22}$$

In (8.22) the probability density $f(\cdot)$ can be constructed from the posterior simulator output as described in Theorem 8.1.2. The density ratio marginal likelihood approximation was proposed by Gelfand and Dey (1994), and the implementation with $f(\cdot)$ constructed from the posterior simulator output is due to Geweke (1999). The advantage of the method is that it is generic. It applies to any posterior simulator, no matter what algorithm is used, and since it approximates the marginal likelihood, it can be applied in approximating Bayes factors for any two models that pertain to the same data $\mathbf{y}^o$. If the evaluation $p(\boldsymbol{\theta}_A^{(m)} \mid A) p(\mathbf{y}^o \mid \boldsymbol{\theta}_A^{(m)}, A)$ is recorded in each iteration, along with $\boldsymbol{\theta}_A^{(m)}$, the density ratio marginal likelihood approximation can be computed after the posterior simulation has been completed. Computation time is typically negligible. Its limitation is that approximations can be poor when $\boldsymbol{\theta}_A$ is of very high dimension, as discussed following Theorem 8.1.2. This limitation applies to all the other known methods of approximating the marginal likelihood as well.

BACC uses the approximation (8.22) for many models, and permits the user to specify the truncation parameter α of the density f_α of Theorem 8.1.2, as well as alternative values of L/M in the tapering function (4.46) for the computation of numerical standard errors.

It is often the case that part of the integration in (8.21) can be carried out analytically, thereby greatly reducing the dimension of the space over which the simulation approximation must be made. Let $\boldsymbol{\theta}'_A = (\boldsymbol{\theta}'_{A(1)}, \boldsymbol{\theta}'_{A(2)})$, where $\boldsymbol{\theta}_{A(2)}$ is of high dimension whereas the order of $\boldsymbol{\theta}_{A(1)}$ is small. Suppose that

$$p(\mathbf{y} \mid \boldsymbol{\theta}_{A(1)}, A) = \int_{\Theta_{A(2)}} p(\boldsymbol{\theta}_{A(2)} \mid \boldsymbol{\theta}_{A(1)}, A) p(\mathbf{y} \mid \boldsymbol{\theta}_{A(1)}, \boldsymbol{\theta}_{A(2)}, A)\, d\nu(\boldsymbol{\theta}_{A(2)})$$

can be evaluated analytically. Then the whole marginal likelihood evaluation problem may be cast in terms of $\boldsymbol{\theta}_{A(1)}$ rather than $\boldsymbol{\theta}_{A(2)}$.

Example 8.2.4 Evaluating the Marginal Likelihood in the Probit Model The posterior simulator constructed in Section 6.2 has two blocks: $\boldsymbol{\beta}$, the $k \times 1$ vector of covariate coefficients, and $\widetilde{\mathbf{y}}$, the $T \times 1$ vector of latent probits. A direct application of the density ratio method of marginal likelihood approximation would require tailoring $f(\cdot)$ to approximate the distribution of the $(k+T) \times 1$ vector $(\boldsymbol{\beta}', \widetilde{\mathbf{y}}')'$. For large T (applications with $T > 1000$ are common), this becomes impractical. Instead, we may exploit the essentially closed-form representation

$$\begin{aligned}
\int_{\mathbb{R}^T} p(\mathbf{y}^o \mid \boldsymbol{\beta}, X, A) p(\mathbf{y}^o \mid \widetilde{\mathbf{y}}, \boldsymbol{\beta}, \mathbf{X}, A)\, d\widetilde{\mathbf{y}} &= \int_{\mathbb{R}^T} p(\mathbf{y}^o \mid \boldsymbol{\beta}, \mathbf{X}, A) p(\mathbf{y}^o \mid \widetilde{\mathbf{y}}, A)\, d\widetilde{\mathbf{y}} \\
&= p(\mathbf{y}^o \mid \boldsymbol{\beta}, \mathbf{X}, A) = \prod_{t=1}^{T} p(y_t^o \mid \boldsymbol{\beta}, \mathbf{x}_t, A) \\
&= \prod_{t=1}^{T} \Phi(-\boldsymbol{\beta}'\mathbf{x}_t)^{(1-y_t^o)} \Phi(\boldsymbol{\beta}'\mathbf{x}_t)^{y_t^o} \qquad (8.23)
\end{aligned}$$

where $y_t^o = 0$ for the first outcome and $y_t^o = 1$ for the second. (The univariate normal cdf Φ cannot be expressed in closed form but can be computed to machine accuracy very rapidly.) We record the product of (8.23) and $p(\boldsymbol{\beta}^{(m)} \mid A)$, along with $\boldsymbol{\beta}^{(m)}$, each iteration, and then apply the density ratio approximation method over the k-dimensional space.

Exercise 8.2.1 Completing the Argument Derive (8.19).

Exercise 8.2.2 Bayes Factor with Common Likelihoods Example 5.1.1 employed a conditionally conjugate prior distribution in which $\boldsymbol{\beta} \sim N(\underline{\boldsymbol{\beta}}, \underline{\mathbf{H}}^{-1})$. Beyond the original specification of the prior, the example considered a "weak" prior that reduced $\underline{\mathbf{H}}$ by a factor of 5, and a "strong" prior that increased it by a factor of 5.

(a) Letting A_1 denote the model with the original prior and A_2 the model with the strong prior, use the method of Section 8.2.1 to approximate the logarithm of the Bayes factor in favor of the model with the strong prior. Find the numerical standard error of approximation, and verify that the approximation is consistent with the exact result reported in Example 5.1.1.

(b) Repeat part (a), but reverse the roles of A_1 and A_2. Use BACC to compute the numerical standard error of the logarithm of the Bayes factor. Then, increase the number of iterations used by a factor of 10, and again compute the numerical standard error. Discuss.

Exercise 8.2.3 Marginal Likelihood Approximation Using Importance Sampling Consider the importance sampling algorithm for the stationary first-order Markov finite state model described in Section 7.2.2, and let $w(\mathbf{P}^{(m)}) = \prod_{j=1}^{m} \pi_j(\mathbf{P}^{(m)})^{n_j^o}$ denote the weight at iteration m.

(a) Show that an almost sure limit $\overline{w}$ of $M^{-1} \sum_{m=1}^{M} w(\mathbf{P}^{(m)})$ exists.

(b) Derive a closed-form expression for the marginal likelihood of the stationary first-order Markov finite state model, expressed in terms of $\overline{w}$, the prior hyperparameters a_{ij}, and the sufficient statistics n_{ij}^o.

Exercise 8.2.4 Marginal Likelihood Approximation Using Gibbs Sampling This exercise is an extension of Example 5.1.1. Use the method of Section 8.2.3 to approximate the log marginal likelihood of the model in Example 5.1.1, and find the numerical standard error of this approximation. Verify that the result is consistent with the exact value reported in Example 5.1.1, and that its numerical standard error is smaller than the numerical standard error of .0030 for the density ratio approximation of the log marginal likelihood reported in that example.

Exercise 8.2.5 Inference about Shape Exercise 5.4.3(b) found the posterior probability that the shape constraints were valid over the range of the data (11.4 to 27), conditional on the regression function for the student : teacher ratio being polynomials of order 2, 3, 4 or 5. Consider, instead, models that constrain the function to have this shape by appropriate truncation of the prior distribution used in Example 5.4.3 and Exercise 5.4.3(b). Find the posterior odds ratio in favor of this model, versus the model of Example 5.4.3 and Exercise 5.4.3(b). Contrast the result with the posterior probabilities found in Exercise 5.4.3(b). (*Hint*: Consider simulating from the prior distribution of Example 5.4.3 in order to find the required normalizing constant for the prior density in the new model.)

8.3 MODEL SPECIFICATION

Creating a complete model entails specification of the prior density of unobservables

$$p(\boldsymbol{\theta}_A \mid A), \tag{8.24}$$

the conditional density of observables

$$p(\mathbf{y} \mid \boldsymbol{\theta}_A, A), \tag{8.25}$$

and the composition of a vector of interest $\boldsymbol{\omega}$ and its conditional density

$$p(\boldsymbol{\omega} \mid \boldsymbol{\theta}_A, \mathbf{y}, A). \tag{8.26}$$

To this point we have taken as given all of these distributions. Yet a sensible specification of all three is essential to the investigator's task of informing and improving decisions. To a substantial extent, success in this endeavor arises from the investigator's creativity, experience, skill, and understanding of the client's situation. These characteristics cannot be endowed analytically. However, the investigator can employ certain systematic procedures that are very useful tasks in the process of creating and improving models. These procedures are known collectively as *Bayesian specification analysis*.

It is useful to distinguish between two kinds of specification analysis. *Prior predictive analysis* takes place when the investigator is considering alternative variants of (8.24)–(8.26). It requires only forward simulation. *Posterior predictive analysis* takes place after the investigator has conditioned on the data $\mathbf{y}^o$ and is considering possible changes in the complete model. It requires a posterior simulator. We distinguish between prior and posterior specification analysis in part because the former is less costly than the latter; if a prior predictive specification analysis indicates serious problems, it may be prudent to consider alternatives to (8.24), (8.25), and (8.26), before undertaking full implementation of the model as outlined in Section 8.1. There is a substantial literature on both of these approaches. The seminal work of Box (1980), and the comments of discussants published with that paper, still provide deep and useful perspectives on specification analysis. For a similar more recent symposium, see Bayarri and Berger (1999) and their discussants.

8.3.1 Prior Predictive Analysis

The objective of prior predictive analysis is to ascertain the prior distribution of functions of the form $h(\mathbf{y}, \boldsymbol{\omega})$ that are interesting and relevant to the problem. This can be accomplished through the forward simulation

$$\boldsymbol{\theta}_A^{(m)} \sim p(\boldsymbol{\theta}_A \mid A), \tag{8.27}$$

$$\mathbf{y}^{(m)} \sim p(\mathbf{y} \mid \boldsymbol{\theta}_A^{(m)}, A), \tag{8.28}$$

$$\boldsymbol{\omega}^{(m)} \sim p(\boldsymbol{\omega} \mid \boldsymbol{\theta}_A^{(m)}, \mathbf{y}^{(m)}, A), \tag{8.29}$$

$$h^{(m)} = h(\mathbf{y}^{(m)}, \boldsymbol{\omega}^{(m)}). \tag{8.30}$$

The forward simulation of h may reveal deficiencies in the model, in the sense that the distribution of $h(\mathbf{y}, \boldsymbol{\omega})$ coincides poorly with beliefs about this function. For instance, in the value at risk example, the function h might be the sample correlation coefficient between squared returns on successive days over a hypothetical sample of (say) 100 days' duration. If the prior distribution of h places probability .99 on the interval $(-.01, .01)$, most clients would regard the model as an unreliable basis on which to assess value at risk, given the well-documented persistence in squared returns on financial assets. They would hesitate to proceed with such a model, because it does not reflect their prior beliefs. More generally, prior predictive analysis interprets the model specification—both the distribution of observables conditional on parameters, and the prior distribution of parameters—in terms of observables and vectors of interest that are usually easier to understand than the parameters themselves.

If the prior predictive analysis reveals deficiencies, then the investigator may revise (8.24), (8.25), or (8.26), or may decide to begin anew with a completely different specification. An investigator with strong insights into the substance of alternative models and the beliefs of clients may find that the prior distribution of h provides useful clues in revising the model or constructing a new one.

Example 8.3.1 Prior Predictive Analysis in the Earnings Example Recall the distribution of earnings conditional on age and education studied in Example 6.4.1 using the normal mixture linear model. In that model the disturbances $\varepsilon_t = y_t - \boldsymbol{\beta}'\mathbf{x}_t$ are independent and identically distributed. Conditional on a latent state $\widetilde{s}_t = j$, $\varepsilon_t \sim N[\alpha_j, (h \cdot h_j)^{-1}]$ $(j = 1, \ldots, m)$. (Section 6.4.2 provides the complete specification of the model.) The prior distribution in Example 6.4.1 had five independent components:

$$\mathbf{X}\boldsymbol{\beta} \sim N[10.5\boldsymbol{\iota}_T, (0.7^2 T)\mathbf{I}_T], \quad (8/3)h \sim \chi^2(10),$$
$$\alpha_j \sim N(0, 2.5), \quad 3h_j \sim \chi^2(3), \quad \mathbf{p} \sim \text{Dirichlet } (1, \ldots, 1). \tag{8.31}$$

Example 6.4.1 considered the cases of mixtures of $m = 2$ and $m = 3$ normal distributions. Since the Bayes factor favored the mixture of three normals, this example pertains to that case.

The prior distribution (8.31) can be understood through its implications for observables $\mathbf{y}$ by means of various functions $h(\mathbf{y})$. The observables we consider here are the log earnings in a population of T men with the same ages and levels of education as in the sample—that is, the $T \times 1$ random vector $\mathbf{y}$. Functions $h(\mathbf{y})$ can be chosen to summarize various aspects of $\mathbf{y}$. Thus the forward simulation (8.27)–(8.30) amounts to drawing a set of parameters from the prior distribution and then, using the $T \times k$ covariate matrix $\mathbf{X}$, simulating values of the corresponding $T \times 1$ log earnings vector $\mathbf{y}$, and then finding the corresponding functions $h(\mathbf{y})$.

These functions $h(\mathbf{y})$ can be chosen to study the implications of the prior distribution for the regression of log wages on age and education. One such function is the difference between the average log earnings of college graduates ($e = 16$) and high school graduates ($e = 12$) in a population of men defined by the covariates in the sample; of course, these men do not all have the same age, and the distribution of ages within each of the two groups tends to differ. Another is the difference between the average log earnings of high school graduates and men with 8 years of education. Turning to systematic differences by age, one function is the difference between the average log earnings of men age 45 and those age 25, while another is the difference between the average log earnings of men age 60 and those age 45.

The prior distribution expresses the variety of conditional distributions of log earnings, as well as conditional means, that are plausible. We cannot observe these conditional distributions, but we can observe closely related sample counterparts. A sample counterpart of the population conditional distribution coefficient of skewness, for example, is the sample skewness coefficient of the least-squares residuals (LSRs) in the population. Construction of this function $h$amounts to drawing parameters from the prior distribution, and then for each drawing simulating the log earnings sample $\mathbf{y}$, computing the least-squares residuals, and forming the sample skewness coefficient of the residuals. Similar exercises can be conducted for the coefficient of kurtosis and for the inequality measures G, P, and R defined in Example 6.4.1.

The results of this prior predictive analysis can be summarized in several ways. One useful summary is the following table, which indicates the observed value $h(\mathbf{y}^o)$, together with the fraction of 2000 draws from the prior distribution that were less than $h(\mathbf{y}^o)$ for each function h. For each function h a centered 95% prior credible interval includes the observed value $h(\mathbf{y}^o)$. In this sense, the prior distribution accommodates the characteristics of the data well:

	$h(\mathbf{y}^o)$	$P^{-1}[h(\mathbf{y}^o) \mid A]$
Average($y \mid e = 12$) − average($y \mid e = 8$)	0.610	0.587
Average($y \mid e = 16$) − average($y \mid e = 12$)	0.491	0.645
Average($y \mid a = 45$) − average($y \mid a = 30$)	0.459	0.579
Average($y \mid a = 60$) − average($y \mid a = 45$)	−0.242	0.478
Variance of LSRs	0.539	0.289
Coefficient of skewness, LSRs	−0.974	0.061
Coefficient of kurtosis, LSRs	6.220	0.904
Gini coefficient, LSRs	0.346	0.186
Proportion below half median, LSRs	0.148	0.309
Earnings fraction to highest decile, LSRs	0.266	0.223

The summary in the foregoing table proceeds one dimension at a time. We can work two dimensions at a time, as illustrated in Figure 8.1. Panels (a) and (b) indicate that the prior distribution of observed systematic differences in earnings by

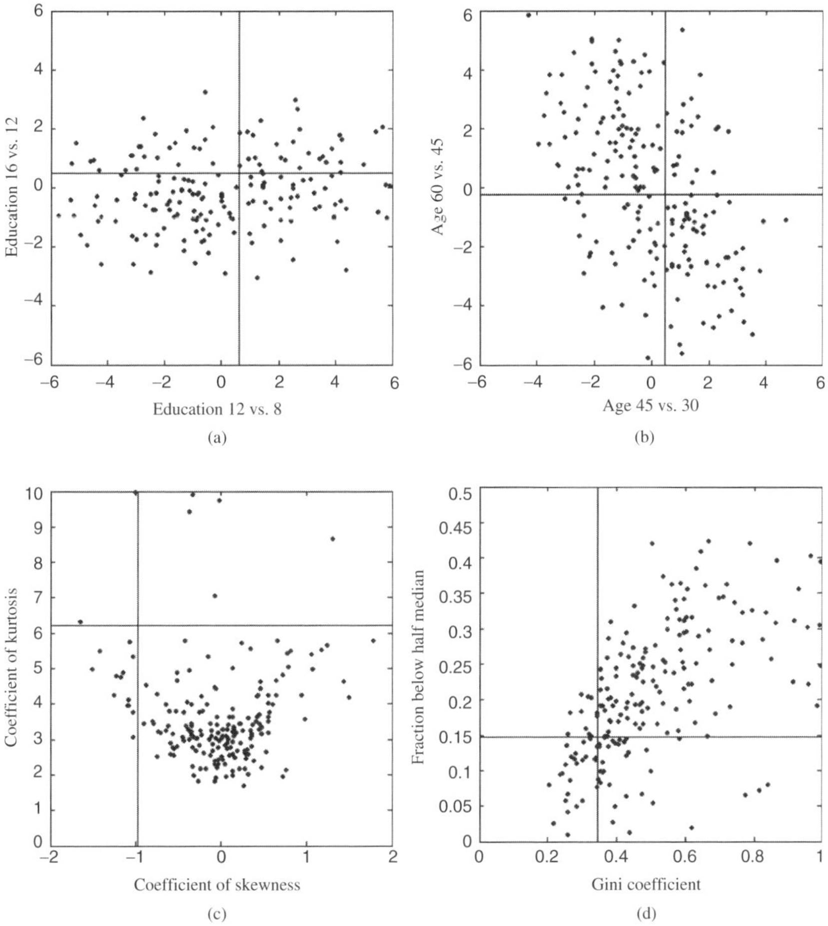

Figure 8.1. In each panel the scatterplots show values of two functions of 200 independent draws from the prior—the observed value is the intersection of the horizontal and vertical lines: (a,b) differences in sample average log earnings; (c) moments from least-squares residuals; (d) inequality measures from least-squares residuals.

age and education is extremely diffuse. They show that the polynomial specification is sufficiently flexible to simultaneously capture the observed difference in the earnings of college and high school graduates and the observed difference in the earnings of high school graduates and those completing eighth grade, in the sample, for example. What is important, here, as well as more generally, is that (8.24) and (8.25) do not appear to systematically exclude any plausible combinations.

The prior distribution of the coefficients of skewness and kurtosis in the least-squares residuals indicates that the model is flexible in some ways; for example, it permits both platykurtic and leptokurtic distributions, including some that are highly leptokurtic. However it provides much less support at the observed values than it does at values that are closer to the moments of a normal distribution. Panel (d) of Figure 8.1 is based on the exponentiated least-squares regression of log wages on the polynomial in age and education. As the Gini coefficient approaches zero, the fraction of men with earnings below half the median must also approach zero, but as the Gini coefficient increases, the prior allows substantial flexibility in the distribution of earnings. The distribution is diffuse but appears to provide greatest support to points near the observed values.

The preceding example illustrates how an investigator makes sense of a complex model, including the prior distribution (8.31). The components of the model have meaning taken together, but not in isolation. The objective of prior predictive analysis is to summarize the model by means of functions $h(\mathbf{y}, \boldsymbol{\omega})$ to which the client can assign prior probability directly. [There is a substantial literature about prior elicitation that builds on this fact, e.g., Garthwaite and Dickey (1988), Kadane (1994), and Kadane and Wolfson (1998).] The investigator's task is to create models (8.24) that reproduce the priors on $h(\mathbf{y}, \boldsymbol{\omega})$ in as plausible a fashion as possible, given the choices of (8.24), (8.25), and (8.26) that are feasible. The next example applies this idea to the specification of the Markov normal mixture linear model in Example 7.3.2.

Example 8.3.2 Prior Predictive Analysis in the S&P 500 Example Recall Example 7.3.2. In their application of the Markov normal mixture linear model to stock returns, Ryden et al. (1998) were motivated, in part, by the challenge of reproducing some stylized facts about observed returns on financial assets identified by Granger and Ding (1995a, 1995b). These characteristics include the following:

1. Returns y_t are not autocorrelated (except possibly at lag one).
2. The autocorrelation functions of $|y_t|$ and y_t^2 decay slowly. The decay is much slower than the exponential ratio of an autoregressive model, like the one considered in Section 7.1.
3. The correlation of $|y_t|^\theta$ and $|y_{t-1}|^\theta$ is highest when $\theta = 1$.
4. Correlations between $\text{sign}(y_t)$ and $\text{sign}(y_{t-k})$ are negligibly small.
5. The correlation between $|y_t|$ and $\text{sign}(y_t)$ is negligibly small.
6. $|y_t|$ has the same mean and standard deviation.

Note that these properties refer to observed returns, not their population counterparts. They indicate features that would be desirable in complete models of financial returns, and strongly suggest functions of observables that can be used to interpret and evaluate any complete model of financial returns before it is applied. The following table indicates the values of the sample statistics $\mathbf{s}^o = h(\mathbf{y}^o)$ from the sample of Example 7.3.2 ($T = 1700$) and the corresponding point in the prior cdf of $s = h(\mathbf{y})$ using the model and prior distribution described in that example. The evaluation of the prior cdf is based on 1000 draws from the prior distribution, followed by a sample of size $T = 1700$ conditional on the parameters drawn.

Property	Observable $s = h(\mathbf{y})$	s^o	$P(s \leq s^o \mid A)$
1	$\text{corr}(y_t, y_{t-1})$	0.1029	0.34
1	$\text{corr}(y_t, y_{t-2})$	0.0338	0.17
2	$\text{corr}(\|y_t\|, \|y_{t-1}\|)$	0.0531	0.20
2	$\text{corr}(\|y_t\|, \|y_{t-2}\|)$	0.0104	0.07
2	$\text{corr}(\|y_t\|, \|y_{t-5}\|)$	0.1107	0.95
2	$\text{corr}(\|y_t\|, \|y_{t-1}\|)^2/\text{corr}(\|y_t\|, \|y_{t-2}\|)$	0.2716	0.69
2	$\text{corr}(\|y_t\|, \|y_{t-1}\|)^5/\text{corr}(\|y_t\|, \|y_{t-5}\|)$	4×10^{-6}	0.26
3	$\text{corr}(y_t^2, y_{t-1}^2)$	0.0697	0.32
3	$\text{corr}(y_t^2, y_{t-1}^2)/\text{corr}(\|y_t\|, \|y_{t-1}\|)$	1.3134	0.93
4	$\text{corr}[\text{sign}(y_t), \text{sign}(y_{t-1})]$	0.0728	0.23
4	$\text{corr}[\text{sign}(y_t), \text{sign}(y_{t-2})]$	0.0031	0.05
5	$\text{corr}[\|y_t\|, \text{sign}(y_t)]$	0.0114	0.55
6	$\overline{\|y\|}_T = \sum_{t=1}^T \|y_t\| / T$	0.0068	0.40
6	$\text{SD}(\|y_t\|) = \sum_{t=1}^T (\|y_t\| - \overline{\|y\|}_T)^2/(T-1)$	0.0059	0.48
6	$\overline{\|y\|}_T/\text{SD}(\|y_t\|)$	1.1589	0.28
	$\left[\sum_{t=1}^T (y_t - \overline{y}_T)^3/(T-1)\right]/\text{SD}(\|y_t\|)^3$	0.227	0.70
	$\left[\sum_{t=1}^T (y_t - \overline{y}_T)^4/(T-1)\right]/\text{SD}(\|y_t\|)^4$	4.4595	0.70

The complete model accounts well for each statistic, taken in isolation. These results, by themselves, say nothing about the ability of the model to account simultaneously for two or more of these statistics, or of the stylized facts identified by Granger and Ding (1995a, 1995b) and Ryden et al. (1998). That possibility can be investigated using the graphical methods of the previous example.

8.3.2 Posterior Predictive Analysis

Consider the following conceptual experiment. We have an observed outcome $\mathbf{y}^o$ from an experiment that can be repeated. Our vector of interest $\boldsymbol{\omega}$ is observable outcomes in independent repetitions of the same experiment. The observed outcome can be summarized in any one of several ways, typically a scalar function $h(\mathbf{y}^o)$, and the outcomes of repetitions of the experiment can be summarized in the same ways, typically $h(\boldsymbol{\omega})$. The predictive density for these outcomes in repetitions of the experiment is $p[h(\boldsymbol{\omega}) \mid \mathbf{y}^o, A]$. The observed $h(\mathbf{y}^o)$, in the context of this distribution, tells us much about the model A. It may turn out that the observed $h(\mathbf{y}^o)$ is implausible, in the sense that $h(\mathbf{y}^o)$ is not an element of a $100(1-\alpha)\%$ highest posterior density credible set for $h(\boldsymbol{\omega})$ with respect to $p[h(\boldsymbol{\omega}) \mid \mathbf{y}^o, A]$. If the predictive density of $h(\boldsymbol{\omega})$ is unimodal, this is equivalent to $h(\mathbf{y}^o)$ being in the extreme tails of $p[h(\boldsymbol{\omega}) \mid \mathbf{y}^o, A]$. This idea goes back to the notion of "surprise" discussed by Good (1956), and its essentials were further developed by Rubin (1984) in what he termed "model monitoring by posterior predictive checks." This idea builds on that of the probability integral transform stressed by Dawid (1984) in prequential forecasting, and formalized by Meng (1994); see also the comprehensive survey of Gelman et al. (1996) and the more recent work of Bayarri and Berger (1999) and the references cited therein.

This conceptual experiment is the motivation for posterior predictive analysis, whose mechanics are the same as those in (8.27)–(8.30), except that we replace (8.27) with $\boldsymbol{\theta}_A^{(m)} \sim p(\boldsymbol{\theta}_A \mid \mathbf{y}^o, A)$. In many cases the posterior predictive analysis may parallel the prior predictive analysis, and in this case the posterior predictive analysis involves almost no additional effort given the output of the posterior simulator.

Example 8.3.3 Posterior Predictive Analysis in the Earnings Example A posterior predictive analysis of the earnings model of Example 6.4.1, using the same functions $h(\cdot)$ as in the prior predictive analysis of Example 8.3.1, yields the results shown below. The posterior cdf evaluated at the observed values is in the interval (0.05, 0.95) in every case. The four functions based on conditional means of earnings indicate observed values near the median of the posterior cdf in two cases, and in the upper quartile in two others. The six functions based on the least squares residuals, which indicate the ability of the model to capture the scale and shape of the conditional distribution, all yield observed values in the interquartile range of the posterior cdf.

	$h(\mathbf{y}^o)$	$P[h(\boldsymbol{\omega}) \leq h(\mathbf{y}^o) \mid \mathbf{y}^o, A]$
Average($y \mid e = 12$) − average($y \mid e = 8$)	0.610	0.940
Average($y \mid e = 16$) − average($y \mid e = 12$)	0.491	0.474
Average($y \mid a = 45$) − average($y \mid a = 30$)	0.459	0.889
Average($y \mid a = 60$) − average($y \mid a = 45$)	−0.242	0.552
Variance of LSRs	0.539	0.442
Coefficient of skewness, LSRs	−0.974	0.619
Coefficient of kurtosis, LSRs	6.220	0.275
Gini coefficient, LSRs	0.346	0.519
Proportion below half-median, LSRs	0.148	0.418
Earnings fraction to highest decile, LSRs	0.266	0.514

These findings are reinforced in the pairwise posterior predictive analysis in Figure 8.2, parallel to the pairwise prior predictive analysis in Figure 8.1. The scatterplots in panels (c) and (d) of Figure 8.2 indicate that the mixture of three normals model accounts well for non-normality, as reflected in the coefficients of skewness and kurtosis and measures of inequality in the least squares residuals.

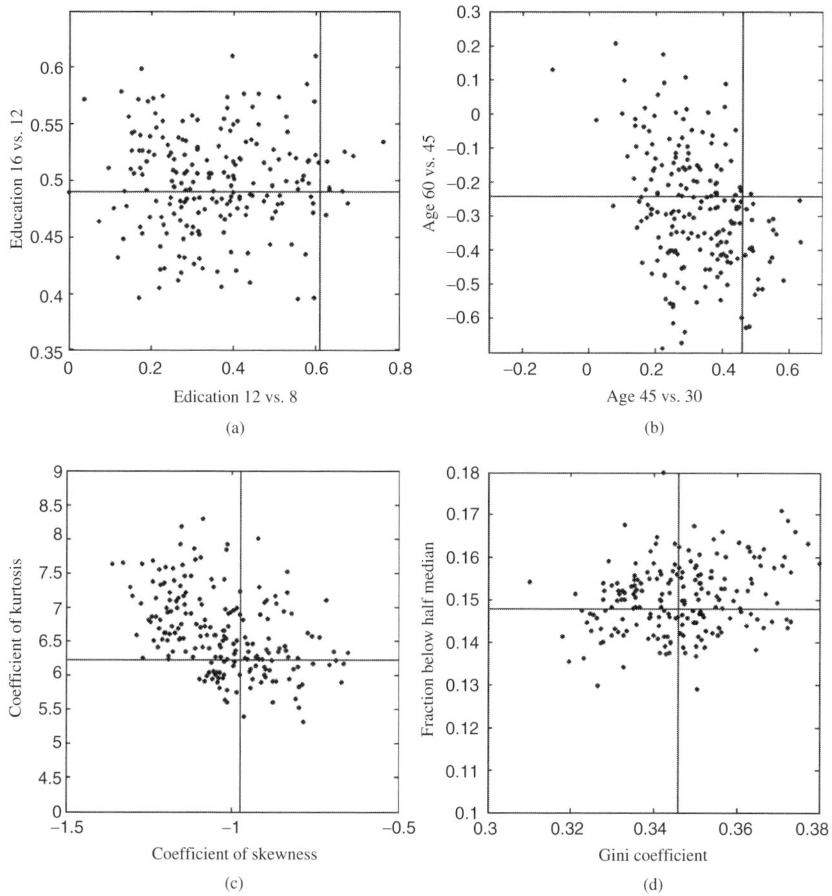

Figure 8.2. In each panel the scatterplots show values of two functions of 200 draws from the posterior—the observed value is the intersection of the horizontal and vertical line: (a, b) difference in sample average log earnings; (c) moments from least-squares residuals; (d) inequality measures from least-squares residuals.

This is consistent with the findings in the foregoing table. The scatterplot in panel (a) reveals that the two returns to education functions are nearly uncorrelated in the posterior distribution, and so this panel adds little to first two lines of the foregoing table. The scatterplot in panel (b) shows that the two returns to age functions are negatively correlated in the posterior distribution. As a consequence, the posterior distribution captures the difference in the returns over these two age ranges, but it is less successful in changes in log earnings between age 30 and age 60.

This example illustrates how Bayesian specification analysis can be used to capture the implications of models for observables. In general, the goal of posterior specification analysis is to highlight inconsistencies between models and observed data, thus increasing our understanding of models and sowing the seeds for the development of better models and improved decisionmaking.

Example 8.3.4 Posterior Predictive Analysis in the S&P 500 Example Repeating the analysis of Example 8.3.2, using the posterior distribution in place of the prior distribution, produces the following results. They are based on the 20,000 iterations of the posterior simulator described in Example 7.3.2.

Property	Observable $s = h(\mathbf{y})$	s^o	$P(s \le s^o \mid \mathbf{y}^o, A)$
1	$\text{corr}(y_t, y_{t-1})$	0.1029	0.964
1	$\text{corr}(y_t, y_{t-2})$	0.0338	0.618
2	$\text{corr}(\lvert y_t\rvert, \lvert y_{t-1}\rvert)$	0.0531	0.141
2	$\text{corr}(\lvert y_t\rvert, \lvert y_{t-2}\rvert)$	0.0104	0.014
2	$\text{corr}(\lvert y_t\rvert, \lvert y_{t-5}\rvert)$	0.1107	0.898
2	$\text{corr}(\lvert y_t\rvert, \lvert y_{t-1}\rvert)^2/\text{corr}(\lvert y_t\rvert, \lvert y_{t-2}\rvert)$	0.2716	0.939
2	$\text{corr}(\lvert y_t\rvert, \lvert y_{t-1}\rvert)^5/\text{corr}(\lvert y_t\rvert, \lvert y_{t-5}\rvert)$	4×10^{-6}	0.107
3	$\text{corr}(y_t^2, y_{t-1}^2)$	0.0697	0.223
3	$\text{corr}(y_t^2, y_{t-1}^2)/\text{corr}(\lvert y_t\rvert, \lvert y_{t-1}\rvert)$	1.3134	0.779
4	$\text{corr}[\text{sign}(y_t), \text{sign}(y_{t-1})]$	0.0728	0.985
4	$\text{corr}[\text{sign}(y_t), \text{sign}(y_{t-2})]$	0.0031	0.392
5	$\text{corr}[\lvert y_t\rvert, \text{sign}(y_t)]$	0.0114	0.670
6	$\overline{\lvert y\rvert}_T = \sum_{t=1}^T \lvert y_t\rvert / T$	0.0068	0.357
6	$\text{SD}(\lvert y_t\rvert) = \sum_{t=1}^T (\lvert y_t\rvert - \overline{\lvert y\rvert}_T)^2/(T-1)$	0.0059	0.627
6	$\overline{\lvert y\rvert}_T/\text{SD}(\lvert y_t\rvert)$	1.1589	0.071
	$\left[\sum_{t=1}^T (y_t - \bar{y}_T)^3/(T-1)\right]/\text{SD}(\lvert y_t\rvert)^3$	0.227	0.781
	$\left[\sum_{t=1}^T (y_t - \bar{y}_T)^4/(T-1)\right]/\text{SD}(\lvert y_t\rvert)^4$	4.4595	0.844

The posterior distribution accounts reasonably well for the properties of interest identified in Example 8.3.2, at least taken one at a time. The most difficult features appear to be the autocorrelation functions of $\lvert y_t\rvert$ and $\text{sign}(y_t)$. Exercise 8.3.3 takes up these and some other possibilities for posterior predictive analysis.

Exercise 8.3.1 Predictive Analyses Suppose that $p[h(\mathbf{y}) = h(\mathbf{y}^o) \mid \mathbf{y}^o, A] = 0$.

(a) At what point in a prior predictive analysis would an investigator become aware of this fact?

(b) At what point in a posterior predictive analysis would an investigator become aware of this fact?

Exercise 8.3.2 Predictive Analyses of the Normal Linear Model for Earnings Repeat the analysis of Examples 8.3.1 and 8.3.3, using a normal distribution in place of the mixture of normals distribution. In how many of the functions $h(\mathbf{y})$ do you find evidence of serious misspecification?

Exercise 8.3.3 Further Predictive Analyses of the S&P 500 Returns Model This exercise entails a more extensive specification analysis of the Markov normal mixture model for the S&P 500 returns.

(a) The analysis in Example 8.3.4 suggested potential difficulties in the distribution of sample autocorrelations of the absolute returns and signs of returns. By modifying the code in the online appendix, use graphical methods to undertake prior and posterior predictive analyses appropriate to each of these features of observed returns.

(b) This model was motivated by the value at risk problem first introduced in Section 1.1.2. Extend the prior and posterior predictive analysis of Examples 8.3.2 and 8.3.4, using a vector of interest $\boldsymbol{\omega}$ suggested by this problem.

Exercise 8.3.4 Predictive Analyses in the Class Size Examples The original work with the Massachusetts test score data (Example 5.1.1) assumed a normal linear regression.

(a) Examples 5.4.1–5.4.4 investigated alternative models in which the regression function is nonlinear in the covariates. Before proceeding to this step, we might wish to perform a posterior predictive analysis, in the context of Example 5.1.1, that would be sensitive to nonlinearity. Carry out such an analysis, using functions $h(\mathbf{y})$ that are sensitive to nonlinearity.

(b) All of Examples 5.1.1 and 5.4.1–5.4.4 assumed that the distribution of the regression disturbances is normal. Design and execute a posterior predictive analysis that provides information on whether this aspect of the model specification appears to be reasonable.

8.4 BAYESIAN COMMUNICATION

An investigator cannot anticipate the uses to which her work will be put, or the variants on her model that may interest a client. Different uses call for different vectors of interest $\boldsymbol{\omega}$. Variants on her model will often revolve around changes in the prior distribution. Yet any investigator who has publicly reported results has confronted the constraint that only a few representative findings can be conveyed in written work.

Posterior simulators provide a clear answer to the question of what the investigator should report, and in the process remove the constraint that only a few representative findings can be communicated. The investigator can provide electronic access to the following $M \times (k+3)$ simulator output matrix:

$$\begin{bmatrix} w(\boldsymbol{\theta}_A^{(1)}) & p(\boldsymbol{\theta}_A^{(1)}) & p(\mathbf{y}^o \mid \boldsymbol{\theta}_A^{(1)}) & \boldsymbol{\theta}_A^{(1)\prime} \\ \vdots & \vdots & \vdots & \vdots \\ w(\boldsymbol{\theta}_A^{(M)}) & p(\boldsymbol{\theta}_A^{(M)}) & p(\mathbf{y}^o \mid \boldsymbol{\theta}_A^{(M)}) & \boldsymbol{\theta}_A^{(M)\prime} \end{bmatrix}.$$

Even very large simulator output matrices can be stored at negligible cost and communicated quickly over the Internet.

Given the simulator output matrix, the client can immediately compute approximations to posterior moments not reported or even considered by the investigator, using spreadsheet arithmetic. For example, a client reading a research report might be skeptical that the investigator's model, prior, and data set provide much information about the effects of an interesting change in a policy variable on the outcome in question. In many cases the client may be able to answer such questions in less time than required to read the paper.

With a small amount of additional effort, the client can modify many of the investigator's assumptions. Suppose that the client wishes to evaluate $E[h(\boldsymbol{\omega}) \mid \mathbf{y}^o, A_2]$ using the same likelihood function as the investigator, but using his own prior density $p(\boldsymbol{\theta}_A \mid A_2)$ in place of the investigator's prior density $p(\boldsymbol{\theta}_A \mid A_1)$. Suppose further that the support of the investigator's prior distribution includes the support of the client's prior. Then the investigator's posterior distribution may be regarded as an importance sampling distribution for the client's posterior. The client reweights the investigator's posterior simulation, using the function

$$w(\boldsymbol{\theta}_A) = \frac{p(\boldsymbol{\theta}_A \mid \mathbf{y}^o, A_2)}{p(\boldsymbol{\theta}_A \mid \mathbf{y}^o, A_1)} = \frac{p(\boldsymbol{\theta}_A \mid A_2)}{p(\boldsymbol{\theta}_A \mid A_1)}. \tag{8.32}$$

The client thus approximates the posterior moment $E[h(\boldsymbol{\omega}) \mid \mathbf{Y}_T^o, A_2]$ by

$$\overline{h}_{A_2}^{(M)} = \frac{\sum_{m=1}^{M} w(\boldsymbol{\theta}_A^{(m)}) w_*(\boldsymbol{\theta}_A^{(m)}) h(\boldsymbol{\omega}^{(m)})}{\sum_{m=1}^{M} w(\boldsymbol{\theta}_A^{(m)}) w_*(\boldsymbol{\theta}_A^{(m)})}. \tag{8.33}$$

In (8.33) $\boldsymbol{\omega}^{(m)}$ is drawn from the conditional density $p(\boldsymbol{\omega} \mid \mathbf{y}^o, \boldsymbol{\theta}_A^{(m)}, A_2)$ independently of any other draws, $w_*(\boldsymbol{\theta}_A)$ is any weighting that accompanies the posterior simulation provided by the investigator, and $w(\cdot)$ is the function in (8.32). If the investigator has employed an importance sampling algorithm, then Theorem 4.2.2 implies $\overline{h}_{A_2}^{(M)} \xrightarrow{\text{a.s.}} E[h(\boldsymbol{\omega}) \mid \mathbf{y}^o, A_2] = \overline{h}_{A_2}$. The same result holds for MCMC algorithms, as formalized in the following result.

Theorem 8.4.1 Simulation-Consistent Approximation of Posterior Moments with a Reweighted Posterior Distribution Suppose that two models, A_1 and A_2, share the same parameter vector $\boldsymbol{\theta}_A$ and observables density $p(\mathbf{y} \mid \boldsymbol{\theta}_A)$, but have respective prior densities $p(\boldsymbol{\theta}_A \mid A_1)$ with $\boldsymbol{\theta}_A \in \Theta_{A_1}$ and $p(\boldsymbol{\theta}_A \mid A_2)$ with $\boldsymbol{\theta}_A \in \Theta_{A_2}$. Suppose $\Theta_{A_2} \subseteq \Theta_{A_1}$ and define

$$w(\boldsymbol{\theta}_A) = p(\boldsymbol{\theta}_A \mid A_2)/p(\boldsymbol{\theta}_A \mid A_1).$$

Let $\{\boldsymbol{\theta}_A^{(m)}\}$ be an ergodic Markov chain with invariant density $p(\boldsymbol{\theta}_A \mid \mathbf{y}^o, A_1)$. Suppose further that the distribution $\boldsymbol{\omega} \mid (\mathbf{y}^o, \boldsymbol{\theta}_A, A)$ is the same in the two models, and that $\boldsymbol{\omega}^{(m)}$ is drawn independently from $p(\boldsymbol{\omega} \mid \mathbf{y}^o, \boldsymbol{\theta}_A^{(m)}, A)$. If $\overline{h}_{A_2} = E[h(\boldsymbol{\omega} \mid \mathbf{y}^o, A_2)]$ exists, then

$$\overline{h}_{A_2}^{(M)} = \sum_{m=1}^{M} w(\boldsymbol{\theta}_A^{(m)})h(\boldsymbol{\omega}_A^{(m)}) / \sum_{m=1}^{M} w(\boldsymbol{\theta}_A^{(m)}) \xrightarrow{\text{a.s.}} \overline{h}_{A_2}.$$

Proof: From Theorem 4.5.2 $\{\boldsymbol{\theta}_A^{(m)}, \boldsymbol{\omega}^{(m)}\}$ is ergodic with invariant density

$$p(\boldsymbol{\theta}_A \mid \mathbf{y}^o, A_1)p(\boldsymbol{\omega} \mid \mathbf{y}^o, \boldsymbol{\theta}_A, A).$$

Since, as shown in Section 8.2.1

$$E[w(\boldsymbol{\theta}_A) \mid \mathbf{y}^o, A_1] = p(\mathbf{y}^o \mid A_2)/p(\mathbf{y}^o \mid A_1)$$

and, by the same argument

$$E[w(\boldsymbol{\theta}_A)h(\boldsymbol{\omega}_A) \mid \mathbf{y}^o, A_1] = [p(\mathbf{y}^o \mid A_2)/p(\mathbf{y}^o \mid A_1)]\overline{h}_{A_2}, \tag{8.34}$$

the result follows. ∎

Note that the prior densities in $w(\boldsymbol{\theta}_A) = p(\boldsymbol{\theta}_A \mid A_2)/p(\boldsymbol{\theta}_A \mid A_1)$ can be replaced with arbitrary kernels in this result.

In Theorem 4.7.1 uniform ergodicity was one of the sufficient conditions for a central limit theorem. If the investigator's algorithm produces uniformly ergodic $\{\boldsymbol{\theta}_A^{(m)}\}$, and if the ratio of the client's prior density to the investigator's prior density is bounded, then there is a central limit theorem under the client's prior as well, as long as the client's function of interest has finite posterior variance using this prior. This condition is strikingly similar to the sufficient condition discussed below Theorem 4.2.2. This is not surprising; the client is using the investigator's posterior as his importance sampling distribution.

Theorem 8.4.2 A Central Limit Theorem for Weighted MCMC Simulators Given the notation and assumptions of Theorem 8.4.1, suppose also that $\{\boldsymbol{\theta}_A^{(m)}\}$ is uniformly ergodic, that $\text{var}[h(\boldsymbol{\omega}) \mid \mathbf{Y}_T^o, A_2]$ exists and is finite, and $w(\boldsymbol{\theta}_A) \leq \overline{w} < \infty$ $\forall\ \boldsymbol{\theta}_A \in \Theta_{A_2}$. Then there exists $\tau^2 > 0$ such that

$$M^{1/2}(\overline{h}_{A_2}^{(M)} - \overline{h}_{A_2}) \xrightarrow{d} N(0, \tau^2).$$

Proof: Arguing exactly as in the proof of Theorem 4.5.2, uniform ergodicity of $\{\boldsymbol{\theta}_A^{(m)}\}$ implies uniform ergodicity of $\{\boldsymbol{\theta}_A^{(m)}, \boldsymbol{\omega}^{(m)}\}$. Let λ_1 and λ_2 be any two real numbers, not both 0. Define $q_{\boldsymbol{\lambda}}(\boldsymbol{\theta}_A, \boldsymbol{\omega}) = \lambda_1 w(\boldsymbol{\theta}_A) + \lambda_2 w(\boldsymbol{\theta}_A) h(\boldsymbol{\omega})$, and observe that $E[q_{\boldsymbol{\lambda}}(\boldsymbol{\theta}_A, \boldsymbol{\omega}) \mid \mathbf{y}^o, A_1] = \lambda_1 \overline{w} + \lambda_2 \overline{wh}_{A_2} = \overline{q}_{\boldsymbol{\lambda}}$ exists and is finite. From Theorem 4.7.1, we obtain

$$M^{1/2}\left[M^{-1}\sum_{m=1}^{M} q_{\boldsymbol{\lambda}}(\boldsymbol{\theta}_A^{(m)}, \boldsymbol{\omega}^{(m)}) - \overline{q}_{\boldsymbol{\lambda}}\right] \xrightarrow{d} N(0, \tau_{\boldsymbol{\lambda}}^2).$$

Since this is true for all $\boldsymbol{\lambda} \neq \mathbf{0}$, it follows [see Rao (1965), Theorem 2c.5(iv)] that

$$M^{1/2}\left\{\begin{bmatrix} M^{-1}\sum_{m=1}^{M} w(\boldsymbol{\theta}_A^{(m)}) h(\boldsymbol{\omega}^{(m)}) \\ M^{-1}\sum_{m=1}^{M} w(\boldsymbol{\theta}_A^{(m)}) \end{bmatrix} - \begin{pmatrix} \overline{wh}_{A_2} \\ \overline{w} \end{pmatrix}\right\} \xrightarrow{d} N(\mathbf{0}, \mathbf{V}).$$

A standard application of the delta method, as in the proof of Theorem 4.2.2, yields the result. ■

Approximating the numerical accuracy of $\overline{h}^{(M)}$ is especially important when the reweighting method of Theorem 8.4.1 is used. An ill-behaved weighting function $w(\boldsymbol{\theta}_A)$ will lead to poor approximation of $\overline{h}_{A_2}$. If this is the case, it is important that the client be aware of this fact. This problem can be approached in a fashion similar to the methods employed in Theorems 4.7.1 and 4.7.3.

Theorem 8.4.3 Variance of the Approximation Error in Weighted MCMC Simulators In addition to the assumptions of Theorem 8.4.2, suppose that

$$\mathbf{z}_m' = [w(\boldsymbol{\theta}_A^{(m)}), w(\boldsymbol{\theta}_A^{(m)}) \cdot h(\boldsymbol{\omega}^{(m)})]$$

is a stationary process with autocovariance function $\mathbf{C}_j = \operatorname{cov}(\mathbf{z}_m, \mathbf{z}_{m+j})$ and spectral density function $\mathbf{S}_{\mathbf{z}}(\lambda) = \sum_{j=-\infty}^{\infty} \mathbf{C}_j \exp(-i\lambda j)$, and that the eigenvalues of $\mathbf{S}_{\mathbf{z}}(\lambda)$ are bounded uniformly both above and away from zero. Then

$$\tau^2 = r^{-2}[\mathbf{S}_{\mathbf{z}_{22}}(0) - 2\overline{h}_{A_2}\mathbf{S}_{\mathbf{z}_{12}}(0) + \overline{h}_{A_2}^{2}\mathbf{S}_{\mathbf{z}_{11}}(0)],$$

where $r = p(\mathbf{y}^o \mid A_2)/p(\mathbf{y}^o \mid A_1)$.

Proof: Straightforward application of the delta method. ■

Corollary 8.4.1 Numerical Standard Errors for Weighted MCMC Simulators In addition to the assumptions of Theorem 8.4.3, suppose also that $\widehat{\mathbf{S}}_{\mathbf{z}}(0) \xrightarrow{\text{a.s.}} \mathbf{S}_{\mathbf{z}}(0)$. Then

$$\widehat{\tau}^{2(M)} = (\overline{w}^{(M)})^{-2}[\widehat{\mathbf{S}}_{\mathbf{z}_{22}}(0) - 2\overline{h}_{A_2}^{(M)}\widehat{\mathbf{S}}_{\mathbf{z}_{12}}(0) + \overline{h}_{A_2}^{2(M)}\widehat{\mathbf{S}}_{\mathbf{z}_{11}}(0)] \xrightarrow{\text{a.s.}} \tau^2,$$

where $\overline{w}^{(M)} = M^{-1}\sum_{m=1}^{M} w(\boldsymbol{\theta}_A^{(m)})$.

The numerical standard error of $\overline{h}_{A_2}^{(M)}$ is then $(\widehat{\tau}^{2(M)}/M)^{1/2}$, whose computation is guided by the following result.

Theorem 8.4.4 Simulation Consistency of Numerical Standard Errors for Weighted MCMC Simulators Define $\{\mathbf{z}_m\}$ as in Theorem 8.4.3. Let $\overline{\mathbf{z}}^{(M)} = M^{-1}\sum_{m=1}^{M}\mathbf{z}_m$ and

$$\mathbf{C}_j^{(M)} = M^{-1}\sum_{m=j+1}^{M}(\mathbf{z}_m - \overline{\mathbf{z}}^{(M)})(\mathbf{z}_{m-j} - \overline{\mathbf{z}}^{(M)})' \; (j = 0, 1, \ldots).$$

Let $L(M)$ satisfy the same conditions as in Theorem 4.7.3. Then

$$\widehat{\mathbf{S}}_{\mathbf{z}}(0) = \mathbf{C}_0^{(M)} + \sum_{s=1}^{L-1}[(L-s)/L]\cdot(\mathbf{C}_s^{(M)} + \mathbf{C}_s^{(M)\prime}) \stackrel{\text{a.s.}}{\rightarrow} \mathbf{S}_{\mathbf{z}}(0).$$

Proof: See Newey and West (1987). ■

BACC permits arbitrary weighting in the computation of posterior moments. Of course, this may also be accomplished with standard mathematical applications software or even spreadsheet arithmetic. However, BACC also utilizes Theorems 8.4.2–8.4.4 to provide numerical standard errors, using either default values of the tapering function $L(M)$ of Theorem 8.4.4, or values of L/M specified by the user.

Example 8.4.1 Bayesian Communication in the Class Size Example (The online appendix contains data, annotated code, and output for this example.) To illustrate the process of Bayesian communication about priors and posterior moments outlined in this section, return to the normal linear model of Example 5.1.1. Suppose that the client's prior is the original prior distribution set out in that example. The investigator, however, has used a prior distribution of the coefficient vector $\boldsymbol{\beta}$ with prior precision $\underline{\mathbf{H}}$ that is 5 times larger than the precision in the client's prior distribution. Using the algorithm of Example 4.3.1 with $M = 100{,}000$ iterations, the investigator, if asked, would report prior and posterior moments as follows:

Investigator's Model: Prior and Posterior Moments, Numerical Accuracy

	Prior		Posterior			
Coefficient	Mean	SD	Mean	SD	NSE	RNE
`intercept`	710	268.84	682.42	10.52	.0329	1.02
`str` coefficient	0	6.7447	−0.6853	0.2644	.00085	1.07
`el` coefficient	0	7.1414	−0.4098	0.2800	.00079	1.25
`lunch` coefficient	0	1.7321	−0.5190	0.0678	.00020	1.12
`income` coefficient	0	75.620	16.517	2.955	.00091	1.04

In these results "NSE" is the numerical standard error computed according to Theorem 4.7.3, with $L/M = 0.02$, and "RNE" is the relative numerical efficiency of the approximation: the ratio of the posterior variance to the product of M and the squared numerical standard error. If RNE is close to 1, then the efficiency of the MCMC algorithm is close to that of a hypothetical simulator making i.i.d. drawings from the posterior distribution itself. That is the case here because there is very little serial correlation in the algorithm of Example 4.3.1.

If the client were to execute the algorithm of Example 4.3.1 with his prior—that is, to repeat the exercise reported in Example 5.1.1—then the client, if asked, would report prior and posterior moments as follows:

Client's Model: Prior and Posterior Moments, Numerical Accuracy

	Prior		Posterior			
Coefficient	Mean	SD	Mean	SD	NSE	RNE
`intercept`	710	120.23	682.61	10.49	.0310	1.15
`str` coefficient	0	3.1063	−0.6757	0.2635	.00081	1.07
`el` coefficient	0	3.1937	−0.3992	0.2783	.00086	1.04
`lunch` coefficient	0	0.7746	−0.5098	0.0675	.00020	1.12
`income` coefficient	0	33.818	16.415	2.956	.00097	0.93

The posterior moments are not exactly the same as those in Example 5.1.1, because the client has used a different seed for the random-number generator. Alternatively, the client might choose to reweight the simulations of the investigator. Denoting the investigator's prior precision matrix by $\underline{\mathbf{H}}_1$ and the client's prior precision matrix by $\underline{\mathbf{H}}_2$, the client will employ the weighting function

$$w(\boldsymbol{\theta}_A) = w(\boldsymbol{\beta}) = \exp[-(\boldsymbol{\beta} - \underline{\boldsymbol{\beta}})'(\underline{\mathbf{H}}_2 - \underline{\mathbf{H}}_1)(\boldsymbol{\beta} - \underline{\boldsymbol{\beta}})/2].$$

Then the client, if asked, would report prior and posterior moments as follows:

Reweighted Model: Prior and Posterior Moments, Numerical Accuracy; Comparison

	Prior		Posterior				Test
Coefficient	Mean	SD	Mean	SD	NSE	RNE	p
`intercept`	710	120.23	682.57	10.50	.0382	0.76	0.383
`str` coefficient	0	3.1063	−0.6751	0.2638	.00092	0.96	0.663
`el` coefficient	0	3.1937	−0.4009	0.2795	.00090	0.96	0.192
`lunch` coefficient	0	0.7746	−0.5098	0.0677	.00022	0.94	0.958
`income` coefficient	0	33.818	16.429	2.949	.00106	0.78	0.328

The rightmost column reports the outcome of a formal test that the posterior mean when the investigator's simulations are reweighted is the same as the posterior mean in the client's simulation, for each of the five coefficients. We expect the p values to be uniformly distributed in the unit interval, and these results are consistent with that hypothesis. There is a loss in efficiency in reweighting the investigator's simulations, apparent in RNE values that are somewhat (but not much) lower than those that result if the client carries out the simulation directly.

Exercise 8.4.1 Completing the Argument Derive (8.34).

Exercise 8.4.2 An informal examination of the three sets of moments reported in Example 8.4.1 might suggest that although the moments in the reweighted model are quite similar to those in the client's model, they are not that different from those in the investigator's model, either. Test this conjecture formally.

Exercise 8.4.3 Recall that both the "weak" and "strong" priors in Example 5.1.1 produced marginal likelihood values that were lower than that of the original prior.

(a) If the investigator were to steadily decrease the precision of her prior, in Example 8.4.1, the corresponding marginal likelihood would also decrease. Pick an even lower precision than the investigator used in Example 8.4.1, and reweight the output of the investigator's simulator to obtain posterior moments of the coefficients under the client's posterior. Verify that the quality of the approximation of the client's posterior moments is essentially as effective as it was in Example 8.4.1.

(b) Using the investigator's posterior from Example 8.4.1, reweight to obtain posterior moments under the "strong" prior of Example 5.1.1. Examine the relative numerical efficiency of the approximation.

(c) Repeat part (b), but substitute the original prior of Example 5.1.1 for the investigator's prior. Examine the relative numerical efficiency, and explain why it is not much higher than that in part (b).

8.5 DENSITY RATIO ROBUSTNESS BOUNDS

An alternative to providing the posterior simulation matrix is to report a range of posterior moments, corresponding to all possible prior distributions within a specified class of distributions. One such class is the density ratio class, discussed in Section 3.3. It consists of all prior density kernels $k(\boldsymbol{\theta}_A \mid A)$ bounded above and below by kernels $a(\boldsymbol{\theta}_A)$ and $b(\boldsymbol{\theta}_A)$. Corresponding to each posterior moment of interest, $E[g(\boldsymbol{\theta}_A) \mid \mathbf{y}^o, A]$, the investigator reports the smallest and largest possible values over all prior density kernels satisfying $0 \le a(\boldsymbol{\theta}_A) \le k(\boldsymbol{\theta}_A \mid A) \le b(\boldsymbol{\theta}_A)$ $\forall$ $\boldsymbol{\theta}_A \in \Theta_A$.

It turns out that it is easy to approximate these bounds using the output of a posterior simulator. The details of the simulation algorithm are unimportant; the method is the same whether the simulator utilizes a variant of MCMC, importance sampling, or any other method that provides a weighted or unweighted sample from the posterior distribution. This is a substantial practical attraction of the density ratio class, because the same procedures and software can be used for all posterior simulation methods. It is necessary only to know that $E[g(\boldsymbol{\theta}_A) \mid \mathbf{y}^o, A]$ exists, and that for simulator output $\{\boldsymbol{\theta}_A^{(m)}\}$, function of interest $g(\boldsymbol{\theta}_A)$ and known weighting function $w(\boldsymbol{\theta}_A)$

$$\frac{\sum_{m=1}^{M} w(\boldsymbol{\theta}_A^{(m)})p(\mathbf{y}^o \mid \boldsymbol{\theta}_A^{(m)}, A)k(\boldsymbol{\theta}_A^{(m)} \mid A)g(\boldsymbol{\theta}_A^{(m)})}{\sum_{m=1}^{M} w(\boldsymbol{\theta}_A^{(m)})p(\mathbf{y}^o \mid \boldsymbol{\theta}_A^{(m)}, A)k(\boldsymbol{\theta}_A^{(m)} \mid A)} \overset{\text{a.s.}}{\to} \frac{\int_{\Theta_A} p(\mathbf{y}^o \mid \boldsymbol{\theta}_A, A)k(\boldsymbol{\theta}_A \mid A)g(\boldsymbol{\theta}_A)\, d\boldsymbol{\theta}_A}{\int_{\Theta_A} p(\mathbf{y}^o \mid \boldsymbol{\theta}_A, A)k(\boldsymbol{\theta}_A \mid A)\, d\boldsymbol{\theta}_A} \tag{8.35}$$

for all prior density kernels $k(\boldsymbol{\theta}_A \mid A)$ in the density ratio class. As indicated in Section 3.3, the problem of finding the lower bound $\underline{E}[g(\boldsymbol{\theta}_A) \mid \mathbf{y}^o, A]$ on the posterior moment is the same as finding the upper bound $\overline{E}[g(\boldsymbol{\theta}_A) \mid \mathbf{y}^o, A]$, and so this section concentrates on the latter question.

To simplify the notation, let $g_m = g(\boldsymbol{\theta}_A^{(m)})$ $(m = 1, \ldots, M)$, and suppose that the simulator output is ordered so that g_m is monotone nondecreasing. Let $\boldsymbol{\iota}_M$ denote an $M \times 1$ vector of ones and consider approximating $\overline{E}[g(\boldsymbol{\theta}_A) \mid \mathbf{y}^o, A]$ by maximizing

$$Q(\mathbf{r}) = \mathbf{g}'\mathbf{r}/\boldsymbol{\iota}_M'\mathbf{r} = \sum_{m=1}^{M} g_m r_m \Big/ \sum_{m=1}^{M} r_m$$

subject to $\mathbf{u} \le \mathbf{r} \le \mathbf{v}$, where $u_m = w(\boldsymbol{\theta}_A^{(m)})a(\boldsymbol{\theta}_A^{(m)})/k(\boldsymbol{\theta}_A^{(m)} \mid S)$ and $v_m = w(\boldsymbol{\theta}_A^{(m)})$ $b(\boldsymbol{\theta}_A^{(m)})/k(\boldsymbol{\theta}_A^{(m)} \mid S)$ $(m = 1, \ldots, M)$ and where $k(\boldsymbol{\theta}_A \mid S)$ is the prior density kernel used in the simulation. We first develop an efficient solution of this discrete maximization problem, and then show that the solution provides a simulation-consistent approximation of $\overline{E}[g(\boldsymbol{\theta}_A) \mid \mathbf{y}^o, A]$.

The solution of the discrete problem parallels the solution of the exact problem given in Theorem 3.3.1.

Theorem 8.5.1 All solutions $\overline{Q}_M$ of $\max_{\mathbf{r}} Q(\mathbf{r})$ subject to $\mathbf{u} \le \mathbf{r} \le \mathbf{v}$ are of the form

$$\mathbf{r} = (u_1, \ldots, u_s, v_{s+1}, \ldots, v_m)', \tag{8.36}$$

where s has the property

$$g_s \le Q_s \le g_{s+1} \tag{8.37}$$

and

$$\overline{Q}_M = Q_s = \frac{N_s}{D_s} = \frac{\sum_{m=1}^{s} g_m u_m + \sum_{m=s+1}^{M} g_m v_m}{\sum_{m=1}^{s} u_m + \sum_{m=s+1}^{M} v_m}. \tag{8.38}$$

Proof: Because the sign of

$$\frac{\partial Q}{\partial r_m} = \frac{\left(\sum_{m=1}^{M} r_j\right) g_m - \sum_{j=1}^{M} g_j r_j}{\left(\sum_{m=1}^{M} r_j\right)^2} = \frac{g_m - Q(r_1, \ldots, r_M)}{\left(\sum_{j=1}^{M} r_j\right)}$$

does not involve r_m, a necessary condition for maximization is $r_m = u_m$ if $g_m < Q(\mathbf{r})$ and $r_m = v_m$ if $g_m > Q(\mathbf{r})$. Hence all solutions $\overline{Q}_M$ are characterized by (8.36)–(8.38). ■

Some relations between $\{Q_s\}$ and $\{g_s\}$ are useful in computing $\max_{\mathbf{r}} Q(\mathbf{r})$ and in showing that $\overline{Q}_M \overset{\text{a.s.}}{\rightarrow} \overline{E}[g(\boldsymbol{\theta}_A) \mid \mathbf{y}^o, A]$.

Theorem 8.5.2 For $Q_1, \ldots, Q_M$ defined in (8.38), and any s, we obtain

(a) $Q_s < g_s \Leftrightarrow Q_s < Q_{s-1} \Longrightarrow Q_{s+1} < Q_s$
(b) $Q_s > g_{s+1} \Leftrightarrow Q_{s+1} > Q_s \Longrightarrow Q_s > Q_{s-1}$
(c) $Q_s = Q_{s+1} \Leftrightarrow g_s \leq Q_s \leq g_{s+1} \leq Q_{s+1} \leq g_{s+2}$

Proof: Simple algebra shows

$$Q_{s+1} - Q_s = (Q_s - g_{s+1})(v_{s+1} - u_{s+1})/D_{s+1} \tag{8.39}$$

$$= (Q_{s+1} - g_{s+1})(v_{s+1} - u_{s+1})/D_s. \tag{8.40}$$

Because $a(\boldsymbol{\theta}_A) < b(\boldsymbol{\theta}_A) \forall \boldsymbol{\theta}_A \in \Theta_A$, it follows that $v_s - u_s > 0 \;\forall\; s = 1, \ldots, M$.

(a) From (8.40), $Q_s < g_s \Longleftrightarrow Q_s < Q_{s-1}$; from (8.39), $Q_s < g_s \leq g_{s+1} \Rightarrow Q_{s+1} < Q_s$.
(b) From (8.39), $Q_s > g_{s+1} \Leftrightarrow Q_{s+1} > Q_s$; from (8.40), $Q_s > g_{s+1} \geq g_s \Longrightarrow Q_s > Q_{s-1}$;
(c) From (8.39) and (8.40), $Q_s = Q_{s+1} \Longrightarrow Q_s = g_{s+1} = Q_{s+1}$, and $g_s \leq g_{s+1} \leq g_{s+2}$; from (8.40), $g_{s+1} \leq Q_{s+1} \Longrightarrow Q_{s+1} \geq Q_s$, and from (8.39), $Q_s \leq g_{s+1} \Longrightarrow Q_{s+1} < Q_s$. ■

Theorem 8.5.2 shows that (8.38) and (8.37) are sufficient for $Q_s = \overline{Q}_M$. It also provides a computationally efficient solution of the discrete problem.

Corollary 8.5.1 $\overline{Q}_M = \max_{\mathbf{r}} Q(\mathbf{r})$ subject to $\mathbf{u} \leq \mathbf{r} \leq \mathbf{v}$ may be computed as follows:

1. Sort $g(\boldsymbol{\theta}_A^{(1)}), \ldots, g(\boldsymbol{\theta}_A^{(M)})$ so that $g_m = g(\boldsymbol{\theta}_A^{(m)})$ is monotone nondecreasing in m.
2. Using successive bisection (Press et al. 1992, Section 3.4), find an index s such that $g_s \leq Q_s \leq g_{s+1}$.
3. Set $\overline{Q}_M = Q_s$.

The sufficiency result in Theorem 8.5.2 means that no further search is required once the condition in step 2 of Corollary 8.5.1 is satisfied. Computation time for step 2 is of the order $\log_2(M)$, whereas computation time for step 1 is of the order $M\log_2(M)$ (Press et al. 1992, Section 8.3). Step 1 is essential to any approach, so as $M \to \infty$, Corollary 8.5.1 provides the most efficient algorithm possible.

Theorem 8.5.3 Suppose that the convergence condition (8.35) is true for all prior density kernels in the density ratio class

$$k(\boldsymbol{\theta}_A \mid A) : 0 < a(\boldsymbol{\theta}_A) < k(\boldsymbol{\theta}_A \mid A) < b(\boldsymbol{\theta}_A) \ \forall \ \boldsymbol{\theta}_A \in \Theta_A,$$

and define $\overline{Q}_M$ as in Corollary 8.5.1. Then

$$\overline{Q}_M \overset{\text{a.s.}}{\to} \overline{E}[g(\boldsymbol{\theta}_A) \mid \mathbf{y}^o, A].$$

Proof: For any real c, define $k(\boldsymbol{\theta}_A; c)$ as in Theorem 3.3.1 and let $s = s(M)$ be the integer s of Corollary 8.5.1. Let

$$s_1 = s_1(M) = \max\{m : g_m \leq \overline{E}[g(\boldsymbol{\theta}_A) \mid \mathbf{y}^o, A]\}.$$

Then $Q_{s_1} \overset{\text{a.s.}}{\to} \overline{E}[g(\boldsymbol{\theta}_A) \mid \mathbf{y}^o, A]$ by assumption.

For any $c < \overline{E}[g(\boldsymbol{\theta}_A) \mid \mathbf{y}^o, A]$, let $s_2 = s_2(M) = \max\{m : g_m \leq c\}$. By assumption and Theorem 3.3.1, we obtain

$$Q_j \overset{\text{a.s.}}{\to} \frac{\displaystyle\int_{\Theta_A} p(\mathbf{y}^o \mid \boldsymbol{\theta}_A, A) k(\boldsymbol{\theta}_A; c) g(\boldsymbol{\theta}_A)\, dv(\boldsymbol{\theta}_A)}{\displaystyle\int_{\Theta_A} p(\mathbf{y}^o \mid \boldsymbol{\theta}_A, A) k(\boldsymbol{\theta}_A; c)\, dv(\boldsymbol{\theta}_A)} < \overline{E}[g(\boldsymbol{\theta}_A) \mid \mathbf{y}^o, A].$$

Hence $\lim_{M\to\infty} P[s_1(M) > s_2(M)] = 1$. By the monotonicity properties of $\{Q_j\}$ established in Theorem 8.5.2, $\lim_{M\to\infty} P[s(M) > s_2(M)] = 1$.

A symmetric argument holds beginning with any $c > \overline{E}[g(\boldsymbol{\theta}_A) \mid \mathbf{y}^o, A]$. ∎

Theorem 8.5.3 applies to any posterior simulation from a model with observables density $p(\mathbf{y}^o \mid \boldsymbol{\theta}_A, A)$ and prior density kernel $k(\boldsymbol{\theta}_A \mid A)$. It does not require $a(\boldsymbol{\theta}_A) < k(\boldsymbol{\theta}_A \mid A) < b(\boldsymbol{\theta}_A)$, but in general, bounds of this type will be of interest to the client. BACC incorporates the yet more specific instance $a(\boldsymbol{\theta}_A) = r^{-1} \cdot k(\boldsymbol{\theta}_A \mid A)$, $b(\boldsymbol{\theta}_A) = r \cdot k(\boldsymbol{\theta}_A \mid A)$, illustrated in Example 5.1.2. The user specifies the function of interest vector $g(\boldsymbol{\theta}_A^{(m)})$, the weight vector $w(\boldsymbol{\theta}_A^{(m)})$ $(m = 1, \ldots, M)$, and one or more values of $r > 1$. BACC computes the lower and upper bounds $\underline{E}[g(\boldsymbol{\theta}_A) \mid \mathbf{y}^o, A]$ and $\overline{E}[g(\boldsymbol{\theta}_A) \mid \mathbf{y}^o, A]$ corresponding to each value of r, using the algorithm described in Corollary 8.5.1.

Exercise 8.5.1 Completing the Argument Derive (8.39) and (8.40).

Bibliography

Ainslie, A. and P. E. Rossi (1998), "Similarities in Choice Behavior across Product Categories," *Marketing Science* **17**: 91–106.

Albano, G. L. and F. Jouneau-Sion (2004), "Bayesian Inference in Repeated English Auctions," *Test* **13**: 193–211.

Albert, J. and S. Chib, (1993a), "Bayes Inference via Gibbs Sampling of Autoregressive Time Series Subject to Markov Mean and Variance Shifts," *Journal of Business and Economic Statistics* **11**: 1–15.

Albert, J. and S. Chib (1993b), "Bayesian Analysis of Binary and Polychotomous Response Data," *Journal of the American Statistical Association* **88**: 669–679.

Allenby, G. M. and P. E. Rossi (1999), "Marketing Models of Consumer Heterogeneity," *Journal of Econometrics* **89**: 57–78.

Amemiya, T. (1985), *Advanced Econometrics*. Cambridge, MA: Harvard University Press.

Anderson, T. W. (1984), *An Introduction to Multivariate Statistical Analysis*, 2nd ed. New York: Wiley.

Bajari, P. and L. X. Lee (2003), "Deciding between Competition and Collusion," *Review of Economics and Statistics* **85**: 971–989.

Barnard, G. A. (1949), "Statistical Inference," *Journal of the Royal Statistical Society Series B* **11**: 115–149.

Barnett, G., R. Kohn, and S. Sheather (1996), "Bayesian Estimation of an Autoregressive Model Using Markov Chain Monte Carlo," *Journal of Econometrics* **74**: 237–254.

Barnett, W. A., and A. Jonas (1983), "The Muntz-Szatz Demand System: An Application of a Globally Well Balanced Series Expansion," *Economics Letters* **11**: 337–342.

Bartlett, M. S. (1957), "A Comment on D. V. Lindley's Statistical Paradox," *Biometrika* **44**: 533–534.

Bayarri, M. J. and J. O. Berger (1999), "Quantifying Surprise in the Data and Model Verification," in J. M. Bernardo et al., eds., *Bayesian Statistics*, Vol. 6, pp. 53–82. Oxford, UK: Oxford University Press.

Berger, J. O. (1985), *Statistical Decision Theory and Bayesian Analysis*, 2nd ed. New York: Springer-Verlag.

Berger, J. O. (1994), "An Overview of Robust Bayesian Analysis" (with discussion), *Test* **3**: 5–124.

Berger, J. O. and R. L. Wolpert (1988), *The Likelihood Principle*, 2nd ed. Hayward, CA: Institute of Mathematical Statistics.

Bernardo, J. M. and A. F. M. Smith (1994), *Bayesian Theory*. New York: Wiley.

Birnbaum, A. (1962), "On the Foundations of Statistical Inference," *Journal of the American Statistical Association* **57**: 269–306.

Box, G. E. P. (1980), "Sampling and Bayes' Inference in Scientific Modelling" (with discussion and rejoinder), *Journal of the Royal Statistical Society Series A* **143**: 383–430.

Brooks, S. P. and A. Gelman (1998), "General Methods for Monitoring Convergence of Iterative Simulations," *Journal of Computational and Graphical Statistics* **7**: 434–455.

Brooks, S. P. and G. O. Roberts (1998), "Convergence Assessment Techniques for Markov Chain Monte Carlo," *Statistics and Computing* **8**: 319–335.

Brown, P. J., T. Fearn, and M. Vannucci (1999), "The Choice of Variables in Multivariate Regression: A Non-conjugate Bayesian Decision Theory Approach," *Biometrika* **86**: 635–648.

Campolieti, M. (2000), "Bayesian Estimation and Smoothing of the Baseline Hazard in Discrete Time Duration Models," *Review of Economics and Statistics* **82**: 685–694.

Campolieti, M. (2001), "Bayesian Semiparametric Estimation of Discrete Duration Models: An Application of the Dirichlet Process Prior," *Journal of Applied Econometrics* **16**: 1–22.

Carlin, B. P. and S. Chib (1995), "Bayesian Model Choice via Markov Chain Monte Carlo Methods," *Journal of the Royal Statistical Society Series B* **57**: 473–484.

Casella, G. and R. L. Berger (2002), *Statistical Inference*, 2nd ed. Pacific Grove, CA: Duxbury.

Celeux, G., M. Hurn, and C. P. Robert (2000), "Computational and Inferential Difficulties with Mixture Posterior Distributions," *Journal of the American Statistical Association* **95**: 957–970.

Chaloner, K. and R. Brant (1988), "A Bayesian Approach to Outlier Detection and Residual Analysis," *Biometrika* **75**: 651–659.

Chan, K. S. and C. J. Geyer (1994), "Markov Chains for Exploring Posterior Distributions—Discussion," *Annals of Statistics* **22**: 1747–1758.

Chatfield, C. (1995), "Model Uncertainty, Data Mining, and Statistical Inference," *Journal of the Royal Statistical Society Series A* **158**: 419–468.

Chauveau, D. and J. Diebolt (2000), "Estimation of the Asymptotic Variance in the CLT for Markov Chains," *Stochastic Models* **19**: 449–465.

Chen, C.-F. (1985), "On Asymptotic Normality of Limiting Density Functions with Bayesian Implications," *Journal of the Royal Statistical Society Series B* **47**: 540–546.

Chib, S. (1992), "Bayes Inference in the Tobit Censored Regression Model," *Journal of Econometrics* **51**: 79–99.

Chib, S. (1993), "Bayes Regression with Autoregressive Errors: A Gibbs Sampling Approach," *Journal of Econometrics* **58**: 275–294.

Chib, S. (1995), "Marginal Likelihood from the Gibbs Output," *Journal of the American Statistical Association* **90**: 1313–1321.

Chib, S. (1996), "Calculating Posterior Distributions and Modal Estimates in Markov Mixture Models," *Journal of Econometrics* **95**: 79–97.

Chib, S. and E. Greenberg (1994), "Bayes Inference in Regression Models with ARMA(p,q) Errors," *Journal of Econometrics* **64**: 183–206.

Chib, S., and E. Greenberg (1995), "Understanding the Metropolis-Hastings Algorithm," *The American Statistician* **49**: 327–335.

Chib, S. and E. Greenberg (1998), "Analysis of Multivariate Probit Models," *Biometrika* **85**: 347–361.

Chib, S. and I. Jeliazkov (2001), "Marginal Likelihood from the Metropolis-Hastings Output," *Journal of the American Statistical Association* **96**: 270–281.

Chipman, H. A., E. I. George, and R. E. McCulloch (1998), "Bayesian CART Model Search," *Journal of the American Statistical Association* **93**: 935–948.

Cowles, M. K. and B. P. Carlin (1996), "Markov Chain Monte Carlo Convergence Diagnostics: A Comparative Review," *Journal of the American Statistical Association* **91**: 883–904.

Cowles, M. K., B. P. Carlin, and J. E. Connett (1996), "Bayesian Tobit Modeling of Longitudinal Ordinal Clinical Trial Compliance Data with Nonignorable Missingness," *Journal of the American Statistical Association* **91**: 86–98.

Dawid, A. P. (1984), "Statistical Theory: The Prequential Approach," *Journal of the Royal Statistical Society Series A* **147**: 278–292.

DeGroot, M. (1970), *Optimal Statistical Decisions*. New York: McGraw-Hill.

DeJong, D. N. (1993), "Bayesian Inference in Limited Dependent Variable Models: An Application to Measuring Strike Duration," *Journal of Applied Econometrics* **8**: 115–128.

Dellaportas, P. and A. F. M. Smith (1993), "Bayesian Inference for Generalized Linear and Proportional Hazards Models via Gibbs Sampling," *Applied Statistics* **42**: 443–459.

DeRobertis, L. and J. A. Hartigan (1981), "Bayesian Inference Using Intervals of Measures," *The Annals of Statistics* **9**: 235–244.

Devroye, L. (1986), *Non-uniform Random Variate Generation*. New York: Springer-Verlag.

Diebolt, J. and C. P. Robert (1994), "Estimation of Finite Mixture Distributions through Bayesian Sampling," *Biometrika* **56**: 363–375.

Doob, J. L. (1953), *Stochastic Processes*. New York: Wiley.

Draper, D. (1995), "Assessment and Propagation of Model Uncertainty," *Journal of the Royal Statistical Society Series B* **57**: 45–97.

Fisher, R. A. (1956), *Statistical Methods and Scientific Inference*. Edinburgh: Oliver and Boyd.

Friedman, M. (1953), "The Methodology of Positive Economics," in M. Friedman, ed., *Essays in Positive Economics*. Chicago: University of Chicago Press.

Friedman, M. (1957), *A Theory of the Consumption Function*. Princeton: Princeton University Press.

Friedman, M. and L. J. Savage (1948), "The Utility Analysis of Choices Involving Risk," *Journal of Political Economy* **56**: 279–304.

Friedman, M. and L. J. Savage (1952), "The Expected-Utility Hypothesis and the Measurability of Utility," *Journal of Political Economy* **60**: 463–474.

Fruhwirth-Schnatter, S. (2001), "Markov Chain Monte Carlo Estimation of Classical and Dynamic Switching and Mixture Models," *Journal of the American Statistical Association* **96**: 194–209.

Fudenberg, D. and J. Tirole (1995), *Game Theory*. Cambridge, MA: MIT Press.

Gallant, A. R. (1981), "On the Bias in Flexible Functional Forms and an Essentially Unbiased Form: The Fourier Flexible Form," *Journal of Econometrics* **15**: 211–245.

Garthwaite, P. H. and J. M. Dickey (1988), "Quantifying Expert Opinion in Linear-Regression Problems," *Journal of the Royal Statistical Society Series B* **50**: 462–474.

Geisel, M. S. (1975), "Bayesian Comparison of Simple Macroeconomic Models," in S. E. Fienberg and A. Zellner, eds., *Studies in Bayesian Econometrics and Statistics: In Honor of Leonard J. Savage*, pp. 227–256. Amsterdam: North-Holland.

Gelfand, A. E. and D. K. Dey (1994), "Bayesian Model Choice: Asymptotics and Exact Calculations," *Journal of the Royal Statistical Society Series B* **56**: 501–514.

Gelfand, A. E. and A. F. M. Smith (1990), "Sampling Based Approaches to Calculating Marginal Densities," *Journal of the American Statistical Association* **85**: 398–409.

Gelman, A. (1996), "Inference and Monitoring Convergence," in W. R. Gilks, S. Richardson, and D. J. Spiegelhalter, eds., *Markov Chain Monte Carlo in Practice*. Chapman and Hall, London.

Gelman, A., J. B. Carlin, H. S. Stern, and D. B. Rubin (1995), *Bayesian Data Analysis*. London: Chapman and Hall.

Gelman, A., X. L. Meng, and H. S. Stern (1996), "Posterior Predictive Assessment of Model Fitness via Realized Discrepancies," *Statistica Sinica* **6**: 733–760.

Gelman, A. and D. B. Rubin (1992), "Inference from Iterative Simulation Using Multiple Sequences," *Statistical Science* **7**: 457–472.

Geman, S. and D. Geman (1984), "Stochastic Relaxation, Gibbs Distributions and the Bayesian Restoration of Images," *IEEE Transactions on Pattern Analysis and Machine Intelligence* **6**: 721–741.

George, E. I. and R. E. McCulloch (1993), "Variable Selection Via Gibbs Sampling," *Journal of the American Statistical Association* **88**: 881–889.

George, E. I. and R. E. McCulloch (1997), "Approaches for Bayesian Variable Selection," *Statistica Sinica* **7**: 339–373.

Geweke, J. (1986), "Exact Inference in the Inequality Constrained Normal Linear Regression Model," *Journal of Applied Econometrics* **1**: 127–142.

Geweke, J. (1988), "Antithetic Acceleration of Monte Carlo Integration in Bayesian Inference," *Journal of Econometrics* **38**: 73–90.

Geweke, J. (1989a), "Bayesian Inference in Econometric Models Using Monte Carlo Integration," *Econometrica* **57**: 1317–1340.

Geweke, J. (1989b), "Exact Predictive Densities in Linear Models with ARCH Disturbances," *Journal of Econometrics* **40**: 63–86.

Geweke, J. (1991), "Efficient Simulation from the Multivariate Normal and Student-t Distributions Subject to Linear Constraints," in E. M. Keramidas, ed., *Computing Science and Statistics: Proceedings of the Twenty-Third Symposium on the Interface*, pp. 571–578. Fairfax, VA: Interface Foundation of North America, Inc.

Geweke, J. (1992), "Evaluating the Accuracy of Sampling-Based Approaches to the Calculation of Posterior Moments," in J. M. Bernardo et al., eds., *Bayesian Statistics*, Vol. 4. Oxford, UK: Clarendon Press.

Geweke, J. (1996a), "Variable Selection and Model Comparison in Regression," in J. O. Berger, J. M. Bernardo, A. P. Dawid, and A. F. M. Smith, eds., *Bayesian Statistics*, Vol. 5, pp. 609–620. Oxford, UK: Oxford University Press.

Geweke, J. (1996b), "Bayesian Inference for Linear Models Subject to Linear Inequality Constraints," in W. O. Johnson, J. C. Lee, and A. Zellner, eds., *Modeling and Prediction: Honoring Seymour Geisser*, pp. 248–263. New York: Springer-Verlag.

Geweke, J. (1999), "Using Simulation Methods for Bayesian Econometric Models: Inference, Development and Communication" (with discussion and rejoinder), *Econometric Reviews* **18**: 1–126.

Geweke, J. (2004), "Getting It Right: Joint Distribution Tests of Posterior Simulators," *Journal of the American Statistical Association* **99**: 799–804.

Geweke, J. and M. Keane (2001), "Computationally Intensive Methods for Integration in Econometrics," in J. Heckman and E. E. Leamer, eds., *Handbook of Econometrics*, Vol. 5, pp. 3463–3568. Amsterdam: North-Holland.

Geweke, J., M. Keane, and D. Runkle (1994), "Alternative Computational Approaches to Inference in the Multinomial Probit Model," *Review of Economics and Statistics* **76**: 609–632.

Geweke, J., M. Keane, and D. Runkle (1997), "Statistical Inference in the Multinomial Multiperiod Probit Model," *Journal of Econometrics* **80**: 125–166.

Geweke, J., R. Marshall, and G. Zarkin (1986), "Mobility Indices in Continuous Time Markov Chains," *Econometrica* **54**: 1407–1424.

Geweke, J. and L. Petrella (1998), "Prior Density-Ratio Class Robustness in Econometrics," *Journal of Business and Economic Statistics* **16**: 469–478.

Geweke, J. and C. Whiteman (in press), "Bayesian Forecasting," in G. Elliott, C. Granger, and A. Timmermann, eds., *Handbook of Economic Forecasting*. Amsterdam: North-Holland.

Geyer, C. J. (1992), "Practical Markov Chain Monte Carlo," *Statistical Science* **7**: 473–481.

Geyer, C. J. (1996), "Estimation and Optimization of Functions," in W. R. Gilks, S. Richardson, and D. J. Spiegelhalter, eds., *Markov Chain Monte Carlo in Practice*. London: Chapman and Hall.

Goldberger, A. S. (1974), "Unobservable Variables in Econometrics," in P. Zarembka, ed., *Frontiers in Econometrics*. New York: Academic Press.

Good, I. J. (1956), "The Surprise Index for the Multivariate Normal Distribution," *Annals of Mathematical Statistics* **27**: 1130–1135.

Granger, C. W. J. and Z. Ding (1995a), "Some Properties of Absolute Returns and Alternative Measure of Risk,"*Annales d'Economie et de Statistique* **40**: 67–91.

Granger, C. W. J. and Z. Ding (1995b), "Stylized Facts on the Temporal and Distributional Properties of Daily Data from Speculative Markets," Department of Economics, University of California—San Diego, unpublished paper.

Green, P. J. (1995), "Reversible Jump Markov Chain Monte Carlo Computation and Bayesian Model Determination," *Biometrika* **82**: 711–732.

Greene, W. H. (2003), *Econometric Analysis*, 5th ed. Upper Saddle River, NJ: Prentice-Hall.

Griliches, Z. (1977), "Errors in Variables and Other Unobservables," in D. Aigner and A. S. Goldberger, eds., *Latent Variables in Socio-economic Models*. Amsterdam: North-Holland.

Hamilton, J. D. (1994), *Time Series Analysis*. Princeton: Princeton University Press.

Hammersly, J. M. and D. C. Handscomb (1964), *Monte Carlo Methods*. London: Methuen and Company.

Hammersly, J. M. and K. H. Morton (1956), "A New Monte Carlo Technique: Antithetic Variates," *Proceedings of the Cambridge Philosophical Society* **52**: 449–474.

Han, C. and B. P. Carlin (2001), "Markov Chain Monte Carlo Methods for Computing Bayes Factors: A Comparative Review," *Journal of the American Statistical Association* **96**: 1122–1132.

Hastings, W. K. (1970), "Monte Carlo Sampling Methods Using Markov Chains and Their Applications," *Biometrika* **57**: 97–109.

Hildreth, C. (1954), "Point Estimates of Concave Functions," *Journal of the American Statistical Association* **49**: 598–619.

Hildreth, C. (1963), "Bayesian Statisticians and Remote Clients," *Econometrica* **31**: 422–438.

Hoeting, J. A., D. Madigan, A. E. Raftery, and C. T. Volinsky (1999), "Bayesian Model Averaging: A Tutorial," *Statistical Science* **14**: 382–401.

Jeffreys, H. (1939), *Theory of Probability*. Oxford, UK: Clarendon Press.

Jeffreys, H. (1961), *Theory of Probability*, 3rd ed. Oxford, UK: Clarendon Press.

Johnson, N. L., S. Kotz, and N. Balakrishnan (1994), *Continuous Univariate Distributions*, Vol. 1, 2nd ed. New York: Wiley.

Johnson, N. L. and S. Kotz (1972), *Distributions in Statistics: Continuous Multivariate Distributions*. New York: Wiley.

Kadane, J. B. (1994), "An Application of Robust Bayesian Analysis to a Medical Experiment," *Journal of Statistical Planning and Inference* **40**: 221–228.

Kadane, J. B. and L. J. Wolfson (1998), "Experiences in Elicitation," *Journal of the Royal Statistical Society Series D* **47**: 3–19.

Kass, R. E. and A. E. Raftery (1995), "Bayes Factors," *Journal of the American Statistical Association* **90**: 773–795.

Kim, J., G. M. Allenby, and P. E. Rossi (2003), "Modeling Consumer Demand for Variety," *Marketing Science* **21**: 229–250.

Kloek, T. and H. K. van Dijk (1978), "Bayesian Estimates of Equation System Parameters: An Application of Integration by Monte Carlo," *Econometrica* **46**: 1–20.

Koop, G. (2003), *Bayesian Econometrics*. West Sussex, UK: Wiley.

Kotz, S., N. Balakrishnan, and N. L. Johnson (2000), *Continuous Multivariate Distributions*, Vol. 1. New York: Wiley.

Lancaster, T. (2004), *An Introduction to Modern Bayesian Econometrics*. Malden, MA: Blackwell.

Lavine, M. (1991a), "Sensitivity in Bayesian Statistics: The Prior and the Likelihood," *Journal of the American Statistical Association* **86**: 396–399.

Lavine, M. (1991b), "An Approach to Robust Bayesian Analysis for Multidimensional Parameter Spaces," *Journal of the American Statistical Association* **86**: 400–403.

Leamer, E. E. (1973), "Multicollinearity: A Bayesian Interpretation," *Review of Economics and Statistics* **55**: 371–380.

Leamer, E. E. (1978), *Specification Searches*. New York: Wiley.

Leamer, E. E. (1982), "Sets of Posterior Means with Bounded Variance Priors," *Econometrica* **50**: 725–736.

Lindgren, G. (1978), "Markov Regime Models for Mixed Distributions and Switching Regressions," *Scandinavian Journal of Statistics* **5**: 81–91.

Lindley, D. V. (1957), "A Statistical Paradox," *Biometrika* **44**: 187–192.

Lindley, D. and A. F. M. Smith (1972), "Bayes Estimates for the Linear Model," *Journal of the Royal Statistical Society Series B* **34**: 1–41.

Little, R. J. A. and D. B. Rubin (2002), *Statistical Analysis with Missing Data*, 2nd ed. New York: Wiley.

Liu, J. S., W. H. Wong, and A. Kong (1994), "Covariance Structure of the Gibbs Sampler with Applications to the Comparison of Estimators and Augmentation Schemes," *Journal of Statistical Planning and Inference* **81**: 27–40.

McCulloch, R. E. and P. E. Rossi (1994), "An Exact Likelihood Analysis of the Multinomial Probit Model," *Journal of Econometrics* **64**: 207–240.

McCulloch, R. E., N. G. Polson, and P. E. Rossi (2000), "A Bayesian Analysis of the Multinomial Probit Model with Fully Identified Parameters," *Journal of Econometrics* **99**: 173–193.

McCulloch, R. and R. Tsay (1994), "Statistical Analysis of Macroeconomic Time Series via Markov Switching Models," *Journal of Time Series Analysis* **15:** 523–539.

McKeague, I. W. and W. Wefelmeyer (2000), "Markov Chain Monte Carlo and Rao-Blackwellization," *Journal of Statistical Planning and Inference* **85**: 171–182.

Meng, X. L. (1994), "Posterior Predictive p-values," *Annals of Statistics* **22**: 1142–1160.

Mengersen, K. L. and R. L. Tweedie (1996), "Rates of Convergence of the Hastings and Metropolis Algorithms," *Annals of Statistics* **24**: 101–121.

Metropolis, N., A. W. Rosenbluth, M. N. Rosenbluth, A. H. Teller, and E. Teller (1953), "Equation of State Calculations by Fast Computing Machines," *The Journal of Chemical Physics* **21**: 1087–1092.

Mincer, J. (1958), "Investment in Human Capital and Personal Income Distribution," *Journal of Political Economy* **66**: 281–302.

Mittelhammer, R. C., G. G. Judge, and D. J. Miller (2000), *Econometric Foundations*. Cambridge, UK: Cambridge University Press.

Mykland, P., L. Tierney, and B. Yu (1995), "Regeneration in Markov-chain Samplers," *Journal of the American Statistical Association* **90**: 233–241.

Newey, W. and K. West (1987), "A Simple Positive Semi-definite, Heteroskedasticity and Autocorrelation Consistent Covariance Matrix," *Econometrica* **55**: 703–708.

Percy, D. F. (1992), "Prediction for Seemingly Unrelated Regressions," *Journal of the Royal Statistical Society, Series B* **54**: 243–252.

Peskun, P. H. (1973), "Optimum Monte-Carlo Sampling using Markov Chains," *Biometrika* **60**: 607–612.

Poirier, D. J. (1988), "Frequentist and Subjectivist Perspectives on the Problems of Model Building in Economics," *Journal of Economic Perspectives* **2**(1): 121–144.

Poirier, D. J. (1995), *Intermediate Statistics and Econometrics: A Comparative Approach*. Cambridge, MA: MIT Press.

Poirier, D. J. (1998), "Revising Beliefs in Nonidentified Models," *Econometric Theory* **14**: 483–509.

Pratt, J. W., H. Raiffa, and R. Schlaifer (1995), *Introduction to Statistical Decision Theory*. Cambridge, MA: MIT Press.

Press, W. H., S. A. Teukolsky, W. T. Vetterling, and B. P. Flannery (1992), *Numerical Recipes: The Art of Scientific Computing*, 2nd ed. Cambridge, UK: Cambridge University Press.

Raftery, A. E. (1996), "Hypothesis Testing and Model Selection via Posterior Simulation," in W. R. Gilks et al., eds., *Markov Chain Monte Carlo in Practice*. London: Chapman and Hall.

Raiffa, H. and R. Schlaiffer (1961), *Applied Statistical Decision Theory*. Cambridge, MA: Harvard Business School.

Rao, C. R. (1965), *Linear Statistical Inference and Its Applications*. New York: Wiley.

Richardson, S. and P. J. Green (1997), "On Bayesian Analysis of Mixtures with an Unknown Number of Components," *Journal of the Royal Statistical Society Series B* **59**: 731–758.

Ripley, R. D. (1987), *Stochastic Simulation*. New York: Wiley.

Robert, C. P. (1995), "Convergence Control Methods for Markov Chain Monte Carlo Algorithms," *Statistical Science* **10**: 231–253.

Roberts, H. V. (1965), "Probabilistic Prediction," *Journal of the American Statistical Association* **60**: 50–62.

Roberts, G. O. and A. F. M. Smith (1994), "Simple Conditions for the Convergence of the Gibbs Sampler and Metropolis-Hastings Algorithms," *Stochastic Processes and Their Applications* **49**: 207–216.

Roeder, K. and L. Wasserman (1997), "Practical Bayesian Density Estimation Using Mixtures of Normals," *Journal of the American Statistical Association* **92**: 894–902.

Rossi, P. E. and G. M. Allenby (2003), "Bayesian Statistics and Marketing," *Marketing Science* **22**: 304–328.

Rossi, P. E., Z. Gilula, and G. M. Allenby (2001), "Overcoming Scale Usage Heterogeneity: A Bayesian Hierarchical Approach," *Journal of the American Statistical Association* **96**: 20–31.

Rossi, P. E., R. E. McCulloch, and G. M. Allenby (1996), "The Value of Purchase History Data in Target Marketing," *Marketing Science* **15**: 321–340.

Royden, H. L. (1968), *Real Analysis*, 2nd ed. New York: Macmillan.

Rubin, D. B. (1984), "Bayesianly Justifiable and Relevant Frequency Calculations for the Applied Statistician," *Annals of Statistics* **12**: 1151–1172.

Ryden, T., T. Terasvirta, and S. Asbrink (1998), "Stylized Facts of Daily Return Series and the Hidden Markov Model," *Journal of Applied Econometrics* **13**: 217–244.

Sareen, S. (1999), "Posterior Odds Comparison of a Symmetric Low-Price, Sealed-Bid Auction within the Common-Value and the Independent-Private-Values Pardigms," *Journal of Applied Econometrics* **14**: 651–676.

Sareen, S. (2003), "Reference Bayesian Inference in Non-regular Models," *Journal of Econometrics* **113**: 265–288.

Savage, L. J. (1951), "The Theory of Statistical Decision," *Journal of the American Statistical Association* **46**: 55–67.

Schott, J. R. (1997), *Matrix Analysis for Statistics*. New York: Wiley.

Shao, J. (1989), "Monte Carlo Approximations in Bayesian Decision Theory," *Journal of the American Statistical Association* **84**: 727–732.

Shiller, R. J. (1984), "Smoothness Priors and Nonlinear Regression," *Journal of the American Statistical Association* **79**: 609–615.

Simon, H. A. (1956), "Dynamic Programming under Uncertainty with a Quadratic Criterion Function," *Econometrica* **24**: 74–81.

Sims, C. A. and H. Uhlig (1991), "Understanding Unit Rooters: A Helicopter Tour," *Econometrica* **59**: 1591–1599.

Smith, A. F. M. (1973), "A General Bayesian Linear Model," *Journal of the Royal Statistical Society Series B* **35**: 67–75.

Smith, M. and R. Kohn (1996), "Nonparametric Regression Using Bayesian Variable Selection," *Journal of Econometrics* **75**: 317–343.

Sweeting, T. J. and A. O. Adekola (1987), "Asymptotic Posterior Normality for Stochastic Processes Revisited," *Journal of the Royal Statistical Soceity Series B* **49**: 215–222.

Tanner, M. A. and W. H. Wong (1987), "The Calculation of Posterior Distributions by Data Augmentation," *Journal of the American Statistical Association* **82**: 528–550.

Theil, H. and A. S. Goldberger (1961), "On Pure and Mixed Statistical Estimation in Economics," *International Economic Review* **2**: 65–78.

Tierney, L. (1991), "Exploring Posterior Distributions Using Markov Chains," in E. M. Keramidas, ed., *Computing Science and Statistics: Proceedings of the 23rd Symposium on the Interface*, pp. 563–570. Fairfax, VA: Interface Foundation of North America, Inc.

Tierney, L. (1994), "Markov Chains for Exploring Posterior Distributions" (with discussion and rejoinder), *Annals of Statistics* **22**: 1701–1762.

Tierney, L. and J. B. Kadane (1986), "Accurate Approximations for Posterior Moments and Marginal Densities," *Journal of the American Statistical Association* **81**: 82–86.

Tierney, L. and J. B. Kadane (1989), "Fully Exponential Laplace Approximations to Expectations and Variances of Nonpositive Functions," *Journal of the American Statistical Association* **84**: 710–716.

Varian, H. (1992), *Microeconomic Analysis*, 3rd ed. New York: Norton.

von Neumann, J. and O. Morgnestern (1944), *Theory of Games and Economic Behavior*. Princeton: Princeton University Press.

Wasserman, L. (1992), "Recent Methodological Advances in Robust Bayesian Inference," in J. M. Bernardo et al., eds., *Bayesian Statistics*, Vol. 4, pp. 483–502. Oxford, UK: Oxford University Press.

Wasserman, L. and J. B. Kadane (1992), "Computing Bounds on Expectations," *Journal of the American Statistical Association* **87**: 516–522.

Wasserman, L., M. Lavine, and R. L. Wolpert (1993), "Linearization of Bayesian Robustness Problems," *Journal of Statistical Planning and Inference* **37**: 307–316.

West, M. and J. Harrison, (1989), *Bayesian Forecasting and Dynamic Models*. New York: Springer-Verlag.

Zellner, A. (1962), "An Efficient Method of Estimating Seemingly Unrelated Regressions and Tests for Aggregation Bias," *Journal of the American Statistical Association* **57**: 348–368.

Zellner, A. (1971), *An Introduction to Bayesian Inference in Econometrics*. New York: Wiley.

Zellner, A. (1986a), "Bayesian Estimation and Prediction Using Asymmetric Loss Functions," *Journal of the American Statistical Association* **81**: 446–451.

Zellner, A. (1986b), "On Assessing Prior Distributions and Bayesian Regression Analysis with g-Prior Distributions," in P. K. Goel and A. Zellner, eds., *Bayesian Inference and Decision Techniques: Essays in Honour of Bruno de Finetti*. Amsterdam: North-Holland.

Zellner, A. and C. K. Min (1995), "Gibbs Sampler Convergence Criteria," *Journal of the American Statistical Association* **90**: 921–927.

Author Index

Contemporary Bayesian Econometrics and Statistics, by John Geweke

Subject Index

Contemporary Bayesian Econometrics and Statistics, by John Geweke

WILEY SERIES IN PROBABILITY AND STATISTICS

ESTABLISHED BY WALTER A. SHEWHART AND SAMUEL S. WILKS

The ***Wiley Series in Probability and Statistics*** is well established and authoritative. It covers many topics of current research interest in both pure and applied statistics and probability theory. Written by leading statisticians and institutions, the titles span both state-of-the-art developments in the field and classical methods.

Reflecting the wide range of current research in statistics, the series encompasses applied, methodological and theoretical statistics, ranging from applications and new techniques made possible by advances in computerized practice to rigorous treatment of theoretical approaches.

This series provides essential and invaluable reading for all statisticians, whether in academia, industry, government, or research.

ABRAHAM and LEDOLTER · Statistical Methods for Forecasting
AGRESTI · Analysis of Ordinal Categorical Data
AGRESTI · An Introduction to Categorical Data Analysis
AGRESTI · Categorical Data Analysis, *Second Edition*
ALTMAN, GILL, and McDONALD · Numerical Issues in Statistical Computing for the Social Scientist
AMARATUNGA and CABRERA · Exploration and Analysis of DNA Microarray and Protein Array Data
ANDĚL · Mathematics of Chance
ANDERSON · An Introduction to Multivariate Statistical Analysis, *Third Edition*
* ANDERSON · The Statistical Analysis of Time Series
ANDERSON, AUQUIER, HAUCK, OAKES, VANDAELE, and WEISBERG · Statistical Methods for Comparative Studies
ANDERSON and LOYNES · The Teaching of Practical Statistics
ARMITAGE and DAVID (editors) · Advances in Biometry
ARNOLD, BALAKRISHNAN, and NAGARAJA · Records
* ARTHANARI and DODGE · Mathematical Programming in Statistics
* BAILEY · The Elements of Stochastic Processes with Applications to the Natural Sciences
BALAKRISHNAN and KOUTRAS · Runs and Scans with Applications
BARNETT · Comparative Statistical Inference, *Third Edition*
BARNETT and LEWIS · Outliers in Statistical Data, *Third Edition*
BARTOSZYNSKI and NIEWIADOMSKA-BUGAJ · Probability and Statistical Inference
BASILEVSKY · Statistical Factor Analysis and Related Methods: Theory and Applications
BASU and RIGDON · Statistical Methods for the Reliability of Repairable Systems
BATES and WATTS · Nonlinear Regression Analysis and Its Applications
BECHHOFER, SANTNER, and GOLDSMAN · Design and Analysis of Experiments for Statistical Selection, Screening, and Multiple Comparisons

† BELSLEY, KUH, and WELSCH · Regression Diagnostics: Identifying Influential Data and Sources of Collinearity
BENDAT and PIERSOL · Random Data: Analysis and Measurement Procedures, *Third Edition*
BERRY, CHALONER, and GEWEKE · Bayesian Analysis in Statistics and Econometrics: Essays in Honor of Arnold Zellner
BERNARDO and SMITH · Bayesian Theory
BHAT and MILLER · Elements of Applied Stochastic Processes, *Third Edition*
BHATTACHARYA and WAYMIRE · Stochastic Processes with Applications
† BIEMER, GROVES, LYBERG, MATHIOWETZ, and SUDMAN · Measurement Errors in Surveys
BILLINGSLEY · Convergence of Probability Measures, *Second Edition*
BILLINGSLEY · Probability and Measure, *Third Edition*
BIRKES and DODGE · Alternative Methods of Regression
BLISCHKE AND MURTHY (editors) · Case Studies in Reliability and Maintenance
BLISCHKE AND MURTHY · Reliability: Modeling, Prediction, and Optimization
BLOOMFIELD · Fourier Analysis of Time Series: An Introduction, *Second Edition*
BOLLEN · Structural Equations with Latent Variables
BOROVKOV · Ergodicity and Stability of Stochastic Processes
BOULEAU · Numerical Methods for Stochastic Processes
BOX · Bayesian Inference in Statistical Analysis
BOX · R. A. Fisher, the Life of a Scientist
BOX and DRAPER · Empirical Model-Building and Response Surfaces
* BOX and DRAPER · Evolutionary Operation: A Statistical Method for Process Improvement
BOX, HUNTER, and HUNTER · Statistics for Experimenters: Design, Innovation, and Discovery, *Second Editon*
BOX and LUCEÑO · Statistical Control by Monitoring and Feedback Adjustment
BRANDIMARTE · Numerical Methods in Finance: A MATLAB-Based Introduction
BROWN and HOLLANDER · Statistics: A Biomedical Introduction
BRUNNER, DOMHOF, and LANGER · Nonparametric Analysis of Longitudinal Data in Factorial Experiments
BUCKLEW · Large Deviation Techniques in Decision, Simulation, and Estimation
CAIROLI and DALANG · Sequential Stochastic Optimization
CASTILLO, HADI, BALAKRISHNAN, and SARABIA · Extreme Value and Related Models with Applications in Engineering and Science
CHAN · Time Series: Applications to Finance
CHARALAMBIDES · Combinatorial Methods in Discrete Distributions
CHATTERJEE and HADI · Sensitivity Analysis in Linear Regression
CHATTERJEE and PRICE · Regression Analysis by Example, *Third Edition*
CHERNICK · Bootstrap Methods: A Practitioner's Guide
CHERNICK and FRIIS · Introductory Biostatistics for the Health Sciences
CHILÈS and DELFINER · Geostatistics: Modeling Spatial Uncertainty
CHOW and LIU · Design and Analysis of Clinical Trials: Concepts and Methodologies, *Second Edition*
CLARKE and DISNEY · Probability and Random Processes: A First Course with Applications, *Second Edition*
* COCHRAN and COX · Experimental Designs, *Second Edition*
CONGDON · Applied Bayesian Modelling
CONGDON · Bayesian Statistical Modelling
CONOVER · Practical Nonparametric Statistics, *Third Edition*
COOK · Regression Graphics
COOK and WEISBERG · Applied Regression Including Computing and Graphics

COOK and WEISBERG · An Introduction to Regression Graphics
CORNELL · Experiments with Mixtures, Designs, Models, and the Analysis of Mixture Data, *Third Edition*
COVER and THOMAS · Elements of Information Theory
COX · A Handbook of Introductory Statistical Methods
* COX · Planning of Experiments
CRESSIE · Statistics for Spatial Data, *Revised Edition*
CSÖRGŐ and HORVÁTH · Limit Theorems in Change Point Analysis
DANIEL · Applications of Statistics to Industrial Experimentation
DANIEL · Biostatistics: A Foundation for Analysis in the Health Sciences, *Eighth Edition*
* DANIEL · Fitting Equations to Data: Computer Analysis of Multifactor Data, *Second Edition*
DASU and JOHNSON · Exploratory Data Mining and Data Cleaning
DAVID and NAGARAJA · Order Statistics, *Third Edition*
* DEGROOT, FIENBERG, and KADANE · Statistics and the Law
DEL CASTILLO · Statistical Process Adjustment for Quality Control
DEMARIS · Regression with Social Data: Modeling Continuous and Limited Response Variables
DEMIDENKO · Mixed Models: Theory and Applications
DENISON, HOLMES, MALLICK and SMITH · Bayesian Methods for Nonlinear Classification and Regression
DETTE and STUDDEN · The Theory of Canonical Moments with Applications in Statistics, Probability, and Analysis
DEY and MUKERJEE · Fractional Factorial Plans
DILLON and GOLDSTEIN · Multivariate Analysis: Methods and Applications
DODGE · Alternative Methods of Regression
* DODGE and ROMIG · Sampling Inspection Tables, *Second Edition*
* DOOB · Stochastic Processes
DOWDY, WEARDEN, and CHILKO · Statistics for Research, *Third Edition*
DRAPER and SMITH · Applied Regression Analysis, *Third Edition*
DRYDEN and MARDIA · Statistical Shape Analysis
DUDEWICZ and MISHRA · Modern Mathematical Statistics
DUNN and CLARK · Basic Statistics: A Primer for the Biomedical Sciences, *Third Edition*
DUPUIS and ELLIS · A Weak Convergence Approach to the Theory of Large Deviations
* ELANDT-JOHNSON and JOHNSON · Survival Models and Data Analysis
ENDERS · Applied Econometric Time Series
ETHIER and KURTZ · Markov Processes: Characterization and Convergence
EVANS, HASTINGS, and PEACOCK · Statistical Distributions, *Third Edition*
FELLER · An Introduction to Probability Theory and Its Applications, Volume I, *Third Edition,* Revised; Volume II, *Second Edition*
FISHER and VAN BELLE · Biostatistics: A Methodology for the Health Sciences
FITZMAURICE, LAIRD, and WARE · Applied Longitudinal Analysis
* FLEISS · The Design and Analysis of Clinical Experiments
FLEISS · Statistical Methods for Rates and Proportions, *Third Edition*
FLEMING and HARRINGTON · Counting Processes and Survival Analysis
FULLER · Introduction to Statistical Time Series, *Second Edition*
FULLER · Measurement Error Models
GALLANT · Nonlinear Statistical Models
GEWEKE · Contemporary Bayesian Econometrics and Statistics
GHOSH, MUKHOPADHYAY, and SEN · Sequential Estimation
GIESBRECHT and GUMPERTZ · Planning, Construction, and Statistical Analysis of Comparative Experiments

GIFI · Nonlinear Multivariate Analysis
GIVENS and HOETING · Computational Statistics
GLASSERMAN and YAO · Monotone Structure in Discrete-Event Systems
GNANADESIKAN · Methods for Statistical Data Analysis of Multivariate Observations, *Second Edition*
GOLDSTEIN and LEWIS · Assessment: Problems, Development, and Statistical Issues
GREENWOOD and NIKULIN · A Guide to Chi-Squared Testing
GROSS and HARRIS · Fundamentals of Queueing Theory, *Third Edition*
† GROVES · Survey Errors and Survey Costs
* HAHN and SHAPIRO · Statistical Models in Engineering
HAHN and MEEKER · Statistical Intervals: A Guide for Practitioners
HALD · A History of Probability and Statistics and their Applications Before 1750
HALD · A History of Mathematical Statistics from 1750 to 1930
† HAMPEL · Robust Statistics: The Approach Based on Influence Functions
HANNAN and DEISTLER · The Statistical Theory of Linear Systems
HEIBERGER · Computation for the Analysis of Designed Experiments
HEDAYAT and SINHA · Design and Inference in Finite Population Sampling
HELLER · MACSYMA for Statisticians
HINKELMANN and KEMPTHORNE · Design and Analysis of Experiments, Volume 1: Introduction to Experimental Design
HINKELMANN and KEMPTHORNE · Design and Analysis of Experiments, Volume 2: Advanced Experimental Design
HOAGLIN, MOSTELLER, and TUKEY · Exploratory Approach to Analysis of Variance
HOAGLIN, MOSTELLER, and TUKEY · Exploring Data Tables, Trends and Shapes
* HOAGLIN, MOSTELLER, and TUKEY · Understanding Robust and Exploratory Data Analysis
HOCHBERG and TAMHANE · Multiple Comparison Procedures
HOCKING · Methods and Applications of Linear Models: Regression and the Analysis of Variance, *Second Edition*
HOEL · Introduction to Mathematical Statistics, *Fifth Edition*
HOGG and KLUGMAN · Loss Distributions
HOLLANDER and WOLFE · Nonparametric Statistical Methods, *Second Edition*
HOSMER and LEMESHOW · Applied Logistic Regression, *Second Edition*
HOSMER and LEMESHOW · Applied Survival Analysis: Regression Modeling of Time to Event Data
† HUBER · Robust Statistics
HUBERTY · Applied Discriminant Analysis
HUNT and KENNEDY · Financial Derivatives in Theory and Practice
HUSKOVA, BERAN, and DUPAC · Collected Works of Jaroslav Hajek—with Commentary
HUZURBAZAR · Flowgraph Models for Multistate Time-to-Event Data
IMAN and CONOVER · A Modern Approach to Statistics
† JACKSON · A User's Guide to Principle Components
JOHN · Statistical Methods in Engineering and Quality Assurance
JOHNSON · Multivariate Statistical Simulation
JOHNSON and BALAKRISHNAN · Advances in the Theory and Practice of Statistics: A Volume in Honor of Samuel Kotz
JOHNSON and BHATTACHARYYA · Statistics: Principles and Methods, *Fifth Edition*
JOHNSON and KOTZ · Distributions in Statistics
JOHNSON and KOTZ (editors) · Leading Personalities in Statistical Sciences: From the Seventeenth Century to the Present

JOHNSON, KOTZ, and BALAKRISHNAN · Continuous Univariate Distributions, Volume 1, *Second Edition*
JOHNSON, KOTZ, and BALAKRISHNAN · Continuous Univariate Distributions, Volume 2, *Second Edition*
JOHNSON, KOTZ, and BALAKRISHNAN · Discrete Multivariate Distributions
JOHNSON, KOTZ, and KEMP · Univariate Discrete Distributions, *Second Edition*
JUDGE, GRIFFITHS, HILL, LÜTKEPOHL, and LEE · The Theory and Practice of Econometrics, *Second Edition*
JUREČKOVÁ and SEN · Robust Statistical Procedures: Aymptotics and Interrelations
JUREK and MASON · Operator-Limit Distributions in Probability Theory
KADANE · Bayesian Methods and Ethics in a Clinical Trial Design
KADANE AND SCHUM · A Probabilistic Analysis of the Sacco and Vanzetti Evidence
KALBFLEISCH and PRENTICE · The Statistical Analysis of Failure Time Data, *Second Edition*
KASS and VOS · Geometrical Foundations of Asymptotic Inference
† KAUFMAN and ROUSSEEUW · Finding Groups in Data: An Introduction to Cluster Analysis
KEDEM and FOKIANOS · Regression Models for Time Series Analysis
KENDALL, BARDEN, CARNE, and LE · Shape and Shape Theory
KHURI · Advanced Calculus with Applications in Statistics, *Second Edition*
KHURI, MATHEW, and SINHA · Statistical Tests for Mixed Linear Models
* KISH · Statistical Design for Research
KLEIBER and KOTZ · Statistical Size Distributions in Economics and Actuarial Sciences
KLUGMAN, PANJER, and WILLMOT · Loss Models: From Data to Decisions, *Second Edition*
KLUGMAN, PANJER, and WILLMOT · Solutions Manual to Accompany Loss Models: From Data to Decisions, *Second Edition*
KOTZ, BALAKRISHNAN, and JOHNSON · Continuous Multivariate Distributions, Volume 1, *Second Edition*
KOTZ and JOHNSON (editors) · Encyclopedia of Statistical Sciences: Volumes 1 to 9 with Index
KOTZ and JOHNSON (editors) · Encyclopedia of Statistical Sciences: Supplement Volume
KOTZ, READ, and BANKS (editors) · Encyclopedia of Statistical Sciences: Update Volume 1
KOTZ, READ, and BANKS (editors) · Encyclopedia of Statistical Sciences: Update Volume 2
KOVALENKO, KUZNETZOV, and PEGG · Mathematical Theory of Reliability of Time-Dependent Systems with Practical Applications
LACHIN · Biostatistical Methods: The Assessment of Relative Risks
LAD · Operational Subjective Statistical Methods: A Mathematical, Philosophical, and Historical Introduction
LAMPERTI · Probability: A Survey of the Mathematical Theory, *Second Edition*
LANGE, RYAN, BILLARD, BRILLINGER, CONQUEST, and GREENHOUSE · Case Studies in Biometry
LARSON · Introduction to Probability Theory and Statistical Inference, *Third Edition*
LAWLESS · Statistical Models and Methods for Lifetime Data, *Second Edition*
LAWSON · Statistical Methods in Spatial Epidemiology
LE · Applied Categorical Data Analysis
LE · Applied Survival Analysis
LEE and WANG · Statistical Methods for Survival Data Analysis, *Third Edition*
LEPAGE and BILLARD · Exploring the Limits of Bootstrap
LEYLAND and GOLDSTEIN (editors) · Multilevel Modelling of Health Statistics

LIAO · Statistical Group Comparison
LINDVALL · Lectures on the Coupling Method
LINHART and ZUCCHINI · Model Selection
LITTLE and RUBIN · Statistical Analysis with Missing Data, *Second Edition*
LLOYD · The Statistical Analysis of Categorical Data
LOWEN and TEICH · Fractal-Based Point Processes
MAGNUS and NEUDECKER · Matrix Differential Calculus with Applications in Statistics and Econometrics, *Revised Edition*
MALLER and ZHOU · Survival Analysis with Long Term Survivors
MALLOWS · Design, Data, and Analysis by Some Friends of Cuthbert Daniel
MANN, SCHAFER, and SINGPURWALLA · Methods for Statistical Analysis of Reliability and Life Data
MANTON, WOODBURY, and TOLLEY · Statistical Applications Using Fuzzy Sets
MARCHETTE · Random Graphs for Statistical Pattern Recognition
MARDIA and JUPP · Directional Statistics
MASON, GUNST, and HESS · Statistical Design and Analysis of Experiments with Applications to Engineering and Science, *Second Edition*
McCULLOCH and SEARLE · Generalized, Linear, and Mixed Models
McFADDEN · Management of Data in Clinical Trials
* McLACHLAN · Discriminant Analysis and Statistical Pattern Recognition
McLACHLAN, DO, and AMBROISE · Analyzing Microarray Gene Expression Data
McLACHLAN and KRISHNAN · The EM Algorithm and Extensions
McLACHLAN and PEEL · Finite Mixture Models
McNEIL · Epidemiological Research Methods
MEEKER and ESCOBAR · Statistical Methods for Reliability Data
MEERSCHAERT and SCHEFFLER · Limit Distributions for Sums of Independent Random Vectors: Heavy Tails in Theory and Practice
MICKEY, DUNN, and CLARK · Applied Statistics: Analysis of Variance and Regression, *Third Edition*
* MILLER · Survival Analysis, *Second Edition*
MONTGOMERY, PECK, and VINING · Introduction to Linear Regression Analysis, *Third Edition*
MORGENTHALER and TUKEY · Configural Polysampling: A Route to Practical Robustness
MUIRHEAD · Aspects of Multivariate Statistical Theory
MULLER and STOYAN · Comparison Methods for Stochastic Models and Risks
MURRAY · X-STAT 2.0 Statistical Experimentation, Design Data Analysis, and Nonlinear Optimization
MURTHY, XIE, and JIANG · Weibull Models
MYERS and MONTGOMERY · Response Surface Methodology: Process and Product Optimization Using Designed Experiments, *Second Edition*
MYERS, MONTGOMERY, and VINING · Generalized Linear Models. With Applications in Engineering and the Sciences
† NELSON · Accelerated Testing, Statistical Models, Test Plans, and Data Analyses
† NELSON · Applied Life Data Analysis
NEWMAN · Biostatistical Methods in Epidemiology
OCHI · Applied Probability and Stochastic Processes in Engineering and Physical Sciences
OKABE, BOOTS, SUGIHARA, and CHIU · Spatial Tesselations: Concepts and Applications of Voronoi Diagrams, *Second Edition*
OLIVER and SMITH · Influence Diagrams, Belief Nets and Decision Analysis
PALTA · Quantitative Methods in Population Health: Extensions of Ordinary Regressions
PANKRATZ · Forecasting with Dynamic Regression Models

* PARZEN · Modern Probability Theory and Its Applications
PEÑA, TIAO, and TSAY · A Course in Time Series Analysis
PIANTADOSI · Clinical Trials: A Methodologic Perspective
PORT · Theoretical Probability for Applications
POURAHMADI · Foundations of Time Series Analysis and Prediction Theory
PRESS · Bayesian Statistics: Principles, Models, and Applications
PRESS · Subjective and Objective Bayesian Statistics, *Second Edition*
PRESS and TANUR · The Subjectivity of Scientists and the Bayesian Approach
PUKELSHEIM · Optimal Experimental Design
PURI, VILAPLANA, and WERTZ · New Perspectives in Theoretical and Applied Statistics
† PUTERMAN · Markov Decision Processes: Discrete Stochastic Dynamic Programming
QIU · Image Processing and Jump Regression Analysis
* RAO · Linear Statistical Inference and Its Applications, *Second Edition*
RAUSAND and HØYLAND · System Reliability Theory: Models, Statistical Methods, and Applications, *Second Edition*
RENCHER · Linear Models in Statistics
RENCHER · Methods of Multivariate Analysis, *Second Edition*
RENCHER · Multivariate Statistical Inference with Applications
* RIPLEY · Spatial Statistics
RIPLEY · Stochastic Simulation
ROBINSON · Practical Strategies for Experimenting
ROHATGI and SALEH · An Introduction to Probability and Statistics, *Second Edition*
ROLSKI, SCHMIDLI, SCHMIDT, and TEUGELS · Stochastic Processes for Insurance and Finance
ROSENBERGER and LACHIN · Randomization in Clinical Trials: Theory and Practice
ROSS · Introduction to Probability and Statistics for Engineers and Scientists
† ROUSSEEUW and LEROY · Robust Regression and Outlier Detection
* RUBIN · Multiple Imputation for Nonresponse in Surveys
RUBINSTEIN · Simulation and the Monte Carlo Method
RUBINSTEIN and MELAMED · Modern Simulation and Modeling
RYAN · Modern Regression Methods
RYAN · Statistical Methods for Quality Improvement, *Second Edition*
SALTELLI, CHAN, and SCOTT (editors) · Sensitivity Analysis
* SCHEFFE · The Analysis of Variance
SCHIMEK · Smoothing and Regression: Approaches, Computation, and Application
SCHOTT · Matrix Analysis for Statistics, *Second Edition*
SCHOUTENS · Levy Processes in Finance: Pricing Financial Derivatives
SCHUSS · Theory and Applications of Stochastic Differential Equations
SCOTT · Multivariate Density Estimation: Theory, Practice, and Visualization
* SEARLE · Linear Models
SEARLE · Linear Models for Unbalanced Data
SEARLE · Matrix Algebra Useful for Statistics
SEARLE, CASELLA, and McCULLOCH · Variance Components
SEARLE and WILLETT · Matrix Algebra for Applied Economics
SEBER and LEE · Linear Regression Analysis, *Second Edition*
† SEBER · Multivariate Observations
† SEBER and WILD · Nonlinear Regression
SENNOTT · Stochastic Dynamic Programming and the Control of Queueing Systems
* SERFLING · Approximation Theorems of Mathematical Statistics
SHAFER and VOVK · Probability and Finance: It's Only a Game!
SILVAPULLE and SEN · Constrained Statistical Inference: Inequality, Order, and Shape Restrictions

SRIVASTAVA · Methods of Multivariate Statistics
STAPLETON · Linear Statistical Models
STAUDTE and SHEATHER · Robust Estimation and Testing
STOYAN, KENDALL, and MECKE · Stochastic Geometry and Its Applications, *Second Edition*
STOYAN and STOYAN · Fractals, Random Shapes and Point Fields: Methods of Geometrical Statistics
STYAN · The Collected Papers of T. W. Anderson: 1943–1985
SUTTON, ABRAMS, JONES, SHELDON, and SONG · Methods for Meta-Analysis in Medical Research
TANAKA · Time Series Analysis: Nonstationary and Noninvertible Distribution Theory
THOMPSON · Empirical Model Building
THOMPSON · Sampling, *Second Edition*
THOMPSON · Simulation: A Modeler's Approach
THOMPSON and SEBER · Adaptive Sampling
THOMPSON, WILLIAMS, and FINDLAY · Models for Investors in Real World Markets
TIAO, BISGAARD, HILL, PEÑA, and STIGLER (editors) · Box on Quality and Discovery: with Design, Control, and Robustness
TIERNEY · LISP-STAT: An Object-Oriented Environment for Statistical Computing and Dynamic Graphics
TSAY · Analysis of Financial Time Series
UPTON and FINGLETON · Spatial Data Analysis by Example, Volume II: Categorical and Directional Data
VAN BELLE · Statistical Rules of Thumb
VAN BELLE, FISHER, HEAGERTY, and LUMLEY · Biostatistics: A Methodology for the Health Sciences, *Second Edition*
VESTRUP · The Theory of Measures and Integration
VIDAKOVIC · Statistical Modeling by Wavelets
VINOD and REAGLE · Preparing for the Worst: Incorporating Downside Risk in Stock Market Investments
WALLER and GOTWAY · Applied Spatial Statistics for Public Health Data
WEERAHANDI · Generalized Inference in Repeated Measures: Exact Methods in MANOVA and Mixed Models
WEISBERG · Applied Linear Regression, *Third Edition*
WELSH · Aspects of Statistical Inference
WESTFALL and YOUNG · Resampling-Based Multiple Testing: Examples and Methods for *p*-Value Adjustment
WHITTAKER · Graphical Models in Applied Multivariate Statistics
WINKER · Optimization Heuristics in Economics: Applications of Threshold Accepting
WONNACOTT and WONNACOTT · Econometrics, *Second Edition*
WOODING · Planning Pharmaceutical Clinical Trials: Basic Statistical Principles
WOODWORTH · Biostatistics: A Bayesian Introduction
WOOLSON and CLARKE · Statistical Methods for the Analysis of Biomedical Data, *Second Edition*
WU and HAMADA · Experiments: Planning, Analysis, and Parameter Design Optimization
YANG · The Construction Theory of Denumerable Markov Processes
*ZELLNER · An Introduction to Bayesian Inference in Econometrics
ZHOU, OBUCHOWSKI, and McCLISH · Statistical Methods in Diagnostic Medicine